MATHEMATICS ACROSS CULTURES

SCIENCE ACROSS CULTURES: THE HISTORY OF NON-WESTERN SCIENCE

VOLUME 2
MATHEMATICS ACROSS CULTURES

Editor:

HELAINE SELIN, *Hampshire College, Amherst, Massachusetts, USA*

MATHEMATICS ACROSS CULTURES

The History of Non-Western Mathematics

Editor

HELAINE SELIN

Hampshire College, Amherst, Massachusetts, USA

Advisory Editor

UBIRATAN D'AMBROSIO

State University of Campinas/UNICAMP, São Paulo, Brazil (Emeritus)

KLUWER ACADEMIC PUBLISHERS

DORDRECHT/BOSTON/LONDON

A catalogue record for this book is available from the Library of Congress.

ISBN 1-4020-0260-2
Transferred to Digital Print 2001

Published by Kluwer Academic Publishers,
PO Box 17, 3300 AA Dordrecht, The Netherlands.

Sold and distributed in North, Central and South America
by Kluwer Academic Publishers,
101 Philip Drive, Norwell, MA 02018, USA

In all other countries, sold and distributed
by Kluwer Academic Publishers,
PO Box 322, 3300 AH Dordrecht, The Netherlands

Printed on acid-free paper

Printed and bound in Great Britain by Antony Rowe Limited.

INTRODUCTION TO THE SERIES: SCIENCE ACROSS CULTURES: THE HISTORY OF NON-WESTERN SCIENCE

In 1997, Kluwer Academic Publishers published the *Encyclopaedia of the History of Science, Technology, and Medicine in Non-Western Cultures.* The encyclopedia, a collection of almost 600 articles by almost 300 contributors, covered a range of topics from Aztec science and Chinese medicine to Tibetan astronomy and Indian ethnobotany. For some cultures, specific individuals could be identified, and their biographies were included. Since the study of non-Western science is not just a study of facts, but a study of culture and philosophy, we included essays on subjects such as Colonialism and Science, Magic and Science, the Transmission of Knowledge from East to West, Technology and Culture, Science as a Western Phenomenon, Values and Science, and Rationality, Objectivity, and Method.

Because the encyclopedia was received with critical acclaim, and because the nature of an encyclopedia is such that articles must be concise and compact, the editors at Kluwer and I felt that there was a need to expand on its success. We thought that the breadth of the encyclopedia could be complemented by a series of books that explored the topics in greater depth. We had an opportunity, without such space limitations, to include more illustrations and much longer bibliographies. We shifted the focus from the general educated audience that the encyclopedia targeted to a more scholarly one, although we have been careful to keep the articles readable and keep jargon to a minimum.

Before we can talk about the field of non-Western science, we have to define both non-Western and science. The term non-Western is not a geographical designation; it is a cultural one. We use it to describe people outside of the Euro-American sphere, including the native cultures of the Americas. The power of European and American colonialism is evident in the fact that the majority of the world's population is defined by what they are not. And in fact, for most of our recorded history the flow of knowledge, art, and power went the other way. In this series, we hope to rectify the lack of scholarly attention paid to most of the world's science.

As for defining science, if we wish to study science in non-Western cultures, we need to take several intellectual steps. First, we must accept that every culture has a science, that is, a way of defining, controlling, and predicting events in the natural world. Then we must accept that every science is legitimate in terms of the culture from which it grew. The transformation of the word science as a distinct rationality valued above magic is uniquely European. It is not common to most non-Western societies, where magic and science and religion can easily co-exist. The empirical, scientific realm of understanding and inquiry is not readily separable from a more abstract, religious realm.

The first two books in the series are *Astronomy Across Cultures: the History of Non-Western Astronomy*, and *Mathematics Across Cultures: the History of Non-Western Mathematics.* Each includes about 20 chapters. Most deal with the topic as it is perceived by different cultures: Australian Aboriginal Astronomy, Native American Mathematics, etc. Each book also contains a variety of essays on related subjects, such as Astronomy and Prehistory, or East and West. The next four in the series will cover Medicine, Nature and the Environment, Chemistry and Alchemy, and Physics and Optics.

We hope the series will be used to provide both factual information about the practices and practitioners of the sciences as well as insights into the world views and philosophies of the cultures that produced them. We hope that readers will achieve a new respect for the accomplishments of ancient civilizations and a deeper understanding of the relationship between science and culture.

TABLE OF CONTENTS

ACKNOWLEDGMENTS

I would like to thank Maja de Keijzer and her staff at Kluwer Academic Publishers. I never believe that Kluwer is a large company, because of all the individual attention the staff gives me. I would also like to thank my colleagues at Hampshire College; I am so fortunate to be able to work with them. Special thanks go to Aaron Berman, Brian Schultz, and Gai Carpenter for recognizing the value of my work and giving me time to do it. And mostly I offer innumerable, unquantifiable thanks to Bob and Lisa and Tim.

ABOUT THE CONTRIBUTORS

HELAINE SELIN (Editor) is the editor of the *Encyclopaedia of the History of Science, Technology, and Medicine in Non-Western Cultures* (Kluwer Academic Publishers, 1997) and Science Librarian and Faculty Associate at Hampshire College in Amherst, Massachusetts. In addition to editing the new series, *Science Across Cultures*, she has been teaching a course on the Science and History of Alternative and Complementary Medicine.

UBIRATAN D'AMBROSIO (Advisory Editor; A Historiographical Proposal for Non-Western Mathematics) is Emeritus Professor of Mathematics at the State University of Campinas/UNICAMP in São Paulo, Brazil. He is President of the Brazilian Society of History of Mathematics/SBHMat, President of the International Study Group on Ethnomathematics/ISGEm, and President of the Institute for Future Studies/I.E.F. He is also a Visiting Professor and graduate advisor at several universities. He was the Director of the Institute of Mathematics, Statistics and Computer Science of UNICAMP, Brazil from 1972–80, Chief of the Unit of Curriculum of the Organization of American States, Washington DC (1980–82), and Pro-Rector [Vice-President] for University Development, of UNICAMP, Brazil from 1982–90. His most recent publications include *Ethnomathematics. The Art or Technique of Explaining and Knowing and History of Mathematics in the Periphery: The Basin Metaphor* (Berlin: Max-Planck-Institute für Wissenschaftsgeschichte, 1999).

MICHAEL CLOSS (Mesoamerican Mathematics) is a Professor in the Department of Mathematics and Statistics at the University of Ottawa, Ottawa, Canada. He is editor of the volume, *Native American Mathematics* (University of Texas Press, 1986, 1996). In addition to his work on mathematical development in different Native American cultures, he has written many research articles on the mathematics, chronology, astronomy and hieroglyphic writing of the ancient Maya.

RON EGLASH (Anthropological Perspectives on Ethnomathematics) holds a B.S. in Cybernetics and a M.S. in Systems Engineering, both from the

University of California at Los Angeles. Following a year as human factors engineer at National Semiconductor, he returned to school for a doctorate in cultural studies of science and technology through the University of California at Santa Cruz History of Consciousness program. He received a Fulbright grant for his work on ethnomathematics, which included sites in Senegal, Gambia, Mali, Burkina-Faso, Cameroon, Benin, and Ghana. The resulting book, *African Fractals: Modern Computing and Indigenous Design*, has just been published by Rutgers University Press. Dr Eglash's software simulations, which allow students to learn about fractal geometry from African designs, is available from Dynamic Software. His other research areas include history of information technology, community computer networks, and anthropology of communication systems. For more see:
http://www.cohums.ohio-state.edu/comp/eglash.htm

PAULUS GERDES (On Mathematical Ideas in Cultural Traditions of Central and Southern Africa) is Professor of Mathematics at the Universidade Pedagogica (Maputo, Mozambique) and Director of Mozambique's Ethnomathematics Research Center – Culture, Mathematics, Education. Since 1986, Dr Gerdes has been the chair of the African Mathematical Union Commission on the History of Mathematics in Africa (AMUCHMA). Among his books published in English are: *Geometry from Africa: Mathematical and Educational Explorations* (The Mathematical Association of America, 1999); *Culture and the Awakening of Geometrical Thinking* (MEP-Press, 1999); *Women, Art, and Geometry in Southern Africa* (Africa World Press, 1998); *Lusona: Geometrical Recreations from Africa* (L'Harmattan, 1997); *Ethnomathematics and Education in Africa* (IIE/University of Stockholm, 1995); and *African Pythagoras: a Study in Culture and Mathematics Education* (Ethnomathematics Research Project, Maputo, 1994).

THOMAS GILSDORF (Inca Mathematics) is Associate Professor and Chair of the Department of Mathematics at the University of North Dakota. He holds a Ph.D. in mathematics from Washington State University. In 1992–1993, he held the Solomon Lefschetz Postdoctoral Fellowship in mathematics at the Centro de Estudios Avanzados in Mexico City. During his stay there he became interested in the mathematics of indigenous groups of the Americas. That interest increased when he returned and began working with several minority groups, especially Native Americans, at the University of North Dakota. His current research interests include functional analysis and Native American mathematics. Most of his publications have been in the area of locally convex spaces. The article in this book is his second on Native American mathematics; the first, 'Native American number systems', will appear in the proceedings of the Midwest History of Mathematics conference, October, 1998.

JOCHI SHIGERU (The Dawn of Wasan (Japanese Mathematics)) is a historian of science and technology in Eastern Asia. His work involves analyzing traditional mathematics and astronomy and their relationship to Eastern Asian culture. He was born in Tokyo, and obtained the Ph.D. at the University of London. He is now an Associate Professor at the National Kaohsiung First University of Science and Technology, Taiwan, and a research commissioner of the Institute of Wasan in Tokyo. His publications include *The Influence of Chinese Mathematical Arts on Seki Kowa (Takakazu)*, Ph.D. Thesis, University of London, 1993, and 16 papers and 11 international conference papers. He has also contrbuted to *the Encyclopaedia of the History of Science, Technology, and Medicine in Non-Western Cultures* (Helaine Selin, ed. Kluwer, 1997) and the *Encyclopedia of Folklore* (Kano, Masanao, Shunsuke Tsurimi and Nakayama Shigeru, eds. Sanseido, 1997).

KIM, SOO-HWAN (Development of Materials for Ethnomathematics in Korea) is Head of the Department of Mathematics Education at Chongju National University of Education in Korea. He holds a D.Ed. and M.Ed. from the Korea National University of Education. He is also the Financial Director of the Korea Society of Mathematical Education. His current research interests include ethnomathematics, the history and pedagogy of mathematics, and assessment in mathematics education. He has published articles on ethnomathematics and mathematics education.

Y. TZVI LANGERMANN (The Hebrew Mathematical Tradition) is Associate Professor of Arabic, Bar Ilan University, Ramat Gan, Israel and Senior Research Associate, Institute of Microfilmed Hebrew Manuscripts, Jerusalem. He holds a Ph.D. in the History of Science from Harvard University. His current research interests center around Science and Philosophy in Jewish and Muslim cultures. His latest books are *Yemenite Midrash* (HarperCollins, 1997) and *The Jews and the Sciences in the Middle Ages* (Ashgate-Variorum, 1999).

JEAN-CLAUDE MARTZLOFF (Chinese Mathematical Astronomy) is Directeur de Recherche and belongs to the Chinese Civilization Research Center of the C.N.R.S. (Centre National de la Recherche Scientifique/French National Center for Scientific Research) Paris. He has studied mathematics, astronomy and Chinese at Paris University and classical Chinese at the University of Liaoning Province, Shenyang, China. He became a full time C.N.R.S. researcher in 1980 and has published numerous articles on the history of mathematics, astronomy, and the Jesuit mission in China (17–18th centuries). After many research sojourns in Mainland China (Academia Sinica), Japan (Japanese Society for the Promotion of Science) and Taiwan, he is now writing a book on traditional Chinese calendrical computations and mathematical astronomy.

DANIEL CLARK OREY (The Ethnomathematics of the Sioux Tipi and Cone) earned a Ph.D. in Curriculum and Instruction in Multicultural Education from the University of New Mexico in 1988. He is currently Professor of Mathematics and Multicultural Education at California State University, Sacramento. In 1998, he served as a Fulbright visiting scholar at the Pontifícia Universidade

Católica de Campinas in Brazil, and has served as an adjunct faculty member at D-Q University, in Davis, California. He has been a Tinker International Field Research Grant recipient for ethnographic field research using children and computers in Puebla, México; a Mellon Interamerican Field Research Grant recipient for ethnographic field research using Logo programming language in a highland Maya school in Patzún, Chimaltenango and the Colegio Americano de Bananera, Izabál, Guatemala; and a Title VII Fellow at the University of New Mexico. His primary research interests and objectives stress the use of an ethnomathematical perspective which seeks to facilitate the documentation, development, study, and evaluation of diverse learning environments, with special emphasis on issues of access and equity in mathematics for all learners.

T. K. PUTTASWAMY (The Mathematical Accomplishments of Ancient Indian Mathematicians) is Professor Emeritus of Mathematical Sciences at Ball State University in Indiana. He also taught at the University of Mysore and Madras University in India. He has received M.Sc. (University of Mysore), M.S. (University of Minnesota), and Ph.D. (State University of New York at Buffalo) degrees in mathematics. He has published several papers on the global behavior of solutions of ordinary differential equations in the complex plane. His research papers have earned him regular invitations to speak at various international conferences around the world. Ball State University recognized his earlier research accomplishments in 1976–77, when he received its Researcher of the Year award. He has served as chair for contributed papers at the regional and national meetings of the American Mathematical Society and various international conferences including the International Congress of Mathematicians held at Berkeley, California during the summer of 1986.

JAMES RITTER (Egyptian Mathematics) is co-director of the program in History and Philosophy of Science in the Department of Mathematics at the Université de Paris 8 and of the collection *Histoires de science*, published by Presses Universitaires de Vincennes. His work is concerned with the history of 'rational practices' in the ancient world (Egypt and Mesopotamia) and with the development of general relativity and unified theories in the twentieth century. In the former guise he has published on mathematics, medicine and divination in *A History of Scientific Thought* (ed. Michel Serres, Blackwell) and co-edited *Histoire des fractions, fractions d'histoire* (Birkhäuser) and *Mathematical Europe* (Presses de la Maison de l'Homme). In the latter, he has written on unified theories in *Albert Einstein: Œuvres choisies* (Le Seuil) and on cosmology and literature in *Melancholies of Knowledge* (SUNY). He has also co-edited *The Expanding Worlds of General Relativity* (Birkhäuser).

ELEANOR ROBSON (The Uses of Mathematics in Ancient Iraq, 3000 BCE–300 CE) is a British Academy Postdoctoral Fellow at the University of Oxford and a Research Fellow of Wolfson College, Oxford. She has a degree in mathematics and a doctorate in Oriental Studies, which was published in 1999 as *Mesopotamian Mathematics, 2100–1600 BC* (Oxford: Clarendon Press). She has taught the archaeology, history, and languages of the ancient Middle East at Oxford since 1995. Her research interests are in the social and intellectual history of mathematics in the pre-Islamic Middle East, focussing especially on the scribal culture of early Mesopotamia (Iraq). She is currently writing a study of an 18th century scribal school, provisionally titled *The Tablet House*.

JACQUES SESIANO (Islamic Mathematics) studied theoretical physics at the Swiss Federal Institue of Technology in Zurich and has a Ph.D. in History of Mathematics from Brown University. He is a Lecturer in the history of mathematics at the Swiss Federal Institute of Technology in Lausanne.

SHAI SIMONSON (The Hebrew Mathematical Tradition) is Professor of Mathematics and Computer Science at Stonehill College in North Easton, Massachusetts. He earned his B.A. in mathematics from Columbia University in 1979 and his Ph.D. in computer science from Northwestern University in 1986. He has taught at the University of Illinois, Northwestern University, and Tel Aviv and Hebrew Universities. His research interests are in theoretical computer science, computer science education, and history of mathematics. He has published a number of papers in and received NSF funding in all these areas. His most recent work is about Levi ben Gershon, a medieval Jewish mathematician. 'Gems of Levi ben Gershon' will appear in a special history focus issue of *Mathematics Teacher*, 'The Missing Problems of Levi ben Gershon, A Critical Edition, Parts I and II' will appear in *Historia Mathematica*, and 'The Mathematics of Levi ben Gershon' in *Bekhol Derakhekha Daehu*. His work on medieval Jewish mathematics and its use in the classroom, was sponsored by Grant STS-9872041 from the National Science Foundation.

WALTER SIZER (Traditional Mathematics in Pacific Cultures) is a Professor of Mathematics at Moorhead State University in Minnesota. He has also taught in Malaysia and as a Fulbright lecturer in Ghana. He received a Ph.D. from the University of London, doing work in abstract linear algebra, and has been indulging an interest in the history of mathematics for fifteen years. Publications include papers in mathematics, the teaching of mathematics, and the history of mathematics, particularly the mathematical concepts of Pacific islanders.

DAVID TURNBULL (Rationality and the Disunity of the Sciences) is a senior lecturer in the School of Social Inquiry at Deakin University working in Science Studies. He has published *Maps are Territories: Science is an Atlas* (University of Chicago Press, 1993), and *Masons, Tricksters and Cartographers: Makers of Knowledge and Space* (Harwood Academic Press, 1999, in press). He is currently working on encounters between knowledge traditions with special interests in Geographic Information Systems (GIS) and indigenous knowledge; theories of cultural and technical change; and space, place and narrative in the creation and performance of knowledge.

EDWIN J. VAN KLEY (East and West) is currently Professor of History, emeritus, at Calvin College, where he taught from 1961 to 1995 after receiving his Ph.D. from the University of Chicago. Most of his research and writing explore aspects of the impact of Asia on Europe during the seventeenth century, most of it published in articles such as 'Europe: Discovery of China and the Writing of World History', (*The American Historical Review* 76(2): 358–385, April, 1971). His major publication, co-authored with Donald F. Lach, is *Asia in the Making of Europe; A Century of Advance*, Books 1–4 (Chicago: University of Chicago Press, 1993). He continues to work on this project.

HELEN VERRAN (Logics and Mathematics; Australian Aboriginal Mathematics; Accounting Mathematics in West Africa) is senior lecturer in the Department of History and Philosophy of Science at the University of Melbourne, Australia. For six years in the 1980s she worked as lecturer/senior lecturer in the Institute of Education, Obafemi Awolowo University, Ile-Ife, Nigeria. On returning to Australia in the late 1980s she took up a position with Deakin University training members of Australia's traditional Aboriginal communities as primary school teachers. The course was delivered to students in their own communities. This was a research based course in which curriculum development in Aboriginal schools was a focus. It was the starting point for twelve years of research on logic with the Yolngu community; this research continues today. Her most recent work is *African Logics: Towards Ontology for Postcolonial Times and Places* (forthcoming). She is currently working on *Logics and Places: Learning from Aboriginal Australia.*

LEIGH WOOD (Communicating Mathematics across Culture and Time) is the Director of the Mathematics Study Centre at the University of Technology, Sydney. The Centre works with students individually and in small groups to help them reach their academic potential. Leigh teaches a range of subjects to mathematics majors, including differential equations and mathematical communication. Her interest in mathematical discourse as part of mathematics started in 1985 as a result of working with refugees from Southeast Asia. Leigh has developed many teaching and learning packages including 2 textbooks and 10 videos, all of which have a strong mathematical communication component. Her research areas are language in mathematics and curriculum design.

INTRODUCTION

Every culture has mathematics. That is not to say that every culture has forms of deductive reasoning or even that every culture has counting. In most places, mathematics grew from necessity, and not every culture had reasons to ask 'How many?' or 'How much?' Without trade, and without herding and agriculture, societies saw little need to enumerate. But enumeration and calculation are only parts of mathematics; a broader definition that includes 'the study of measurements, forms, patterns, variability and change'[1] encompasses the mathematical systems of many non-Western cultures.

I was a Peace Corps volunteer in Central Africa when I first realized that there were other forms of mathematics. I noticed that people counted differently on their fingers. Whereas I would signal one using my index finger, Malawians used their thumbs. To indicate four, I would use all fingers except the thumb; Malawians used those same fingers, but arranged them so that the two adjacent fingers touched. When they got to five, they made a fist instead of extending all their fingers. At the time this seemed like another fascinating cultural distinction, part of the whole process of learning to see myself as a product of my society and to realize that things I thought were universal human activities were in fact American ones. What intrigued me more was that Malawians indicated 100 by shaking their fists twice, as if knocking on a door. This seemed to me an unusual mathematical organization and was my first hint that numbers and how we express them are connected to our culture and upbringing.

Until the past few decades, histories of mathematics have virtually ignored the mathematics of non-European cultures, even after Egyptian mathematical papyri and Babylonian clay tablets illustrating complex mathematical problems were discovered. This neglect grew from the colonial mentality, which ignored or devalued the contributions of the colonized peoples as part of the rationale for subjugation and dominance. The disparagement by European colonial societies is particularly ironic because the early Europeans, the ancient Greeks themselves, acknowledged their intellectual debt to the Egyptians. Herodotus believed that geometry had originated in Egypt, emerging from the practical need for re-surveying after the annual flooding of the Nile. Aristotle thought that the existence of a priestly leisure class in Egypt prompted the pursuit of

H. Selin (ed.), Mathematics Across Cultures: The History of Non-Western Mathematics, xvii–xx.

geometry. Modern Euro-American scholars, blinded by their belief in their own concept of rationality, disregarded the origins of their own discipline by ignoring Aristotle and Herodotus.

The views of Herodotus and Aristotle also illustrate two theories concerning the beginnings of mathematics, one arising from practical necessity and the other from priestly leisure and ritual. This dichotomy appears in many of the articles in this collection, along with a third element that is also present. Mathematics is either closely connected to religion and ritual or to the needs of daily life, but occasionally it seems to be a purely intellectual activity, as in the case of magic squares.

Eurocentric views of scholarship have changed radically in recent years, and newer histories of mathematics, such as George Ghevergese Joseph's *The Crest of the Peacock: Non-European Roots of Mathematics* (London: I. B. Tauris, 1991) and Frank J. Swetz's *From Five Fingers to Infinity: A Journey Through the History of Mathematics* (Chicago: Open Court, 1994), reflect this trend. They illustrate the debt we owe to the Arabs in bringing together the technique of measurement, evolved from its Egyptian roots to its final form in the hands of the Alexandrians, and our remarkable number system, which originated in India. Although these books and other new scholarship have contributed significantly to the study of the mathematics of non-Western cultures, they focus almost entirely on cultures whose mathematical systems are precursors to our own. The mathematics of other cultures is still largely overlooked.

The *Encyclopaedia of the History of Science, Technology, and Medicine in Non-Western Cultures*, published by Kluwer Academic Publishers in 1997, contained many articles on world mathematics. *Mathematics Across Cultures* is an extension of that encyclopedia. Our aim is to add depth to the articles that, as encyclopedia articles must, covered the subjects as broadly as possible. The collection is divided into two sections. Essays in the first section discuss the connection between mathematics and culture, the field of ethnomathematics, the concept of rationality and how it applies to the study of culture, and the transmission of knowledge from East to West. The chapters in the second part describe individual cultures and their mathematics.

The book begins with Leigh Wood's essay, 'Communicating Mathematics across Culture and Time.' Condensing a great deal of history and philosophy in her discussion, Wood analyzes mathematical communication in societies without written language. She goes on to examine the challenges to successful communication even within cultures that have left written records. Ron Eglash's essay on ethnomathematics compares ways of looking at mathematics in different cultures. In one view, ethnomathematics is the study of mathematical concepts in indigenous cultures; in the other, the word applies to the connections between mathematics and culture regardless of geography or type of societal structure. Edwin Van Kley's paper, distilled from his massive and impressive work, *Asia in the Making of Europe*, describes the centuries-long movement of science and technology from East to West. Among the scientific achievements that the West adopted was the Indian system of arithmetical notation, trigonometry, and the system of calculating with nine Arabic numbers and a zero. Van Kley sees the dominance of Western culture, especially in the last century, as a threat both to the world's traditional cultures and to traditional Christian Western culture.

David Turnbull's provocative essay in the sociology of scientific knowledge, 'Rationality and the Disunity of the Sciences', proposes rethinking the notion of the unity and universality of science and mathematics. He sees the concept of local knowledge as a means of comparing knowledge production systems across cultures; he goes on to dissect the concepts of rationality, objectivity, and scientific method. In his view, 'rather than rejecting universalizing explanations what is needed is a new understanding of the dialectical tension between the local and the global.' In her paper, 'Logics and Mathematics: Challenges Arising in Working across Cultures', Helen Verran also tackles the connection between a universal concept – logic – and its effect on cultural misperceptions. By looking at Australian Aboriginal and Nigerian classrooms, she demonstrates how logic is perceived and carried out in dissimilar societies.

In the final chapter of the first section, 'A Historiographical Proposal for Non-Western Mathematics,' Ubiratan D'Ambrosio shows how colonialism led to a disparaging and belittling of the colonized cultures and their mathematical and scientific achievements. He proposes a new way of looking at mathematics that recognizes other systems of intellectual and social knowledge.

Fifteen culture areas are studied in depth in the second section. Eleanor Robson explores Mesopotamian mathematics in her article on the uses of mathematics in ancient Iraq. James Ritter examines Egyptian mathematics, urging us to study what it meant and how it was used by the Egyptians themselves. Jacques Sesiano describes the extraordinary synthesis of Arab mathematics from its Mesopotamian, Indian, and Greek heritage. He analyzes arithmetic, algebra, geometry, and number theory and ends with a detailed look at magic squares. Y. Tzvi Langermann and Shai Simonson look at two different branches of mathematics in their essay on the Hebrew mathematical tradition. Langermann investigates the arithmetic and numerology of Abraham Ibn Ezra and goes on to describe Hebrew contributions in geometry; Simonson focuses on algebra and its evolution from a geometric to a combinatorial subject.

Thomas Gilsdorf considers several environmental and cultural influences on Inca mathematics, emphasizing the interdependence of mathematics and non-mathematical factors. Michael Closs uses glyphs and available texts to interpret the mathematics of several Mesoamerican people, including the Olmec, Zapotec, Maya, Mixtec, and Aztec. He examines their mathematical astronomy and the connection between the calendar and the state. For North America, Daniel Orey investigates one group of native Americans, the Sioux, and the mathematics used in building and setting up their tipis.

Walter Sizer's essay on traditional mathematics in Pacific cultures examines numeration, geometry, games, and kinship relations in Polynesia, Melanesia, and Micronesia. Helen Verran explores a different part of the Pacific region in her article on land ownership mathematics among Australian Aboriginal people. She believes that questions concerning different forms of logic and mathematics are central to cross-cultural understanding.

In his paper on Central and Southern Africa, Paulus Gerdes presents evidence for early mathematical activity and discusses details of geometrical ideas as expressed in mat weaving, house wall decoration and sand drawings. Helen Verran explores the West African tradition as she relates stories of Yoruba number use. She shows how the properties of a number system reflect the needs of the culture that creates it and stresses the importance of the link between a society and its workings for understanding its number system.

Jean-Claude Martzloff condenses hundreds of years of history in his study of Chinese mathematical astronomy. He discusses the sources themselves and places them in their historical, political, and epistemological context. The second part of his paper explores in depth the mathematics of the Chinese calendar. T. K. Puttaswamy's chapter on India describes the richness and complexity of the mathematics of ancient Indians by focusing on early works such as the *Sulbasütras* and great mathematicians such as Āryabhaṭa I and Brahamagupta. Jochi Shigeru's essay, 'The Dawn of *Wasan* (Japanese Mathematics),' provides us with an excellent illustration of a way to do mathematics that is unique to the Japanese culture. He explains its evolution, relates the stories of some important mathematicians, and also studies magic squares in detail. Kim Soo Hwan takes a completely different approach when he shows how to use games and flags to reveal Korean ethnomathematics.

There are many reasons to study other systems of mathematics, including their potential contribution to our own mathematical discourse and to our appreciation of the richness of cultural and scientific diversity. We invite you to explore different ways of doing mathematics with us.

Helaine Selin
Amherst, Massachusetts
Winter 2000

NOTE

[1] Australian Academy of Sciences, *Mathematical Sciences: Adding to Australia* (1996: ix), cited in Wood, Leigh, 'Communicating Mathematics across Culture and Time', in this volume.

LEIGH N. WOOD

COMMUNICATING MATHEMATICS ACROSS CULTURE AND TIME

> The truth is that history, as we commonly conceive of it, is not what happened, but what gets recorded and told. Most of what happens escapes the telling because it is too common, too repetitious to be worth recording...
>
> The business of making accessible the richness of the world we are in, of making dense and substantial our ordinary, day-to-day living in a place, is the real work of culture.
>
> (Malouf, 1998: 17)

Mathematics is a method for communicating ideas between people about concepts such as numbers, space and time. In any culture there is a common, structured system for such communication, whether it be in unwritten or written forms. These systems can form bridges of communication across culture and across time.

What is communication? Crowley and Heyer (1995: 7) describe communication as '... an exchange of information and messages. It is a process. About one hundred thousand years ago our early ancestors communicated through nonverbal gestures and an evolving system of spoken language. As their world became more complex, they needed more than just a shared memory' ... this led to 'the development of media to store and retrieve the growing volume of information.' Communication is about an exchange of information and the techniques humans have developed to store and retrieve that information. It is not necessary to have writing, computer disks and so on: communication can be verbal (oral traditions), non-verbal (gestures, *quipu*), temporary (sand drawing) or more permanent (clay tablet, woven cloth).

Communicating in or about the field of mathematics involves taking part in mathematical discourse, whether by reading, writing, listening or speaking. Discourse is a broader concept than language because it also involves all the activities and practices that are used to make meaning in a particular profession. The discourse of mathematics includes all the ways that mathematics is done: through language, textbooks, mathematicians talking to each other and to a

H. Selin (ed.), Mathematics Across Cultures: The History of Non-Western Mathematics, 1–12.

wider public, and through popularisation and application of mathematical knowledge (Wood and Perrett, 1997).

Discourse does not see the terms *language* and *communication* as synonymous, although it certainly recognises language as a central resource for the communication process. This view sees language as a resource for making meanings within different situations (Halliday, 1978; Kamp and Reyle, 1993). Communication is seen as a complex process within which natural language supplies a major (but seldom the only) resource for making specific meanings within the framework of specific social practices or areas of knowledge. Readers who wish to know more about the complexities of natural language may consult texts such as Finegan and Besnier, 1989.

When we speak of making meaning within different areas of knowledge we are referring to different discourse practices. The language used in mathematical discourse is natural language but it is a language that has evolved in specialised ways to deal with the demands of expressing mathematical concepts. The most striking difference between mathematical writing and most other types of writing is the extensive use that mathematics makes of symbols and numbers. These are sometimes seen as alternatives to natural language, alternatives that mathematicians value as being more precise and more elegant than natural language. However the symbolic language of mathematics is an extension of natural language as much as a replacement for it, just as a pile of cowrie shells or a gesture can represent a number word. Natural language is used to complement the symbolic language of mathematics when mathematicians talk and write to each other and it has to replace the symbolic language when mathematicians communicate with non-mathematicians. Thus we would argue that natural language plays an essential part in doing mathematics.

All cultures are mathematised, in that people within any culture use ideas of mathematics in their everyday life. For this I will take a wide definition of mathematics to include concepts of number, space, chance, and time. A report of the Australian Academy of Sciences, *Mathematical Sciences: Adding to Australia* (1996: ix), gives this description of modern mathematics:

> Mathematics is the study of measurements, forms, patterns, variability and change. It evolved from our efforts to understand the natural world ...
>
> Over the course of time, the mathematical sciences have developed a rich and intrinsic culture that feeds back into the natural sciences and technology, often in unexpected ways. The mathematical sciences now reach far beyond the physical sciences and engineering; they reach into medicine, commerce, industry, the life sciences, the social sciences and to every other application that needs quantitative analysis.

This description takes mathematics back to its roots. Mathematics has been part of all societies, a part of every profession as well as everyday life. Western mathematics became narrower with the insistence that only deductive mathmatics from a set of axioms, following the Greek tradition, was *real* mathematics. The broader view of mathematics, especially with the consideration of computing, validates the work of non-European mathematicians. A majority of mathematicians today are working much closer to the way that Indian,

Chinese and Arab mathematicians worked, as they focus on real applications and better computing algorithms (Horgan, 1993).

Culture too can be viewed broadly. Ascher (1991: 2) says, 'In any culture people share a language; a place; traditions; and ways of organizing, interpreting, conceptualizing and giving meaning to their physical and social worlds.' David Malouf (1998: 17) describes the process of acculturation as the 'business of making accessible the richness of the world we are in, of making dense and substantial our ordinary, day-to-day living in a place ...' The description of mathematics above states that the 'mathematical sciences have developed a rich and intrinsic culture.'

Mathematics has often worked on many levels, as part of everyday culture and also as used by subgroups within the main culture. In the Inca culture only a small group would have been able to construct and interpret the *quipus* (knotted cords used by the Inca to convey data) (Ascher, 1991). In the Babylonian culture of 3000 BC few people would have been skilled in the algebraic and computational techniques required for commerce, as is evidenced by the many clay tablets which can be interpreted as instructional textbooks. In Egypt, the Ahmes papyrus consists of instructional material for a mathematical subculture that calculated the land areas after the annual floods and documented commercial transactions. Indeed in many cultures, the mathematics of calendars and astronomy was in the hands of the priestly classes.

The mathematical subcultures communicated their knowledge to future mathematicians in a similar way to classes or apprentices today. Students worked through sets of paradigm problems designed to develop the calculating skills, ideas and language necessary for their future careers. Some of the ways that mathematics has been taught is remarkably similar across the cultures and centuries. The drill and practice examples on the Ahmes papyrus and some Babylonian clay tablets are close to the way mathematics is taught in many classrooms today (Fowler and Robson, 1998: 369). Teachers wrote commentaries and extensions from previous texts; they improved algorithms and compared methods. Høyrup (1994) concluded that mathematics as a discipline began to be organised systematically with the need to teach it to professional scribes in about 3000 BC. [Editor's note: see the articles by Gildorf , Robson, and Ritter on Inca, Babylonian, and Egyptian mathematics.]

COMMUNICATION WITHOUT WRITTEN LANGUAGE

We will examine some examples of mathematical discourse in cultures with no written language. We are dealing with speaking and listening, but also with art, artefacts and gesture. It is possible to consider the *quipu* (Ascher, 1991) for example, as a 'written' number system, but for this essay I do not wish to include such artefacts as written language.

How were mathematics and mathematical ideas communicated within cultures without writing? Here we make the assumption that cultures that do not have written language (95% of cultures, Ascher, 1991: 2) communicate mathematics by methods that have been documented by outside observers. The reliability of the documentation varies greatly with the biases and the mathematical knowledge of the observer. These snapshots can nonetheless help to extrapolate back into history for cultures that have lived in isolation for some time. For example the counting system of the Gomileroi in South-Eastern Australia (Table 1) has been documented and I suggest that this number system has been *in situ* for thousands of years prior to the arrival of Europeans in Australia in 1788. Similarly the counting systems in Papua New Guinea, documented by David Lean (summarised in Phythian, 1997), would have been used for centuries before the arrival of Europeans and European mathematics. Richard Pankhurst (as cited in Zaslavsky: 89) has made an extensive study of the measures, weights and values in use among the various Ethiopian peoples throughout their history.

Between the years 1968 and 1988, Glendon Lean collected and recorded data on the counting systems of Papua New Guinea and Oceania. There are 1200 languages in the region and Lean collected data on nearly 900. He also discussed how the number systems were communicated through migration, wars and marriage. As these languages are not written, he made use of diagrams to show gesture counting and used phonetics to write down number words. His awareness of mathematics has meant that he was conscious of the 2, 5, 10 and 20 cycles that occur in counting systems and gathered enough data to be able to classify each system. From his study, Lean concluded that the counting and tally systems in Papua New Guinea could have covered a period of thousands of years. Due to the fact that many neighbouring languages have few number words in common, he concluded that there has been little contact. An extreme example of this is two languages, Baruya (6000 speakers) and Yagwoia (5000 speakers). People who live north of Marawaka in the Eastern Highlands speak Baruya, and Yagwoia is spoken to the south. Table 2 gives a

Table 1 Gomileroi counting system (as quoted in Petocz, Petocz and Wood, 1992: 164 after consultation with elders)

Gomileroi words	*English translation*	
mal	finger	1
bular	two fingers	2
guliba	three fingers	3
bularbular	two fingers and two fingers	4
mulanbu	belonging to one hand	5
malmulanbu mummi	one finger and one hand added on	6
bularmulanbu mummi	two fingers and one hand added on	7
gulibamulanbu mummi	three fingers and one hand added on	8
bularbularmulanbu mummi	two and two fingers and one hand added on	9
bulariu murra	belonging to two hands	10
maldinna mummi	one toe added (to two hands)	11
bulardinna mummi	two toes added on	12
gulibadinna mummi	three toes added on	13
bularbulardinna mummi	two and two toes added on	14
mulanbudinna mummi	one foot added on	15
maldinna mulanbu	one toe and a foot added on	16

Table 2 Examples of Baruya and Yagwoia numbers (Phythian, 1997: 66)

Baruya	*Yagwoia*	
da-'	ungwonangi	1
da-waai	huwlaqu	2
da-waai-da	huwlaqungwa	3
da-waai da-waai	hyaqu-hyaqu	4
at-i	hwolyem pu	5
at-iraai	hwolye kaplaqu	10

few examples of differences (Phythian, 1997). Should these two groups meet, they would need to have a good gesture system or concrete materials to come to a mathematical understanding, despite living next to each other.

The detailed work in Papua New Guinea contrasts with the work of linguists in the 1800s who recorded Tasmanian aboriginal languages before they (the aboriginals and the languages) died out. Perhaps due to a lack of mathematical awareness they did not record details of the counting systems. Some words for numbers are recorded but not in a systematic fashion. For example, one Tasmanian aboriginal group use the word *karde* for 5 and *karde karde* for 10 (Roth, 1899: Appendix B, xi, xvii) but only a few other numbers are recorded. So we have a tantalising suggestion of a base five system but insufficient evidence to come to a conclusion. This lack of evidence for numbers contrasts with the detailed coverage of words for the male and female genital areas (Roth, 1899). I think we can make some conjectures about these Victorian scientists and their cultural proclivities.

In *Hidden in Plain View*, Jacqueline Tobin (1999) records the oral history of several African American quilt makers. They described the meanings and detailed communication that could be passed on by the combination of colours, beads and knots. They could give routes, times, maps and instructions all through quilting. Thus they passed on their African history and made plans to escape from slavery, all under the noses of their masters. The following quote (Tobin, 1999: 78) discusses the *lukasa*, the memory boards of the Central African Luba people.

> The use of stitches and knots, as a kind of Morse code in thread, along with fabric color and quilt patterns, made it possible to design a visual language.
>
> The *lukasa*, or memory board, is a mnemonic device used by the highest level of the Luba royal association. The *lukasa* contains secret mythical, historical, genealogical and medical knowledge. Beads on the front ... Engraved geometric patterns on the back ... All these serve to recall aspects of Luba history.

There are mathematical implications. The quilts are like computer programs displaying information graphically and can be read by those who understand the code. Fortunately the code in this case has been recorded. It is an example of how mathematical (and other) ideas can be hidden in plain view.

Aboriginal art is another area where mathematics has been overlooked by Western eyes. Michael Cooke (1991) spent 10 years in the Yolngu community in northern Australia and his principal sources were Yolngu teachers. He acknowledges that by taking some of the Yolngu world and fitting it with a Western idea of mathematics, he has lost the full significance of the meanings and some of the intricacy of the Yolngu world. His paper covers many mathematical aspects including how kinship relationships are depicted in song and painting (Cooke, 1991: 38).

> For the Yolngu artist, painting is a means of schematising Yolngu world order in a way parallel to the Western mathematical theorist who constructs graphs and diagrams. Just as the Western mathematician seeks elegance, symmetry and aesthetic satisfaction in such work, so does the Yolngu artist. Both rely on extensive use of systems of symbolic representation in their abstract modelling of order. ...
>
> For the Yolngu it is the system of song cycles which provides the theoretical basis and rationale for the Yolngu system of order and relationship.

There are many other examples of mathematical discourse in cultures without a written language:

1. *Finger reckoning*. Hand signals for numbers and number operations are used in many cultures. Examples in Africa are given in Zaslavsky (1973: 239–253) and for Papua New Guinea in Phythian (1997).
2. *Weaving or patterns*. Northern Australian aboriginals use painting to illustrate kinship patterns (Cooke, 1991; Harris, 1991).
3. *Cowrie shells* for currency. Again there are many examples in Zaslavsky.
4. *Knotting, quipu of the Incas*. Much has been written on the quipu (e.g., Ascher, 1991: 16–27; Smith, 1925: 196; Joseph, 1992: 28–37) and it remains an important example of how mathematics and mathematical ideas can be communicated effectively without a written language or number system.
5. *Classification*. For example, Yolngu Aboriginal culture divides the world into two parts, *Yirritja* and *Dhuwa* (Cooke, 1991). [Editor's note: see Verran's article on Aboriginal mathematics in this volume.]

All these cultures communicated mathematical ideas without written language. Some of the ideas would have been accessible to all people within the culture, such as the counting systems in Papua New Guinea, and some of the ideas would be restricted to a privileged group, such as the quipu makers in South America.

Making meaning from artefacts left to us presents a difficult task. What meaning can a modern observer make of the Tasmanian Aboriginal rock carvings in Figures 1 and 2? Is this a number system, a pattern of moon phases, a written language? Even the arrangement by the modern artist may influence our ideas. We are using our western-trained minds to classify and order the images to fit with our concepts of logic and pattern. Unfortunately there are no Rosetta stones or living speakers to assist with translation.

CULTURES WITH WRITTEN LANGUAGE

Making meaning from history involves examining the discourse of cultures from the information that is recorded and passed on. Written forms of math-

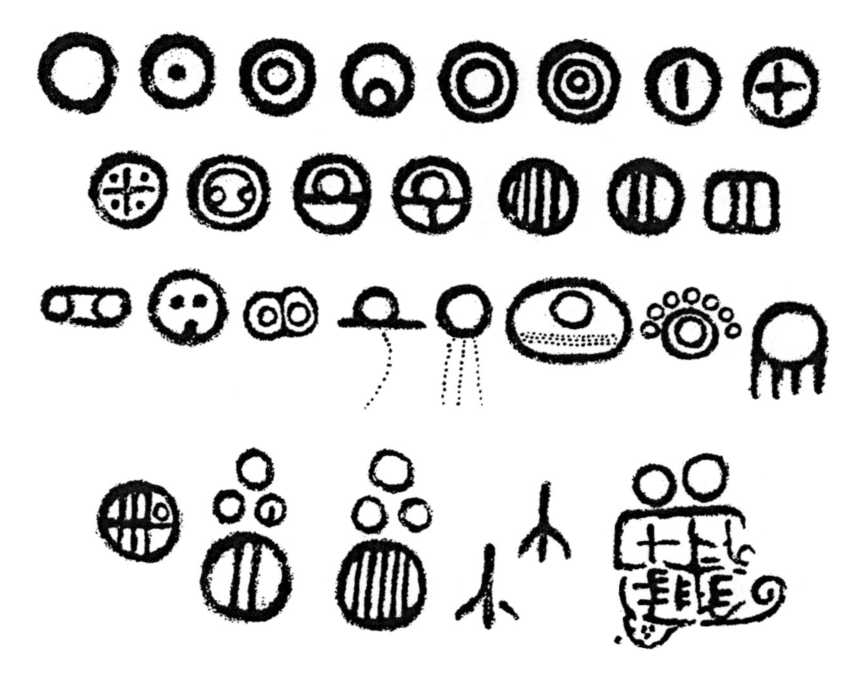

Figure 1 Symbols used in Tasmanian Aboriginal art. Clark, 1986: 32.

ematical discourse have largely formed the basis of modern mathematics: thus mathematical ideas have communicated across time. However, even with written records there are hazards to successful communication.

It is in part good management and mostly luck that written materials are available to modern readers, even though writers, such as Rashīd al Dīn (1971 translation) in 13th century Persia went to great lengths to increase the probability that his work would not be lost. He made two complete copies of each work; each was translated into Persian and Arabic. Only the best quality paper and only scribes with the best handwriting were used. Even so, not all of his works exist today, but those that do give an insight into the mathematics in the lives of everyday people of that era. New materials such as a letter of al-Kāshī on scientific life in Samarkand, described and translated in Bagheri (1997), are being added to our collection of records of mathematics. Translations, for example al Dīn (1971) make original works accessible to readers who are unable to read ancient languages.

Significant communication of mathematical ideas and techniques occurred between cultures in the past. An example of how commentaries and improved algorithms occurred between Arabic and Indian mathematicians is explained in Rashed (1994: 143–148) who showed that al-Bīrūnī (10th century AD) was aware of Brahmagupta's methods of quadratic interpolation, used for trigonometry and astronomy.

However, as David Malouf (1988) states, much of what really happened is not recorded, and therefore must be inferred in the same way as for cultures

Figure 2 Tasmanian Aboriginal rock carvings *in situ*. Clark, 1986: 33.

with non-written languages. The information that is recorded is filtered through the eyes of the observer, who has his or her own cultural prejudices. Such cultural prejudices have caused criticism of non-European mathematics, in particular with the idea of proof. Modern mathematicians are in considerable conflict about the status and usefulness of proof (Horgan, 1993) but some commentators feel free to criticise other cultures for not fitting within their own narrow definition.

Much has been made of the Greek concept of proof as the basis of modern mathematics to the extent that many writers (for example Kline, 1972: 190, quoted in Joseph, 1994: 194) have disparaged the Indians for their supposed haphazard ideas of proof. But as Joseph (1994) demonstrates, Indian mathematics does prove theorems though not in the same way as Greek deductions from axioms – in fact the Indian demonstrations are very similar to the way proofs are presented to students studying secondary and early tertiary mathematics. A broader idea of proof, such as that advocated by Mason, Burton and Stacey (1985), amongst others, would include the demonstrations of Indian and Chinese mathematicians.

Another reason for miscommunication is error of interpretation. We mainly rely on translations of ancient texts. Translations are open to interpretation, compounded by time and cultural differences. Added to this, many translators are not mathematicians and so may miss subtle points which may be critical mathematically. This is not a criticism of translators but a criticism of commentators who do not take these difficulties into account when interpreting discourse over time and culture.

The mistranslation of one word can have critical effects. For example, the mistranslation of the word *asanna* in the translation of an Indian text has led to considerable misrepresentation of Indian mathematics. In verse 10 of the section entitled *Ganiṭapāda* from Āryabhaṭa's *Āryabhaṭīya* (AD 499) the following appears: '100 plus 4, multiplied by 8, and added to 62,000: this will be the "asanna" value of the circumference of a circle of diameter 2000.' The Sanskrit word *asanna* has been translated as approximate, inexact, rough, crude and so on. On the basis of this translation, some Western historians of mathematics have concluded that Indian mathematicians did not realise the irrational nature of π

George Gheverghese Joseph believes that the word *asanna* has a more fruitful meaning. It incorporates two different meanings that overlap with one another: 'inexact' and 'unattainable'. The second meaning, which is closer to the idea of irrationality, is what Joseph thinks Āryabhaṭa had in mind, and it was an extremely creative concept that gave the motivation to Kerala mathematicians to work on infinite series (Joseph, personal communication).

Reza Hatami, an Arab-speaking scholar working in Sweden, has translated sections of the work of al-Khwārizmī and Omar Khayyām and has found similar difficulties with some of the standard English and Swedish translations (Hatami, 1999).

Many translations are written in the modern discourse of mathematics and include use of symbols and algebra. The clarity of the symbolism in use today can cause us to underestimate the difficulties that mathematicians faced in the past. There is a good example of a theorem from Cardano (1545) that shows how mathematical writing has changed in the last 400 years. A comparison of this theorem with its modern equivalent shows how useful symbolic notation can be, and also illustrates the relationship of such notation to ordinary language (Smith, 1997: 6). Cardano's intention was to provide a solution to the equation $x^3 + mx = n$, where m and n were implicitly assumed to be positive. However, even the statement of the theorem was different to modern language. Renaissance Italians referred to this equation as *a cube plus a first power equal to a number*. Cardano gave the solution as:

> Cube one-third of the coefficient of the unknown; add to it the square of one half the constant of the equation; and take the square root of the whole. You will duplicate this, and to one of the two you add one half the number you have already squared and from the other you subtract one half the same. Then, subtracting the cube root of the first from the cube root of the second, the remainder left is the value of the unknown.

Today we would write something like the following:

Theorem: A solution of the equation $x^3 + mx = n$ is given by

$$x = \sqrt[3]{\frac{n}{2} + \sqrt{\frac{n^2}{4} + \frac{m^3}{27}}} - \sqrt[3]{-\frac{n}{2} + \sqrt{\frac{n^2}{4} + \frac{m^3}{27}}}.$$

These examples show some of the difficulties of interpreting works from other cultures and times. The observer brings his or her knowledge and culture into the equation. I take issue with commentators, such as Kline (1980: 111), who make strident assertions about the mathematics of another culture.

> It is fairly certain that the Hindus did not appreciate the significance of their own contribution. The few good ideas they had, such as separate symbols for the numbers 1 to 9, the conversion from positional notation in base 60 to base 10, negative numbers, and the recognition of 0 as a number, were introduced casually with no apparent realization that they were valuable innovations. They were not sensitive to mathematical values.

This is based on flimsy evidence and a very narrow view of mathematics. Notice the use of 'mathematical values'. Whose values? It may be that future commentators claim that strictly holding to the Greek view of deductive proof has seriously curtailed the mathematics of the 20th century.

A TENUOUS LINK

> This art originated with Mahomet the son of Moses the Arab (al-Khowarizmi). Leonardo of Pisa (Fibonacci) is a trustworthy source for this statement.
>
> (Cardano, 1545, translation Whitmer, 1968).

This is a lovely example of communication. Cardano has acknowledged the originator of his work, al-Khwārizmī, and the conduit of this knowledge, Leonardo Fibonacci.

The work of al-Khwārizmī (*ca.* 780–*ca.* 850), through the channel of *Liber Abaci* (1202) by Leonardo Fibonacci became well known in Europe and was the basis of Cardano's work in 1545. This is just one example of how European mathematics was influenced by Arab mathematics, but it also shows how the European mathematicians readily acknowledged the prior work of non-European mathematicians.

Leonardo Fibonacci made considerable contributions to mathematics himself, notably in solving cubic equations, but it is his 1202 *Liber abaci* that was the most influential. Fibonacci travelled widely through Egypt, Syria, Sicily and Greece and came into contact with Arabic mathematics. He was convinced that the Indian-Arabic numerals and methods of calculation were vastly superior to the current methods in Italy (Eves, 1983). *Liber abaci* can be considered the main reason why Indian-Arabic numerals and the achievements of Arabic mathematicians had spread through Europe by the start of the Renaissance. How much would Cardano and others have achieved without al-Khwārizmī and Indian-Arabic numerals?

MAKING MEANING

Making meaning from the sources available to us leaves most unsaid and unwritten. Much of the ordinary day-to-day practices of mathematics are not recorded. The mathematical contributions of women, artisans and many workers are not represented. Many records have been lost. Of those discovered, there are risks of mistranslation, misinterpretation, or, in the case of the Tasmanian Aboriginal rock carvings, no basis for interpretation.

It is reasonable to expect difficulties with communication across time and cultures. Even within the same culture, recorders and observers often disagree. However there have also been triumphs of communication across time and culture – Fibonacci's *Liber abaci* brought the ideas of algebra and Indian-Arabic numerals to Europe giving an excellent starting point for the work of Cardano and others. Fibonacci displayed what many commentators have not: an open mind. He was open to the work of other cultures and actively disseminated their ideas. Glendon Lean and Richard Parkhurst are more modern observers who have shown cultural sensitivity as well as keen mathematical knowledge.

As Zeilberger (1993) stated during a lively debate in the American Mathematical Society, 'Although there will always be a small group of "rigorous" old-style mathematicians ... who will insist that the true religion is theirs and that the computer is a false Messiah, they may be viewed by future mainstream mathematicians as a fringe sect of harmless eccentrics ...' (Zeilberger as cited in Cohen, 1997).

Mathematics itself is not one culture with one discourse. The dominant paradigm in Western mathematics is shifting from Greek deductive proof to a more experimental and applied mathematics and with this comes a new discourse. A reappraisal of the contributions of non-European mathematics under the new paradigm will see a better valuing of their mathematical achievements and consequent changes in forms of mathematical discourse.

Let us look forward to a time when we can appreciate the mathematics of all cultures and the contribution of mathematical ideas to the 'business of making accessible the richness of the world we are in, of making dense and substantial our ordinary, day-to-day living in a place – the real work of culture'.

BIBLIOGRAPHY

Ascher, Marcia. *Ethnomathematics: A Multicultural View of Mathematical Ideas.* Belmont: Wadsworth, 1991.

Bagheri, Mohammad. 'A newly found letter of al-Kāshī on scientific life in Samarkand.' *Historia Mathematica* 24: 241–256, 1997.

Benson, James D. and William S. Greaves, eds. *Systemic Perspectives on Discourse.* Norwood, New Jersey: ABLEX, 1985.

Cardano, G. *Ars Magna or the Rules of Algebra*, trans. T. Richard Witmer. Mineola, New York: MIT, 1968, original 1545.

Clark, Julia. *The Aboriginal People of Tasmania*, 2nd ed. Hobart: Tasmanian Museum and Art Gallery, 1986.

Cohen, G. L. 'Speculation and rigour in modern mathematics.' In *Advanced Mathematical Discourse*, Leigh N. Wood and Gillian Perrett, eds. Sydney: UTS, 1997, pp. 61–69.

Cooke, Michael. *Seeing Yolngu, Seeing Mathematics.* Ngoonjook Occasional Paper Series. Batchelor, Northern Territory: Batchelor College, 1991.

Crowley, David and Paul Heyer. *Communication in History: Technology, Culture, Society*, 2nd ed. White Plains, New York: Longman, 1995.

al Dīn, Rashīd. *The Successors of Genghis Khan.* Translated from the Persian by John Andrew Boyle. New York: Oxford University Press, 1971.

Eves, Howard. *Great Moments in Mathematics before 1650.* Washington, DC: The Mathematical Society of America, 1983.

Fowler, David and Eleanor Robson. 'Square root approximations in Old Babylonian mathematics.' *Historia Mathematica* 25: 366–378, 1998.

Gregory, Michael and Peter Fries, eds. *Discourse in Society: Functional Perspectives.* Norwood, New Jersey: ABLEX, 1995.

Halliday, Michael. A. K. *Language as Social Semiotic.* Baltimore, Maryland: University Park Press, 1978.

Harris, Pam. *Mathematics in a Cultural Context: Aboriginal Perspectives on Space, Time and Money.* Geelong, Australia: Deakin University Press, 1991.

Hatami, Reza. *Från Al jábr till Ars magna.* M.S. Thesis, University of Växjö, 1999.

Horgan, John. 'The death of proof.' *Scientific American* 269(4): 92–102, 1993.

Høyrup, Jens. *Measure, Number and Weight: Studies in Mathematics and Culture.* Albany, New York: State University of New York Press, 1994.

Joseph, George Gheverghese. 'Different ways of knowing: contrasting styles of argument in Indian and Greek mathematical traditions.' In *Mathematics, Education and Philosophy*, Paul Ernest, ed. London: The Farmer Press, 1994.

Kamp, Hans and Uwe Reyle. *From Discourse to Logic.* Dordrecht: Kluwer Academic Publishers, 1993.

Kline, Morris. *Mathematics: The Loss of Certainty.* New York: Oxford University Press, 1980.

Malouf, David. 'Imagination puts us in our place.' *The Sydney Morning Herald.* 23rd November, 1998, p. 17.

Mason, John, Leonie Burton and Kaye Stacey. *Thinking Mathematically.* Bristol: The Bath Press, 1985.

National Committee for Mathematics of the Australian Academy of Science. *Mathematical Sciences: Adding to Australia.* Canberra: Australian Academy of Science, 1996.

Petocz, Peter, Dubravka Petocz and Leigh N. Wood. *Introductory Mathematics.* Melbourne: Nelson, 1992.

Phythian, Ted J. 'Counting systems of Papua New Guinea.' *PNG Journal of Mathematics, Computing & Education* 3(1): 65–78, 1997.

Roth, H. Ling. *The Aborigines of Tasmania*, 2nd ed. Halifax: F. King & Sons, 1899.

Smith, David Eugene. *History of Mathematics.* New York: Ginn and Company, 1925; reprinted New York: Dover, 1958.

Smith, Geoffrey Howard. 'Reading and writing mathematics.' In *Advanced Mathematical Discourse*, Leigh N. Wood and Gillian Perrett, eds. Sydney: UTS, 1997, pp. 5–17.

Tobin, Jacqueline L. and Raymond G. Dobard. *Hidden in Plain View.* New York: Oxford University Press, 1999.

Wood, Leigh Norma and Gillian Perrett, eds. *Advanced Mathematical Discourse.* Sydney: UTS, 1997.

Zaslavsky, Claudia. *Africa Counts.* New York: Lawrence Hill Books, 1973.

Zeilberger, Doron, 'Theorems for a price: tomorrow's semi-rigorous mathematical culture.' *Notices of the American Mathematical Society* 40: 978–981, 1993.

RON EGLASH

ANTHROPOLOGICAL PERSPECTIVES ON ETHNOMATHEMATICS

The term 'ethnomathematics' has two distinct meanings currently in use. D'Ambrosio (1990), who coined the term, takes it to mean a general anthropology of mathematical thought and practice. In that sense, the word applies to the connections between mathematics and culture in every geographic area, every human group, and in every historical moment on earth. Ascher (1990), on the other hand, has defined ethnomathematics as the study of mathematical concepts in small-scale or indigenous cultures; 'by and large, the indigenous people of the places that were "discovered" and colonized by Europeans.' This essay will explore the motivation behind these two definitions, and discuss the ways in which the discipline of anthropology can help to illuminate the need for both approaches.

ANTHROPOLOGICAL PERSPECTIVES ON CULTURE

Nineteenth century anthropologists, such as Edward Tylor and Lewis H. Morgan, mistakenly assigned cultures to various 'levels of advancement'. The most popular version of this theory was Morgan's *Ancient Society*, published in 1877, which posited that human society had passed through three major stages: savagery, barbarism, and civilization. Thus cultural attributes which were different from European ones were explained as coming from societies which were still at a previous level along this unilineal chain – a very convenient excuse for colonialism (Adas, 1989). While Morgan's view of unilineal development has long been abandoned by contemporary anthropologists, it lives on in the popular imagination and even in some contemporary social ideologies. On the one hand, right-wing neo-nazi groups continue to make use of the unilineal model in their insistence on white cultural superiority. On the other hand, this model was used by Friedrich Engels and Karl Marx to develop a historical basis for their social critique of capitalism and can thus be seen in some contemporary left-wing ideologies as well.

The unilineal development model actually had its start in biological evolutionary theory. Lovejoy (1922) notes that the image of a 'great chain of being'

H. Selin (ed.), Mathematics Across Cultures: The History of Non-Western Mathematics, 13–22.

from animals to people to God is quite ancient in European history and was quickly (and mistakenly) applied by early biologists. Steven J. Gould (1996: 348) describes how contemporary advocates, such as Arthur Jensen, of the idea that there are racial differences in intelligence still depend on this myth of a single ladder of progress:

> As a paleontologist, I am astounded. Evolution forms a copiously branching bush, not a unilineal progressive sequence. Jensen speaks of 'different levels of the phyletic scale – that is, earthworms, crabs, fishes, turtles, pigeons, rats, and monkeys.' Doesn't he realize that modern earthworms and crabs are descendants of lineages that have evolved separately from vertebrates for more than 500 million years? They are not our ancestors ... [t]hey represent good solutions for their own way of life. ... Does Jensen really think that pigeon-rat-monkey-human represents an evolutionary sequence among warm-blooded vertebrates?

With the exception of pseudo-scientific claims such as those of Jensen, all modern biological accounts confirm evolutionary development in terms of Gould's 'copiously branching bush.' Similarly, cultural diversity is seen by anthropologists not in terms of unilineal development, but rather as 'solutions for their own way of life.' Just as organs for sensing the external environment have a wide variety of different evolutionary developments in biology (from echolocation in the blind river dolphin to the compound eye of the house fly), the conceptual developments of different cultures should be viewed as a branching diversity of forms rather than ranked on a non-existent 'ladder of progress'.

CROSS-CULTURAL COMPARISONS OF MATHEMATICS

The mythology of a unilineal development sequence has been particularly damaging when it comes to cross-cultural comparisons of mathematics, because Europeans have habitually viewed their mathematical development in the same way Jensen viewed human evolution: as the apex of a unilineal sequence. This framework invites a poor characterization of mathematics in non-European cultures, because it blinds us to the possibility that a cultural group which has not developed mathematical tools that came early in European history could still have investigated areas of mathematics that Europeans came to more recently, or even have yet to discover.

The unilineal development model has also been reinforced by assumptions of a deterministic relationship between political structure, economy, and mathematics. Anthropologists note that societies which practice gather-hunter economies tend to have developed band structures in which there is little political hierarchy or labor specialization, while those with horticultural and pastoral economies tend to develop tribal structures in which there is only a slight political hierarchy and weak labor specialization. Those with agricultural and industrial economies tend to develop state structures in which there are strong political hierarchies and a great deal of labor specialization. Actually, the relationships are not deterministic; for example several of the Pacific Northwest Native American societies had a gather-hunter economic base, but used a hierarchical tribal political system. Conversely, anthropologists were surprised to find 'acephalous' African societies with large-scale agricultural production (such as wet rice) but little or no political hierarchy. But reductive theorists and the popular imagination have been hasty to ignore this diversity and place the economic and political categories along a unilineal development ladder.

Thus non-western state empires such as the ancient Chinese, Hindu and Muslim societies are most often used for examples of 'non-western mathematics'. This has led to the mistaken impression that mathematical inventions in indigenous band and tribal societies such as sub-Saharan Africa, North and South America, and the South Pacific are absent. It is indeed true that the mathematics of those literate, non-western empires provides a close match to the mathematics of Europe, but that is only because these European cultures are themselves state societies. Once we rid ourselves of the unilineal development model, we can then see that a diversity of mathematical ideas occurs in all cultures, regardless of their economic or political structure. We may have difficulty in *translating* the mathematics of non-state societies to the forms of Western math we are familiar with, but there is no *a priori* reason to see that difficulty as one of inferiority.

Ascher's definition of ethnomathematics can thus be seen as a critique of this misuse of cross-cultural comparison. The point is not simply one of opposition to biological determinism or ethnocentrism, because both of those oppositional intellectual movements have at times succumbed to the same unilineal development model. For example, in racist theories that European brains are biologically superior to those of non-European ethnic groups, it was often claimed that only Europeans developed state societies. Louis Agassiz (1850: 143–144), a Swiss naturalist who became America's leading proponent of white biological superiority, stated the matter specifically for the case of Africans: 'This compact continent of Africa exhibits a population which had been in constant intercourse with the white race, which had enjoyed the benefit of the example of the Egyptian civilization ... and nevertheless there has never been a regulated society of black men on that continent.' It never occurred to Agassiz that ancient Egypt could have been founded by Africans, and indeed much of what is today referred to as the Afrocentric intellectual movement has been founded on the premise of a black African origin for ancient Egypt. But by focusing on a direct contradiction – 'you say we don't have state societies? Ha! We do!' – we are forced into remaining within the unilineal development model. Although the existence of such state empires in non-western societies does indeed contradict the racist claims that only Europeans were capable of such 'regulated societies', it also plays into the assumption that the kinds of intellectual achievements which can contradict white supremacy can only be found in state societies. By exalting the non-western empires of the ancient Egyptians, Maya, Chinese, Hindus, and Muslims, we risk further considering the tribal and band societies that surrounded them as primitive.

THE ETHNOMATHEMATICS APPROACH TO INDIGENOUS SOCIETIES

While Ascher's focus on the mathematics of indigenous societies was thus a much-needed response to the over-emphasis on non-western state empires (cf.

Katz, 1992 for a critique), the methodologies for such investigations are by no means self-evident. What would it mean to discuss mathematics in a culture that did not have a writing system? What does the presence of a particular geometric design mean in terms of knowledge of geometry? The answers ethnomathematics provides depend on establishing intention – the opposite emphasis provided by the discipline of 'mathematical anthropology' (Eglash, 1997). By showing conscious intent in the utilization of symbolic and geometric patterns, ethnomathematics can make a distinction between phenomena that are merely the result of unconscious social dynamics and those which are active mathematical ideas and practices. Working in many different areas of the world, Ascher (1990), Closs (1986), Crump (1990), D'Ambrosio (1990), Eglash (1999), Gerdes (1991), Lipka *et al.* (1999), Njock (1979), Rauff, (1993), Washburn and Crowe (1988), Zaslavsky (1973), and many others (see Fisher, 1992, Shirley, 1995 for reviews), have provided mathematical analyses of a variety of indigenous patterns and abstractions, while drawing attention to the role of conscious intent in these designs. Many of these authors have chapters in this volume.

For example, Gerdes (1991) used the *lusona* sand drawings of the Tchokwe people of Northeastern Angola to demonstrate indigenous mathematical knowledge. His analysis showed the constraints necessary to define an 'Eulerian Path' (the stylus never leaves the surface and no line is re-traced), and a recursive generation system (increasingly complex forms are created by successive iterations through the same geometric algorithm). While it would have been possible to limit description of these features to the physical construction only (and thus hypothesize – as a mathematical anthropologist might have – that they are the result of an unconscious social process) these were, rather, placed in the context of indigenous concepts and activities. The Eulerian constraint, for example, was critical for not only definitions of drawing skill within their society, but also externally when the lusona were deployed by the Tchokwe as a way to deflate the ego of over-confident European visitors.

Ascher (1990) notes the same type of Eulerian path drawings in the South Pacific, but these tend to be less recursive (i.e. requiring combinations of different geometric algorithms, which Ascher likens to algebraic systems, rather than the fractal-like iterations through the same algorithm that dominate the African versions). Ascher describes the South Pacific drawings as primarily motivated by symbolic narratives, in particular that of the Malekula islanders as an abstract mapping of kinship relations. Again, this is in strong contrast to the tradition of mathematical anthropology, where kinship algebra was considered a triumph of western analysis (cf. Kay, 1971, and Helen Verran's essay on Yolngu kinship mathematics in this volume).

Ascher's description of the Native American game of Dish shows this contrast in a more subtle form. In the Cayuga version of the game six peach stones, blackened on one side, are tossed, and the numbers landing black side or brown side recorded as the outcome. The traditional Cayuga point scores for each outcome are (to the nearest integer value) inversely proportionate to the probability. Ascher does not posit an individual Cayuga genius who discovered probability theory, nor does she explain the pattern as merely an unintentional epiphenomenon of repeated activity. Rather, her description (p. 93) is focused on how the game is embedded in community ceremonials, spiritual beliefs, and healing rituals, specifically through the concept of 'communal playing' in which winnings are attributed to the group rather than the individual player. Juxtaposing this context with detailed attention to abstract concepts of randomness and predictability in association with the game – in particular the idea of 'expected values' associated with successive tosses – has the effect of attributing the invention of probability assignments to collective intent.

At the skeptical extreme in ethnomathematics, Donald Crowe has refrained from making any inferences about intentionality, and insists that his studies of symmetry in indigenous pattern creations (cf. Washburn and Crowe, 1988) are simply examples of applied mathematics. But since Crowe has restricted his work to only those patterns which *could* be attributed to conscious design (painting, carving, and weaving), it creates the opposite effect of mathematical anthropology's attempt to eliminate indigenous intent. This is evidenced by Crowe's dedication to using these patterns in mathematics education (particularly his teaching experience in Nigeria during the late 1960s, which greatly contributed to Zaslavsky's (1973) seminal text, *Africa Counts*).

Thus the ethnomathematics of indigenous societies is epistemologically distinguished from the mathematics of non-western state societies in that it is not limited to direct translations of western forms, but rather can be open to any mathematical pattern discernable to the researcher. In fact, even that description might be too restrictive: previous to Gerdes's study there was no western category of 'recursively generated Eulerian paths'; it was only in the act of applying a western analysis to the lusona that Gerdes (and the Tchokwe) created that hybrid.

ETHNOMATHEMATICS AS THE ANTHROPOLOGY OF MATHEMATICS

In addition to its epistemological conflicts, ethnomathematics is immersed in sociopolitical struggles as well; indeed it was in the efforts to overcome the colonial legacies of pedagogy in the Third World that Brazilian educator Ubiratan D'Ambrosio coined the term. D'Ambrosio was inspired by a UNESCO project he attended in Mali in 1970, and was later influenced by the social critique of fellow Brazilian Paulo Freire. D'Ambrosio's decision to define ethnomathematics in terms of the mathematical practices of any cultural group, rather than specifically indigenous societies, was an outcome of this context. While wary of the non-western state society trap that Ascher had avoided, the post-colonial context also required recognition that denial of mathematical capability was a problem for many groups other than indigenous societies, and that much of the Third World was populated by people whose mixed heritage and history could not be so easily distinguished. Illusions of cultural purity and organic innocence are too easily projected onto traditional cultures. For that reason many researchers (and cultural workers, particularly artists) have developed a postmodern emphasis on hybridity, creolization, and other impure identities (cf. Minh-ha, 1986; Anzaldúa, 1987; Bhabba, 1990). In addition, non-ethnic identities of marginalization – e.g. sex/gender and class/caste systems – can also have profound cultural relations to technological knowledge, including that of mathematics.

Ethnomathematics in this broad definition can be broken into four categories. In addition to the previous distinction between the non-western empire civilizations and indigenous societies, one would also include vernacular knowledge systems, such as the 'street mathematics' of Nunes *et al.* (e.g., calculation by peasant pushcart venders), the 'situated cognition' of Lave (e.g., calculations by homemakers at a supermarket), and the working class mathematics of carpet-layers studied by Masingila (1994). Gerdes (1994) lists similar designations under titles such as 'folk mathematics,' 'informal mathematics', 'everyday mathematics', 'out-of-school mathematics', and 'non-standard mathematics'. Finally, the connections between culture and mathematics in Western societies would also fall under this broad definition of ethnomathematics. This fourth category derives from the type of research conducted under the social studies of science rubric (Restivo, 1993), such as the influence of British class structure on mathematical debates over statistical measures in the early 20th century (Barnes and MacKenzie, 1979), or the history of women in western mathematics (Olsen, 1974).

The Ascher and D'Ambrosio definitions are not mutually exclusive. The definition of indigenous can be blurred – Ascher herself began with a study of the Inca *quipu*, which was developed in a state society – and in many instances the work inspired by D'Ambrosio's postcolonial context has been focused on indigenous knowledge systems. In Africa for example, Gerdes (1982) and Gay and Cole (1967) were both motivated by local attempts to develop a new post-colonial basis for pedagogy. In the United States, Claudia Zaslavsky, whose 1973 *Africa Counts* is often regarded as the first of its genre, attributes her project to the civil rights activities of the 1950s, which resulted in an increase in African studies materials in her school and thus alerted her to the conspicuous absence of material on indigenous African mathematics. Gloria Gilmer, past president of the International Study Group on Ethnomathematics, cites her identity as an African American mathematician in the 1950s as fundamental to her own inclusion of African cultural resources (Gilmer, 1989).

THE PROBLEM OF PEDAGOGY

While ethnomathematics has struggled to define itself as an anthropological field practice, its use as applied anthropology – ethnomathematics in the classroom – has engendered even more controversy. In some of these discussions (cf. Jackson, 1992) any tie to political motivations is described as an inherent defect, a loss of scholarly status, and thus (unless one is willing to deny the kinds of historical connections mentioned in the previous section) ethnomathematics can be eliminated out of hand. Moreover, it is indeed possible to cite cases in ethnosciences where counter-hegemonic political motivations are at fault (cf. discussion of the 'Portland Baseline Essays' in Oritz de Montellano,

1993; Martel, 1994). Given the *a priori* hostility to ethnomathematics, and its own potential flaws, its application to education has been understandably difficult. Nevertheless, there are several reasons such efforts are worthwhile.

The education reform efforts which consider ethnomathematics include multicultural mathematics (Nelson *et al.*, 1991), critical mathematics (Skovsmose, 1985; Frankenstein and Powell, 1997), humanist mathematics (White, 1986), and situated cognition (Lave, 1988) among others. In addition to the politics of post-colonial education (Knijnik, 1993), these approaches generally cite cultural alienation from standard mathematics pedagogy for minority ethnic groups as well as other identities (see Keitel *et al.*, 1989 for a detailed listing). Another important motivation is the idea that individuals from dominant groups will tend to have better relations with subordinate groups if they are exposed to more egalitarian presentations of the other's culture. Finally, there is also the contention that racist and other views of biological determination of intelligence can be combated by presenting mathematical knowledge generated through these groups.

The problem of cultural alienation does find support in field research. Powell (1990), for example, notes that pervasive mainstream stereotypes of scientists and mathematicians conflict with certain aspects of African-American cultural orientation. Similar disjunctures between African-American identity and mathematics education in terms of self-perception, course selection and career guidance have been noted (cf. Hall and Postman-Kammer, 1987; Boyer, 1983). One critique maintains that if there is alienation, then the solution should lie in making teaching materials more universal rather than more local. A similar suggestion has been employed in response to sexism in the word problems of math textbooks, but research reviewed in Nibbelink *et al.* (1986) indicates that gender-neutral examples have been inadequate, and they recommend reinstating gender with more balanced presentation of both male and female figures. Similarly, attempting to get rid of all cultural reference would reduce the quality of the textbook for everyone. Concrete examples are important for learning application skills, enhancing general interest, and reaching a wider range of cognitive styles. And there are many culture-specific elements, such as the Greek names of Euclid and Pythagoras, which would be absurd to eliminate, suggesting that cultural balance is a better strategy than cultural obliteration.

In support of the theory that over-emphasis on biological determinism creates a learning deterrent, Geary (1994) reviews cross-cultural studies which indicate that while children, teachers and parents in China and Japan tend to view difficulty with mathematics as a problem of time and effort, their American counterparts attribute differences in mathematics performance to innate ability (which thus becomes a self-fulfilling prophecy). Thus it is possible that even if the cultural alienation theory is incorrect, the opposition to biological determinism provided by ethnomathematics would be of strong benefit to the students. While no formal studies have yet been carried out, Anderson (1990), Frankenstein (1990), Gerdes (1994), Moore (1994), and Zaslavsky (1991) have given anecdotal reports of positive results in using ethnomathematics to teach minority students.

Despite these optimistic outlooks, there are still many potential difficulties in applications to pedagogy. Williams (1994) suggests that any multicultural science teaching implies that minority students have less aptitude than white students, since it gives them special treatment. Although this sounds similar to politically conservative critiques of affirmative action, the accusation of a patronizing stance has also been made from the opposite end of the political spectrum:

> Where there is 'multicultural' input into the science curriculum it tends to focus on so-called 'Third World Science' and involves activities like making salt from banana skins. ... The patronizing view of the 'clever and resourceful native' which underlies such practice is not far removed from the racist views of 'other peoples and cultures' which pervade attempts at multicultural education (Gill *et al.*, 1987).

This critique touches on several difficulties. There is a danger of singling out minority students and increasing their 'otherness', of reductive presentations of minority cultures, and, perhaps most pointedly, an ahistoricizing effect in which romantic portrayals of a mythically pure tradition overshadow the political actualities of Third World experience.

Finally, there have been certain attempts to write under the multicultural mathematics rubric which are far less attentive to educational requirements, sacrificing mathematical content for a third-world cultural gloss. What is missing from these failed attempts at inclusiveness is in part the insistence on intentional indigenous mathematics. Most students and teachers are delighted to find real examples of African geometric algorithms or Native American applications of probability. But these examples take a great deal of skill to discover, analyze, and combine with standard mathematics curricula. What goes under the name of multicultural mathematics is too often a cheap short-cut that merely replaces Dick and Jane counting marbles with Tatuk and Esteban counting coconuts. Of the few texts that do use indigenous math, almost all examples are restricted to primary school level. Again, this restriction might unintentionally imply primitivism (e.g. that mathematical concepts from African culture are only child-like).

While there are many worthwhile projects that have used the 'multicultural mathematics' label, far too many have shown a lack of reflection on these pedagogical issues. Rather than working with the cultural concepts held by the students themselves – either by researching the students' own cultural self-identity or attempting to educate the students better about the heterogeneous complexities of their local heritage – these poorly applied versions assume a static, essentialist racial identity. Under this assumption, a child from Puerto Rico may find herself confronted with Inca llamas, as if she should automatically be familiar with any artifact from the universe of Latin American societies simply because she is 'Latina'.

In addition, this essentialist approach leans too heavily on the crutch of self-esteem, as if all cultural barriers could be reduced to a self-imposed shame (ignoring, for example, the dangers of primitivist romanticism, as noted previously, or the systematically misdirected pride recently exposed in John Hoberman's excellent text, *Darwin's Athletes*). In contrast, ethnomathematics directly addresses the over-emphasis on biological determinism that creates a learning deterrent for students of all social groups.

BIBLIOGRAPHY

Adas, M. *Machines as the Measure of Men*. Ithaca, New York: Cornell University Press, 1989.

Agassiz, L. 'The diversity of origin of the human races.' *Christian Examiner* 49: 110–145, 1950.

Anzaldúa, G. *Borderlands, La Frontera*. San Francisco: Spinsters, 1987.

Ascher, M. *Ethnomathematics: a Multicultural View of Mathematical Ideas*. Pacific Grove: Brooks/Cole Publishing, 1990.

Barnes, B. and MacKenzie, D. 'On the role of interests in scientific change.' In *On the Margins of Science*, Roy Wallis, ed. Sociological Review Monograph no. 27, Keele, Staffordshire: University of Keele, 1979.

Bhabba, H. *Nation and Narration*. London: Routledge, 1990.

Closs, M. P., ed. *Native American Mathematics*. Austin: University of Texas Press, 1986.

Crowe, D. W. 'Ethnomathematics reviews.' *Mathematical Intelligencer* 9(2): 68–70, 1987.

Crump, T. *The Anthropology of Numbers*. Cambridge: Cambridge University Press, 1990.

D'Ambrosio, U. *Etnomatematica*. Sao Paulo: Editora Atica, 1990.

Eglash, R. 'When math worlds collide: intention and invention in ethnomathematics.' *Science, Technology and Human Values* 22(1): 79–97, Winter, 1997.

Eglash, R. *African Fractals: Modern Computing and Indigenous Design*. New Brunswick: Rutgers University Press, 1999.

Fisher, W. H. 'Review of *Etnomatematica*.' *Science, Technology, and Human Values* 17(4): 425, 1992.

Gay, J. and Cole, M. *The New Mathematics and an Old Culture*. New York: Holt, Rinehart & Wilson, 1967.

Gerdes, P. *Lusona: Geometrical Recreations of Africa*. Maputo: E.M. University Press, 1991.

Gerdes, P. 'Reflections on ethnomathematics.' *For the Learning of Mathematics* 14(2): pp. 19–22, June, 1994.

Gill, D., Singh, E., and Vance, M. 'Multicultural versus anti-racist science teaching.' In *Anti-racist Science Teaching*, D. Gill and L. Levidow, eds. London: Free Association Books, 1987, pp. 124–135.

Gilmer, Gloria. 'Making mathematics work for African Americans from a practitioner's perspective.' *Making Mathematics Work for Minorities*. Washington, DC: Mathematical Sciences Education Board, 1989, pp. 100–104.

Joseph, G. G. *The Crest of the Peacock*. London: I.B.Tauris & Co., 1991.

Katz, V. J. Book review of *Ethnomathematics*. *Historia Mathematica* 19(3): 310–314, 1992.

Kay, P. *Explorations in Mathematical Anthropology*. Cambridge, Massachusetts: MIT Press, 1971.

Keitel C., Damerow P., Bishop A. and Gerdes P. 'Mathematics, education and society.' *Science and Technology Education Document Series*, #35. Paris: UNESCO, 1989.

Knijnik, G. 'An ethnomathematical approach in mathematical education: A matter of political power.' *For the Learning of Mathematics* 13(2): 23–25, 1993.

Lave, J. *Cognition in Practice*. New York: Cambridge University Press, 1988.

Lerman, S. 'Vedic Multiplication.' *Critical Mathematics Educators Group Newsletter* 2: 4, 1992.

Lipka, Jerry, Mohatt, Gerald V. and The Cuilistet Group. *Transforming the Culture of Schools: Yup'ik Eskimo Examples*. Mahwah, New Jersey: Lawrence Erlbaum Associates, 1999.

Martel, Erich. '... And how not to: a critique of the Portland baseline essays.' *American Educator* 18(1): 33–5, 1994.

Masingila, J. O. 'Mathematics practice in carpet laying.' *Anthropology & Education Quarterly* 25(4): 430–462, 1994.

Minh-ha, T. 'She, the inappropriated other.' *Discourse* 8: 11–37, 1986.

Moore, C. G. 'Research in Native American mathematics education.' *For the Learning of Mathematics* 14(2): 9–14, 1994.

Njock, E. 'Langues Africaines et non Africaines dans l'enseignement des mathématiques en Afrique.' *Seminaire Interafricain sur l'enseignement des mathématiques*. Accra, Ghana, May, 1979.

Osen, Lynn M. *Women in Mathematics*. Cambridge, Massachusetts: MIT Press, 1974.

Oritz de Montellano, B. 'Melanin, Afrocentricity, and pseudoscience.' *Yearbook of Physical Anthropology* 36: 33–58, 1993.
Powell, L. 'Factors associated with the underrepresentation of African Americans in mathematics and science.' *Journal of Negro Education* 59(3): 292–8, 1990.
Powell, Arthur B. and Frankenstein, Marilyn, eds. *Ethnomathematics: Challenging Eurocentrism in Mathematics Education.* Albany: State University of New York Press, 1997.
Rauff, J. 'Algebraic structures in Walbiri iconography.' *Mathematical Connections* 1(2): 5–11, 1993.
Restivo, S. *Math Worlds.* Albany: State University of New York Press, 1993.
Shirley, L. 'Using ethnomathematics to find multicultural mathematical connections.' *National Council of Teachers of Mathematics Yearbook*, 1995, pp. 34–43.
Skovmose, O. 'Mathematical education versus critical education.' *Educational Studies in Mathematics* 16: 337–354, 1985.
Washburn, D. K. and Crowe, D. W. *Symmetries of Culture.* Seattle: University of Washington Press, 1988.
White, A. *Humanistic Mathematics Network Newsletter.* Claremont, California: Harvey Mudd College, 1986–.
Williams, H. 'A critique of Hodson.' *Science Education* 78(5): 515–519, 1994.
Zaslavsky, Claudia. *Africa Counts.* Boston: Prindle, Weber & Schmidt, 1973.
Zaslavsky, C. *Fear of Math.* New Brunswick, New Jersey: Rutgers University Press, 1994.

EDWIN J. VAN KLEY

EAST AND WEST

As traditionally used in the West, the terms East and West imply that the two are somehow of equal importance. While in some senses that might be arguable in the nineteenth and twentieth centuries it certainly was not true during the long reaches of human history prior to the nineteenth century. By about 500 BC, following the scheme developed in William McNeill's *The Rise of the West*, the globe supported four major centers of civilization: the Chinese, the Indian, the Near Eastern, and the Western, considering Greek culture as antecedent to what eventually became the West. Of the four the West was probably the least impressive in terms of territory, military power, wealth, and perhaps even traditional culture. Certainly this was the case after the fall of the western Roman Empire in the fifth century AD. From that time until about AD 1500 the West probably should be regarded almost as a frontier region compared to the other centers of civilization (Hodgeson, 1993: 27).

From roughly 500 BC to AD 1500 a cultural balance obtained between the four major centers of civilization. During these millennia each center continued to develop its peculiar style of civilized life and each continued to spread its culture and often its political control to peoples and lands on its periphery. While the inhabitants of each center of civilization were aware of the other centers, sometimes traded with them, and occasionally borrowed from them, the contacts were too thin to permit any one center to threaten – commercially, militarily, politically, or culturally – the existence of the others. During this long period, that is through most of recorded human history, there was no possibility of Western superiority or hegemony. No visitor from Mars would likely have predicted that the West would eventually dominate the planet.

Obviously the Greeks and Romans knew quite a bit about the Near-Eastern world, especially about the Persian Empire of Alexander the Great and the successor states formed after its collapse. About India and China they knew much less, and what they knew was much less accurate. Although Herodotus (*ca.* 494–425 BC) reported some things about India, most of the information available to the Greeks came from the writers who described Alexander's campaigns in the Indus Valley (326–234 BC) and from Megasthenes. They described India as fabulously rich, the source of much gold and precious stones.

H. Selin (ed.), Mathematics Across Cultures: The History of Non-Western Mathematics, 23–35.

It was hot; the sun stood directly overhead at midday and cast shadows toward the south in summer and toward the north in winter. They described huge rivers, monsoons, tame peacocks and pheasants, polygamy, and the practice of *suttee* (widow burning). But they also reported fantastic things such as gold-digging ants, cannibals, dog-headed people, and people with feet so large that they served as sun shades when lying supine. The Romans knew that India was the source of spices and that China, which they called Serica as well as Sinae, was the source of silk. There are possible traces of Indian influence in some Roman silver and ivory work, some indication that Plotinus and other Neo-Platonists were familiar with the *Upanishads*, and possibly some influence of Buddhism on Manichaeanism (Lach, 1965: I, 5–19).

From about the fourth century, even before the fall of the western Roman Empire, until the return of Marco Polo from China in the late thirteenth century, Europe or the West added little factual information to its understanding of India and China. After the fall of the Roman Empire there was no direct trade between Europe and Asia and thus there were no opportunities to test the stories by observation. The rise of Islam in the seventh and eighth centuries completed Europe's isolation. During these centuries the old stories inherited from the Greeks were retold and embellished with little effort to distinguish fact from myth. To these were added three legends of more recent origin: the stories celebrating the heroic exploits of the mythical Alexander, those rehearsing Saint Thomas the Apostle's missionary journey to India and his subsequent martyrdom, and those describing the rich, powerful, Christian kingdom of Prester John located somewhere to the east of the Islamic world with which European rulers dreamed of allying against the Muslims. Even the trickle of precious Asian products brought to Europe by intermediaries seemed only to confirm the image of Asia as an exotic and mysterious world, exceedingly rich and exceedingly distant (Lach, 1965: I, 20–30).

The rise of the Mongol empire in Asia during the thirteenth century resulted in direct overland travel between Europe, or the West, and China. The Mongols' success also revived hopes among European rulers of finding a powerful ally to the east of the Muslims. Even the devastating Mongol incursions into Poland and Hungary in 1240 and 1241 scarcely dampened their enthusiasm. Already in 1245 the Pope sent an embassy led by John of Plano Carpini to the Mongol headquarters near Karakorum. He was followed during the ensuing century by a fairly large number of envoys, missionaries, and merchants, several of whom wrote reports of what they saw and did in Eastern Asia. Marco Polo's was the most comprehensive and reliable, and the most widely distributed, of the Medieval reports. By the time the Polos first arrived at Kublai Khan's court in 1264 it was newly established at Cambaluc (Beijing), from which the khan ruled the newly conquered Cathay (China). Like many other foreigners during the Mongol period (the Yuan Dynasty in China) the Polos were taken into the Khan's service. They were employed in the Mongol administration for seventeen years during which time Marco appears to have traveled extensively throughout China. On his return to Europe he produced the first detailed description of China in the West based primarily on first-hand observation and experience, although his failure to mention details such as footbinding has led scholars like Frances Wood to question how much of his account is actually first-hand (Wood, 1996). Whether based on his own observations or those of someone else, however, no better account of China appeared in Europe before the middle of the sixteenth century. Marco Polo described China as the wealthiest, largest, and most populous land in the thirteenth-century world. While his understanding of Chinese culture was apparently minimal he accurately and admiringly described Chinese cities, canals, ships, crafts, industries, and products. He noted the routes, topography, and people encountered in his travels, including his voyage home through Southeast Asia to Sumatra, Ceylon, and along the west coast of India (Lach, 1965: I, 30–48).

The decline of the Mongol Empire and the establishment of the Ming Dynasty in China in 1368 severed the direct connection between Europe and China. The fall of Constantinople in 1454 and the establishment of the Turkish empire in the Near East disrupted Europe's older connections with the Near East and India. Europe's isolation from the outside world was complete, not to be restored until the opening of the sea route around the tip of Africa in the waning years of the fifteenth century. During this period no European appears to have traveled to China, although a few travel reports refer to India and Southeast Asia. Of them only that written in 1441 by the renowned humanist and papal secretary, Poggio Bracciolini, based on Nicolò de' Conti's travels, added to the West's store of knowledge about India and confirmed some of the more accurate of the ancient Greek reports (Lach, 1965: I, 59–65). In fact amid the Renaissance humanists' enthusiasm for the rediscovery of ancient Greek literature, the old Greek writings on India received new respect and attention.

During the long era of cultural balance before AD 1500 many important technological and scientific inventions and innovations appear to have migrated to the West from the other centers of civilization, more often from China than from the others. The migration of technology was usually gradual, involving one or more intermediaries, the inventions usually being established in the West without any clear ideas about their origins. Much of the basic technology that enabled the Europeans to sail directly to Asia around 1500 and later to begin their march towards global domination, was known earlier in the Asian centers and only later adopted or separately invented in Europe.

Among the more important technological borrowings were gunpowder, the magnetic compass, printing, and paper, all apparently originating in China. For none of them is the path of migration entirely clear, and thus for none of them can the possibility of independent invention by Europeans be entirely ruled out. Gunpowder, for example, was known in China by 1040 and did not appear in Europe until the middle of the thirteenth century. The magnetic compass was fully described in an eleventh-century Chinese book, *Meng Qi Bitan* (Dream Pool Essays), written by Shen Gua in 1088. It began to be used in Europe during the late twelfth or early thirteenth century. Most likely Europeans learned about it from the Arabs. The case for moveable-type printing having been borrowed from the Chinese is more hotly debated than that for gunpowder or the compass. Wood-block printing was used in China by the seventh century, and paper was invented much earlier; the first printed books appeared there during the ninth century, six centuries before the invention of printing in the West by Johannes Gutenberg in 1445. Block printing probably became known in Europe through the introduction of printed playing cards and paper money during the Mongol period; medieval travelers frequently mentioned these. Because of the large number of Chinese characters, the Chinese continued to prefer printing from page-sized blocks of wood carved as a single unit; European printing almost immediately employed moveable type, thus convincing some scholars that it was a separate invention. But while they may have preferred block printing the Chinese also developed moveable type as early as the eleventh century. For none of these basic inventions taken separately – gunpowder, the compass, and printing – is the case for its diffusion from China to Europe indisputably demonstrated. Taken together, however, along with a rather large number of other technological and scientific innovations such as paper, the stern-post rudder, the segmental arch-bridge, canal lock-gates, and the wheelbarrow, which all appear to have migrated from China to the West, it becomes apparent that the general flow of technology and science in pre-modern times was from East to West, which would seem to increase the likelihood that these basic innovations also migrated to Europe from Asia. Those who used the technological innovations probably cared little about their ultimate origin and apparently did not seek it out. Nevertheless, European mariners after 1500, confronted first hand with evidence that printing, gunpowder, the mariners' compass, and the like had been in use much longer in Asia than in Europe, frequently suggested that they had been borrowed from the Asians (Lach, 1977: II, Bk. 2, 81–84).

Even before gunpowder a group of military innovations found their way to Europe from China and India, again through intermediaries and apparently without Europeans being aware of their origins. The Chinese form of the Indic stirrup was the most important of these and may have been as important to European military development in the eighth century as gunpowder was later. The Javan fiddle bow and the Indian Buddhist pointed arch and vault were acclimated in Europe before 1100. The traction trebuchet along with the compass and paper appeared in the twelfth century. Much more important for subsequent Western scientific achievements was the adoption of Hindu-Arabic mathematics in the twelfth and thirteenth centuries: the Indian system of arithmetical notation, trigonometry, and the system of calculating with nine Arabic numbers and a zero, all practiced in India as early as AD 270. Some components of Indian mathematics may have come from the Babylonians or the Chinese. It came to Europe, however, through the translation of Arabic writings, and the European borrowers usually credited India rather than China, Babylonia, or the Arabic intermediaries as the source of the new mathematics. Along with Indian mathematics Europeans learned some elements of Indian astronomy and also became fascinated with the Indian idea of perpetual motion (Lach, 1977: II, Bk. 3, 398–99).

Also before 1500 Europeans sometimes attempted to imitate desirable Asian products, not always successfully. Already in the sixth century the Byzantine emperor Justinian monopolized the silk trade in his realm and expressed his determination to learn the secret of its manufacture. In 553 a monk supposedly smuggled some silkworm eggs into Constantinople carrying them in a hollow stick, perhaps of bamboo. Nothing is said in the story about the importation of silk technology, but less than a century later sericulture had obviously taken root in Syria. From there it spread to Greece, Sicily, Spain, Italy, and France. In Italy during the fourteenth century water power was used in silk spinning, as it had been used in China much earlier. Attempts to imitate Chinese porcelain, however, were much less successful. The best attempts to do so were made in northern Italian cities during the fifteenth century. None, however, approached Chinese porcelain in composition, color, or texture. Nor did the Dutch Delftware of the seventeenth century, another attempt to imitate the Chinese product. Not until the eighteenth century were European craftsmen able to produce a hard-paste porcelain to rival that of China (Lach, 1965; I, 20–21, 81–82). Also appearing in Europe before 1500 were less important devices or techniques such as the Malay blowgun, playing cards, the Chinese helicopter top, the Chinese water-powered trip-hammer, the ball and chain governor, and, according to Miyasita Saburo, perhaps even Chinese techniques of anatomical dissection (Lach, 1977: II, Bk. 3, 399–401; Miyasita, 1967).

While impressive, the West's importation of Asian science and technology before 1500 in no way deflected Western culture from its traditional paths. Seldom was the provenance of the new inventions or techniques known, and they could all rather easily be incorporated into the traditional Christian European worldview. They did not provoke any serious questions about the European way of life, its religious basis, its artistic and cultural traditions, or even its traditional scientific views. This is also true for the artistic and cultural borrowings from Asia prior to 1500. They too were often unconscious, and even when they were not they were regarded as embellishments or decorations, rather than in any way a challenge to traditional themes. For example, the incorporation into the Christian calendar and the corpus of edifying literature stories about Saints Baarlam and Josephat, derived as they were from stories of the life of Buddha, resulted not in a Buddhist challenge to the Christian faith but simply in the addition of two new saints to the growing Christian pantheon (Lach, 1977: II, Bk. 2, 101–05).

Nevertheless, the borrowing and adaptation of Asian science and technology by the West before 1500 was indispensable in making possible the long overseas voyages to Asia and the protection of the European ships and shore installations there. Without gunpowder, cannons, the compass, the stern-post rudder, etc. there would have been no European expansion. While they had lagged well behind the other centers of civilization through most of civilized history, by 1500 European marine and military technology were beginning to surpass that of the Near East, India, and China, and were obviously superior to that of the peripheral areas of Africa, Southeast Asia, and the Americas. The Portuguese voyages down the coast of Africa and Columbus' voyage across the Atlantic attest to Europe's rapidly improving technology.

A small Portuguese fleet under Vasco da Gamma reached Calicut on the Malabar Coast of India in 1498, thus establishing direct contact between Europe and Asia by sea and also inaugurating a new era in the relationships between the four major centers of civilization on the globe. Soon after 1500 the Europeans began to dominate the seas of Asia, the Portuguese in the Indian Ocean being the first. The Portuguese, and following them the Dutch and the English, moved along sealanes, visited seaports, and fit into a trading world which had been developed earlier by Muslim traders and which stretched from eastern Africa to the Philippines. As the Europeans at first tried to compete in that world, and as they later tried to dominate and even to monopolize it, they found Muslim merchants and merchant-princes to be their most formidable opponents. That they were able to move so rapidly and effectively into that international trading system was due not so much to their superior technology and firepower as to the fact that the great Muslim empires of the sixteenth century – the Ottoman Turkish, the Safavid Persian, and the Mughul in northern India – seem to have been too busy consolidating their newly won empires to contest the European intrusion. These Muslim empires, as well as the Southeast and East Asian empires of Siam, Vietnam, China, and Japan, had all as a matter of governmental policy turned away from the sea and looked inward to the control of their land empires and to land taxes as the primary source of their wealth and power. Had any or all of these major Asian and Near-Eastern states seriously resisted the western incursion the story might have had a different conclusion. How, for example, might it have ended had the Chinese not deliberately dismantled their ocean-going fleets after Admiral Zheng He's several formidable incursions into the Indian Ocean and along the eastern coast of Africa between 1405 and 1433? The first of these voyages alone comprised sixty-two ships carrying 28,000 men (Lavathes, 1994). Apart from Muslim traders in the Indian Ocean, however, those who formidably opposed European power in the sixteenth and seventeenth centuries were small Muslim commercial port-city states like Makassar on Celebes and Aceh on Sumatra. The Mughul emperors in northern India usually allowed the Europeans to trade freely in their ports, often exempting them from customs duties. Apart from the illegal but locally tolerated Portuguese settlement in Macao the Europeans were not permitted to trade in China at all. China briefly contested Dutch maritime power only in 1624 after the Dutch had spent two years raiding the coast of Fukien Province, burning villages, seizing junks, enslaving their crews, and constructing a fort on one of the Pescadores Islands – all in an effort to force the Chinese to allow them to trade freely at some port along the Chinese coast. When confronted with a full scale Chinese war fleet, however, the Dutch commander hastily sued for peace and gratefully accepted the Chinese admiral's offer to allow the Dutch to trade on Formosa, which was not yet considered Chinese territory (Groeneveldt, 1898). From 1500 until about the middle of the eighteenth century the Europeans were able to carve out empires in the Americas, insular Southeast Asia, and the Pacific Islands, but they did not threaten the major centers of civilization in Asia.

During the first three centuries of direct maritime contact between Europe and Asia (1500–1800) the Westerners showed little sense of cultural superiority toward the high cultures of Asia. If anything they tended to exaggerate the wealth, power, and sophistication of the Asian centers of civilization. Most Europeans were confident that Christianity was indeed the true religion, and they quickly began to send missionaries to convert Asian peoples. By 1600, if not earlier, they were also justifiably confident that European mathematics, science, and technology were superior to that of the Asians. Those areas aside, however, Europeans were endlessly fascinated with what they discovered beyond the line and realized that they still had much to learn from Asia. Between 1500 and 1800 the currents of cultural influence continued to flow mainly from east to west, but during this era the impact of Asia on the West was usually more conscious and deliberate than previously, and it was primarily in areas other than science and technology.

The new seaborne commerce with Asia almost immediately brought greatly increased quantities and varieties of Asian products into Europe. Pepper and fine spices were the first to appear on the docks in European ports, but they were soon followed by such goods as Chinese porcelain and lacquerware, tea, silks, Indian cotton cloth, and cinnamon. Following these staples of the trade came also more exotic products: Japanese swords, Sumatran or Javanese krisses, jewelry, camphor, rhubarb, and the like; the list is very long. Some of these products, such as tea, provoked striking social changes in Europe. Attempts to imitate others resulted in new industries and in new manufacturing techniques: Delft pottery in imitation of Ming porcelain, for example. Attempts to compete with cheap Indian cotton cloth seem to have touched off a technical revolution in the British textile industry which we customarily regard as the beginning of the industrial revolution (Rothermund, 1978).

Along with the Asian products came descriptions of the places and peoples who had produced the products. The earliest sixteenth-century descriptions usually seem designed to inform other Asia-bound fleets about the conditions of trade. After Christian missions were established the missionaries wrote letters and reports intended to elicit support for the missions. But before long the travel tales and missionaries' descriptions became popular in their own right and profitable to publish. During the seventeenth century what had been a sizeable stream of literature about Asia became a veritable torrent. Hundreds of books about the various parts of Asia, written by missionaries, merchants, mariners, physicians, soldiers, and independent travelers, were published during the period. For example, during the seventeenth century alone there appeared at least twenty-five major descriptions of South Asia, another fifteen devoted to mainland Southeast Asia, about twenty to the Southeast Asian archipelagoes, and sixty or more to East Asia. Alongside these major independent contributions stood scores of Jesuit letterbooks, derivative accounts, travel accounts with brief descriptions of many Asian places, pamphlets, news sheets, and the like. Many of the accounts were collected into the several large multivolume compilations of travel literature published during the period: those of G. B. Ramusio, Richard Hakluyt, Johann Theodor and Johann Israel De Bry, Levinus Hulsius, Samuel Purchas, Isaac Commelin, and Melchisédech

Thévenot. In addition to the missionaries' accounts, the travel tales, and composite encyclopedic descriptions such as those of Johann Nieuhof, Olfert Dapper, and Arnoldus Montanus, several important scholarly studies pertaining to Asia were published during the seventeenth century. There were studies of: Asian medicine such as Jacob Bontius' general work and Michael Boym's on Chinese medicine; botany, for example those of Boym and the massive work on Malabar flora by Hendrik Adriaan van Rheede tot Drakestein; religion in works like those of Philippus Baldaeus and Abraham Rogerius; history, outstandingly Martino Martini's *Sinicae historiae*; and translations of important Chinese and Sanskrit literature such as the Jesuits' *Confucius sinarum philosophus*.

The published accounts range in size from small pamphlets to lavishly illustrated folio volumes. They were published in Latin and in almost all of the vernaculars, and what was published in one language was soon translated into several others, so that determined enthusiasts could probably have read most of them in their own languages. They were frequently reprinted in press runs which ranged from 250 to 1,000 copies. Five to ten editions were not at all uncommon, and some of the more popular accounts would rival modern best sellers. In short the Early-Modern image of Asia was channeled to Europe in a huge corpus of publications which was widely distributed in all European lands and languages. Few literate Europeans could have been completely untouched by it, and it would be surprising if its effects could not have been seen in contemporary European literature, art, learning, and culture (Lach and Van Kley, 1993: III, 301–597).

From this literature European readers could have learned a great deal about Asia and its various parts. Perhaps most obviously their geographic horizons would have been continually expanded. Gradually Europeans gained accurate knowledge about the size and shape of India, China, and Southeast Asia. During the seventeenth century, several puzzles which had plagued earlier geographers were solved: for example, the identification of China with Marco Polo's Cathay, the discovery that Korea was a peninsula and that Hokkaido was an island. By the end of the century Europeans had charted most of the coasts of a real Australia to replace the long-imagined Antipodes as well as the coasts of New Guinea, the Papuas, numerous Pacific islands, and parts of New Zealand. Interior Ceylon and Java, as well as Tibet, were visited and accurately described by Europeans before the end of the century. By 1700 only some areas of continental Asia north of India and China, the interior of Australia, New Zealand, and New Guinea and parts of their coastlines remained unknown to the Europeans. Most of these lacunae were filled in during the eighteenth century. Even more impressive than the greatly expanded geographic knowledge available to European readers in early modern times was the rapidly increasing and increasingly detailed information about the interiors, societies, cultures and even histories of Asia's high cultures. Already during the seventeenth century European readers could have read detailed descriptions and even viewed printed cityscapes and street scenes of scores of Asian cities, interior provincial cities as well as capitals and seaports, and they could have learned countless details about Asia's various peoples, their occupations, appearance, social customs, class structures, education, ways of rearing children, religious beliefs, and the like. Details regarding Asia's abundant natural resources, crafts, and arts were described as well as its commercial practices and patterns of trade. Asian governments were described in exceedingly close detail, especially those of major powerful states such as China, the Mughul Empire, Siam, and Japan. Jesuit missionaries in China, for example, described the awesome power of the emperor, his elaborate court, the complex imperial bureaucracy and its selection through competitive written examinations, and the Confucian moral philosophy on which it all was supposed to have been based. They also described the frequently less orderly and less savory practice of Chinese government, complete with detailed examples of officials' abuse of power and competing factions within the administration. Similar details were reported for the governments of all the major states as well as for countless smaller states. By the end of the seventeenth century European observers had published many sophisticated accounts of Asian religions and philosophies – not only the frequently deplored Hindu 'idolatry' and widow burning, but also the Hindu world view which lay beneath the panoply of deities and temples, the various schools and sects of Hinduism, and the ancient texts of Hindu religion. Similarly sophisticated and detailed accounts of Confucianism and Buddhism were available, as well as descriptions of the more primitive beliefs of peoples like the Formosan aborigines, the Ainu of Hokkaido and the inner Asian and Manchurian tribes. Seventeenth and eighteenth-century readers could also have learned much about Asian history, especially, but not exclusively, that of Asia's high cultures. By the mid-seventeenth century, for example, a very detailed sketch of China's long dynastic history culled from official Confucian histories by Jesuit missionaries had been published. Martino Martini wrote the *Sinicae historiae decas prima*, which was published in Munich in 1658. During the eighteenth century an important Chinese history, the *Tongjian Gangmu* (Outline and Details of the Comprehensive Mirror [for aid in government]) by Zhu Xi, was translated into French in its entirety by Joseph-Anne-Marie de Mailla and published in Paris (*Histoire générale de la Chine, 1777).* Not only history, however, but also news about Asia was reported to early modern readers. Their image of Asia was surely not that of a static world far away. Among the more important events reported in almost newspaper-like detail during the seventeenth century alone were the Mughul emperor's successful campaigns in the south of India, the Maratha challenge to Mughul supremacy, the fall of the Indian states of Golconda and Vijayanagar to the Mughul empire, the Manchu Conquest of China in 1644, the feudal wars and the establishment of the Tokugawa shogunate in Japan (1600), and the internal rivalries and wars in Siam and Vietnam. Natural disasters such as earthquakes, fires, volcanic eruptions, the appearance of comets, etc. were also regularly reported. Readers of this richly detailed, voluminous, and widely distributed literature may well have known relatively more about Asia and its various parts than do most educated westerners today (Lach and Van Kley, 1993: Bks. 2, 3, and 4).

The post-1500 literature on Asia also contains a great many descriptions of Asian science, technology, and crafts: weaving, printing, paper-making, binding, measuring devices, porcelain manufacture, pumps, watermills, hammocks, palanquins, speaking tubes, sailing chariots, timekeepers, astronomical instruments, agricultural techniques and tools, bamboo and other reeds for carrying water, as well as products such as musical instruments, wax, resin, caulking, tung-oil varnish, elephant hooks and bells, folding screens, and parasols. Some, such as Chinese-style ship's caulking, leeboards, strake layers on hulls, lug sails, mat and batten sails, chain pumps for emptying bilges, paddle-wheel boats, wheelchairs, and sulphur matches, were quickly employed or imitated by Europeans. Some provoked documented experimentation and invention: Giambattista della Porta's kite in 1589, and Simon Stevin's sailing chariot in 1600. The effects of the new information on the sciences of cartography and geography are obvious and profound. Simon Stevin, whose sailing chariot probably was inspired by descriptions of similar Chinese devices, and which incidentally inspired two Latin poems by Hugo Grotius, also introduced decimal fractions and a method of calculating an equally-tempered musical scale, both of which might also have been inspired by Chinese examples. The sixteenth-century mariners' cross staff may have been inspired by the Arab navigators' *kamals*. The Western science of botany was profoundly influenced by the descriptions of Asian flora and by the specimens taken back to Europe and successfully grown in European experimental gardens – rice, oranges, lemons, limes, ginger, pepper, and rhubarb being among the most useful. More important for botany, however, the Asian plants provoked comparisons with plants familiar to Europeans, the development of comprehensive classification schemes, and thus the beginnings of modern plant taxonomy (Lach and Van Kley, 1993: 925–27). Some Asian cures (herbs and drugs) were borrowed, especially for tropical medicine. Chinese acupuncture, moxibustion, and methods of diagnosis by taking the pulse were minutely described and much admired by European scholars, but it is not yet clear to what extent they were actually used in Europe.

The flow of cultural influence was not exclusively from East to West between 1500 and 1800. Many Asians became Christian. Able Jesuit missionaries translated scriptures and wrote theological works in Chinese and other Asian languages; they also translated European mathematical and scientific treatises into Chinese. The Kangxi emperor himself studied Euclidian geometry, western astronomy, cartography, geography, the harpsichord, and painting under the guidance of Jesuit missionaries whom he kept in his court. Like many Chinese he was fascinated by European clocks. Many Asians, including the Chinese, learned how to use western firearms and cast western-style cannon. But the cultural consequences of these efforts were disappointingly small and of short duration. In China before the end of the eighteenth century they seem largely to have disappeared along with the Christian mission. The Japanese, however, even during the closed-country period after 1640, were far more curious about Western science. Samurai scholars, for example, studied Dutch medicine and science and in the eighteenth century repeated Benjamin Franklin's kite-flying experiment. Nevertheless through most of the early modern period the Europeans were far more curious about and more open to influence from Asia than were any of the high cultures of Asia to influence from the West.

During the seventeenth and eighteenth centuries the new information about Asia influenced Western culture primarily in areas other than science and technology. The extent of this impact remains to be comprehensively studied, and even of that which is known only a few examples can be mentioned here. Asian events and themes entered European literature in scores of instances from Felix Lope de Vega, Lodovico Ariosto, François Rabelais, and Thomas More in the sixteenth century (Lach, 1977: Book 2) to the several Dutch, German, and English plays and novels depicting the fall of the Ming Dynasty and the triumph of the Manchus in the seventeenth century (Van Kley, 1976), to Voltaire's literary and philosophical works in the eighteenth (Pinot, 1932; Rowbotham, 1932). Even popular literature, seventeenth-century Dutch plays and pious tracts, for example, show surprising familiarity with the new information about Asia (Van Kley, 1976a). Asian influences in European art, architecture, garden architecture and the decorative arts also began in the sixteenth century and culminated in the *chinoiserie* of the eighteenth century (Lach, 1970: II, Bk. 1; Impey, 1977). Confrontation with China's ancient history challenged the traditional European Four-Monarchies framework of universal history and touched off a controversy among European scholars which by the mid-eighteenth century resulted in an entirely new conception of ancient world history (Van Kley, 1971). Some scholars, beginning with Pierre Bayle in the late seventeenth century, have detected a Neo-Confucian influence in the thought of Spinoza and in some aspects of Leibniz's philosophy (Maverick, 1930; Mungello, 1977). Chinese government and especially the system of selecting officials through a series of competitive, state supervised, written examinations was frequently held up for emulation by European states during the seventeenth and eighteenth centuries. It might well be that the institution of written civil-service exams in Western states, beginning in eighteenth-century Prussia, was inspired by the Chinese example. Of more general importance than any single instance of influence, however, was the challenge presented by long-enduring, sophisticated, and successful Asian cultures to traditional European assumptions about the universality of their own. Perhaps here can be found the beginnings of cultural relativism in the West.

By 1600 European science and technology generally and especially marine and military technology outstripped that of any Asian society, whatever had been borrowed earlier from them. And while the fascination with and appreciation of the high cultures of Asia continued through most of the eighteenth century, the West between 1600 and 1800 experienced the radical transmutation of its traditional culture which resulted in the development of a rational, scientific approach to the use of nature and to society – to agriculture, business, industry, politics, and above all warfare – which we have traditionally labeled the Scientific Revolution, Enlightenment, and Industrial Revolution. This transmutation has enabled western nation-states during the past two centuries to establish the world-wide competitive empires which came to dominate all the other centers of civilization and which has resulted in the global dominance of Western culture. The triumph of this transmuted Western culture has reversed the centuries-long East-to-West flow of cultural influence and threatens all of the world's traditional cultures. It should be remembered, however, that this rational, scientific, industrialized Western culture was not an obviously natural outgrowth of traditional Western culture, that in its early development it received important basic components from Asia, and that its triumph threatens traditional Christian Western culture just as seriously as it threatens the traditional cultures of Asia and the Near East. Nor is it necessary that this rational, scientific, industrialized, and now global culture continues to be led and directed by Westerners. Obviously Asians are as able to master its intricacies as are Europeans or Americans; important parts of it originally came from Asia after all.

BIBLIOGRAPHY

Chaudhurii, K. N. *Trade and Civilization in the Indian Ocean: An Economic History from the Rise of Islam to 1750*. Cambridge: Cambridge University Press, 1985.

Cipolla, Carlo M. *Guns, Sails, and Empires: Technological Innovation and the Early Phases of European Expansion, 1400–1700*. New York: Pantheon Books, 1965.

Crosby, Alfred. *The Measure of Reality; Quantification and Western Society, 1250–1600*. Cambridge: Cambridge University Press, 1997.

Groeneveldt, W. P. *De Nederlanders in China*. Vol. XLVIII of *Bijdragen tot de taal-, land- en volkenkunde van Nederlandsch-Indië*. The Hague: Martinus Nijhoff, 1898.

Hodgson, Marshall G. S. *Rethinking World History; Essays on Europe, Islam, and World History*, Edmund Burke, III, ed. Cambridge: Cambridge University Press, 1993.

Hodgson, Marshall G. S. *The Venture of Islam: Conscience and History in a World Civilization*. Chicago: The University of Chicago Press, 1974.

Impey, Oliver. *Chinoiserie; The Impact of Oriental Styles on Western Art and Decoration*. New York: Scribners, 1977.

Lach, Donald F. *Asia in the Making of Europe*. Vol. 1: *The Century of Discovery*. Bks. 1–2. Chicago: The University of Chicago Press, 1965. Vol. II: *A Century of Wonder*. Bks. 1–3. Chicago: The University of Chicago Press, 1970, 1977.

Lach, Donald F. and Van Kley, Edwin J. *Asia in the Making of Europe*. Vol. III: *A Century of Advance*. Bks. 1–4. Chicago: The University of Chicago Press, 1993.

Levathes, Louise. *When China Ruled the Seas; The Treasure Fleet of the Dragon Throne, 1405–1433*. New York: Simon and Schuster, 1994.

Maverick, Lewis A. 'A possible source of Spinoza's doctrine.' *Revue de littérature comparée* 19: 417–428, 1939.

McNeill, William H. *The Pursuit of Power: Technology, Armed Force, and Society Since A.D. 1000*. Chicago: The University of Chicago Press, 1982.

McNeill, William H. *The Rise of the West: A History of the Human Community*. Chicago: The University of Chicago Press, 1963.

Miyasita, Saburo. 'A link in the westward transmission of Chinese anatomy in the Later Middle Ages.' *Isis* 18: 486–490, 1967.

Mungello, David E. *Leibniz and Confucianism; the Search for Accord*. Honolulu: University Press of Hawaii, 1977.

Needham, Joseph. *Science and Civilisation in China*. 6 vols. Cambridge: Cambridge University Press, 1954–1984.

Parker, Geoffrey. *The Military Revolution: Military Innovation and the Rise of the West, 1500–1800*. Cambridge: Cambridge University Press, 1988.

Pinot, Virgile. *La Chine et la formation de l'esprit philosophique en France: 1640–1740*. Paris: P. Geuthner, 1932.

Rowbotham, A. H. 'Voltaire, sinophile.' *Publication of the Modern Language Association* 48: 1050–1065, 1932.

Rothermund, Dietmar. *Europa und Asien im Zeitalter des Merkantilismus*, 'Erträge der Forschung,' 80. Darmstadt: Wissenschaftliche Buchgesellschaft, 1978.

Sanderson, Stephen K. 'East and West in the development of the modern world-system.' *Itinerario* 21(2): 225–41, 1998.

Steensgaard, Niels. *Carracks, Caravans, and Companies; the Structural Crisis in the European-Asian Trade in the Early 17th Century*. Lund: Studentliteratur, 1973.

Van Kley, Edwin J. 'An alternative muse: the Manchu conquest of China in the literature of seventeenth-century Northern Europe.' *European Studies Review* 6: 21–43, 1976.

Van Kley, Edwin J. 'Europe's 'discovery' of China and the writing of world history.' *The American Historical Review* 76: 358–385, 1971.

Van Kley, Edwin J. 'The effect of the discoveries on seventeenth-century Dutch popular culture.' *Terrae Incognitae VIII*: 29–43, 1976.

Wood, Frances. *Did Marco Polo Go to China?* Boulder, Colorado: Westview Press, 1996.

Wolf, Eric R. *Europe and the People without History*. Berkeley: University of California Press, 1982.

DAVID TURNBULL

RATIONALITY AND THE DISUNITY OF THE SCIENCES

The canonical model of science as a unified body of universal, objective truth has now to be set against a variety of studies locating science in historical, philosophical and sociological perspectives. Detailed analyses on the history of epistemology, on the cultural origins of science and on the practice of science reveal science as diverse, heterogeneous and disunited, taking a variety of different forms depending on the discipline, the location and the period in question (Dupré, 1993; Galison, 1997; Galison and Stump, 1996; Law, 1987; Rosenberg, 1994; Star, 1989). The unity, stability and cohesion of science, which according to the standard realist/rationalist view are due to its fidelity with a unified reality, are seen as social and historical accomplishments. Similarly the universality of science has been found to result from specific forms of social organization in coordinating and transmitting scientific knowledge rather than from its inherent truth. These alternative conceptions of the production of scientific knowledge raise crucial problems for our understanding of 'rationality', that is our capacity to choose critically between differing accounts of reality. By implication it also raises problems for the canonical model of mathematics as a paradigmatic cultural universal.

LOCAL KNOWLEDGE

A prime example of the sort of thing which suggests the necessity for rethinking the nature of the unity and universality of science is the concept of 'local knowledge'. This concept has recently come to the fore in the field of the sociology of scientific knowledge, where it is a common empirical finding that knowledge production is an essentially local process (Shapin, 1994, 1998; Turnbull, 1995, 1999). Knowledge claims are not adjudicated by absolute standards; rather their authority is established through the workings of *local* negotiations and judgments in particular contexts. This focus on the localness of knowledge production provides the condition for the possibility for a fully-fledged comparison between the ways in which different modes of understanding the natural world have been produced by different cultures and at different times. Such cross-cultural comparisons of knowledge production systems have

H. Selin (ed.), Mathematics Across Cultures: The History of Non-Western Mathematics, 37–54.

only recently started to emerge in the sociology of science (Agrawal 1995; Turnbull 1993, 1997; Watson-Verran and Turnbull, 1995). It follows that a necessary condition for the possibility of fully equitable comparisons between knowledge traditions is that Western science, rather than being taken as the definition of knowledge, rationality, or objectivity, should be treated as an example of a knowledge system which has developed its own historically specific forms of rationality. The question of how and why such forms have come to dominate then becomes a matter of empirical inquiry not epistemological necessity.

Though knowledge systems may differ in their epistemologies, methodologies, logic, cognitive structures, or in their socio-economic contexts, a characteristic that they all share is their localness. Hence, in so far as they are collective bodies of knowledge, many of their small but significant differences lie in the work involved in creating assemblages from the 'motley' of differing practices, instrumentation, theories, and people (Turnbull, 1993). Much of that work can be seen as strategies and techniques for creating the equivalencies and connections whereby otherwise heterogeneous and isolated forms of knowledge are enabled to move in space and time from the local site and moment of their production and application to other places and times.

Taking this perspective, from which all knowledge systems from whatever culture or time including contemporary technology and science are based on local knowledge, raises two problems. Can the term 'local knowledge' be given adequate specificity, and how do we reconcile it with the strongly held conviction that an essential underpinning of the enlightenment project, modernism and the concept of progress, is that scientific knowledge is universal, objective and true in all times and all places.

One way of drawing the divide between the natural sciences and the human sciences has been to emphasize the distinction between the universal and the contingent, but currently a wide range of disciplines including physics, ecology, history, feminist theory, literary theory, anthropology, geography, economics, politics, and sociology of science have come to focus on the specific, the contingent, and the particular. From meteorology to medicine it is now recognised that we need to include a focus on the particular conditions at specific sites and times rather than losing that specificity in unlocalized generalizations. This may involve looking at the social and historical specificities of a text, a reading, a culture, a population, a period, a site, a region, an electron, or a laboratory, but within this diversity of uses of local two broad and rather different senses can be discerned. On the one hand there is the notion of a 'voice' or a 'reading'. The voice may be purely individual and subjective or may be a collection of voices belonging to a group, class, gender, or culture, but in all cases the notion captures a central postmodernist insight that all texts or cultures are multivocal and polysemous. That is, they are capable of having a multiplicity of competing meanings, readings or voices and are hence subject to 'interpretive flexibility' (Collins, 1985). On the other hand, local is used both in the more explicitly geopolitical sense of place and in the experiential sense of contextual, embodied, partial, or individual.

Science studies have also taken a 'local turn'. Some philosophers of science have come to re-evaluate the role of theory and argue that scientists practicing in the real world do not deduce their explanations from universal laws but rather make do with rules of thumb derived from the way the phenomena present themselves in the operation of instruments and devices (Cartwright, 1983; Hacking, 1983). Similarly philosophers and sociologists of science alike have recognized the lack of absolute standards and the role of tacit knowledge in technoscientific practice, and have sought to display the context in which the practice of science is manifested as negotiated judgments, craft skills and collective work. The recognition of the social and material embodiment of judgments, skills and work in the cultural practice of individuals and groups has coalesced into the general claim that all knowledge is local. Knowledge, from this constructivist perspective, can be local in a range of different senses. 'It is knowledge produced and reproduced in *mutual interaction* that relies on the *presence* of other human beings on a direct, face-to-face basis.' (Thrift, 1985). It is knowledge that is produced in contingent, site, discipline or culture specific circumstances (Rouse, 1987). It is the product of open systems with heterogeneous and asynchronous inputs 'that stand in no necessary relationship to one another' (Pickering, 1992). In sum scientific knowledge is 'situated knowledge' (Haraway, 1991).

Perhaps the most important consequence of the recognition of the localness of scientific knowledge is that it permits parity in the comparison of the production of contemporary technoscientific knowledge with knowledge production in other cultures. Previously the possibility of a truly equitable comparison was negated by the assumption that indigenous knowledge systems were merely local and were to be evaluated for the extent to which they had scientific characteristics. Localness essentially subsumes many of the supposed limitations of other knowledge systems compared with western science. So-called traditional knowledge systems have frequently been portrayed as closed, pragmatic, utilitarian, value laden, indexical, context dependent, and so on. All of which was held to imply that they can not have the same authority and credibility as science because their localness restricts them to the social and cultural circumstances of their production. Science by contrast was held to be universal, non-indexical, value free, and as a consequence floating, in some mysterious way, above culture. Treating science as local simultaneously puts all knowledge systems on a par and renders otiose discussion of their degree of fit with transcendental criteria of rationality and logic. This gives science studies an anthropological dimension which could flag the emergence of a sub-discipline that might be called 'comparative knowledge traditions'.

Emphasizing the local in this way necessitates a re-evaluation of the role of theory, which is often held by philosophers and physicists to provide the main dynamic and rationale of science as well as being the source of its universality. Karl Popper, for example, claims that all science is cosmology and Gerald Holton sees physics as a quest for the Holy Grail, which is no less than the 'mastery of the whole world of experience, by subsuming it under one unified theoretical structure' (Allport, 1991). It is this claim to be able to produce universal theory that Western culture has used simultaneously to promote and reinforce its own stability and to justify the dispossession of other peoples. It constitutes part of the ideological justification of scientific objectivity, the 'god-trick' as Donna Haraway calls it: the illusion that there can be a positionless vision of everything. This allegiance to mimesis has been severely undermined by analysts like Richard Rorty, but theory has also been found wanting at the level of practice, where analytical and empirical studies have shown it cannot provide the sole guide to experimental research and on occasion has little or no role at all. The conception of grand unified theories guiding research is also incompatible with a key finding in the sociology of science: 'consensus is not necessary for cooperation nor for the successful conduct of work'. This sociological perspective is succinctly captured in Leigh Star's description:

> Scientific theory building is deeply heterogeneous: different viewpoints are constantly being adduced and reconciled ... Each actor, site, or node of a scientific community has a viewpoint, a partial truth consisting of local beliefs, local practices, local constants, and resources, none of which are fully verifiable across all sites. The aggregation of all viewpoints is the source of the robustness of science (Star, 1989: 46).

Theories from this perspective have the characteristics of what Star calls 'boundary objects'; that is they are 'objects which are both plastic enough to adapt to local needs and constraints of the several parties employing them, yet robust enough to maintain a common identity across sites.' Thus theorizing, like unity, is the outcome of the assemblage of heterogeneous local practices.

Given the localization of knowledge, how are the universality and connectedness that typify technoscientific knowledge achieved? Given all these discrete knowledge/practices, imbued with their concrete specificities, how can they be assembled into fields or knowledge systems? In Star's terms, 'how is the robustness of findings and decision making achieved?' (Star, 1989). Ophir and Shapin ask, 'How is it, if knowledge is indeed local, that certain forms of it appear global as in the domain of application' (Ophir and Shapin, 1993). How in other words does science travel and how is it given cohesion? The answers are primarily spatial and linguistic and consist in a variety of social, literary, and technical devices that provide for treating instances of knowledge/practice as similar or equivalent and for making connections, that is, in enabling local knowledge/practices to move and to be assembled. These are processes John Law calls 'heterogeneous engineering' (Law, 1987). Among the many social strategies that enable the possibility of equivalence are processes of standardization and collective work to produce agreements about what counts as appropriate forms of ordering and what counts as evidence, facts, proof, etc. Technical and literary devices that provide for connections and mobility may be material or conceptual and include maps, calendars, theories, books, lists, metaphors, narratives and systems of recursion, but their common function is to enable otherwise incommensurable and isolated kinds of knowledge to move in space and time from the local site and moment of their production to other places and times.

One approach to understanding the movement of local knowledge is that of Bruno Latour, for whom the most successful devices in the agonistic struggle for authority are those which are mobile and also 'immutable, presentable, readable and combinable with one another' (Latour, 1986). These immutable mobiles are the kinds of texts and images that the printing press and distant point perspective have made possible. Such small and unexpected differences in the technology of representation are, in his account, the causes of the large and powerful effects of science. That which was previously completely indexical, having meaning only in the context of the site of production, and having no means of moving beyond that site, is standardized and made commensurable and re-presentable within the common framework provided by distant point perspective. Hence that which has been observed, created, or recorded at one site can be moved without distortion to another. At centers of calculation such mobile representations can be accumulated, analyzed, and iterated in a cascade of subsequent calculations and analyses (Latour, 1987).

Latour's account is complemented by the work of Steven Shapin and Simon Schaffer in *Leviathan and The Air Pump* (1985). They have shown that experimental practice in science is sustained and made credible by the adoption of forms of social organization such as professional scientific societies, modes of literary and graphic expression such as the scientific paper, scientific instruments such as the vacuum pump and careful delimitations of public and private spaces such as the laboratory and the university. We now take all of these for granted, but they had to be deliberately created to overcome the fundamentally local character of experimentally derived knowledge claims.

In the seventeenth century, the problem for Robert Boyle, one of the earliest experimentalists, was to counter the arguments of his opponent Thomas Hobbes about the grounding of true and certain knowledge which they both agreed was essential in a country riven by dissent and conflicting opinion. Reliable knowledge of the world for Hobbes was to be derived from self-evident first principles, and anything that was produced experimentally was inevitably doomed to reflect its artifactual nature and the contingencies of its production; its localness would deny it the status of fact or law. Boyle recognized the cogency of these arguments and set out to create the forms of life within which the knowledge created at one site could be relayed to and replicated at other sites. In order for an experimental fact to be accepted as such it had to be witnessable by all, but the very nature of an experimental laboratory restricted the audience of witnesses to a very few. Boyle, therefore, had to create the technology of what Shapin calls virtual witnessing. For this to be possible three general sorts of devices or technologies had to be developed. Socially, groups of reliable witnesses had to be formed. In the seventeenth century such reliability was to be in found in the objectivity of gentlemen who were held to be disinterested because they did not aspire to wealth, property and power since they had already acquired it. These gentlemen witnesses had to be able to communicate their observations to other groups of gentlemen so that they too might witness the phenomena. This required the establishment of journals using clear and unadorned prose that could carry the immutable mobiles, experimental accounts, and diagrams. The apparatus had to be made technically reliable and reproducible, but perhaps most importantly the physical space for such empirical knowledge had to be created.

Hobbes, of course, was right: experimental knowledge is artifactual. It is the product of human labor, of craft and skill, and necessarily reflects the contingencies of the circumstances. It is because craft or tacit knowledge is such a fundamental component of knowledge production that accounts of its generation, transmission, acceptance, and application cannot be given solely in terms of texts and inscriptions. A vital component of local knowledge is moved by people in their heads and hands. Harry Collins, a sociologist of science, has argued that this ineradicable craft component in science is ultimately what makes science a social practice. Because knowledge claims about the world are based on the skilled performance of experiments, their acceptance is a judgment of competence not of truth. An example of the centrality of craft skill is the TEA laser, invented in Canada by Bob Harrison in the late sixties, which British scientists attempted to replicate in the early seventies. 'No scientist succeeded in building a laser using only information found in published or other written sources' and furthermore the people who did succeed in building one were only able to do so after extended personal contact with somebody who had himself built one. Now TEA lasers are blackboxed and their production is routine and algorithmic. But in order to become a matter of routine Harrison's local knowledge had to be moved literally by hand (Collins, 1985; Collins and Pinch, 1998).

Joseph Rouse (1987), in considering the contemporary production process of scientific knowledge, has summarized the implications of this understanding of science:

> Science is first and foremost knowing one's way about in the laboratory (or clinic, field site etc.). Such knowledge is of course transferable outside the laboratory site into a variety of other situations. But the transfer is not to be understood in terms of the instantiation of universally applied knowledge claims in different particular settings by applying bridge principles and plugging in particular local values for theoretical variables. It must be understood in terms of the adaptation of one local knowledge to create another. We go from one local knowledge to another rather than from universal theories to their particular instantiations.

According to the historian of science Thomas Kuhn (1970), the way a scientist learns to solve problems is not by applying theory deductively but by learning to apply theory through recognizing situations as similar. Hence theories are models or tools whose application results from situations being conceived as or actually being made equivalent. This point is implicit in the recognition that knowledge produced in a laboratory does not simply reflect nature because nature as such is seldom available in a form that can be considered directly in the lab. Specially simplified and purified artifacts are the typical subject of instrumental analysis in scientific laboratories. For the results of such an artificial process to have any efficacy in the world beyond the lab, the world itself has to be modified to conform to the rigors of science. A wide variety of institutional structures have to be put in place to achieve the equivalencies needed between the microworld created inside the lab and the macroworld outside in order for the knowledge to be transmittable. The largest and most expensive example of this is the Bureau of Standards, a massive bureaucracy costing six times the R&D budget (Latour, 1987). Without such social institutions the results of scientific research are mere artifacts. They gain their truth, efficacy, and accuracy not through a passive mirroring of reality but through an active social process that brings our understanding and reality into conformity with each other.

Non-western societies have also developed a variety of social and technical devices for coping with their localness and enabling knowledge to move. Some of them are technical devices of representation like the Inca *ceques* and *quipus.* [Editor's note: see the article by Thomas Gilsdorf on Inca mathematics in this volume.] Some of them are abstract cognitive constructs, like the Anasazi and Inca calendars, and the Micronesian navigation system (Turnbull, 1995). All of them also require social organization, rituals, and ceremonies. All of them have proved capable of producing complex bodies of knowledge and in many cases have been accompanied by substantial transformations of the environment. The major difference between Western science and other knowledge systems lies in the question of power. Western science has succeeded in transforming the world and our lives in ways that no other system has. The source of the power of science on this account lies not in the nature of scientific knowledge but in its greater ability to move and apply the knowledge it produces beyond the site of its production. However at the end of the twentieth century we can now perceive that there is a high cost to pay for science's hegemony. Much of that cost in terms of environmental degradation and ethnocide is due not so much to the all-encompassing nature of scientific theories but to the social strategies and technical devices science has developed in eliminating the local.

Some of these devices have been revealed with relatively few problems through direct observation. Others are less susceptible to investigation and analysis, being embodied in our forms of life. One way to catch a glimpse of these hidden presuppositions and taken for granted ways of thinking, seeing, and acting, is to misperceive, to be jolted out of our habitual modes of understanding through allowing a process of interrogation between our knowledge system and others. Such an interrogative process of mutual inter-translation can enable us to catch sight of the cultural glasses we wear instead of looking through them as if they were transparent.

This challenging of the comprehensive discourses of science by other knowledge systems is what Foucault had in mind when he claimed that we are 'witnessing an insurrection of subjugated knowledges' and corresponds to an emphasis on the local that has emerged in anthropology at least since Clifford Geertz's *Interpretation of Cultures.* In his critique of global theories and in his emphasis on 'thick description', Geertz pointed out that cultural meanings cannot be understood at the general level because they result from complex organizations of signs in a particular local context and that the way to reveal the structures of power attached to the global discourse is to set the local knowledge in contrast with it.

However we should not be too easily seduced by the apparently liberating effects of celebrating the local since it is all too easy to allow the local to become a 'new kind of globalizing imperative' (Hayles, 1990). In order for all knowledge systems to have a voice and in order to allow for the possibility of inter-cultural comparison and critique, we have to be able to maintain the local and the global in dialectical opposition to one another. This dilemma is the most profound difficulty facing liberal democracies now that they have lost the convenient foil of communism and the world has Balkanized into special interest groups – genders, races, nationalities, tribes, religions, or whatever. By moving into a comparative mode there is a grave danger of the subsuming of the other into the hegemony of western rationality, but conversely unbridled cultural relativism can only lead to the proliferation of ghettos and dogmatic nationalism. We cannot abandon the strength of generalizations and theories, particularly their capacity for making connections and for providing the possibility of criticism (Nanda, 1997, 1998).

At the same time we need to acknowledge the reflexive point that theory and practice are not distinct. Theorizing is also a local practice. If we do not embrace this joint dialectic of theory and practice, the local and the global, we will not be able to establish the conditions for the possibility of directing the circulation and structure of power in knowledge systems. It is in the light of this recognition that we need to consider how the movement of local knowledge is accomplished in different knowledge systems and the consequent effects on the ways people and objects are constituted and linked together, i.e., the effects on power. The essential strength of the sociology of scientific knowledge is its claim to show that what we accept as science and technology could be other than it is. The great weakness of the sociology of scientific knowledge is the general failure to grasp the political nature of the enterprise and to work towards change. With some exceptions it has had a quiet tendency to adopt the neutral analyst's stance that it devotes so much time to criticizing in scientists. One way of capitalizing on the sociology of science's strength and avoiding the reflexive dilemma is to devise ways in which alternative knowledge systems can be made to interrogate each other.

Without the kinds of connections and patterns that theories make possible we will never be able to perceive the interconnectedness of all things. Without the awareness of local differences we will lose the diversity and particularity of the things themselves. Thus, rather than rejecting universalizing explanations what is needed is a new understanding of the dialectical tension between the local and the global. We need to focus on the ways in which science creates and solves problems through its treatment of the local. Science gains its truthful character through suppressing or denying the circumstances of its production and through the social mechanisms for the transmission and authorization of the knowledge by the scientific community. Both of these devices have the effect of rendering scientific knowledge autonomous, above culture, and hence beyond criticism. Equally problematic is the establishment of the standardization and equivalencies required in order that the knowledge produced in the lab works in the world. The joint processes of making the world fit the knowledge instead of the other way round and immunizing scientific knowledge from criticism are best resisted by developing forms of understanding in which the local, the particular, the specific, and the individual are not homogenized but are enabled to talk back. Such a position implies that there is still hope for the enlightenment project and that some form of rationality is required; otherwise such cross-cultural interrogation and critique will not be possible. However, it can no longer be the kind of simple uniform rationality that has been previously taken as typifying science.

RATIONALITY

Rationality, objectivity, and method are the three central concepts that are frequently invoked as capturing the essence of science and, taken along with unity and universality, provide the mythic structure of the canonical model. Science is the authoritative form of knowledge in the world today. The most convincing proof of its superiority appears to lie in the ever expanding body of knowledge we have about reality. We can predict where, when, and how meteorites will collide with Jupiter, and we can build interplanetary spacecraft to send back signals recording the event. We can explore the atomic structure of chemicals and design new ones to suit whatever purpose we have in mind, ecstasy or health. We can explain the origins of everything in the universe down to the first few nanoseconds. The key, according to the myth, to this unparalleled success lies in science's embodiment of the highest form of rationality and objectivity in the scientific method. This mythical underpinning of science also provides the rationale for the celebration of modernism and the current domination of the West. This view is unselfconsciously exemplified by the philosopher Ernest Gellner who claims, 'If a doctrine conflicts with the acceptance of the superiority of scientific-industrial societies over others, then it really is out' (cited in Salmond, 1985).

Therein lies the first set of intrinsic problems and contradictions this myth conceals. Modernism is supposedly synonymous with development and social improvement, but it has become apparent in recent times that science and technology are no longer unalloyed agents of progress. They now seem to contribute significantly to the difficulties we are facing in environmental degradation, pollution, climate change, and waste disposal. We have then a joint problem. On the one hand the future is not what it used to be, courtesy of the negative effects of science and technology, but nonetheless we will inevitably rely on them for their problem solving capacities. On the other hand it is becoming apparent that the grand project of modernism – a universal scientific culture – is in danger of collapsing from over-exploitation and specialisation and ought to be modified in favor of encouraging cultural diversity. Just as biological diversity has become recognized as an ecological necessity so too has our cultural survival come to be seen as dependent on a diversity of knowledge. Consequently 'the central problem of social and political theory today is to decide the nature of communicatory reason between irreducibly different cultures' (Davidson, 1994). Given this double difficulty we need to explore the possibility of a reconstituted knowledge system with multiple forms of rationality.

Rationality is a deeply problematic concept. It is profoundly embedded in the hidden assumptions of late twentieth century occidentalism about what it is to be a knowing, moral, sane individual. Indeed it is so embedded that to be anything other than rational is to be ignorant, immoral, insane, or a member of an undifferentiated herd. Hence rationality cannot be treated as simply an epistemological concept about the conditions under which one can know something; it also carries ideological overtones, privileging certain ways of knowing over others. Rationality is a constitutive element in the moral economy.

Yet despite, or perhaps because of, this central role of rationality, there are no fully articulated rules or criteria for being rational in the acceptance of beliefs or in the pursuit of knowledge, nor is there a single type of rationality. As with all concepts, when the epistemological is overemphasised at the expense of the practical, rationality has a degree of incoherence which seems to reach total intransigence in the recognition that ultimately there can be no rational (as opposed to aesthetic, moral or social) justification for being rational or choosing between types of rationality. Nonetheless critical rationalism as advocated by Karl Popper has a primal persuasiveness in contemporary western society. Mario Bunge has captured some of that self-evidentiality and variability in his seven desiderata for rationality.

1. Conceptual: minimizing fuzziness, vagueness, or imprecision.
2. Logical: striving for consistency, avoiding contradiction.
3. Methodological: questioning, doubting, criticizing, demanding proof or evidence.
4. Epistemological: caring for empirical support and compatibility with bulk of accepted knowledge.
5. Ontological: adopting a view consistent with science and technology.
6. Valuational: striving for goals which are worthy and attainable.
7. Practical: adopting means likely to attain the goals in view.

Setting the criteria out like this gives them a kind philosophical self-evidence which makes their denial seem irrational. But there are no universal criteria of rationality, and even if the claim for rationality's governing role is weakened to talk of desiderata as Bunge does, Wittgenstein's point prevails: no body of rules can contain the rules for their application. Rationality consists in the application of locally agreed criteria in particular contexts. Hence it should be acknowledged that science, rather than exemplifying some transcendental rationality, has developed its own rationality that has in turn served to create a great divide between science and traditional beliefs.

To talk of rationality as if it were purely a philosophical problem denies its historical and sociological dimensions. The concept of the individual as a rational actor that is now so basic to Western ways of thought arose in conjunction with the development of modern science in the seventeenth century. This was a period which saw great debates over the appropriate forms of rationality between the Cartesian rationalists and the Baconian empiricists. Whether true knowledge was to be derived deductively from self-evident first principles or by observation and experiment, it had already been accepted that the acquisition of such knowledge was within the capacity of human individuals. The recognition that human reason and experience was not inherently limited and could be a source of knowledge re-emerged in the twelfth and thirteenth centuries in the West with the separation of the church from the state and with the development of secular law from the accompanying canon law (Huff, 1993). The development of this concept of rationality was not universal. For example it was not paralleled in Islamic society where men were denied rational agency; they were held to lack the capacity to change nature or to understand it. Knowledge was instead to be derived from traditional authority. This is not to deny that there has been any Islamic science or any Islamic discussions or debates on the nature of rationality. On the contrary, there have been major achievements in Islamic science but in a radically different moral economy.

While the notion of the rational actor as unconstrained by circumstance or authority, moved only by logic and evidence, has become embedded in our legal, economic, and scientific presuppositions, such an idealized conception is at variance with our lived reality both at the societal and the individual level. At the societal level modern western capitalism has become a bureaucratic system which, as Weber pointed out, relies on a calculative rationality. The administration and perpetuation of this system is crucially dependent on a system of rules from which legal and administrative calculations can be derived by professional objective experts. Hence modern science and capitalism are interdependent; they were co-produced on the basis of a calculability derived in part from rational structures of law and administration. In Weber's view it is the specific and peculiar rationalism of western culture that makes science unique to the west. Even if Weber was right, what needs further examination is how specific and peculiar that calculative rationalism is. That form of rationality has a number of interwoven components, for example the acceptance of written documents as opposed to oral testimony as evidence. This transition also occurred in the twelfth and thirteenth centuries but required the development of a 'literate mentality' before it became self-evident that records and archives were more belief worthy than the word of 'twelve good men and true' (Street, 1984). Some, like Goody, have further argued that the accumulation of knowledge and the possibility of criticism and hence rationality are only possible in a literate culture. Similarly vision had to be rationalized to provide grammar or rules for the relationship between the representation of objects and their shapes as located in space. Yet another component of the form of rationality we equate with science was the acceptance of the validity of experimental evidence. This, as we have already seen, had to be accomplished through the institutionalization of historically contingent forms (Shapin and Schaffer, 1985). Similarly what now seems like a natural and self-evident relationship between fact and numbers was forged through a long and uneven historical process (Poovey, 1998: 29; Dear, 1997).

Rationality is not however a singular human capacity, nor is it quite the iron cage Weber described. Rather there are varieties and compounds of rationality which at the societal level, as Foucault has shown, are dependent on particular social and historical institutions, constituted through the interwoven practices, techniques, strategies and modes of calculation that traverse them. Nor, at the individual level, do we behave like the ends/means optimization calculators that economic rationalism would have us believe we are. We are at least as interested in meaning, significance, and personal values as we are in economic concerns (Friedland and Alford, 1991; DiMaggio and Powell, 1991). Though much of our knowledge is tacit and local and derived from our own experience we are not quite the rational individuals that the legal and philosophical theorists claim. A large proportion of our knowledge derives not from personal experience but from a communal stock relayed by books, newspapers, journals, teachers, and experts. But we often find that much of this expert knowledge is contested and stands in contradiction with our own experience (Wynne, 1996). Our individual lived rationality is thus expressed through working and living amongst multiple forms of rationality and is based in a range of social practices, traditions, and moralities that are suppressed and concealed in the portrayal of rationality as an ahistorical, universal form of reasoning exemplified by science. In the actual practice of science and daily life what counts as being rational is an essentially messy business much better described as 'muddling through' (Fortun and Bernstein, 1998).

OBJECTIVITY

Much the same can be said of objectivity. Objective knowledge is held to be the product of science that has established methods to ensure that individual, institutional, and cultural biases are eliminated. On closer examination objectivity is not characteristic of one special kind of knowledge – science. Rather it is the result of whatever institutionalized practices serve in a particular culture to create self-evident validity. Objective knowledge, in modern terms, is held to contrast with subjective knowledge. It is knowledge that is not local, that is not contingent on the circumstance, authority, or the perspective of the individual knower. However, the concept of objectivity, like that of rationality, is not immutable; it is an historic compound. In the seventeenth century objectivity meant 'the thing insofar as it is known'. The concept of aperspectival objectivity emerged in the moral and aesthetic philosophy of the late eighteenth century and spread to the natural sciences only in the mid-nineteenth century as a result of the institutionalization of scientific life as a group rather than an individual activity (Daston, 1992; Dear, 1992).

This characterization of objectivity as the 'view from nowhere' (Nagel, 1986) represents one of the essential contradictions of scientific knowledge production. Knowledge is necessarily a social product; it is the messy, contingent, and situated outcome of group activity. Yet in order to achieve credibility and authority in a culture that prefers the abstract over the concrete and that separates facts from values, knowledge has to be presented as unbiased and undistorted, as being without a place or a knower.

Objectivity, like democracy, is at best a worthy goal but one that is never capable of achievement. Since knowledge is the product of social processes it can never completely transcend the social. Objective knowledge cannot, for example, simply be knowledge which is unaffected by non-rational psychological forces, since scientists always have motivations even if they pursue knowledge for its own sake. Nor can objectivity be restricted to the avoidance of dogmatic commitment, because there have been scientists like Kepler whose obsession with regular solids and cosmic harmony led to his derivation of the laws of planetary motion. While it may be possible to avoid personal idiosyncrasy this can only be achieved through the establishment of communal or public knowledge. If knowledge is a communal product, then the question of how the community should be constituted arises. Should the scientific community be an essentially western institution? Consequently objective knowledge cannot simply be 'value free' knowledge because what is counted as knowledge is itself a value. Similarly the criterion of practical effectiveness cannot determine objectivity since it too is based in community standards (Albury, 1983).

The only remaining possibilities for objective knowledge lie in the notions of correspondence with reality and experimental verification. There are well known difficulties here, since correspondence theories of truth and verification are dependent on empiricism and the scientific method, both of which have been subject to powerful criticism in the last half century or so. Essentially, what philosophers like Pierre Duhem, W. V. O. Quine and Ludwig Wittgenstein have argued is that our ways of knowing about the world are riddled with *indeterminacies*. Which is to say that there is no set of procedures sufficiently powerful to determine which knowledge claims are absolutely true and certain, nor is there a certain way of grounding such claims. There are uncertainties inherent in all our ways of knowing that have to be bridged by a variety of practical and social strategies. Karl Popper and Thomas Kuhn have argued that neither deduction nor induction is capable of providing true and certain knowledge. Furthermore observations and theories are interrelated and hence neither can be an independent foundation for the other. Both Popper and Kuhn recognize that observations have point and meaning within the context or framework of a theory: we do not simply observe natural phenomena, we observe them in the light of some theory we already have in mind, or minimally with some set of expectations about what is interesting or what to look for. In this sense our observations of the world are 'theory dependent' or 'theory laden'.

As Duhem points out, all theories are enmeshed in a web of other theories and assumptions. The apparent conflict between an experimental result and a particular hypothesis cannot conclusively lead to the rejection of that hypothesis, since the strongest conclusion that can be drawn is that the hypothesis under test and the web of theories and assumptions in which it is embedded cannot both be true. Since the experimental result by itself is insufficient to tell us where the flaw lies, we can always maintain a theory in the light of an apparently falsifying experiment if we are prepared to make sufficiently radical adjustments in the web of our assumptions. Conversely it is the case that for any given set of facts there is an indeterminably large number of possible theories that could explain them.

In addition to the problematic relationship of theory and observation there is a difficulty concerning the language in which our claims about the world are expressed. All propositions or observation statements contain descriptive predicates which imply a classification or categorization of the world based on postulated essences or natural kinds. We are stuck with some degree of circularity since we gain our knowledge about natural kinds from theories which are in turn based on observations. There is no neutral observation language; our only option is to recognize and acknowledge the conventional character of our linguistic classifications (Hesse, 1980).

SCIENTIFIC METHOD

Despite all these difficulties scientists are able to reach firm conclusions about the natural world. How is this possible? Many would claim that even though it may be logically true that there are an indefinite number of possible explanations for a given body of facts, in a given case there are typically a very restricted set of alternatives and there are adequate means of selecting the right theory, or at least the best possible theory in the circumstances, given the application of certain criteria. For example, we obviously desire theories which are internally coherent, consistent with other accepted theories, and simple rather than complex. However coherence, consistency, and simplicity as well as other criteria like plausibility have all proved notoriously difficult to express in a way which can be used to measure all theories in all circumstances. It is, nonetheless, very tempting from our twentieth century standpoint, imbued as we are with the scientific ethos, to suppose that there must be a particular set of rules, procedures, and criteria to which all scientists adhere. Taken together they should constitute the scientific method, and by diligent application of this method we should be able to arrive at all the scientific discoveries of our age. However, Paul Feyerabend in *Against Method*, for example, argues that no proposed set of rules and procedures has survived criticism or has been universally adopted by all scientists in all circumstances. Likewise he claims that in no case can it be shown that the success of science can be solely attributed to its adherence to the scientific method.

There can of course be endless debates about such claims, but so far no one has been able to identify the one scientific method which has been adopted in all the sciences. Compare for example theoretical physics (mainly mathematical) and biology (mainly observational). Nor is there a method which has been unilaterally accepted in a particular discipline. Compare again, Newton's espoused methodology ('I feign no hypotheses') with Einstein's (bold hypotheses and deductive tests). Further, particular instances of scientific practice under close examination reveal a pragmatic willingness to suspend or modify any particular version of the scientific method if necessary, as Feyerabend has shown in his analysis of Galileo and Copernicus. It seems then that there is no single, invariant methodology of science. Instead there are series of complex interactions between method and practice. As science and technology develop, so the practitioners develop and negotiate the rules for doing them in a local and contingent fashion. New methodologies are propounded in order to provide support and credibility to a newly preferred scientific theory (Schuster and Yeo, 1986).

SCIENTIFIC PRACTICE

It could be argued that the indeterminacies of science and the lack of a specific scientific method are merely philosophical and theoretical problems, and that science is firmly grounded in experimental practice. The best way to know whether a particular knowledge claim about the world is true or false is to subject it to experimental test and then have somebody else repeat the experiment.

There are two kinds of related difficulties with this empirical approach. Experiments are inevitably performed on a simplified, artifactual, and isolated portion of reality. Hence the universal generalizations drawn from them are not indubitable. Cartwright goes so far as to claim that the Laws of Physics are, in effect, lies. Equally the effectiveness of experimental replication as the litmus test of truth is somewhat undermined by the role of skill or tacit knowledge in scientific practice. 'The problem being that, since experimentation is a matter of skillful practice, it can never be clear whether a second experiment has been done sufficiently well to count as a check on the results of the first. Some further test is needed to test the quality of the experiment – and so forth' (Collins, 1985). This results in what Collins calls *the experimenter's regress.* In the normal course this is resolved by social processes in which the judgment about whether to accept a particular result is based on the relevant community's evaluation of the skills of the experimenter in question. It becomes deeply problematic when the existence of the phenomenon itself is at issue, as in the case of gravity waves. An experiment showing the existence of such waves was followed by others seemingly denying their existence, or at least failing to detect them. Which was correct? The existence of gravity waves turns on the judgment of the community about competently performed experiments, and those judgments of competence are based on the accepted community knowledge about the nature of gravity waves. Replication then is not the test of their existence; rather it reflects the ability of the experimenter to achieve community standards of experimental practice.

Thus it would seem that conceptions of rationality, objectivity, and the scientific method cannot be derived from self-evident epistemological principles. They are instead embedded in the historically contingent processes of scientific practice whereby the resistances and limitations of reality are encountered and accommodated. Science is a social activity that is essentially dependent on community and tradition, and the practice of science is governed by concrete, discrete, local traditions which resist rationalization.

* * *

The notion of a great divide between western and so-called primitive knowledge systems has turned crucially on the question of the rationality of science. If as the arguments above suggest science has a rationality of its own, but not one that is especially privileged, how do we both account for and deal with similarities and differences between cultures? Given that the peoples of the world are sufficiently alike to have universally developed complex languages, and that those languages and their accompanying knowledge systems have produced profoundly different cultures, how are we to ensure communication and preserve cultural diversity?

In the earlier section on local knowledge it was claimed that the common element in all knowledge systems is their localness, and that their differences lie in the way that local knowledge is assembled through social strategies and technical devices for establishing equivalencies and connections. It is no small reflexive irony that this discussion of rationality and science is dependent on unspoken assumptions concerning credibility and authority which constitute a form of rationality. It is assumed that this chapter, like the others in the book, dependent as it is on evidence, analysis, and argument, is capable of being read, understood, and utilized by readers from all cultures. In other words there is a strong resemblance between the assumed rationality of this book and the practice of science. Science is dependent on the assemblage of heterogeneous inputs, but that assemblage is not achieved by the application of logical and rational rules or conformity to a method or plan. Indeed it is not even dependent on a clearly articulated consensus. Rather the assemblage results from the work of negotiation and judgment that each of the participants puts in to create the equivalencies and connections that produce order and meaning. Perhaps then it has to be acknowledged that there is a minimal rationality assumption and that links between kinds of rationality can be created by common human endeavor in a practical process of muddling through. Given the lack of universal criteria of rationality, what is needed to handle the problem of working disparate knowledge systems together is the creation of a shared knowledge space in which equivalencies and connections between differing forms of rationality can be constructed. Communication, understanding, equality, and diversity will not be achieved by others' simply adopting western information, knowledge, science, and rationality. It will only come from finding ways to work together in joint forms of rationality.

BIBLIOGRAPHY

Agrawal, A. 'Dismantling the divide between indigenous and scientific knowledge.' *Development and Change* 26(3): 413–439, 1995.

Albury, R. *The Politics of Objectivity*. Geelong: Deakin University Press, 1983.

Allport, P. 'Still searching for the Holy Grail.' *New Scientist* 132: 51–2, Oct 5th, 1991.

Bunge, M. 'Seven desiderata for rationality.' In *Rationality: The Critical View*, J. Agassi and I. C. Jarvie, eds. Dordrecht: Martinus Nijhoff Publishers, 1987, pp. 5–17.

Cartwright, N. *How The Laws of Physics Lie*. Oxford: Clarendon Press, 1983.

Collins, H. *Changing Order: Replication and Induction in Scientific Practice*. London: Sage, 1985.

Collins, H. and T. Pinch. *The Golem at Large: What You Should Know About Technology*. Cambridge: Cambridge University Press, 1998.

Collins, H. and T. Pinch. *The Golem: What You Should Know About Science*. Cambridge: Cambridge University Press, 1998.

Daston, L. 'Objectivity and the escape from perspective.' *Social Studies of Science* 22: 597–618, 1992.

Daston, L. 'The moral economy of science.' *Osiris* 10 (Constructing Knowledge in the History of Science): 1–26, 1995.

Davidson, A. 'Arbitrage.' *Thesis Eleven* 38: 158–62, 1994.

Dean, M. *Critical and Effective History: Foucault's Methods and Historical Sociology*. London: Routledge, 1994.

Dear, P. 'From truth to disinterestedness in the Seventeenth Century.' *Social Studies of Science* 22: 619–31, 1992.

Dear, P. *Discipline and Experience: The Mathematical Way in the Scientific Revolution*. Chicago: University of Chicago Press, 1995.

DiMaggio, P. and W. Powell. 'The iron cage revisited: institutional isomorphism and collective rationality.' In *The New Institutionalism in Organizational Analysis*, W. Powell and P. DiMaggio, eds. Chicago: University of Chicago Press, 1991, pp. 63–82.

Duhem, P. *The Aim and Structure of Physical Theory*. Princeton: Princeton University Press, 1954.

Dupré, J. *The Disorder of Things: Metaphysical Foundations of the Disunity of Science*. Cambridge, Massachusetts: Harvard University Press, 1993.

Feyerabend, P. *Against Method: Outline of an Anarchist Theory of Knowledge*. London: Verso, 1978.

Finnegan, R. *Literacy and Orality: Studies in the Technology of Communication*. Oxford: Basil Blackwell, 1988.

Fortun, M. and H. Bernstein. *Muddling Through: Pursuing Science and Truths in the 21st Century*. Washington, DC: Counterpoint, 1998.

Foucault, M. *Power/Knowledge: Selected Interviews and Other Writings 1972–77*. New York: Pantheon Books, 1980.

Friedland, R. and R. Alford. 'Bringing society back in: symbols, practices and institutional contradictions.' In *The New Institutionalism in Organizational Analysis*, W. Powell and P. DiMaggio, eds. Chicago: University of Chicago Press, 1991, pp. 232–266.

Galison, P. *Image and Logic: A Material Culture of Microphysics*. Chicago: University of Chicago Press, 1997.

Galison, P. and D. Stump, eds. *The Disunity of Science: Boundaries, Contexts and Power*. Stanford: Stanford University Press, 1996.

Geertz, C. *The Interpretation of Cultures: Selected Essays*. New York: Basic Books, 1973.

Hacking, I. *Representing and Intervening: Introductory Topics in the Philosophy of Natural Science*. Cambridge: Cambridge University Press, 1983.

Hacking, I. 'The self-vindication of the laboratory sciences.' In *Science as Practice and Culture*, A. Pickering, ed. Chicago: University of Chicago Press, 1992, pp. 29–64.

Haraway, D. *Simians, Cyborgs and Women: The Reinvention of Nature*. London: Free Association Books, 1991.

Hayles, K. *Chaos Bound: Orderly Disorder in Contemporary Literature and Science*. Ithaca: Cornell University Press, 1990.

Hesse, M. *Revolutions and Reconstructions in the Philosophy of Science*. Sussex, Harvester Press, 1980.

Hollis, M. and S. Lukes, eds. *Rationality and Relativism*. Oxford: Basil Blackwell, 1982.

Huff, T. *The Rise of Early Modern Science; Islam, China and the West*. Cambridge: Cambridge University Press, 1993.

Ivins, W. M. *Prints and Visual Communication*. Cambridge, Massachusetts: Harvard University Press, 1953.

King, M. D. 'Reason, tradition, and the progressiveness of science.' In *Paradigms and Revolutions: Applications and Appraisals of Thomas Kuhn's Philosophy of Science*, G. Gutting, ed. South Bend, Indiana: University of Notre Dame Press, 1980, pp. 97–116.

Kuhn, T. *The Structure of Scientific Revolutions*. Chicago: University of Chicago Press, 1970.

Latour, B. 'Visualisation and cognition: thinking with eyes and hands.' *Knowledge and Society* 6: 1–40, 1986.

Latour, B. *Science In Action*. Milton Keynes: Open University Press, 1987.

Lave, J. *Cognition in Practice*. Cambridge: Cambridge University Press, 1988.

Law, J. 'Technology and heterogeneous engineering: the case of Portuguese expansion.' In *The Social Construction of Technological Systems; New Directions in the Sociology and History of Technology*, W. Bijker, T. Hughes and T. Pinch, eds. Cambridge: MIT Press, 1987, pp. 111–34.

Marcus, G. E. and M. M. J. Fischer. *Anthropology as Cultural Critique: An Experimental Moment in the Human Sciences*. Chicago: University of Chicago Press, 1986.

Nader, L., ed. *Naked Science: Anthropological Inquiry into Boundaries, Power and Knowledge.* New York: Routledge, 1996.

Nanda, M. 'The science question in postcolonial feminism.' In *The Flight From Science and Reason,* P. Gross, N. Levitt and M. Lewis, eds. New York: New York Academy of Sciences, 1997, pp. 420–36.

Nanda, M. 'The epistemic charity of the constructivist critics of science and why the third world should refuse the offer.' In *A House Built on Sand: Exposing Postmodernist Myths about Science,* N. Koertge, ed. New York: Oxford University Press, 1998, pp. 286–311.

Ophir, A. and S. Shapin. 'The place of knowledge: a methodological survey.' *Science in Context* 4: 3–21, 1991.

Pickering, A. 'Objectivity and the mangle of practice.' *Annals of Scholarship* 8: 409–425, 1991.

Pickering, A., ed. *Science as Practice and Culture.* Chicago: University of Chicago Press, 1992.

Pickering, A. *The Mangle of Practice: Time, Agency, and Science.* Chicago: University of Chicago Press, 1995.

Poovey, M. *A History of the Modern Fact: Problems of Knowledge in the Sciences of Wealth and Society.* Chicago: University of Chicago Press, 1998.

Popper, K. *Conjectures and Refutations: The Growth of Scientific Knowledge.* London: Routledge Kegan Paul, 1963.

Quine, W. V. O. *From a Logical Point of View: Logico-Philosophical Essays.* New York: Harper Torchbooks, 1963.

Rosenberg, A. *Instrumental Biology or the Disunity of Science.* Chicago: University of Chicago Press, 1994.

Rouse, J. *Knowledge and Power: Towards a Political Philosophy of Science.* Ithaca: Cornell University Press, 1987.

Salmond, A. 'Maori epistemologies.' In *Reason and Morality,* J. Overing, ed. London: Tavistock, 1985, pp. 240–63.

Schuster, J. A. and R. Yeo, eds. *The Politics and Rhetoric of Scientific Method.* Dordrecht: Reidel, 1986.

Shapin, S. *A Social History of Truth: Civility and Science in 17th Century England.* Chicago: University of Chicago Press, 1994.

Shapin, S. 'Placing the view from nowhere: historical and sociological problems in the location of science.' *Transactions of the Institute of British Geographers* 23: 5–12, 1998.

Shapin, S. and S. Schaffer. *Leviathan and the Air Pump: Hobbes, Boyle and the Experimental Life.* Princeton: Princeton University Press, 1985.

Star, S. L. 'The structure of ill-structured solutions: boundary objects and heterogeneous distributed problem solving.' In *Distributed Artificial Intelligence,* L. Gasser and N. Huhns, eds. New York: Morgan Kauffman Publications, 1989, pp. 37–54.

Street, B. V. *Literacy in Theory and Practice.* Cambridge: Cambridge University Press, 1984.

Thrift, N. 'Flies and germs: a geography of knowledge.' In *Social Relations and Spatial Structures,* D. Gregory and J. Urry, eds. London: Macmillan, 1985, pp. 366–403.

Turnbull, D. 'Local knowledge and comparative scientific traditions.' *Knowledge and Policy* 6(3/4): 29–54, 1993.

Turnbull, D. 'Reframing science and other local knowledge traditions.' *Futures* 29(6): 551–62, 1997.

Turnbull, D. *Masons, Tricksters and Cartographers: Makers of Knowledge and Space.* Reading: Harwood Academic Publishers, 1999, in press.

Watson-Verran, H. and D. Turnbull. 'Science and other indigenous knowledge systems.' In *Handbook of Science and Technology Studies,* S. Jasanoff, G. Markle, T. Pinch and J. Petersen, eds. Thousand Oaks, Sage Publications, 1995, pp. 115–139.

Weber, M. 'Law, rationalism and capitalism.' In *Law and Society,* C. M. Campbell and P. Wiles, eds. Oxford: Martin Robertson, 1979, pp. 51–89.

Wittgenstein, L. *Philosophical Investigations.* Oxford: Blackwell, 1958.

Wynne, B. 'Misunderstood misunderstandings: social identities and public uptake of science.' In *Misunderstanding Science: The Public Reconstruction of Science and Technology,* A. Irwin and B. Wynne, eds. Cambridge: Cambridge University Press, 1996, pp. 19–46.

HELEN VERRAN

LOGICS AND MATHEMATICS: CHALLENGES ARISING IN WORKING ACROSS CULTURES

Scene One

A mathematics lesson in a school recently moved to new, cyclone-proofed buildings at the top of a hill behind the Yolngu Aboriginal settlement of Yirrkala in the far northeast of Australia's Northern Territory. The teacher, Mandawuy Yunupingu, a young man in his final year of BA(Ed) studies, is set to become, in 1989, the first Aboriginal school principal in Australia. He has chosen to teach this lesson in the school hall, which doubles as a basketball court. He wanted a space not cluttered by school desks. Watching an edited excerpt of a video (Yirrkala Community School, 1996c) of this lesson in 1998, I am struck again by the certainty that Mandawuy and his pupils evince over the subject matter of this lesson. It is a certainty arising from familiarity. The children, learning the details of the system of rules being presented to them, are certain enough to ask questions. Familiarity with the subject matter can be seen clearly in the body language – the postures, the ease of the interactions – even if sometimes the correct answers escape the children. The edited fragment of this lesson begins with Mandawuy seated on the concrete floor of the school hall – an open, roofed area – and the children grinning in delight, sitting around him in a circle.

Girls, boys and teacher are all barefoot and cross-legged on the floor. Mandawuy is giving this lesson in Gumatj, a Yolngu language which is also the first language of many of the pupils, and certainly understood fully by all the children even if some, in answering his questions, reply in Rirratjingu, another of the Yolngu languages. Mandawuy asks a question which is translated in a subtitle in the video, 'What are some other Yirritja trees?' We hear a chorus of children's voices; thanks to the subtitle we pick out the word 'Gadayka'. The next subtitle reduces Mandawuy's long response to this (wrong) answer as, 'Gadayka tree is sung by Yirritja people when it grows on Yirritja land. But it is a Dhuwa tree'. Mandawuy is gentle in his correction. 'Gungurru', volunteers a small girl. 'Gungurru is one Yirritja tree', Mandawuy affirms. 'Do you see the Gungurru tree?', he asks, pointing outside. 'Dhurrtji', another child

H. Selin (ed.), Mathematics Across Cultures: The History of Non-Western Mathematics, 55–78.

volunteers. 'What is Dhurrtji? Is it Dhuwa?' 'Dhuwa, yes it is'. 'So Gungurru, dhumulu, are some Yirritja trees'.

'What about other living things such as fish?', the lesson continues. 'Yo, Nguykal', Mandawuy agrees, vigorously patting the little boy sitting next to him on the back. 'Nguykal; yes that's right; it's Yirritja. He thinks really fast'. 'What else?' 'Yo, bäru, crocodile'. 'Yes that right; that's Yirritja'. 'What about Dhuwa?' 'Manda, yes that's right. What else?' 'Goanna'. 'Yes you're right but there are two kinds of goanna. One is Yirritja and one is Dhuwa. The Dhuwa one is called Djanda. The Yirritja one is called Beyay'. 'Are they male and female?' asks a shirtless little boy. The other children squeal with laughter, falling about in hilarity. But Mandawuy rubs him on the shoulder, 'That's a good question', he says admiringly. And in the lesson (although not in this edit of the video) he goes on to explain the theory of formal opposites: Yirritja and Dhuwa in contrast to the empirical opposites: male and female (Watson-Verran, 1997b; Verran, this volume).

The embodied certainty which suffuses this lesson and underlies the laughter is also evident in an interview where Mandawuy justifies this particular mathematics lesson. The certainty is evident despite the fact that here Mandawuy is speaking English (Yirrkala Community School, 1996c).

> Yes, there's rules about all these ways of interaction with other members of the community. And those rules are, say, universal, in Yolngu thinking. And it's adhered to very strongly by the elder people, by the people who keep social practices going in the community correctly. And that is the sort of situation we have to establish and continue for the young people to learn at school. So that they get familiar with their own *gurrutu* system, which is the kinship system, and at the same time can be able to draw in the *Balanda* [white Australian] concepts, which is mathematics. But the main purpose is to emphasise the fact that there is a Yolngu system that's there, that needs to be looked at first in a formal way of learning.

Scene Two

In February 1991, the entire primary section of Yirrkala School is assembled on the grassy square at the school's centre. On the walls of the classrooms which bound the square are examples of work the children have produced in mathematics lessons over the past two weeks. The entire school has participated in a maths workshop on '*Gurrutu*, The Yolngu Kinship System'. The teaching was supervised by a group of elders who are clan leaders in the community.

In the booklet (Yirrkala Community School, 1991) reporting this episode, objectives of this teaching/learning episode are identified.

Week One:
1. To develop children's understanding about gurrutu, and its reciprocal nature;
2. To develop children's spoken language using correct reciprocal gurrutu terminology.

Week Two:
1. Children will learn the system of mäl names;
2. Children will learn that gurrutu has a recursive pattern;
3. Children will understand why gurrutu and mäl have always been important to Yolngu.

One class produced a chart listing some of the multiple categories of names that the children in the class had for their teacher: for example some children stood in the gurruṯu position which called her 'child' (*waku*), while others were in the 'sister-in-law' (*galay*) position. (See Watson-Verran, 1997b for an elaboration of this system.) The more senior classes developed an exhaustive list of the nine reciprocal gurruṯu positions (2^3 positions plus the zero point position) (see Verran, this volume) which they tabulated. In the lower classes children had drawn diagrams which showed both egocentric and generalised diagrammatic maps of gurruṯu and mäl recursions.

Perhaps a first response to these stories is to question whether these lessons are mathematics education. It is a legitimate response. It might seem to many that what is happening here is a form of 'cultural education', and of course they would be correct. But this alone does not disqualify it as mathematics education. Who would deny that a class focussing on multiplication, or how to measure with length, is a form of cultural education? However that might not be a very satisfying answer, for what motivates the question is likely to be a feeling of unease: to include these classes as mathematics education might be pushing the notions of logic and mathematics too far.

Recognising this unease, in this chapter I explore the grounds on which these lessons, so confidently understood as mathematics by those involved in them, can be generally regarded as legitimate instances of mathematics education in a Yolngu Aboriginal school. Let me put these lessons in context by pointing to other instances of what my Yolngu friends understand as mathematics lessons: a workshop where children learn to use number in a quantification exercise; a class focussing on the recursive nature of number names; an excursion to the beach to measure the direction of the wind.

All these lessons can be viewed in the videos (Yirrkala Community School, 1996a–e) which elaborate what the Yolngu teachers, whose work I am presenting here, call a 'Garma Mathematics Curriculum'. *Garma* can be translated

Table 1 A chart produced by Yolngu children in their classroom study of gurruṯu (Yirrkala Community School, 1991)

Gurrutu category name that children call their teacher	*Names of children in the class who call their teacher this name*
Ngäṉḏi (mother)	Winimbu, Lirrpuma, Gapanbula, Gatang, Baninggirr, Djälang Wulu
mukul-bapa (aunty, mother's brother)	Yarrmiya, Larrtjpira, Baraltji, Larrandangu
gutha	Baringguma, Tony
yapa (sister)	Gapiny
momu (father's mother)	
märi (mother's mother)	Garul
mumalkur	Bandil
waku (child)	Banatha
gaminyarr (son's child)	Waninggurr, Muyulyun, Mathuray, Yangarryangrr
mukul-rumaru	Djawulu
galay (brother's wife)	Baymala

as 'common ground'. It incorporates the sense that what is presented (that is, in the curriculum) is something unified enough to be taken as an entity. It is a thing that has been generated in painstaking negotiation and experimentation over a long period of time, and is thus (almost) stable. It represents a publicly agreed upon position in the community. The inclusion of garma overtly recognises this curriculum as a negotiated outcome of (often opposed) interests, in this case the officials of Northern Territory mathematics education and the relevant Yolngu officials.

While the issue has been satisfactorily, although possibly only temporarily, settled well enough for some officials in Australia's Northern Territory to agree, for many others it remains controversial. Having such lessons as those I have just described as mathematics education raises the challenging question of what mathematics and logics *are* in contexts of working across cultures. My view is that taking this challenge seriously is important. The issue might seem to be a problem only for a few 'others', and not the concern of the mainstream academy. But as we shall see such challenges contain within them the possibility for the mainstream to see itself in new way. With this comes the possibility that we might make our futures different from our pasts. Specifically, we might find ways to go beyond colonialism in dealing with difference in doing mathematics and logics.

In taking up this challenge I argue here that we need to go beyond conventional understandings of logic as an *a priori* entity, somehow lying behind and within mathematics and other forms of certain knowledge. My argument is implied in using the term 'logics' in my title, a usage which might seem a contradiction. For many people logic is by definition singular. There can only be one; it is the basis of human reasoning and knowledge. Kant is useful to elaborate the conventional notion of logic I am rejecting. By introducing Kant I do not wish to suggest that somehow his writings have caused us to mistake the nature of logic, but rather that the account of logic he gives us is by now, some two hundred years later, inadequate. For me the Kantian account of logic is emblematic of what, near the end of the twentieth century, we need to go beyond. For Kant logic is given, pre-contextual, and it is normative.

> ... [T]his science of the necessary laws of the understanding and reason in general, or – which is the same – the mere form of thinking, we call logic ... [i]t must contain nothing but laws a priori that are necessary and concern the understanding in general ... [L]ogic ... must be taken to be a canon of the use of the understanding and of reason (Kant, 1974: 16–17).

In contrast, I point to the notion of garma as instructive: actively generated orderings robustly hanging together as outcomes of past collective acting. These are generated as thoroughly embedded, and embodied as routine practices. I take the position that to work logics and mathematics across cultures in ways that are valid and ethical we need to wean ourselves from foundationism. This is the notion that the world (or a world) has a given and set logic taken to be 'out there' as a complete and given entity, and that certain knowledge, like mathematics, is a symbolic representation of this.

I turn now to Nigeria in the early 1980s. Some of the mathematics lessons given by the teachers I worked with there might similarly be styled as non-standard. They did not abide strictly by Nigeria's national curriculum.

Scene Three

Mr Ojo was a poor organiser, but he had good rapport with children. This morning he had excelled himself in his preparations; he had assembled about twenty small cards, thick cardboard 10 cm long and about 5 cm wide, marked off in 1 cm divisions along the length. There was one card for each group of two or three children. To go with this he had twenty lengths of string about 2 m long. It was a lesson we had prepared as a group back in the laboratory in the Institute of Education at University of Ife (now Obafemi Awolowo University) in Nigeria. We had begun with one of the 'Measuring Ourselves' pamphlets produced by the African Primary Science Program – a large and prestigious USAID project which sponsored science and maths curriculum development with a focus on practical work in many countries in Africa in the 1960s–70s (Yoloye, 1978). Discussing how the lesson might be modified so that it would be suitable for a group of 50 or so Yoruba children, and classrooms quite devoid of resources, helped students prepare for their practical teaching exercise in the schools around Ile-Ife in Nigeria.

Like the rest of students at the Institute of Education, Mr Ojo was around my own age and a far more experienced teacher. As a lecturer, I was responsible for at least part of their retraining, but my complete inexperience when it came to Nigerian classrooms meant that this course in science education was, by necessity, very much a two way program of training. As a group, students and lecturer, we worked out a way of negotiating the curriculum we were developing.

The lesson Mr Ojo was to teach, 'Length in Our Bodies', involved children's using string to record another child's height, leg length, arm length, etc., then a metre rule to report the length in metric units. In the lab we had measured each other: we used the string to represent height, lay the string on the floor and used chalk to record the length. Then when one of the few metre rulers became available, we measured the distance between chalk marks and recorded the measurement in a chart.

Name	*Height*
Mrs Taiwo	1 m 62 cm
Mr Ojerinde	1 m 70 cm
Dr (Mrs) Watson-Verran	1 m 50 cm

We had also devised a means for evaluating the effectiveness of the lesson: Given a chart with fictitious children's names and heights, and using the process of the lesson in reverse, children would show a particular height using a length of string to demonstrate whether or not they had understood. The students were nervous about teaching this way. It meant getting children out of their desks and putting materials other than pencils and exercise books into their hands. It meant children talking to and working with one another instead of working only from the blackboard and speaking only in reply to the teacher. The children were liable to become unruly and noisy at such a departure from the norm, and this could be a serious problem with 45–50 in a small, enclosed space.

Speaking in Yoruba, Mr Ojo demonstrated the procedure. He called a small boy to the front: with the end of string just under the boy's heel, he held his finger at the point on the string which matched the top of the boy's head. Tying a loose knot at this point, he took the other end of the string from under the boy's foot; holding this at one end of a card he wound the length of string around until he came to the knot. Then he instructed: 'Count the number of strings around the card, e.g. "9" (i.e., 10 cm lengths) and write down the number. Multiply by ten. How do we multiply by ten? ... ninety ... now add the bit of string left over ... Yes, we have 96 cm'.

The children set to work in pairs or threes; soon the chart Mr Ojo had drawn on the blackboard was full of names and numbers. Several children very efficiently used card and string to show how tall the fictitious Dupe, Tunde and Bola of our evaluation exercise were. The lesson could be judged a complete success; the children had obviously got hold of using metric units to express a value. The children were pleased with their accomplishment; Mr Ojo was certain of his success.

Mr Ojo had deviated significantly from the exemplary lesson we had devised in our lab sessions back at the University and from the prescribed way of teaching length outlined in the Nigerian Primary Mathematics Curriculum. Yet his lesson was successful, unlike many of those which followed the prepared script more closely (Verran, 1999). For the prescribed stretched string and the extending metre ruler, he had substituted a small card wound around with string. Disconcertingly he had measurement of length beginning in the plurality of the strands of wound-up string rather than in the singularity of an extension. Should he have been corrected? Or should we trust the sanctioning of his approach by a class of fifty or so restless and difficult little children? Having listened to Mr Ojo's brief presentation these children were certain of what to do and did it, enjoying an experience which trained them in the bodily routines of quantifying in the process.

Scene Four

The school of my next story was located on one of the narrow roads radiating off from the Ooni's palace in Ile-Ife, Nigeria. The road was usually unaccountably clear and clean, and the school grounds always seemed neat; rake marks showed in the dust when we arrived in the morning. On Mondays the children were well scrubbed; by Friday their navy uniforms would be drab and dusty. Today Mr Ojeniyi was dressed as usual in a clean crisply ironed white shirt.

These were older children, who were quiet and responsive to Mr Ojeniyi, sitting in orderly rows, their exercise books open before them. A lesson on division was scheduled, and I knew from painful experience that division lessons could be excruciating as teachers and children struggled with the mechanics of long division, tying themselves in knots with 'the carrying over' and the 'bringing down of the next column'. But I was expecting something good today, I always enjoyed Mr Ojeniyi's lessons; he was one of those people excited by the aesthetic of maths. Numbers gave him joy and pleasure which he communicated easily to the children.

Mr Ojeniyi began in English, but after a few sentences shifted to Yoruba as he got warmed up in his explanation. I closed my eyes to try to pick up the Yoruba better and to follow his explanation. I was expecting an account of division as some sort of serial process, something like the reverse of multiplication, understood as serial addition. I was astonished to hear Mr Ojeniyi identify division as definitive of whole number. 'You will not understand a number unless you understand the many ways it can be divided'. He emphasised the point by repeating it in English, a common strategy in these classrooms.

First he presented a Yoruba number and showed it as a product of twenty plus or minus various factors of twenty, in translating it into a base ten English language type number. Then he did the same process in reverse: using English number names and the base ten system, he converted it into a Yoruba number, using division into sets of twenty as the first and defining process. The children followed his explanation and then wrote in their books the series he had elaborated for each translation on the blackboard. After two more such demonstrations on the board, one a Yoruba number translated to English base ten number, the other an English number translated to Yoruba, Mr Ojeniyi wrote ten Yoruba and ten English number names on the blackboard, all between one hundred and four hundred. He instructed the children to make similar translations for those twenty numbers, and to my amazement most of the children completed the exercise with some facility.

Then came the fun. Each translation could be done in more than one way, yet clearly some ways were more elegant than others. Mr Ojeniyi asked for volunteers, and the children loudly suggested alternatives. Amid much laughter and shouting, children jumping out of their seats to rush to the blackboard to demonstrate an alternative, a generally agreed upon best translation was gradually reached for each of the twenty. All thought of serious focus on the process of division vanished in the delight of the game, yet the game was all about division.

I had been told of and read about the base twenty Yoruba number system (see Abraham, 1962; Armstrong, 1962; Johnson, 1921; Watson, 1986) and I saw that this was an inspiration in Mr Ojeniyi's brilliant lesson. I was reminded too of the multiple kinds of arithmetic that Liberian tailors have access to (Reed and Lave, 1979). The brilliance of Yoruba children over number manipulations had been noted before.

> Yoruba conservation [of number] results are as good or better than those on American children [tested in the same way]. Comparisons with other African groups were favorable though Yoruba performance did not match the near-errorless results of Tiv children (Lloyd, 1981).

Yet, according to some theorists, such brilliance would not necessarily imply a 'full understanding of number ... [for] a true grasp of the logical foundations of number extends beyond mere arithmetic and is integrally related to measurement [i.e. conception of qualitative extension]' (Hallpike, 1979: 252).

From this position these lessons would be regarded as shot through with failure. There would be failure on the part of the children but more importantly failure by the teacher and the teacher educator. In contrast, there are other theorists who might praise these lessons, seeing them as based in an indigenous mathematics. Those observers would praise the teachers for working out of their own traditions, and the teacher educator would be approved for providing gallant support.

These opposed explanations grow from alternative accounts of logic and mathematics. The opposed judgements of failure or success, and in turn the different apportionment of blame and praise, result from alternative visions of the nature of the logical foundation. Engaging one or another of these visions, not only can we distribute praise and blame, but also we can come up with certain judgements we can trust. These are accounts of logic and mathematics which give us a basis for going on, justifiably certain of our correctness. The point I want to make is that while they do achieve all this, they also have us remaking our colonial pasts and prescribing particular moral orders. For this reason I argue that both these accounts of logic and mathematics are equally flawed as explanations and as the basis for judgement.

I now present two foundationist accounts of logic which might ground such judging explanations. I consider how they apportion praise and blame, bringing with them a particular moral universe. The first account I offer might be understood as orthodoxy; its alternative is the 'loyal opposition'. Actually, within each of the two traditions I identify there are labyrinthine and multiple distinctions and oppositions. I am presenting cartoon versions here, for once in the maze of realism-relativism it is very difficult to find a way out. The two interpretive traditions I present reach back through the modernising discourses of the academy on the one hand through the logical positivists directly to Kant, and on the other through the sociology of scientific knowledge, only indirectly mobilising Kant. Both give a vision of knowledge as symbolising, growing on a material foundation, but mediated in differing ways by experience.

The first tradition emphasises the given structure of the physical world as the foundation of symbolising which represents the structure. So, for example, numbers as symbols in some sense show the structure of the physical world. There are multiple versions of the mechanisms whereby the connection between the physical and the symbolic might be achieved. I call this cluster of explanations 'universalism'; others might call it 'realism'. Exploring its many versions is not my interest here. The alternative, 'relativism', also sometimes called 'constructivism', has the foundation of the social practices of ordering, sorting, arranging and patterning using given symbolising resources.[1] For those who identify J. S. Mill as an intellectual ancestor, the structure of the physical world is the stimulus behind a historical, possibly culturally specific, ordering, sorting, arranging and patterning which has generated the categories of the foundation. Other strands in this tradition point to alternative origins; they all agree in having the foundation as a historical and social product. Again the alternatives will not detain me. In one case – universalism – the foundation originates in a physical structure (a found, natural order); in the other – relativism – it is in a set of conventional practices (an order constructed in social activity).

When things are presented in this way each foundation seems eminently likely and we immediately begin to wonder how we could possibly go about choosing one or the other. But that too is not my focus. What I want to point to is that choosing either, for whatever reason, has significant consequences. At first we see that going for one or the other determines who gets praised or blamed for what in some contemporary classrooms. But more significantly we then see that irrespective of those details, both prescribe a particular moral order.

UNIVERSALISM: LOGIC REPRESENTS THE STRUCTURE OF THE MATERIAL WORLD – A PHYSICAL FOUNDATION

For the epistemic tradition which prefers to adopt a physical foundation, which sees that symbolic renderings like numbers, for example, reflect given categorisations in the real world, I turn to the logical positivists. I take Carnap's *Philosophical Foundations of Physics* (1966) as an exemplary account. Here we see quantifying, and by implication mathematics in general, as an expression of a singular, universal logic, taken to be a found universal form embedded in the material world.

In elaborating this logic Carnap identifies three kinds of concepts: classificatory, comparative and quantitative (Carnap, 1966: 51–52). By a classificatory concept he means simply 'a concept that places an object within a certain class' – what we would ordinarily call a proper name. In science they are the concepts of the taxonomies of zoology and botany, (cool blooded or warm blooded for example) but in ordinary life the earliest words a child learns – dog, cat, house, tree – are of this kind. Comparative concepts play a kind of intermediate role between classificatory and quantitative concepts. As Carnap presents it, comparative concepts are always qualities or properties of objects. For Carnap a quantitative concept is a particular way of representing a particular extent of a quality – naturally occurring abstract objects to be found in all physical objects. Quantification reports the extent of a quality, and so characterises a particular body.

The underlying assumption here is that the world naturally presents as spatial extensions which endure over time – things – defined as matter situated in space and time. Here the world is assumed to be an array of spatio-temporal entities, and this is the natural physical foundation for the symbolic practices of quantification. Quantifying here points first to the universal qualities in these spatio-temporal entities, the first abstract entities of quantifying. Then numbers, a second-order abstract entity, taken to be analogous in form to the extension of qualities held by the concrete spatio-temporal entities, are used to represent a real value of the entity through a given quality. Spatio-temporal entities are the natural foundation and quantifying is a dual level qualitative abstraction generating abstract entities-numbers. The logic in numbers lies in the material world. Mathematics is of course far more than numbers, but just as logic is embedded in numbers in completely accountable ways, so too all mathematical objects originate in the logic embedded in the world. Mathematics is singular and symbolic, its abstract objects grounded in a material reality.

According to this account of logic and mathematics, Mr Ojo is wrong to present an image to the children where length is portrayed as a bundle of string whose many strands can be counted and manipulated to come up with a single value for length. An uninterrupted straight line, sectioned in a subsequent action, portrays the proper, and only correct way to do length. Linear extension is understood to constitute a true image of how qualities really exist in the world, and measuring *should* begin with that. It is as important to get hold of the abstraction as to get the right answer. Mr Ojo has the children getting the right answer in the wrong way. As for Mr Ojeniyi, with his suggestion that division is constitutive of number, he has failed to communicate the correct linear account of the number array. And the lessons of the Yolngu Aboriginal teachers are quite beyond the pale. In this view of the matter there is nothing in those lessons which can be considered either logical or mathematical.

This way of distributing praise and blame grows from a particular account of logic committed to a vision of a physical foundation for the symbolising project of mathematics. Within this interpretive frame we can tell an orthodox story of institutional power relations which explains these mathematics lessons as failure and inadequacy. The lecturer, and in turn the teachers, are accredited with institutional power to teach the truth. The teachers are under the jurisdiction of the lecturer who has the right and the responsibility to tell them they were wrong and pass or fail them. This account of logic engenders a story which exalts the institutional power of the lecturer as the agent of a colonising modernity and sees only failure in these stories. This is failure on the part of Africans or Aborigines – children, teachers and whole societies – who in one case adopt 'primitive' ways of quantifying, and in the other are so primitive as to fail to organise their lives through logic. It also sees failure on the part of the lecturer who did not uphold the standards of modern university teaching and insist on the logical ways of teaching mathematics.

Out of this account can be generated legislating explanations about 'their' primitiveness, and the consequent need for 'their' uplifting through development and education. Difference is ruled out; policies are devised to abolish it. This story legislates a particular moral order: you should give up your non-modern Yolngu or Yoruba ways to become full knowing subjects in the process of making modernity. Colonialism here is seen as an agent of modernity and any notion of postcolonialism is necessarily neocolonialism, a continuing struggle to roll back the darkness through learning to use the given categories of the universal logic.

RELATIVISM: LOGIC REPRESENTS THE STRUCTURE OF A PARTICULAR CULTURE'S SYMBOLIC SYSTEM – A SOCIAL FOUNDATION

The alternative foundationist tradition considers social practices as the origin of a foundation for knowledge. My past writings can be taken as an example of this tradition (Watson, 1987; Watson with the Yolngu Community at Yirrkala and Chambers, 1989; Watson, 1990). In that work we can recognise links to Bloor's programme in the sociology of knowledge (Bloor, 1991) which identifies originary ordering practices as the (social) foundation of knowledge. Extending the ways that is usually understood, I added language use as one such constitutive practice. The pictures of Yolngu, Yoruba and English logics I presented each had three sets of social practices as foundational: practices of referring in ordinary language use, practices of unitising the material world, and recursive practices. Although my adoption of the relativist frame was somewhat ambiguous (I emphasised the physical nature of these practices and their here-and-nowness) in the end I point to them as historically set social methods of constituting symbols.[2]

The major difference between my accounts of a logic founded on historical, social practices, and the foundationists who take the given structure of the physical world as a foundation, lies in my disputing the naturalness of spatio-temporal entities – things defined by their situation in space and time – as a universal foundation. Making a comparative analysis of the grammars of English and Yolngu languages and English and Yoruba language, and taking language use to be a symbolising practice, I showed the possibility of disparate methods in the (universal) social practice of referring in language use. I had this creating alternative starting points in logic, generating for example in the Yoruba case, numbers which symbolise in different ways, and in the Yolngu case, quite differently constituted organising recursions. The explanation presents 'doing number' or more generally 'doing recursion', as constituting separate symbolising domains. The account implies that children might learn either one or the other, or sometimes maybe both if they are profoundly bilingual.

In relativisms, mathematics can be plural. Necessarily abstract mathematical objects represent the categories of a system of ordering happened upon historically as an act of social construction. It might seem also that logics can be plural; however in practice logic can only be singular. Because we can never give an account of how each system of symbols relates to a world which may, or may not, have given categories, the worlds variously represented by various systems of abstract mathematical objects are incommensurable; only one logic is ever accessible (Quine, 1960). We might however come up with various *ad hoc* ways the plural mathematics, the plural systems of symbols, can be translated.

On this basis Mr Ojo is correct and due for praise for doing a good job in coming up with a neat picture of number in Yoruba quantifying. He might come in for approbation on the basis that he failed to recognise and acknowledge the logical domain of English number, for this failure will serve children ill in the modern world. But in a relativist scheme this is a political (social) mistake, not a logical failing. In assessing Mr Ojeniyi's lesson we might argue that his arithmetic shows the juxtaposition of two systems of symbolising number, suggesting how each set of symbols might be translated into the other. His lesson can be judged both logically and pedagogically sound. The Yolngu teachers too come in for praise for their focus on the general notion of recursion as a characteristic common to the two symbolising domains.

This is a relativist account where logics, in symbolising foundations given in disparate conventional social practices, result in a pluralist world where alternative forms of knowledge compete, and where a deep incommensurability underlies a superficial capacity to translate between symbolic systems. It is an equally orthodox story of powerful Yoruba/Yolngu resistance and Western impotence. In this version teachers should heroically resist Western incursion and teach an indigenous version of logic in the school curriculum. Using indigenous language they resist and challenge the white lecturer. This story has the school teachers as agents of local resistance against the impotent university lecturer – the agent of a rejected and resisted Western imposition. We can begin to see how it too legislates in a moral arena; only this prescribes the heroism of resistance set against the tragedy of alienation. It is a vision of fragmentation, not unity.

Colonialism here oppresses and destroys the indigenous forms of knowledge using powerful Western forms of knowledge. This is a pluralist world of mutual incomprehension, incommensurability, and 'other' forms of knowledge, where colonial oppression and domination are pursued through competition on an uneven playing field, so that one knowledge (that of the coloniser) dominates. This is just another way of ruling out difference and denying any possibility of shared knowledge making. This relativist account often goes along with an 'othering', where 'our' (modern) base and spiritually deficient reductionism is compared to 'their' wholeness, emphasising what 'we' (moderns) have lost compared to 'them' (traditionals). Postcolonialism on this account is the expulsion of invading Western forms by a renaissance and resurgence of indigenous forms of knowledge; fragmentation and separate development (sometimes called apartheid) are prescribed.

The judgements affected by universalist and relativist accounts of logic and mathematics differ – they distribute praise and blame differently. Yet the moral orders they prescribe (Pyne Addelson, 1994: 137) have much in common. Relativism and universalism are opposed, effecting opposed judgements, but they both further the moral order of colonialism. I argue that both are morally untenable, and for this reason we must find ways to go beyond foundationist accounts of logics and mathematics. In summarising my argument here, I list what I see as several characteristics of the moral order prescribed by these foundationist ways of understanding logic(s) and mathematics.

1. Paradoxically we have denied difference (albeit in alternative ways) at the same time as we regenerate the boundary between Yoruba, Yolngu and English, between the traditional and the modern or between the one and the other.
2. The individual knower of logic and mathematics must choose to be one or the other – rational or primitive, with us or with them.
3. We are re-making the naturalness of domination. A particular institutional set of oppositional and colonial power relations is regenerated either by justifying it in universalism or by justifying its inversion, in relativism.
4. Most immediate and painful for me, foundationism is re-inscribing the

researcher as disembodied authority who participates as an (academic) legislator. Obliged only to adopt valid reasoning procedures, this disembodied researcher cannot participate in the times and places of those who her stories are about. Even if in presenting her work the researcher does include others, this can never be more than a rhetorical flourish (Pyne Addelson, 1994: xi). In a foundationist framework, my participation, my being caught up in the complex power relations and reciprocal indebtednesses of being both teacher and learner, at the same time as being a theorist (a storyteller) of the episodes, must be denied. I must adopt one or other of these cause and effect stories which both foreclose and legislate from a position of privilege.

5. Foundationist accounts of logic and mathematics fail to recognise Yolngu and Yoruba communities as arenas generating new ways to go on, and regenerating old ways of going on together. They fail to recognise the creativity of members of these communities as they proceed logically in their times and places. These are times and places where, for example Yoruba numbers and English numbers are equally likely to feature in ways of going on in Nigeria. Or they are times and places where gurruṯu and number both contribute to the making of a particular contemporary form of Australian life.
6. Foundationist accounts misconstrue the generative nature of 'doing logic' and 'doing mathematics' in mathematics teaching and learning, as much as in other contexts. This misrepresentation hides the nature of the struggle that mathematics teachers and learners everywhere experience, but particularly so in multicultural situations. It hides from teachers the nature of and the outcomes of their work, leading to degraded accounts of teaching and learning.
7. Foundationist ways of doing knowledge misconstrue the nature of colonialism and the place of knowledge in making and re-making colonialist moral orders. Hence they also misrepresent what a struggle for postcolonialism might be and where such struggles might be waged.

Stories such as those I told in beginning this chapter were grist for both realists and relativists from the 1960s to the 1980s.[3] What was at issue in that discourse was the true nature of the foundation, the debate proceeding quite conventionally within a foundationist frame. The rival claims for either a physical or a social foundation for logic, and by derivation mathematics, came with well established ways to evaluate and dispute them as knowledge claims. Argument was over the criteria by which claims might be judged, as much as endless dispute over the proper nature of the foundation.

As foundationist ontologies and epistemologies, universalism and relativism necessarily and unproblematically take logic – the foundation – as an entity, an object, a complete set of categories which can be expressed as a set of symbolising processes generating abstract concepts. Mathematics is necessarily also an object in this scheme of things – the set of mathematical concepts and ways of manipulating the concepts. As objects, logic and mathematics have particular properties, and the properties or characteristics of logic, and by extension mathematics, differ depending on which foundation you accept. We might experience logic and mathematics as sets of procedures and related concepts, but in the foundationist scheme of explaining what they are, they are first and foremost *things*. Their properties as things can be characterised through description.

Necessarily descriptions of the thing that is logic arise in particular instances; they are particular instances of the object logic or by derivation the object mathematics. Yet the descriptions make a claim about logic or mathematics in *general*. As knowledge claims they can be evaluated as examples of inductive reasoning, 'Is the evidence sufficient?', 'Are the reasoning processes valid?', and so on. This is the knowledge project, the way of 'doing knowledge' in the modern academy. Knowledge claims, in this case claims about the true nature of logic and mathematics, are held up for scrutiny, to be commented upon, condemned or commended as deemed fit, by judging observers, remote from the particularities on which the (generalised) knowledge claims are based. The foundationist metaphor brings with it a particular style of reasoning which establishes its particular propositional notion of truth and falsity (Hacking, 1982). This is a notion of true–false which is timeless and placeless, expressing verities of all times and all places. The reasoning goes this way irrespective of whether the foundation is taken as physical entities or social practices. The picture of a foundation has us adopting a particular mode of ontology/epistemology, a mode fitted to an object world. This foundationist mode can be engaged to make general claims about the nature of the foundation object.

The objectification of logic is the first step in foundationism. The object logic has, as Kant for example sees it, some general properties, and his texts walk us through the difficulties of seeing those properties, as well as making arguments about what those properties *must* be. For Kant the object logic is set, stable and complete in its form and determining of knowing.

I go on now to look at the way the *assumption* of the object 'logic', which results unproblematically in an object world, is expressed in Kant's most famous text, *Critique of Pure Reason*. This exercise has significant outcomes. I show that the way this is expressed in the text alerts us to a complexity we might otherwise miss in the seeming obviousness of the assumption that the world *is* objects. Going back to *The Critique of Pure Reason* to investigate how logic emerges unproblematically as a modern object, I follow feminist philosopher Michèle Le Dœuff (1989: 5–6). As part of her feminist critique of contemporary philosophy she went looking for some defining moments in modern philosophy. Such moments would contain prescriptions of how we should see things, among them being the proper position of women. More importantly from our point of view is the proper way of figuring the known and the knower.

That others would be sceptical about this enterprise Le Dœuff freely acknowledges. Surely picturing and prescribing ways of seeing is not what modern philosophy does. Every schoolgirl knows that philosophy is not a story or a pictorial description. Philosophy declares itself as philosophy through a break with the domain of the image. Looking in philosophy's seminal texts Le Dœuff found something quite surprising for philosophers: convincing evidence of what she called 'a properly philosophical imaginary', images thoroughly specific to philosophy. Her conclusion was that

> imagery has a relation to what we call 'conceptualized' intellectual work, or at least that it occupies the place of theory's impossible ... imagery copes with the problems posed by the theoretical enterprise ... The imagery which is present in theoretical texts stands in a relation of solidarity with the theoretical enterprise itself (and with its troubles) that is, in the strict sense of the phrase imagery is *at work* in these productions (Le Dœuff, 1989: 5–6).

Because of its strident denial of its images, and because philosophy claims to be a self-founding discipline, making recourse to myth inevitable, most or all the images in philosophy are about philosophy itself:

> Philosophy defines and designs its own myths making use of spatial and narrative places and layouts ... [Spatial] images ... form part of the language of the corporation ... which may be, just as much as certain principles, rules and misapprehensions, structuring elements of the philosophical position itself (Le Dœuff, 1989: 5–6).

Despite its denials philosophy *does* give us the pictures. And we should not be surprised. Common sense tells us that a prescription of how to do something necessarily contains instructions for how to see it. In presenting an account of how to do ontology and epistemology, philosophy must prescribe how to envisage the relation between known and knower. In particular Le Dœuff points to Kant whose 'work stands as a watershed in the self-understanding of modernity' (Rundell, 1994: 87), as someone whose texts can be usefully read in this way. She shows how the metaphor of ordered and ordering space goes right through Kant's *Critique*, and yet the way it is treated in the text effectively denies and hides the work it does in the reasoning.

In explaining his fundamental distinction between the analytic, concerned with the *a priori*, and the synthetic, the domain of the *a posteriori*, a spatial metaphor is implicit:

> In the analytic judgement we *keep to* the given concept and seek to extract something from it ... But in synthetic judgements I have to *advance beyond* the given concept, viewing as in relation to the concept something altogether different from what was thought in it. ...
>
> Granted then that we must *advance beyond* a given concept in order to compare it synthetically with another, a third something is necessary. ... *inner* sense and its *a priori* form, time (Kant, 1929: 171) (emphasis added).

Seventy pages later we get the full picture from which his account of the nature of logic is drawn.

> We have now not merely explored the territory of pure understanding, and carefully surveyed every part of it, but have also measured its extent, and assigned to everything in it its rightful place. This domain is an island enclosed by nature itself within unalterable limits. It is the land of truth – enchanting name! – surrounded by a wide and stormy ocean, the native home of illusion, where many a fog bank and many a swiftly melting iceberg give the deceptive appearance of farther shores, deluding the adventurous seafarer ever anew with empty hopes, and engaging him in enterprises which he can never abandon and yet is unable to carry to completion (Kant, 1929: 171).

Notwithstanding the intimate role of the spatial image in his own text, Kant casts imagery out of reason. With his island Kant paradoxically uses a metaphor to outlaw metaphor. Kant sees the exclusion of the image, of metaphor, as the defining property of logic which is rules and laws. He gives us a striking picture of a proper mode of ontology/epistemology, and to do it he uses the very method which his method outlaws!

Others characterise this shocking paradox in Kant's work as follows:

> Kant confronts an abyss, where were he to fall into it, he would confront chaos and uncertainty. He pulls back onto the ground of certitude. In doing so he circumscribes the nature and role of the imagination, especially its synthesising power, making it dependent on understanding. ... This has the effect, as the *Critique of Pure Reason* unfolds, not only of reducing the nature of the imagination to that of cognition, ... but more importantly of driving a wedge between reason and imagination. Reason contains no creative power, only a regulative power which gives rules and standards (Rundell, 1994: 95).

The island is the emblem of the Kantian enterprise, yet an emblem unthinkable in the logic of the program it seeks to establish. The image of the bleak ordered northern isle is thus a precondition of the foundationist mode of ontology and epistemology which Kant prescribes, and is systematically denied in elaborating that prescription.

The paradox accomplishes logic as an object. Logic is an immaculate conception – the first object of modernity. Its characteristics, its pure, clean lines, are boundaries which establish an object world which can be known, and by default a knower who may be on the island but is not of it. The world is pictured and thus made picturable through the work of the metaphor of the island.

This immaculate conception of logic as an object and denial of the resource which enabled the conception has several consequences. It gives knowers the possibility of specific categories to know with. It is these specificities that universalists and relativists have been disagreeing about ever since. It creates the possibility for further objects which might then be pictured using the lineaments of the island's order. Through knowing, through the image of the island, myriad other objects in the world can be recognised. These objects, we understand, were waiting there unrecognised. Seeing the object logic with its characteristics is the key to the process of now bringing many further objects to light. The island both creates and prescribes a way of doing knowledge which is by now thoroughly pervasive, but at the time Kant was writing this way was still very much a project in the making.

All modern objects from stones to human bodies, from numbers to nuclear bombs, and including mathematics and culture, the two objects assumed in the title of this collection of essays, are pictured and justified through the object logic. All are unproblematically bits of matter set in space and time with specific properties, or understood as abstract objects by analogy to that picture. All follow in the image of Kant's found, stable and orderly island (logic) set in treacherous seas (metaphor), whose role in generating the island *as* island is denied.

There is no denying that the paradox has been infinitely productive. The usefulness of the figure of objects – bits of ordered matter set in space and time – is evident. While accepting this, it is instructive to ask *why* Kant did not notice the figuring role of the island. What hid the paradox by which the *image* of the modern object was established? What made it appear self-evident that logic is an object imaged in this way? To answer these questions we need to consider what was at stake here. According to Le Dœuff, the textual conflict expressed in Kant's treatment of the metaphor of the island is 'a sign that something important and troubling is seeking utterance – something which cannot be acknowledged but is keenly cherished' (Le Dœuff, 1989: 9).

What is this something, that cannot be acknowledged, but which generates the prolonged metaphor of *The Critique*? Le Dœuff identifies it as the need to close the question of philosophy's role in the global project of the Great Instauration, as Kant, following Bacon, calls the making of modernity. Kant must characterise the work of doing knowledge, set down the proper mode of ontology and epistemology, by showing us how to imagine it, *and* at the same time make it appear as found. It must seem that this method of doing knowledge has all this time been there waiting for recognition.

Excluding the seas, fixing our gaze solely on the island, defines logic as an object. In turn something very important is ensured: the project of modernity. The possibility of completed knowledge and hence of absolute certainty, of redemption from the state of uncertainty, is ensured through the work of the figure. Allow metaphor and picturing into the project of doing knowledge and you can never complete anything. The island and seas become recognisable as foreground and background, and a further step of foregrounding/backgrounding is always possible. Things can always be re-imagined.

Having logic as an object through the aegis of Kant's island generates and justifies the possibility of completed knowledge and absolute certainty. And two hundred years ago that was the only legitimate form of knowledge. It was not something that Kant and his compatriots would consider as needing any justification. It would have been surprising if Kant had noticed the paradox.

Before I go on to consider where we might go from here, I want to point to several further things that come along with the modern objects so helpfully pictured for us in the island. The first of these is the image of the knower of those objects, and the role for that knower, an adjudicating and legislative role. Along with the ordered island to be known, the removed judging observer, with specific and particular characteristics, comes into existence. The properties of the universal knower in this scheme are the characteristics of a way of knowing evinced by white, higher class men of modern education.

It is the feminist philosophers (in particular Haraway, 1997: 22–24; Pyne Addelson, 1994; see also Code, 1991, 1995; Harding, 1986, 1991, 1998) like LeDœuff who have alerted us to what was/is taken for granted about the nature of knowers. They, resisting the prescription to become, symbolically at least, white, higher class men of modern education, began the process of revealing some of what is embedded in the orthodox mode of ontology/epistemology.

Less often recognised, yet equally clearly located in the modern knowledge project, is another characteristic of European societies of the time. The elaboration of a knowledge project in which completed knowledge and absolute certainty were the justified ends, implied a need for, and legitimated, European colonising. Colonising is both necessary and enabling in the struggle for completion which *must* involve a recognising, and a going beyond, the bounds of the currently known. At the same time the possibility of completing knowledge both enables and motivates the colonising impulse – one must both imagine there is something out there, and consider it worth collecting, to set off on a journey of appropriation (Shapin, 1996: 19).

The task of examining the working of Kant's text has brought us a long way. Recognising the work of the figure of matter set in space and time, introduced as Kant's ordered island set in treacherous seas, has enabled us to see how the foundationist mode of ontology/epistemology is a moral project of European colonising. It effects a re-making of the community life of eighteenth century northern Europe. Having seen this, what we need to ask now is, can we refuse the fantasies of completeness and absolute certainty without foregoing the benefits which come with the modern knowledge project? Can we use Kant's image of an ordered island as emblem of our objects without taking on all the baggage that comes along with it? In the context of this collection of essays we can specifically ask, can we begin to extricate ourselves from the projects of European colonising, without foregoing the considerable benefits brought about by taking logic as an object?

The benefits accomplished through the myriad modern objects that derive from the object 'logic', working through an object centred ontology/epistemology, are tangible, and it would be foolish to suggest that we can or should do without them. Can we rework Kant's paradox and still benefit from his insights? I suggest that we can. What we need to do is take Kant's metaphor in full, to keep it in full view, and to treat it seriously. My suggestion is that it is the lack of recognition of the work the image is doing, its deleting and hiding, that is the problem here. It is the *denial* which effects such a tight linkage of a particular, and now recognisably abhorrent, moral order, to a beneficent mode of ontology and epistemology.

My suggestion is that we take Kant's metaphor in full and take its role as a figure seriously. To take the metaphor in full, and to celebrate it, is to see the island and the seas which surround it as integral to each other. We can understand this as changing our mode of ontology and epistemology.

The first things to go with this move are the dual fantasies of completable knowledge and absolute certainty. A recognition of Kant's treacherous seas as part of the ordered island has us recognising our use of pictures and narratives, as much as following rules, as part of doing logic as infinitely generative. The foregrounded/backgrounded figure no longer effects or accomplishes limitation. Or rather, any limitation becomes recognisable as a contrivance; the limitation that the world *is* objects is seen as an outcome.

Reconciling ourselves to uncompletability is perhaps not so difficult. Unlike Kant's Europe, it is (almost) self-evident in our times and places two to three centuries into modernity that certainty based on completability *is* unattainable. In our times and places we are perhaps more inclined to feel the ineffable complexity of the world. That notions of completable knowledge, and certainty in consequence of that, featured strongly in establishing the modern project is quite understandable. Now we have come to know the modern project too well for the fantasy of completion to be sustainable. Of course we need to go on to ask what are the grounds of *that* certainty over (un)certainty, and my argument there is for a form of embodied certainty (Verran, forthcoming). There is not room here to pursue that question.

The move to include the seas as part of the island has further consequences. Taking Kant's island seriously amounts to an ontic/epistemic commitment to knowledge as primarily enacted (rather than as propositional which becomes recognisable as a particular form of enacting). It involves a different way of dealing with the paradox inherent in ontic/epistemic commitment, in assuming meaning. Including the treacherous seas with the island we see straight away that an ordered island set against treacherous seas is an accomplishment. Objectification thus becomes recognisable as an accomplishment. Getting comfortable with this way of seeing things, we then come to recognise that we must *continue* to enact the accomplishment of the ordered island. An object world is effected in our enacting it. The inherent uncertainty, messiness, complexity, and multiplicity of this necessarily on-going action is what was hidden when Kant failed to notice his paradox.

Including the seas, recognising the primacy of enacting, does not necessarily mean that we must give up objects and give up all hope of an ordered island. We can keep logic as an exemplary object. But now it is an object that is openly recognised as accomplished in an on-going, continual and never-ending project of objectification. We can have our metaphors and pictures, jettison completability and achievable certainty, and keep our objects if we no longer hide from ourselves the work we do in our myriad and vast modern projects of objectification. This is the constant working of the mess and complexity in embodied being – the thankless task of continually ordering, subduing and controlling. It keeps in view all the time the continuing and continual doing of logic in doing routine practices – including the routines through which we do logic as an object.

When we take Kant's island seriously as a *figure* it immediately loses its usefulness as a justification, as grounding the possibility of completable knowledge and absolute certainty. Including the seas makes Kant's island a recursion – foreground and background. As part of a recursion of foreground and background, the figure of object it establishes is altered, so that now it must be seen as an accomplishment – the temporary halting of a recursion achieved only with an on-going effort. The point is *not* that logic, as not a given object, should now be understood instead as processes. It would be a mistake to understand that what I am saying here is that logic is processes which might be adumbrated along with the telling of its metaphors, which somehow lie behind or inside everyday life, and alternatively behind or inside the more restricted arenas like mathematics, science, and economics. That would be just another way of taking logic as prescriptive and *de facto* as a completed and given object. That is a version of the foundation envisaged by relativism. It is more complicated than that.

Logic(s) in this picture are both and neither objects and sets of routines (adumbrated or not) – and hence both and neither singular or plural. Logic(s) do not lie behind or inside anything, but are done in various contingent, partial and uncompletable projects. Logics and mathematics are done in the routines of myriad times and places, and include routines which do logics and mathematics as objects. These are the multiple routines acquired by those who have participated in education in mathematics and philosophy.

The routines embodied in doing logics or mathematics in one time and place are to varying extents the same as, or different from, the routines in another, and the same can be said for the images. We all know, for example, that the routines we learn in mathematics classes in university lecture halls and laboratories are rather different from the routines and imagings of mathematics lessons in primary schools. But there are enough continuities for there to be no problems with calling them both mathematics. Similarly, the content of mathematics lessons in Yoruba and Yolngu schools is to some extent shared and to some extent quite distinct.

Some places and time do logics and mathematics as objects; some do not. Connecting up the logics done in these different times and places, or not, depends on translating the practices. And the capacity to translate and to justify the translations depends on an acceptance of and a familiarity with figures which are constitutive of the doing of logics and mathematics. *That* is the significance of including the seas and taking the metaphor of the island in full.

* * *

Mandawuy, his fellow Yolngu teachers and other members of Yolngu communities, along with Balanda teachers experienced in the difficulties of teaching mathematics to Yolngu Aboriginal children, have over a number of years developed routines and practices which gradually coalesced as a new sort of mathematics curriculum. It is a curriculum in which the translating role of recursion is central. To that extent it is a curriculum that is both very similar and very different to other mathematics curricula.

In their classrooms, getting children doing the routines in the precise ways they must be done, we can imagine the Yolngu teachers disciplining and training the children's bodies. Inculcating the small routines and gestures of doing number or doing gurrut̲u, they are developing the grounds of a collective embodied certainty – making the island. In this difficult, tedious, creative and imaginative work they are dealing actively with pictures and stories, evoking the possibility of the ordered island. Their struggle is creating not only knowers of mathematics; the grounds for this Yolngu Aboriginal community to continue are being (re)generated. Mathematics education here enables a going on *as* a contemporary Yolngu Aboriginal community. Mathematics lessons are no longer sites of colonialism but rather sites of community (re)generation.

So too Mr Ojo and Mr Ojeniyi, in such lessons as those I described, are expanding the grounded certainty of being contemporary Yoruba. They are arranging quantifying experiences, and children's participation consolidates the routines of small bodily gestures by which numbers are achieved. Yoruba teachers also arrange other experiences where alternative sets of small bodily gestures and alternative images and stories are rehearsed. These more orthodox lessons embody the routines of measuring length with extension and an additive number line. Yoruba children need to incorporate both sets of routines to become fully prepared as participants in the (re)generation of the life of their times and places.

It is also the case that sometimes Yoruba and Yolngu teachers will do the routines which make mathematics as an object. They do this, for example when they deal with a school inspector, justifying their work, or when they study mathematics education and engage in practices which have logics and mathematics as objects. Failing to recognise the constitutive role of the figure/picture/metaphor/tropes of logic and mistaking logic for a given object, we might imbue the latter objectifying practices with disciplinary capacities. Then we will want to discipline the teachers and the children, distribute praise and blame this way or that, and in that process find ourselves re-making colonialism. But this does not mean that anything goes in a mathematics lesson. The criterion of translatability is stringent and rigorous when we understand certainty as embodied in being in particular times and places.

ACKNOWLEDGMENTS

I am deeply indebted to my co-participants in mathematics lessons, in particular the groups of Yoruba teachers, and Yolngu Aboriginal teachers I worked with. Their creative and imaginative work, and deep commitment to promoting genuine understanding in the children they taught, alerted me to the philosophical and moral problems implicit in conventional mathematics curricula in primary schools. The constancy of these teachers inspired me to question conventional understandings about mathematics and logics.

NOTES

[1] Some readers might find it odd for me to construe relativism as foundationist analysis, reserving that adjective for universalist regimes of interpretation, assuming that relativism is distinguished from universalist analysis by not assuming set foundational categories. A detailed response to the issue would recognise the multiplicity of relativisms which characterise connection between the abstract realm of knowledge and the concrete world of experience in various ways. The global response I propose is that all relativisms are properly identified as foundationist analyses by the very fact of their hypothesising knowledge as a symbolic realm, inevitably setting up a material realm which that abstract/symbolic knowledge is in some sense about, or in which it resides.

[2] In identifying my naivete in framing these studies, I do not mean to imply that they are worthless. I am merely pointing to the ways they are compromised, and urging readers to work against the grain as they read them.

[3] Following the trail of texts we can identify 1967 as a high point in that old debate. In the developing philosophical debate on 'African thought' Robin Horton's 1967 'African Traditional Religion and Western Science' was widely recognised as pivotal. Horton took a Popperian line on

the analysis of scientific rationality and extended it to traditional African religion, purporting to show the 'openness' of science as against the 'closedness' of traditional African religion. Open and closed are terms Popper uses in elaborating his particular brand of postitivistic scientific rationality. Rationality had become the hot topic in philosophy of science following the publication of Kuhn's *The Structure of Scientific Revolutions*; perhaps we can credit Winch with the introduction of Africans into this debate with his 'Understanding a Primitive Society'. In 1970 the edited collection *Rationality* appeared as a contribution to the debate in philosophy of science. Horton continued his approach in his paper in the collection he edited with Ruth Finnegan, *Modes of Thought*, and philosophers at the University of Ife continued to critique Horton's analyses. Anthropologists Geertz and Gellner also had their say. Kwasi Wiredu (1980) was the first African voice to be clearly and widely heard. By 1982, when *Rationality and Relativism*, which includes a re-think from Robin Horton, was published, the debate had largely puffed itself out.

On the cross-cultural psychology side Gay and Cole published their *New Mathematics and an Old Culture* in 1967, which reported their work with the Kpelle people of Liberia in West Africa, following hot on the heels of the 1966 publication of *Studies in Cognitive Growth*, edited by Jerome Bruner among others, in which the content was strongly cross-cultural. We can trace the continuation of this work in Cole and Scribner's *Culture and Thought: A Psychological Introduction* (1974) and the collection edited by Berry and Dasen, *Culture and Cognition: Readings in Cross-Cultural Psychology* (1974) and recognise Hallpike's *Foundations of Primitive Thought* (1979) as both a definitive work and in some senses a final gasp in this phase of the debate.

BIBLIOGRPAHY

Abraham, R. C. *Dictionary of Modern Yoruba*. London: Hodder and Stoughton, 1962.

Armstrong, R. G. *Yoruba Numerals*. Ibadan: Oxford University Press, 1962.

Ascher, Marcia. *Ethnomathematics: A Multicultural View of Mathematical Ideas*. Pacific Grove: Brooks/Cole Publishing Co., 1991.

Berry, J. W. and P. R. Dasen, eds. *Culture and Cognition: Readings in Cross-Cultural Psychology*. London: Methuen, 1974.

Bloor, David. *Knowledge and Social Imagery*. Chicago: University of Chicago Press, 1976; 2nd edition, 1991.

Bruner, Jerome, R. R. Oliver and P. M. Greenfield, eds. *Studies in Cognitive Growth*. New York: Wiley, 1966.

Carnap, Rudolf. *Philosophical Foundations of Physics: An Introduction to the Philosophy of Science*, Martin Gardner, ed. Chicago: University of Chicago Press, 1966.

Code, Lorraine. *What Can She Know? Feminist Theory and the Construction of Knowledge*. Ithaca, New York: Cornell University Press, 1991.

Code, Lorraine. *Rhetorical Spaces: Essays on Gendered Locations*. London: Routledge, 1995.

Cole, Michael and S. Scribner. *Culture and Thought: A Psychological Introduction*. New York: Wiley, 1974.

Finnegan, Ruth and Robin Horton, eds. *Modes of Thought*. London: Faber, 1973.

Gay, John and Michael Cole. *The New Mathematics and an Old Culture*. New York: Holt Rinehart Winston, 1967.

Geertz, Clifford. *The Interpretation of Cultures*. New York: Basic Books, 1973.

Gellner, Ernest. *Legitimation of Belief*. Cambridge: Cambridge University Press, 1974.

Hacking, Ian. 'Styles of reasoning.' In *Rationality and Relativism*, Martin Hollis and Steven Lukes, eds. Cambridge, Massachusetts: MIT Press, 1982, pp. 48–66.

Hallen, Barry. 'Robin Horton on critical philosophy and traditional religion.' *Second Order* 6(1): 81–92, 1977.

Hallen, Barry and J. O. Sodipo. *Knowledge, Belief and Witchcraft*. London: Ethnographica, 1986.

Hallpike, C. R. *The Foundations of Primitive Thought*. Oxford: Clarendon Press, 1979.

Haraway, Donna. *Modest_Witness@Second_Millennium FemaleMan_Meets_OncoMouse*. New York: Routledge, 1997.

Harding, Sandra. *The Science Question in Feminism*. Ithaca: Cornell University Press, 1986.

Harding, Sandra. *Whose Science? Whose Knowledge?* Ithaca: Cornell University Press, 1991.

Harding, Sandra. *Is Science Multicultural? Postcolonialisms, Feminisms, and Epistemologies*. Bloomington: Indiana University Press, 1998.

Horton, Robin. 'African traditional religion and Western science.' *Africa* 37: 50–71, 155–187, 1967.

Johnson, Samuel. *The History of the Yorubas, From the Earliest Times to the Beginning of the British Protectorate*. Lagos: CMS (Nigeria) Bookshops, 1921.

Kant, Immanuel. *Critique of Pure Reason*. (Originally *Critick der reinen Vernunft*. Königsberg: Bey Friedrich Nocolovius, 1st edition, 1781, 2nd edition, 1787.) J. M. D. Meiklejohn, trans. London: Macmillan, 1929.

Kant, Immanuel. *Logic*. (Originally *Logik ein Handbuch zu Vorlesungen*. Königsberg: Bey Friedrich Nocolovius, 1800.) R. Hartman and W. Schwarz, trans. Indianapolis: Bobbs-Merrill Co., 1974.

Kuhn, Thomas. *The Structure of Scientific Revolutions*. Chicago: University of Chicago Press, 1962.

Le Dœuff, Michèle. *The Philosophical Imaginary*. London: The Athelone Press Ltd, 1989.

Lloyd, Barbara. 'Cognitive development, education and social mobility.' In *Universals of Human Thought: Some African Evidence*, Barbara Lloyd and John Gay, eds. Cambridge: Cambridge University Press, 1981, pp. 176–196.

Lukes, S. and M. Hollis, eds. *Rationality and Relativism*. Oxford: Basil Blackwell, 1982.

Mill, J. S. *A System of Logic: Ratiocinative and Inductive*. London: Longmans, 1848.

Popper, Karl. *Conjectures and Refutations: The Growth of Scientific Knowledge*. New York: Basic Books, 1962.

Powell, Arthur B. and Marilyn Frankenstein, eds. *Ethnomathematics: Challenging Eurocentrism in Mathematics Education*. Albany: State University of New York Press, 1997.

Price-Williams, E. 'A study concerning concepts of conservation of quantities among primitive children.' *Acta Psychologica* 18: 297–302, 1961.

Pyne Addelson, Kathryn. *Moral Passages*. New York: Routledge, 1994.

Quine, W. V. O. *Word and Object*. Cambridge, Massachusetts: MIT Press, 1960.

Reed, H. J. and Jean Lave. 'Arithmetic as a tool for investigating relations between culture and cognition.' *American Ethnologist* 6: 568–582, 1979.

Rundell, John. 'Creativity and judgement: Kant on reason and imagination.' In *Rethinking Imagination, Culture and Creativity*, John Rundell and Gillian Robinson, eds. London and New York: Routledge, 1994, pp. 87–117.

Shapin, Steven. *The Scientific Revolution*. Chicago: University of Chicago Press, 1996.

Verran, Helen. 'Staying true to the laughter in Nigerian classrooms.' In *Actor Network Theory and After*, John Law and John Hassard, eds. Oxford: Blackwell Publishers, 1999, pp. 136–155.

Verran, Helen. *Numbers, Judgement, and Certainty. Storytelling about Logics and Mathematics in Africa*. Chicago: University of Chicago Press, forthcoming.

Watson, Helen. 'Applying numbers to nature: a comparative view in English and Yoruba.' *The Journal of Cultures and Ideas* 2(3): 1–26, 1986.

Watson, Helen. 'Learning to apply numbers to nature: a comparison of English speaking and Yoruba speaking children learning to quantify.' *Educational Studies in Mathematics* 18: 339–357, 1987.

Watson, Helen with the Yolngu Community at Yirrkala and D. W. Chambers. *Singing the Land, Signing the Land*. Geelong: Deakin University Press, 1989.

Watson, Helen. 'Investigating the social foundations of mathematics: natural number in culturally diverse forms of life.' *Social Studies of Science* 20: 283–312, 1990.

Watson Verran, Helen and David Turnbull. 'Science and other indigenous knowledge systems'. In *Handbook of Science and Technology Studies*, Sheila Jasanoff, *et al.*, eds, 1995, pp. 115–139.

Watson-Verran, Helen. 'Knowledge systems of the Australian Aboriginal people.' In *Encyclopaedia of the History of Science, Technology, and Medicine in Non-Western Countries*, Helaine Selin, ed. Dordrecht: Kluwer Academic Publishing, 1997a, pp. 490–494.

Watson-Verran, Helen. 'Mathematics of the Australian Aboriginal people.' In *Encyclopaedia of the History of Science, Technology, and Medicine in Non-Western Countries*, Helaine Selin, ed. Dordrecht: Kluwer Academic Publishing, 1997b, pp. 619–622.

Winch, P. 'Understanding a primitive society.' *American Philosophical Quarterly* 1: 307–324, 1964.

Wilson, Bryan, ed. *Rationality*. Oxford: Basil Blackwell, 1970.

Wiredu, Kwasi. *Philosophy and an African Culture*. Cambridge: Cambridge University Press, 1980.

Yirrkala Community School. *Garma Maths Curriculum (Yolngu gali') Primary School Study of Gurrutu (the Yolngu kinship system Feb 11–22, 1991*, Yirrkala Literature Production Centre, Box 896, Nhulunbuy, NT, 0881, Australia, 1991.

Yirrkala Community School. *Living Maths 1: Djalkiri – Space Through Analogs.* Boulder Valley Films. Distributed Australian Film Institute, Melbourne, 1996a.

Yirrkala Community School. *Living Maths 2: Space – The Grid Digitised.* Boulder Valley Films. Distributed Australian Film Institute, Melbourne, 1996b.

Yirrkala Community School. *Living Maths 3: Gurrutu – Recursion Through Kinship.* Boulder Valley Films. Distributed Australian Film Institute, Melbourne, 1996c.

Yirrkala Community School. *Living Maths 4: Tallying Number.* Boulder Valley Films. Distributed Australian Film Institute, Melbourne, 1996d.

Yirrkala Community School. *Living Maths – The Book of the Videos.* Boulder Valley Films. Distributed Australian Film Institute, Melbourne, 1996e.

Yoloye, E. Ayotunde. *Evaluation for Innovation: African Primary Science Programme Evaluation Report.* Ibadan: Ibadan University Press, 1978.

UBIRATAN D'AMBROSIO

A HISTORIOGRAPHICAL PROPOSAL FOR NON-WESTERN MATHEMATICS

Mathematics and other sciences, as we generally understand them today, emerged in a distinctive form in Europe. But every culture generates something equivalent to mathematics and science that works satisfactorily within its own context. These are bodies of knowledge that have been generated in a particular context, with specific motivations, and that have been and are subject to insufficiencies and criticism as well as changes resulting from exposure to other cultures. These are results of the major challenges facing the human species, which is driven to survive through encounters with others and with nature as a whole and to transcend the moment, searching the past and probing into the future.

In search of survival humans developed ways of dealing with the immediate environment, which provided air, water, and nourishment. These ways are techniques and styles of individual and collective behavior, which include communication and language. In search of transcendence they developed perceptions of the past, present and future and means of explaining facts and phenomena. These means are memory, individual and collective, and myths and divinatory arts. In memory and myths are the traditions, which include history, religions and systems of values. In the divinatory arts are the systems of explanation, such as astrology, oracles, logics, numerology and the laws of nature. The sciences may also tell us what might happen.

Knowledge is the response to these drives. How is knowledge generated, organized intellectually and socially, and diffused? Language and mathematics both offer major challenges. Both have grown differently in different cultures. And both have been affected by cultural encounters throughout history. Particularly important for our analysis are the encounters which occurred after the 15th century between European and non-European cultures. I agree with Urs Bitterli (1989) when he explains the encounter of European and non-European cultures as having three basic phases: contact, collision and relationship. He shows that they do not occur necessarily in this order, that they are not mutually exclusive and that all three types have occurred. In some cases the contact led directly to relationship. This classification is convenient for

H. Selin (ed.), Mathematics Across Cultures: The History of Non-Western Mathematics, 79–92.

understanding the cultural dynamics of the encounters, from the beginning of European overseas conquest through early industrial relations. At the start of the conquest, mathematics was beginning to establish itself as a field of knowledge that would be central for the development of industrial civilization.

The history and philosophy of mathematics focuses on the mathematics which synthesizes centuries of ideas developed in the Mediterranean basin and enriched by contacts with Africa and the Far East. Historiography is largely based on written sources and relies on names, epochs, dates and places proposed by early historians.

Although the conquered civilizations had mathematical knowledge, current historiography is inadequate in recognizing it. Its nature and history are practically unknown. The collision phase resulted in the denial of the forms of knowledge of the conquered. The relationship phase was marked by an effort to transfer European mathematics to the colonies. This continued until the transition from the 19th through the 20th century, when local production of mathematics, as originated in Europe, began to be delineated.

This paper focuses on the social, political and cultural factors in the dynamics of the transfer and the production of mathematical knowledge in the colonies, as well as on the recognition of non-European forms of mathematics, extant or buried in the colonial process. It is a historiography which relies on the memory of people and events which survived in a literate era. The methodology puts together scraps of information and recognizes extant mathematical practices, normally called ethnomathematics.

ETHNOSCIENCE AND ETHNOMATHEMATICS

The great navigations that began in the 16th century created an awareness of forms of scientific knowledge from different cultural environments. The several ethnosciences involved in the encounters, which obviously included European science, were subjected to great changes. In this paper I will examine some of the consequences of this mutual exposure.

By ethnosciences I mean the corpora of knowledge established as systems of explanations and ways of doing accumulated through generations in distinct cultural environments. Particularly important for us is ethnomathematics – knowledge derived from quantitative and qualitative practices such as counting, weighing and measuring, sorting and classifying. As with academic western science and mathematics, the two have a symbiotic relation. Both are not new disciplines; they are part of a research program on history and epistemology. Both research and educational programs take into account all the forces that shape a mode of thought, in the sense of looking into the generation, organization (both intellectual and social) and diffusion of knowledge.

The research program, typically interdisciplinary, brings together and interrelates results from the cognitive sciences, epistemology, history, sociology and education. An essential component is the recognition that mathematics and science are intellectual constructs of mankind in response to needs of survival and transcendence. The need for an intellectual framework to organize the

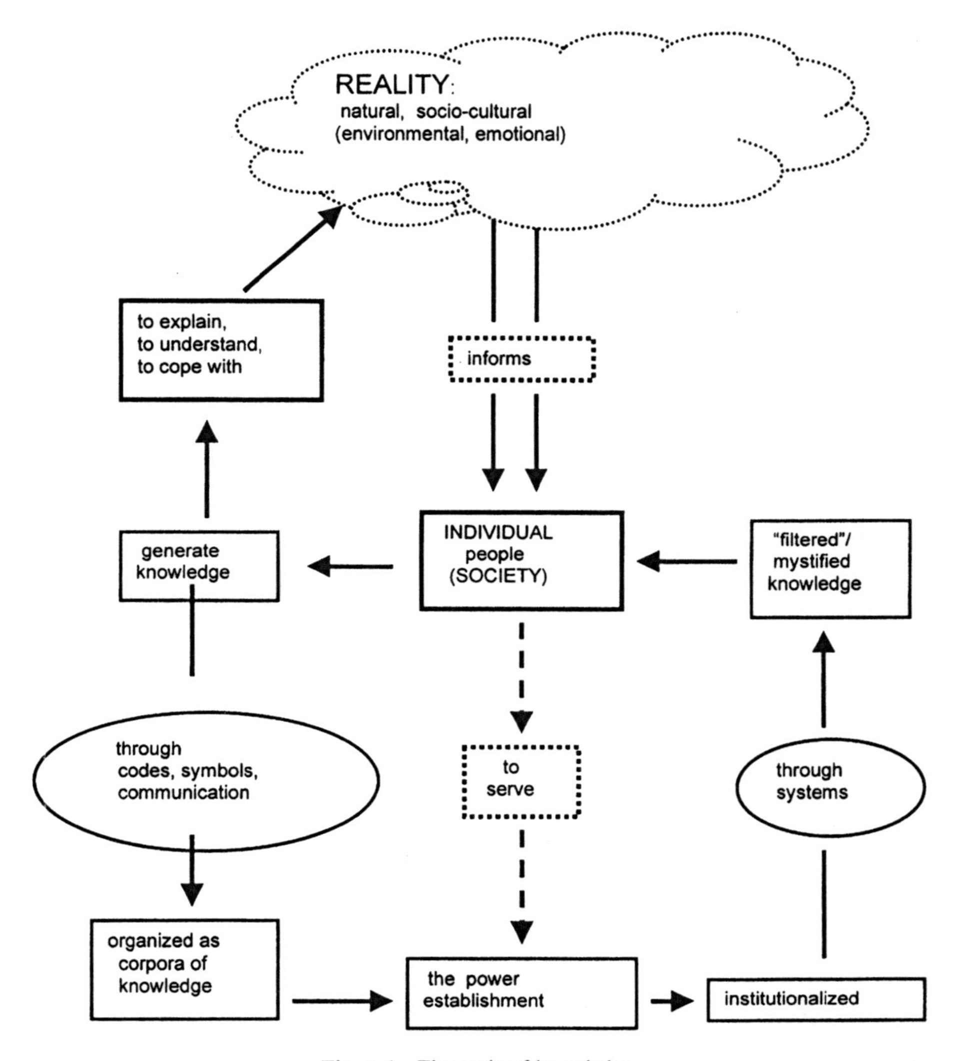

Figure 1 The cycle of knowledge.

corresponding systems of codes, norms and practices gave rise to many aspects of science and mathematics (D'Ambrosio, 1994). In the research program particular attention is given to those dimensions of knowledge which bear some relation to what became known as the several disciplines of science and mathematics in Europe after the 15th century. Ethnoscience, both as body of knowledge and as pedagogical practice, is supported by the history of science and reflects the dynamics of cultural acquisition. Some examples illustrate this.

All over the world, many of the weather explanations and predictions, agricultural practices, health and healing processes, dressing and institutional codes, food, and commerce, came from the European tradition developed in the Middle Ages and the Renaissance. But we also see that these practices are significantly modified and molded to individual cultures. For example, it is common to see indigenous peoples in the Americas using Indo-Arabic numerals but performing the operations from bottom to top, explaining that this is the way trees grow.

The practices and perceptions of learners are the substratum upon which new knowledge is built. Thus new knowledge has to be based on the individual and cultural history of the learner, and it has to recognize the diversity of extant cultures, present in specific communities, all over the world. This is the essence of a new educational posture called multicultural education. It depends on a new historical attitude which recognizes the contribution of past cultures in building the modern world and which avoids the omissions and errors of past treatments of cultural differences.

We easily identify two categories of scientific knowledge: scholarly (or 'formal' or 'academic') science, supported by an epistemology whose practice is restricted to professionals with specialties, and cultural (or 'practical' or 'popular' or 'street') science.[1] These categories are closely related and their main distinction refers to criteria of rigor, to the nature, domain and breadth of scientific pursuits, and to what and how much one can do with them.

For example, pre-Columbian cultures had different styles of measuring and computing, and these practices are still prevalent in some native communities. Most Amazonian tribes count like this: 'one, two, three, four, many'. These numbers are all that are necessary to satisfy their needs (Closs, 1986). We also see ways of dealing with pottery and tapestry that demonstrate strong mathematical characteristics in several cultures (Ascher, 1991; Gerdes, 1995). The people from these cultures have no problems assimilating the European number system and deal perfectly well with counting, measuring and money when trading with Europeans. Land measurement, as practiced by peasants in Latin America, comes from ancient geometry transmitted to medieval surveyors, since land property and measurement (geo-metry) are foreign to pre-Columbian cultures. Another example comes from Africa, where the people deal with numbers and counting according to their specific cultural background (Zaslavsky, 1979).

The prestige of science comes mainly from its being recognized as the basic intellectual instrument of progress. Modern technology depends on science, and the instruments of validation in social, economic and political affairs, mainly through storing and handling data, are based on science and mathematics. Particularly important in this respect is statistics. This brings to science an aura of essentiality. There is a feeling that there are practically no limits to what can be explained by science. Many of the applications which give science such a prestigious position are part of various forms of cultural conflict.

Studies of ethnoscience and ethnomathematics are motivated by the demands of the natural and cultural environment and are present everywhere. It is a fact that, even without recognizing it, just about everybody deals with mathematical practices, incorporated into their daily routines. When walking or driving, people memorize routes, in most cases optimizing trajectories. When dealing with money, with measurements and quantification in general, we recognize an intrinsic mathematical component. The same is true of classifying, ordering, selecting and memorizing routines.

These practices are generated, organized and transmitted informally, as is language, to satisfy the immediate needs of a population. They are incorporated into the pool of common knowledge which keeps a group of individuals together and operational, and this is what is called culture. Culture thus manifests itself in different, interrelated, forms and domains. Cultural forms, such as language, mathematical practices, religious feelings, family structure, and dress and behavior patterns, are diversified. They are associated with the history of the individuals, communities and societies where they are developed. A larger community is partitioned into several distinct cultural variants, each with its own history and responsive to differentiated cultural forms.

SOME REMARKS ON HISTORIOGRAPHY

History, as a major academic discipline, carries with it an intrinsic bias which makes it difficult to explain the ever-present process of cultural dynamics which permeates the evolution of mankind. This paves the way for paternalism and arrogance, for intolerance and intransigence. It interferes with the understanding, for different cultural groups, of others' cultural realities when trying to satisfy their needs of survival and transcendence. These biases have been methodological as well as ideological, particularly in the history of science. Helge Kragh (1987: 11) says, 'History of Science has its own "imperialism" that partly reflects the fact that viewed historically and socially science is almost purely a western phenomenon, concentrated on a few, rich countries. While science may be international, history of science is not.'

This problem seems to be almost unavoidable in the framework of historiographies which rely on reductionist approaches. The mere fact that to pursue historical analyses one talks about the sciences as distinct from religion, art, and politics, impedes our understanding of the processes of the evolution of ideas and methods which underlie our struggle to find explanations, to understand and cope with our environment, and to live well with nature.

The reductionism which characterizes histories based on facts and names, on places and dates, derives from the prevailing ideology and justifies current actions. Even when we move a step further than narrative history and go to historiography, the facts get immersed in the processes, and we may be satisfied with the false impression of having understood the past because we have data verified and facts described and explained. I agree with Armando Saitta that historiography should be focused on a problem, never losing sight of all the forces which play a part in historical reality and avoiding the unilateral approach of the specialist and the reduction of the historical flow to a few elements. Saitta (1955: 12) asks the historian to look into 'What today isn't but tomorrow will be'. He proposes a global history. When he refuses the history of the 'if', he opens the way to an evaluation of all the alternatives that were present in the process, and he claims that the one alternative which may have succeeded should not imply the rejection of the others. E. H. Carr (1968) agrees when he says that the historical moment in which several alternatives were open does not imply abandoning those which did not succeed, but rather one should look into the reason some did not succeed and what was the cost of these decisions.

Paraphrasing Miguel León-Portilla (1985), it is a matter of listening also to the looser. History has been mostly the history of winners. The looser has been marginalized, and this is very noticeable in the histories of modern science[2]. The dawn of modern science is identified with the modern geography of the world, and brings with it privileges for those capable of mastering both science and technology. How did this privileged role come into being? Why do the conquered and colonized still have problems in mastering science and technology? Why have science and technology progressed so rapidly and have left aside social and ethical concerns, thus paving the way for enormous social, political and environmental problems? These questions are germane to the concept of knowledge itself.

CREATING SCIENTIFIC KNOWLEDGE

We see knowledge as emanating from the people, essentially a product of their drive to explain, understand and cope with their immediate environment and with reality in general. This drive is subjected to a process of exposure to other members of society, and, thanks to communication, both immediate and remote in time and space, goes through a process of codification, intertwined with an associated underlying logic, inherent to the people as a form of knowledge that some call wisdom. The modes of communication and the underlying logic are recognized as the result of the prevailing cognitive processes. Cognitive evolution, related to environmental specificity, gives rise to different modes of thought and different underlying logic, communication and codification. Hence knowledge is structured and formalized according to culture. The power structure, which itself grows out of society as a form of political knowledge, appropriates structured knowledge and organizes it into institutions. In this form and under the control of the establishment and the power structure, which mutually support each other, knowledge is given back to the people, who in the first instance generated it, through systems and filters which are designed to keep the established power structure. The generation, transmission, institutionalization and diffusion of knowledge constitute a holistic approach to knowledge and to the dynamics of change. This is the essence of the research program in the history of science which I call 'ethnomathematics' (D'Ambrosio, 1990 and 1998).

The disciplinary approach to knowledge focuses on cognition, epistemology, history and sociology. This makes it difficult to understand the dynamics of change. Mutual exposure of distinct approaches to knowledge, resulting from distinct environmental realities, is global, embracing the entire cycle from the generation through the diffusion of knowledge. The process of cultural dynamics which takes place in the exposure is based on mechanisms which balance the process of change, which I call *acquiescence* – that is, the capability of consciously accepting change (modernity) – and the cultural *ethos* – which acts as a sort of protective mechanism against change that produces new cultural forms. This behavior can be traced back throughout the entire history of mankind. These conceptual tools are close to the ethos introduced by Gregory Bateson (1972) in dealing with cultural contact and enculturation.

In the encounter of the two worlds (Europe and America) there were many instances of cultural violation. The origin of these violations may be related to distinct views of nature. A scientific conceptualization, which resulted from an intertwining of medieval Judeo-Christian and Greco-Arabic thought, led to looking at nature and at the universe as an inexhaustible source of richness and to exploiting these resources with a drive towards power and possession. This behavior towards nature and life led to a single model of development and hence to ignoring the cultural, economical, spiritual and social diversities which constitute the essence of our species.

These reflections question the set of current concepts and models and call for the acceptance of the idea that survival depends on a global and holistic view of reality. This demands a radical change which applies to all levels of knowing and doing. Thus we must look for radical changes in our models of development, of education and of civilization, based on the recognition of a plurality of models, of cultures, of spirituality and of social and economic diversity, with full respect for each one of the distinct options.

VISIONS OF THE WORLD

The European navigators of the end of the 15th and early 16th centuries reached all of America, Africa, India and China. In the case of Africa and in Asia, there had previously been contacts, but these voyages amplified and deepened contact with the rest of the world. Columbus and the Spaniards met the 'new', the unknown, and the unexpected, in 1492 and in subsequent voyages. Earlier contacts with the Americas are known, but the motivations and behavior of earlier navigators were completely different from the Spanish and Portuguese, and afterwards the English, French and Dutch.[3]

The influence of the navigators and chroniclers, particularly the Portuguese, in creating the mode of thought which underlies modern European science is notable. In the words of Joaquim Barradas de Carvalho (1983: 13), 'the authors of the Portuguese literature of navigation made possible the Galileos and the Descartes' essentially through the development of 'objective and serene curiosity, rigorous observations and creative experimentation' (Correia, 1940: 468).

The lack of recognition of Portuguese science in the 15th and 16th centuries illustrates the observations above about biased historiography. The important *Tractatus de sphera* (early 13th century) written by Johannes de Sacrobosco, was recognized as 'the clearest, most elementary, and most used textbook in astronomy and cosmography from the thirteenth to the seventeenth century' (Thorndike, 1949: 1). It received two important translations with commentaries in Portugal, one by Pedro Nunes, in 1537 and one by João de Castro, possibly in 1546. Nunes' translation incorporates much of the observational and experimental science which had been pursued by Portuguese navigators since the early 15th century and recorded in their writings. Curiously enough, neither is recognized in the most important study of Sacrobosco, written by Lynn Thorndike. We might say that Iberian science up to the Enlightenment fits into the characterization of 'non-Western'.

Particularly important are the *Crónica dos feitos de Guiné* of Gomes Eanes de Zurara (1453) and the *Esmeraldo de situ orbis*, by Duarte Pacheco Pereira, written between 1505 and 1508, probably the first major scientific work reporting on what was observed and experimented in the newly 'discovered' environments. In fact, we have to understand the sense of the word 'discovery' among the Portuguese authors of that period to better realize the role of the navigations in paving the way for modern science. In his contribution, Joaquim Barradas de Carvalho (see Note 2) gives both an exhaustive study of the *Esmeraldo de situ orbis* and a discussion of the meaning of the word 'discovery'.

The voyages themselves brought a broader view of the world. Venturing to the Southern Hemisphere demanded two major enterprises: the construction of the caravel, an extremely versatile ship built by the Portuguese in the 15th century as the result of a remarkable engineering project,[4] and novel navigation techniques, relying on tables constructed from systematically recorded observation carried on by the commanders of those ships. The commanders were also responsible for recording the 'different skies' which they were the first Europeans to see. The contributions of Gil Eanes crossing the Bojador Cape in 1434, Nuno Tristão reaching the coast of Mauritania in 1443, and the major achievement of Diogo Cão crossing the Equator in 1483, all paved the way for Bartolomeo Dias to cross the Cape of Hope in 1488 and for Vasco da Gama to reach Calicut in India, in 1498. With all of these voyages, including Columbus' reaching the Western lands in 1492, the view of the world changed. All lands and peoples were within the reach of the navigators. It was the beginning of a new phase in history.

THE 'NEW SCIENCES' AS SEEN IN THE ENCOUNTER

As I said above, America and to some extent Africa, were more surprising to Europeans than what they had seen in lands which had been reached before by land routes. America particularly showed peoples with new forms of explanation, of rituals and of societal arrangements. Reflections on natural philosophy or the physical sciences, particularly astronomy, were part of the overall cosmovision of the pre-Columbian civilizations. But the conquerors did not recognize the scientists that were surely present in the society of the conquered cultures.

In one of the earliest registers of these cultures, Fray Bernardino de Sahagún wrote in the 16th century that, 'The reader will rightfully be bored in reading this Book Seven [which treats astrology and natural philosophy of the natives of this New Spain], ... trying only to know and to write what they understood in the matter of astrology and natural philosophy, what is very little and very low' (1989: 478). Sahagún's report explains much about the flora and fauna, as well as medicinal properties of herbs of Nueva España. But he does not give any credit to indigenous formal structured knowledge. This is typical of what might be called an epistemological obstacle of the encounter.

Another important book is the *Sumario compendioso ... con algunas reglas tocantes al Aritmética* by Juan Diaz Freyle, printed in Mexico in 1556, the first arithmetic book printed in the New World. It has a description of the number system of the Aztecs. But this book soon disappeared from circulation and Aztec arithmetic was replaced by the Spanish system.

Much research is needed on science at the time of the encounter. This requires a new historiography, since names and facts, on which the history of science heavily relies, were not a concern in the registry of these cultures. A history 'from below', which might throw some light on the modes of explanation and of understanding reality in these cultures, has not been common in the history of science, although there are more sources for the history of the natural and health sciences.

THE BASIN METAPHOR AND A SOCIOLOGY OF MATHEMATICS

Mathematics is essential in the modern world. Public opinion is ready to support investment in mathematical research in spite of being absolutely unable to understand it. Parents invest in the mathematical education of their children. We regard those that get good grades in mathematics as potential geniuses, while those that do not do well are regarded as stupid. On the other hand, evidence from research showing that both individual and social creativity is enhanced by self-esteem is not taken into account for those that do well in the arts or in sports but fail in mathematics. Let us introduce at this time some concepts and reflections from the discipline called social studies of science or science policy. This is basically the study of the politics of scientific development, the backbone of funding agencies.[5]

When deciding on investments in science and technology, it is natural to expect social benefits. These investments have been substantial, both through governmental or aid agencies, either bi- (British Council) or multilateral (UNESCO). The outcomes in the so-called Third World have not been encouraging. The gap between central nations and peripheral nations in the production of scientific knowledge is getting larger. Over 80% of the money for scientific and technological research benefits the First World. 'The gap between rich and poor countries is a gap of knowledge,' says Federico Mayor (1994). Scientific productivity is related both to the cultural atmosphere and to self-esteem. Self-esteem can hardly prevail among a population deprived of its history.

We may consider, as is frequent in discussions of policy, especially in the United Nations and other agencies, measurable indices of scientific, mathematical, and technological progress. Scientometrics relies on several indicators, and the studies of quantitative history speak of central nations – those who produce new knowledge – and peripheral nations – those who absorb knowledge. Production and absorption of knowledge are clearly distinguishable. The sad situation is that the peripheral nations have been slow in absorbing new

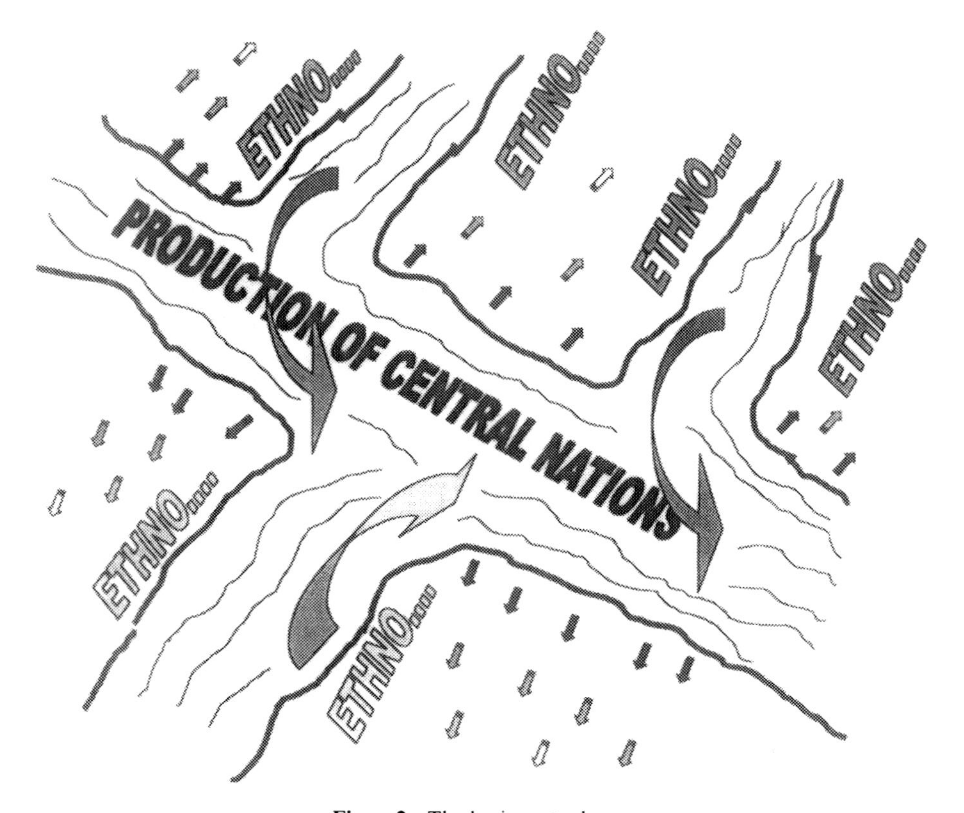

Figure 2 The basin metaphor.

knowledge. The lack of infrastructure acts as a barrier for this process (D'Ambrosio, 1975: 94).

The basin metaphor helps to explain the process. The main producers of knowledge, which are the central nations, are represented by the main stream. The water [knowledge] fertilizes their margins. They will have an effect in the margins of the tributaries [peripheral nations] much later, when the waters have already flown along the stream [thus producing the gap of knowledge]. The water [knowledge] does not flow upstream of the tributaries.

The water of the tributaries [indigenous knowledge] fertilizes their own margins and adds and contributes to the volume of water of the main stream. This is manifest in the emigration of academics and in the subordination of laboratories and research institutions of the peripheral nations to the priorities of their major homologues in the central nations (D'Ambrosio, 1979). As an example of this, we see efforts to entice research institutions in the peripheral nations to join major biotechnology research plans. The co-opting of scientists in the periphery is normally done by sending experts, in many cases highly reputed scientists, to the periphery for short visits and conferences, by offering fellowships, by giving stipends higher than the current national salaries, by sending equipment, in many cases obsolete or already worn out, and by offering international travel to seminars and congresses. This is true in academia and, in the peripheral nations which have a higher degree of development – the so-called emergent nations – also in industry. In mathematics, we have numerous examples of such practices in the post-war period. The presence of monies from the United States Army, Navy and Air Force research agencies, as well as from the American National Science Foundation (NSF), the French Centre National de la Recherche Scientifique (CNRS), the British Council, the DAAD and other agencies, following the pattern mentioned above, is remarkable.

These cases and policies have not yet been studied in detail. They often produce results without analyzing if the peripheral countries can absorb and make these results useful for their major needs. Normally this is the result of a lack of qualitative directives in science policy of the peripheral nations. Practically every scientific development plan in the periphery is a program entirely based in quantitative goals. The World Bank, United National Development Programme (UNDP) and other financing agencies rely on, indeed stimulate, plans based on quantitative goals. Clearly, they are easier to check, but the benefits for the poor populations of the peripheral nations are practically nil.

In the basin metaphor, the sources of the rivers, both the main stream and the tributaries, correspond to ethnomathematical knowledge. Ethnomathematical knowledge, like the waters, keeps flowing and fertilizing the margins of the tributary and eventually mixes in the major stream, contributing to this flow. Waters of the main stream do not go upstream through the tributaries, but the converse is true. This is evident in the life sciences and less so in the exact sciences. The progress carried on by the main stream will benefit the margins of the tributary after a long way through difficult land paths – which correspond to the acquisition of knowledge from other socio-cultural and environmental sources. The needs of the peripheral nations are met by what comes from the tributaries plus what is downloaded in the margins of the main stream. From these margins to the sites where they are needed, the paths go through grounds which are not receptive to alien ideas. An alternative is the production of knowledge which responds to cultural and environmental history and needs. Our hope is the erosion of the basin scenario [a world order based on leavers and takers[6]] and the creation of a great lake [a new planetary order without inequity, arrogance and bigotry (D'Ambrosio, 1997 and 1998)]. Mathematics has everything to do with this.

* * *

Conquest and colonization had an enormous influence in the course of the development of civilization. The chroniclers of the conquest told of what they saw and learnt. They mentioned very different ways of explaining the cosmos and ways of dealing with the surrounding environment. Indigenous religious systems, political structures, architecture and urban arrangements, sciences and values were, in a few decades, suppressed and replaced by those of the conqueror. A few remnants of the original practices of these cultures were, and in some cases still are, outlawed or treated as folklore. But they continue to integrate the cultural memory of the peoples descended from the conquered. Many of these behaviors are easily recognized in everyday life.

Mathematics, as a human endeavor, is the same. This is one focal point of the research program known as ethnomathematics, which deals with the generation, the intellectual and social organization and the diffusion of different ways, styles, and modes (*tics*) of explanation, understanding, learning, coping with and probing beyond (*mathema*) the immediate natural and socio-cultural environment (*ethno*).

Conquest led to colonization. In the Americas, the early colonizers, the Spanish and the Portuguese, paved the way for the French, the English and the Dutch settlers and later on for Africans, Europeans and Asiatic immigrants. With them came new forms of coping with the environment, of dealing with daily life, and new ways of explanation and learning. The result was the emergence of a synthesis of different forms of knowing and explaining which were generated by and available to the different communities. Often these were merged into new religions, new cuisine, new music, new arts and new languages. All of these interrelated as a synthesis of the cultural forms of the ancestors.

The colonization which took place in the Americas during most of the 16th, 17th, and 18th centuries occurred while new philosophical ideas, new sciences, new ways of production and new political arrangements were flourishing in Europe. Cultural artifacts produced in Europe were assimilated in the Americas. There was a co-existence of cultural goods, particularly knowledge, produced in the Americas and abroad. The former was consumed mostly by the lower strata of society and the latter by the upper classes. These boundaries were not clearly defined and there was much mutual influence of the resulting intellectual productions.

This poses the following basic question: What are the relations between the producers and consumers of cultural goods? This guides my proposal for a historiography of mathematics and what I have called 'the basin metaphor'. Although this is a question affecting the relations between academia and society in general, hence between the ruling elite and the population as a whole, it is particularly important for understanding the role of intellectuality in the colonial era. Thus ethnomathematics is a fundamental instrument of historical analysis (D'Ambrosio, 1995 and 1996). Curiously enough, the factors influencing the consumption of what we might call academic mathematics produced in an alien cultural environment, and what non-mathematicians have to say about mathematics, have not been given attention in the prevailing historiography (D'Ambrosio, 1993). It is important to incorporate the views of both mathematicians and 'outsiders' into the history of mathematics. This broader look, suggested by new historical scholarship, came under severe attack in what became to be known as the science wars.[7] This paved the way for another kind of attack, coming from the ultra-conservative groups, on the teaching of evolution and from another sector of academia, fueling what is known as 'math wars' (Klein, 1999).

The recognition of other systems of generation, of intellectual and social organization and of diffusion of modes of explanation, understanding, learning, coping with and probing beyond the immediate natural and socio-cultural environment is the only way we can escape the arrogance associated with the Western concept of truth. As I said in the introductory remarks, this can not be done unless we examine simultaneously categories, techniques, behavior, communication, language, traditions, history, systems of values, religions, and the sciences.

NOTES

1 Many scholars do not agree with the use of 'cultural science'. Ethnoscience might be a better choice.

2 We use the term modern science to denote the set of ideas arising from paradigms established in the 17th and 18th centuries, mainly through the works of Descartes, Newton, Leibniz and their followers.

3 See Ivan Van Sertima's interesting study: *They Came Before Columbus* (New York: Random House, 1976) and the reports on the voyages of the Chinese monk Huei Shen in the 5th century to Mexico. See the communication of Juan Hung Hui: 'Tecnologia Naval China y Viaje al Nuevo Mundo del Monje Chino Huei Shen,' III Congreso Latinoamericano y III Congreso Mexicano de Historia de la Ciencia y la Tecnologia, Ciudad de Mexico, 12–16 Enero 1992.

4 See in this respect Antonio Cardoso: *As Caravelas dos Descobrimentos e os mais Ilustres Caravelistas Portugueses* (Monografia n.7 do Museu de Marinha, Lisboa, 1984).

5 These topics have in the post-war period drawn much attention and generated important studies whose results throw some light on the production of scientific knowledge throughout history. Particularly interesting is the historiography adopted by Harold Dorn in his exciting book *The Geography of Science*, Baltimore: The Johns Hopkins University Press, 1991.

6 I use 'leavers and takers' in the sense introduced by Daniel Quinn in *Ishmael. A romance of Human Condition* (New York: Bantam/Turner, 1992).

7 See the issue devoted to the theme 'Science Wars' of *Social Text*, Spring/Summer 1996, pp. 46–47. The issue has very interesting papers. Regrettably, attention was given only to the hoax of Alan Sokal. As a consequence there was a renewal of attacks on Afrocentrism, the warnings against a 'new dark age of irrationalism' and other controversial disputes in the academic world. All this intellectual fundamentalism is nothing but a defensive posture against the challenge to the current epistemological order.

BIBLIOGRAPHY

Ascher, Marcia. *Ethnomathematics: A Multicultural View of Mathematical Ideas*. Pacific Grove, California: Brooks/Cole Publishing Company, 1991.

Barradas de Carvalho, Joaquim. *À la recherche de la spécifité de la renaissance portugaise*, 2 vols. Paris: Fondation Calouste Gulbenkian/Centre Culturel Portugais, 1983.

Bateson, Gregory. *Steps to an Ecology of Mind*. New York: Ballantine Books, 1972.

Bitterli, Urs. *Cultures in Conflict: Encounters Between European and Non-European Cultures, 1492–1800*. Cambridge: Polity Press, 1989.

Carr, E. H. *What is History?* Harmondsworth: Penguin Books, 1968.

Closs, Michael, ed. *Native American Mathematics*. Austin: University of Texas Press, 1986.

Correia, Mendes. 'Influência da expansão ultramarina no progresso científico.' *História da Expansão Portuguesa no Mundo* 3: 468, 1940.

D'Ambrosio, Ubiratan. 'Adapting the structure of education to the needs of developing countries.' (letter) *Impact of Science on Society* 25(1): 94, 1975.

D'Ambrosio, Ubiratan. 'Knowledge transfer and the universities: a policy dilemma.' *Impact of Science on Society* 29(3): 223–229, 1979.

D'Ambrosio, Ubiratan. 'Mathematics and literature.' In *Essays in Humanistic Mathematics*, Alvin M. White, ed. Washington, DC: The Mathematical Association of America, 1993, pp. 35–47.

D'Ambrosio, Ubiratan. 'Ethno-mathematics, the nature of mathematics and mathematics education.' In *Mathematics, Education and Philosophy: An International Perspective*, Paul Ernest, ed. London: The Falmer Press, 1994.

D'Ambrosio, Ubiratan. 'Ethnomathematics, history of mathematics and the basin metaphor.' In *Histoire et Épistemologie dans l'Education Mathématique/History and Epistemology in Mathematics Education* (Actes de la Première Université d'Été Européenne, Montpellier, 19–23 juillet 1993), F. Lalande, F. Jaboeuf and Y. Nouaze, eds. Montpellier: IREM, 1995, pp. 571–580.

D'Ambrosio, Ubiratan. 'Ethnomathematics: an explanation.' In *Vita Mathematica: Historical Research and Integration with Teaching*, Ronald Calinger, ed. Washington, DC: The Mathematical Association of America, 1996, pp. 245–250.

D'Ambrosio, Ubiratan. 'Diversity, equity, and peace: from dream to reality.' In *Multicultural and Gender Equity in the Mathematics Classroom: The Gift of Diversity*. 1997 Yearbook of the NCTM – National Council of Teachers of Mathematics, Janet Trentacosta and Margaret J. Kenney, eds. Reston, Virginia: NCTM, 1997, pp. 243–248.

D'Ambrosio, Ubiratan. 'Mathematics and peace: our responsibilities.' *Zentralblatt für Didaktik der Mathematik/ZDM* 30(3): 67–73, 1998.

D'Ambrosio, Ubiratan. *Etnomatemática: Arte ou Técnica de Explicar e Conhecer*. São Paulo: Editora Ática, 1990. (*Ethnomathematics: The Art or Technique of Explaining and Knowing*, Patrick B. Scott, trans. Las Cruces, New Mexico: NMSU/ISGEm, 1998).

Gerdes, Paulus. *Ethnomathematics and Education in Africa*. Stockholm: Institute of International Education/Stockholms Universitet, 1995.

Klein, David *et al.* Open letter to the U.S. Secretary of Education Richard Riley (advertisement). *Washington Post*, November 18, 1999.

Kragh, Helge. *An Introduction to the Historiography of Science*. Cambridge: Cambridge University Press, 1987.

León-Portilla, Miguel. 'Visión de los vencidos (crónicas indígenas mexicanas).' *Historia* 16, 1985.

Mayor, Federico. Opening speech at the *Conference on Scientific and Technological Cooperation in Africa*, Nairobi, Kenya, March 1994.

Sahagún. Fray Bernardino de. *Historia General de las Cosas de Nueva España*, 2 vols. Mexico: Alianza Editorial Mexicana, 1989.

Saitta, Armando. *Il Programma della Collezione Storica*. Bari: Laterza, 1955.

Thorndike, Lynn. *The Sphere of Sacrobosco and its Commentators*. Chicago: University of Chicago Press, 1949.

Zaslavsky, Claudia. *Africa Counts: Number and Pattern for Teachers*. New York: Lawrence Hill, 1979.

ELEANOR ROBSON

THE USES OF MATHEMATICS IN ANCIENT IRAQ, 6000–600 BC

Every culture has mathematics, but some have more than others. The cuneiform cultures of the pre-Islamic Middle East left a particularly rich mathematical heritage, some of which profoundly influenced late Classical and medieval Arabic traditions, but which was for the most part lost in antiquity and has begun to be recovered only in the last century or so.

People in the Middle East began to live in cities during the fourth millennium BC, and since then mud-brick, being cheap and plentiful, has been the principal urban building material. But it is not particularly durable or weather resistant if it has not been baked, so that buildings have to be repaired or renewed every few generations. Inevitably, then, long-inhabited settlements gradually rise above their surroundings on a bed of rubbish and rubble as the centuries and millennia progress, forming mounds or 'tells'. These tells are the raw material of archaeology, comprising successive layers of ever-older street plans, houses, domestic objects, waste pits – and written artefacts.

The centre of numerate, literate, urban culture in the pre-Islamic Middle East was southern Iraq, often called now by the Greek-derived term Mesopotamia, 'between the rivers', of the Tigris and Euphrates. The terms Sumer and Babylonia are also used, to refer to the area south of modern-day Baghdad in the third millennium BC and the second and first millennia BC respectively. Sumer was the cultural area of the speakers of a language called Sumerian, related to no other language known, which was first written down in the late fourth millennium and used, for an ever smaller number of functions, until the last centuries BC. It was gradually replaced by Babylonian, the southern dialect of the Semitic language Akkadian, named after the city of Babylon which was the region's capital from the mid-eighteenth century BC for most of the following two millennia. In the north of Iraq, home of the Assyrian dialect of Akkadian, the city of Ashur was the cultural, religious and political centre from the late third millennium onwards, until a succession of more northerly capitals, the most famous of which was Nineveh, replaced it in the early first millennium BC.

Sumerian and Akkadian, although linguistically unrelated, shared a syllabic

H. Selin (ed.), Mathematics Across Cultures: The History of Non-Western Mathematics, 93–113.

script we now call cuneiform, 'wedge-shaped', made by impressing a reed stylus on clay tablets which could range in size from a postage stamp to a laptop computer but were most often designed to be held comfortably in the hand. Clay is essentially an inorganic material and is not subject to the same decay processes as papyrus- or leather-based writing materials. When lost, abandoned, or thrown away in and around the mud-brick buildings of Mesopotamian cities tablets were for the most part preserved intact for millennia. Documents in languages such as Aramaic, on the other hand – written alphabetically with ink on perishable media, from the late second millennium onwards – survive rarely and badly.

We therefore rely primarily on cuneiform sources for our understanding of ancient Middle Eastern literate culture – and there is certainly no shortage of them. Conservative estimates put the number of clay tablets accessible to scholars in museum and university collections at around half a million worldwide. Many times that number are still in the ground. While all clay tablets are archaeological artefacts like pots or bones – for no texts have been passed on in continuous transmission – only a small fraction of those above ground have known archaeological contexts. In the early days of exploration and excavation, in the latter half of the nineteenth century AD, the aim of explorers and scholars was to fill Western collections with artefacts; the skills and methodology involved in the recovery of ancient environments and lifestyles had not yet been developed. Many thousands of tablets were dug up before the days of controlled, stratigraphic archaeology. But even during the twentieth century, excavators were slow to record detailed findspots for written artefacts, assuming that somehow they could 'speak for themselves' and did not need to be contextualised archaeologically. At the same time, particularly before the formation of the Iraqi Antiquities Authority in 1923, many sites were looted for their tablets, which were then sold on to private collectors and public institutions in Europe and America. As archaeological techniques were refined and excavation became both slower and more meticulous, and as specialist equipment became more expensive while budgets got smaller, digs became better managed and more limited in scope and focus. Since the Gulf War in 1990 the illegal trade in looted tablets and other antiquities has massively increased once more (Gibson, 1997). Inevitably, then, the number of tablets found in well-documented archaeological contexts is relatively small – but still can be counted in the tens of thousands.

A large percentage of those tablets, with context and without, deal with numbers. The vast majority are administrative documents, drawn up by bureaucrats working for large temples, palaces, or private enterprises which needed extensive and sophisticated literate and numerate management techniques to control their enormous resources. While these have been the bread and butter of social and economic historians, they have rarely been treated as sources for the history of mathematics. This is because we also have several thousand tablets which can strictly be called mathematical; that is, they deal with number or other mathematical matters for their own sake. While many are simple multiplication tables, others show a conceptual abstraction which is far removed from the practical arithmetical needs of everyday life. These mathematical tablets are known from the earliest phases of cuneiform culture to the latest, but date predominantly from the Old Babylonian period of the early second millennium BC.

Since its discovery in the early twentieth century AD, this mathematics has been treated implicitly as part of the 'Western' tradition; even now one finds 'Mesopotamian' mathematics categorised as 'Early Western mathematics', while Iraqi mathematics in Arabic, some of which is directly related to its compatriot precursors, appears under 'other traditions' (e.g. Cooke, 1997). There is, however, no evidence of Mesopotamian influence on Classical mathematics until 150 years post-Euclid – despite a century of determined attempts to show otherwise. To some extent this slant has been part of the mainstream European colonisation of the ancient civilisations. As Bahrani (1999: 163) puts it:

> In the simplest terms, if the earliest 'signs of civilisation' were unearthed in an Ottoman province inhabited primarily by Arabs and Kurds, how was this to be reconciled with the European notion of the progress of civilisation as one organic whole? Civilisation had to have been passed from ancient Mesopotamia and Egypt to Greece.

But at the same time as 'this unruly ancient time was [being] brought within the linear development of civilisation' (Bahrani, 1999: 163) by dissociating it from the modern Middle East and grafting it instead to early European history, mathematics was itself being divorced from the history of the region for the simple reason that 'few Assyriologists like numbers' (Englund, 1998: 111). History of ancient Middle Eastern mathematics has, by and large, and certainly in the last fifty years, been left to mathematicians and historians of mathematics who have little feel for the culture which produced the mathematics or the archaeology which recovered the artefacts, and no technical training in the languages and scripts in which the mathematics was written (cf. Høyrup, 1996). Not surprisingly, the history of Mesopotamian mathematics has predominantly been a history of techniques and 'facts' about 'what "everyone knew" in Babylon' (Cooke, 1997: 47) – for the most part 'translated' beyond all recognition into modern symbolic algebra.

But the field is at last growing up, and since the late 1980s more serious efforts have been made to understand the language, conceptualisation, and concepts behind the mathematics of ancient Iraq (e.g. Høyrup, 1990a and 1994; Robson, 1999). This article is an attempt to pull the focus back further and to make a first approximation to a description of Mesopotamian numerate, quantitative, or patterned approaches to the past, present, and future; to the built environment and the agricultural landscape; and to the natural and supernatural worlds.

DECORATIVE ARTS

Mathematics contains a strong element of the visual, making decorative designs and motifs one of the most pervasive and demotic sources for the identification of mathematical concepts within non-literate and extra-literate cultures (cf. Washburn and Crowe, 1988). Plotting the changing fashions in pottery design over the millennia is a key tool for the study of ancient Iraq – and indeed most archaeologically recovered societies – because fired clay is one of the most ubiquitous, malleable, replaceable, breakable and yet indestructible resources known to humankind. Changes in pottery style have been used to trace developments in society, technology, and artistic sophistication, but are also crucial witnesses to the place of geometrical concepts such as symmetry, rotation, tessellation, and reflection in the dominant aesthetics at all levels of a society.

Pottery firing technology was in widespread use in northern Iraq from around 6000 BC. The earliest phases (so-called Samarra ware, cf. Figure 1) already exhibit strong geometric stylisation, as Leslie (1952: 60) notes:

> A potter begins to decorate a pot with some notion of how the pot should look and with some ability [to] carry through this notion. Stated in formalistic terms, this notion of the Samarran potter is as follows:
>
> 1. The predominant use of unpaneled elements within enclosed bands.
> 2. The alternations of direction of movement and/or symmetrical motion of contiguous bands.
> 3. The emphasis of fourfold rotation or of quatrofold radial symmetry of finite designs.
> 4. The dominant use of lineal rather than areal design elements.
> 5. The uniformity and precision of draftsmanship.
> 6. The tendency to balance equally painted and ground areas.
>
> Stated in psychological terms, as an aesthetic ideal, the style is neat, busy, and abstract.

These principles are exemplified by the well-preserved bowl shown in Figure 1. Around a central clockwise swastika four stylised herons catch fish in their mouths while eight fish circle round them, also in a clockwise direction. An outer band of stepped lines moves outwards anticlockwise, countering the swirling effect of the animalian figures.

We see similar geometrical concerns in the Halaf ware of the late sixth

Figure 1 Samarra period bowl *ca.* 6000 BC, northern Iraq; painted clay; diameter 27.7 cm (after Caubet and Pouyssegur, 1998: 34).

millennium BC. Figure 2 shows intersecting half- and quarter-arcs of circles forming symmetrical petal-like figures within hatched bands and wavy lines around the rim of the vessel. The overall impression is more static than the earlier Samarra ware. Kilmer (1990: 87) has drawn attention to the similarities with Old Babylonian geometrical figures called 'cargo-boats' composed of intersecting quarter-arcs (cf. Figure 10). Five- and six-fold radial symmetry are also occasionally attested in Samarra and Halaf ware (cf. Goff, 1963: figs. 34–35, fig. 69). Incidentally, potters at this early period are thought to have been almost exclusively women, working at home without potters' wheels, to supply their immediate household's domestic needs.

Another key artefact type in the archaeology of ancient Iraq is the cylinder seal. These small stone objects, usually 3–5 cm in height and 1–3 cm in diameter, bore on their cylindrical surfaces incised designs which served to identify their owner or institutional function when rolled out onto the surface of clay covering vessel necks, knots, and other sealings. The original incisions had to be mirror images of the intended clay impressions. Cylinder seals made their first appearance during the fifth millennium BC and, like pottery designs, showed a strong geometrical aesthetic from the earliest days, as well as a topological understanding of cylindrical surfaces, bounded at the top and bottom edges but unbounded on the curved plane. Most cylinder seal designs exploit this horizontal continuity, resulting in a seamless, never-ending design when rolled out on the more-or-less Euclidean surface of the clay. Figure 3 shows a sophisticated design of mirror-image fantastic animals whose elongated necks and tails intertwine and overlap each other, breaking up the vertical boundaries between the figures. Figure 4, on the other hand, exhibits a purely abstract continuous design whose main components are hatched arches which alternately descend and ascend

Figure 2 Halaf period bowl *ca.* 5000 BC, northern Iraq; painted clay; diameter 14 cm (after Reade, 1991: fig. 15).

Figure 3 Modern impression of cylinder seal *ca.* 3200 BC, unprovenanced; green stone; 4.6 × 4 cm (after Collon, 1987: no. 885).

Figure 4 Modern impression of cylinder seal *ca.* 3000 BC, southwestern Iran; limestone; 4.0 × 1.3 cm (after Collon, 1987: no. 42).

from the edges of the seal, reaching almost to the opposite edge in two-fold rotational near-symmetry.

These devices of symmetry and continuity could equally well be used for abstract or figurative images. In Figure 5 we see pairs of birds (ducks?) with their wings outstretched within panels borders by twisted ropes. The image exhibits both horizontal and vertical symmetry. In Figure 6, water buffalo standing back-to-back drink from flowing water jars held by kneeling curly-haired men facing centre who are naked except for their cummerbunds. Underneath, a river runs continuously through a stylised mountainous terrain.

Figure 5 Ancient impression of cylinder seal on clay, *ca.* 3000 BC, Uruk, southern Iraq; height 6.4 cm (after Collon, 1987: no. 9).

Figure 6 Modern impression of cylinder seal *ca.* 2200 BC, unprovenanced; jasper; 4.0 × 2.7 cm (after Collon, 1987: no. 529).

The horns of the buffalo support a central cuneiform inscription in Sumerian recording the ownership of the seal: 'divine Shar-kali-sharri, king of Akkad: Ibni-sharrum, the scribe [and seal-owner] is your servant'.

Symmetry could also be subverted or used to add meaning to an image. In Figure 7, a male human ruler, standing sideways, is mirrored to the right by a larger female deity facing front. The king pours offerings onto a stylised altar while the goddess offers the rod and ring, symbols of kingship, towards him. Trees and minor deities flank the main protagonists, but while the mountain gods are mirror images one tree is crooked (and deciduous?) while the other is a straight pine. The symmetry is subverted by placing the gods to the left of each tree, but the overall feeling of balance – between human and divine, male and female, etc. – is maintained. The seal also carried a cuneiform inscription in Sumerian, enabling us to identify not only the owner of the seal but also the protagonists of the image. In this instance the text is not an integral part of the image but occupies the remaining third of the curved surface of the object, between pine tree and minor deity. It reads: 'divine Amar-Suena, king

Figure 7 Ancient impression of cylinder seal on clay; *ca.* 2050 BC, Nippur, southern Iraq (after Gibson and Biggs, 1991: cover).

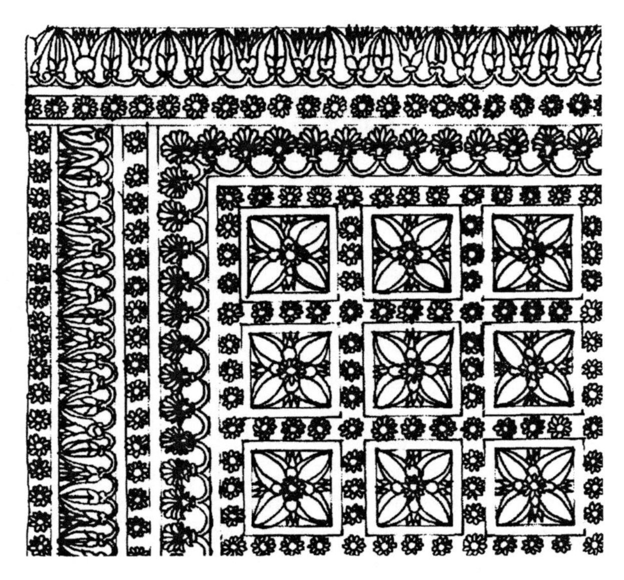

Figure 8 Design on the door sill of a palace throne room, *ca.* 685 BC; Nineveh, northern Iraq; Mosul marble; 45 × 45 cm (after Collon, 1995: fig. 114).

of the four corners, beloved of the goddess Inana: Lugal-engardug – overseer of the temple of Inana, *nu-esh*-priest of the god Enlil, son of Enlil-amakh, overseer of the temple of Inana, *nu-esh*-priest of the god Enlil – is your servant.'

We have little evidence for mathematically inspired design on textiles and other biodegradable objects, except for representations of them in less perishable media. The stone 'carpets' on the floors of major thresholds of the Neo-Assyrian palaces in northern Iraq are some of the most beautiful examples. Both of those shown here (Figures 8 and 9) are bordered with alternating open and closed symmetrical lotuses (giving a tasselled effect) and rows of rosettes. The central designs of Figure 8 are stylised flowers with four-fold symmetry, strongly reminiscent of the bowl in Figure 2, while the main ground of Figure 9 is composed of a series of interlocking circles forming petal-shaped figures from third-circle arcs.

Two-, four- and eight-fold symmetry based on the square, with overlapping figures and shapes within shapes, is best attested for literate mathematics in the geometrical tablet BM 15285 (Figure 10), probably from the eighteenth century city of Larsa in southern Iraq (cf. Kilmer, 1990: 84–86; Robson, 1999: 34–56, 218–230). Geometrical figures based on the equilateral triangle and/or a third of a circle, however, are so far attested only in eighteenth century Eshnuna (east central Iraq) and seventeenth century Susa (south-west Iran) (Robson, 1999: 45–48).

BUREAUCRACY AND ACCOUNTANCY

Quantitative conceptual tools for managing and controlling property and wealth predate literacy by perhaps a thousand years (Englund, 1998: 42–55).

Figure 9 Design on the door sill of a palace throne room, *ca.* 645 BC; Nineveh, northern Iraq; Mosul marble; 65 × 65 cm (after Curtis and Reade, 1995: no. 45).

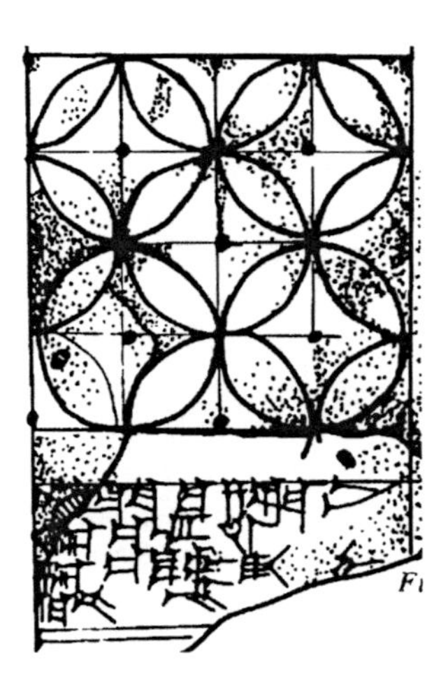

'The side of the square is 60 rods (*ca.* 360 m). [Inside it are] 4 triangles, 16 barges, 5 concave squares. What are their areas?'

Figure 10. The last preserved problem of the geometrical 'text-book' BM 15285 *ca.* 1750 BC; Larsa, southern Iraq (after Robson, 1999: 217).

During the course of the fourth millennium BC, temples were at the social and economic heart of developing urban centres such as Uruk and Susa. Through accumulation and close management of offerings (such as grain, cattle, land, precious stones, and metals) to deities they became wealthy and powerful institutions with interests in maintaining and increasing that wealth and power. Numerical methods of control afforded one such means (along with social methods such as propagation of religious ideologies). Small clay counters a few centimetres high, shaped into crude spheres, cones, and discs, appear to have represented fixed amounts of commodities within an accounting system; those commodities are no longer identifiable to us but the counters must have had context and meaning to the ancient administrators.

We can satisfactorily distinguish these counters from other small clay objects (such as sling-shots, loom weights, and beads) only when they are found archaeologically with other administrative artefacts such as cylinder seals, standard-sized mass-produced ration bowls, and the remains of the clay envelopes in which the counters were stored. More often than not they turn up not in the contexts in which they must have originally been used but in rubbish dumps within or in the vicinity of temple precincts. Small collections of counters are occasionally found within intact envelopes, which have sometimes been marked with cylinder seal impressions – marks of some institutional authority. Impressions of the tokens themselves may also be found on the surface of the envelopes – records of their contents (Nissen *et al.*, 1993: 11–13).

The oldest clay tablets known carry impressions of counters too, with or without cylinder sealings. We can trace a chronological and conceptual development over the fourth millennium: from pre-arithmetical tablets on which a single counter was impressed many times (perhaps as simple tallies), to those which show small numbers of counter-impressions of different shapes and sizes and implicit numerical relationships between them, to similar tablets which have been impressed not with a counter but with a stylus in the shape of a counter (Nissen *et al.*, 1993: 13–14). It would be tempting to posit an evolution from unmarked envelopes to those bearing impressions of counters, and from these to the arithmetical tablets – but the archaeological data does not (at the moment) support such a clear-cut chronological development.

Tablets with word signs as well as number signs probably date from the last third of the fourth millennium. The number signs continue to be made in imitation of the counters, while the words are for the most part pictograms scratched (not impressed) into the surface of the clay. Because each of the incised signs represents a whole word or idea rather than a particular sound, it is difficult (and perhaps inappropriate) to assign a language to these signs – but on the whole it is felt likely that the language is Sumerian. Nearly 4000 early tablets have been recovered from the foundations of the central temple of Uruk, where they had been re-used as building rubble. A Berlin-based team has studied them intensively, showing them to be elements of a sophisticated and complex system of managing the material wealth of the temple (Nissen *et al.*, 1993: 4–6).

The tablets use around a dozen different metrological systems and bases, dependent on the subject of the accounts. For instance, while most discrete objects, including nearly all dairy products, were counted in base 60, many others, including cheese, were counted in base 120. There were also separate methods for counting time, areas, and no less than six different capacity systems for various grains and liquids (Nissen *et al.*, 1993: 28–29). Because these number systems were contextual it was feasible to use visually identical signs in different numerical relationships. For instance, a small circular impression was worth ten small conical impressions in the discrete sexagesimal and bisexagesimal systems but sixteen in the areal system and just six in the barley capacity system. These relative values have been deduced from tablets on which several commodities are totalled together. Circular stylus-impressions on these early tablets are almost identical to the imprints of spherical counters on clay envelopes, while the conical impressions closely resemble the indentations made by conical counters. We can hypothesise, then, that the preliterate counters, like the early number signs, represented not absolute numbers but had the potential to embody a range of different numerical relationships dependent on what was being accounted for – and what those commodities were we are unlikely ever to know.

Even in those first tablets from the late fourth millennium, we find complex summations of different categories of goods, theoretical estimates of raw materials needed for food products, and the continuous tracking of grain harvests over several years (Nissen *et al.*, 1993: 30–46). The accounting year was assumed to be 360 days long, comprising 12 months of 30 days each, with an extra month inserted when needed to realign it with the seasons (Englund, 1988). This ad hoc intercalation was replaced by a standardised 19-year cycle of 235 lunar months only in the late first millennium BC (Rochberg, 1995: 1938). Over the course of the third millennium, writing was still used primarily for quantitative (and increasingly, legal) purposes, and its users were still mostly restricted to the professional managers of institutional (temple and palace) wealth. In southern Iraq we have the names of individual accountants and administrators some 500 years earlier than the first royal inscription (cf. Postgate, 1992: 30, 66)! An important innovation of the later third millennium was the balanced account, in which theoretical outgoings were measured against actual expenditures – reaching a peak of complexity and ubiquity in the twenty-first century under the empire whose capital was the city of Ur (so-called Ur III). Under this highly managerial regime, accounts could be drawn up in silver (for merchants working for the state [Snell, 1982]), grain (for millers), clay vessels (for potters), and even units of agricultural labour. Expected work- and production-rates were set at the upper limit of feasibility so that more often than not team foremen carried over a deficit of labour owed to the state from one accounting period to the next, year in year out (Englund, 1991).

Some time before the end of the third millennium the sexagesimal place value system (SPVS) was developed or invented. Metrologies, though rationalised periodically by administrative reform or royal decree (Powell, 1987–90), had continued to be context-dependent and to utilise many different number bases. This extract from the prologue to a 21st century law-code shows the increasing tendency to sexagesimalisation and cross-metrological relationships, where 1 *sila* $\approx$ 1 litre and 60 *gin* = 1 *mana* $\approx$ 0.5 kg:

> I made the copper *bariga* and standardised it at 60 *sila*. I made the copper *ban* and standardised it at 10 *sila*. I made the regular royal copper *ban* and standardised it at 5 *sila*. I standardised the 1 *gin* metal weight against 1 *mana*. I made the bronze *sila* and standardised it at 1 *mina* (after Roth, 1995: 16).

Nevertheless, it was not always a trivial matter to convert between systems – to calculate the area of field given its length and width, say, or to find the grain capacity of a given volume – and the SPVS seems to have been a calculational device created to overcome these difficulties. Measures were converted to sexagesimal multiples and fractions of a designated base unit and the calculation performed in base 60, the results of which were transformed back into the units of the appropriate metrology. Professional scribes were apparently not supposed to show their arithmetical workings, so that we get only occasional glimpses of the SPVS at work in the late third millennium (Powell, 1976).

By the early second millennium, with the further spread of writing into the personal sphere, we have a good deal of evidence about legal and financial uses of mathematics. The standard units of commercial exchange were barley (for items of low value) and silver (for more expensive goods). Law-codes set out ideal rates of exchange, wages, and professional fees (Roth, 1995: 23–142) but they were often much higher in practice than laid down in theory (Postgate, 1992: 195). Loans of barley were made at an average interest rate of $33\frac{1}{3}\%$, silver at 20%. These rates were not annual but for the duration of the loan, however long it lasted – usually a matter of months. Willed property was measured out and divided in equal portions among the heirs (usually, but not exclusively, sons) with the eldest getting an extra share in recompense for performing *kispum* rituals for dead ancestors. Women could own and manage property too, whether in their own right or on behalf of their families (Postgate, 1992: 88–108). Mathematical problems from the same period, however, do not reflect contemporary practice but rather use inheritance and loan scenarios to set up pseudo-realistic word problems on topics like arithmetic progressions and division by irregular numbers (Friberg, 1987–90: 569–570).

QUANTITY SURVEYING AND ARCHITECTURE

It is a very old chestnut indeed that mathematical innovation was driven in premodern societies by the need to measure fields and predict harvests. Nonetheless, there is a kernel of truth to it. We have already seen that the oldest known literate institution, the temple in Uruk, included quantitative land and crop management amongst its activities.

Once again, though, our best evidence is from the later third millennium. Southern Iraqi fields at this time were typically elongated strips, designed both to minimise the number of turns ploughing teams of oxen had to make and to maximise the number of fields abutting the irrigation channels. Field plans drawn up by the scribes of Ur show that irregularly shaped areas around the main arable lands were divided into approximately right triangles and irregular quadrilaterals, whose areas were calculated as the products of averaged opposite sides (Liverani, 1990). The diagrammatic elements of the plans are never to scale; rather, we might say that they are relational or topological, showing only the spatial relationships between the field elements and their basic geometrical shapes. All quantitative information was contained in the annotations to the plans: the measurements of the fields' sides, their calculated areas, and sometimes one or more cardinal points at the edges of the plan. The text could also contain descriptive information, such as the quality of the soil or the names of the fields. From such plans theoretical harvest yields could be calculated, and compared to those actually achieved.

Agricultural labour was also closely managed from at least the mid-third millennium on, with workers being allotted target work-rates for tasks such as renovating irrigation canals or weeding. Under the Ur III regime, such work-rates were used as units of account in a system of annual double-entry book-keeping. In the region of Umma in southern Iraq teams of 20, including an overseer, were expected to contribute 7,200 working days a year, plus whatever was owed from the year before. Administrators kept records of the work they performed, following the agricultural cycle from the spring-time harvest (reaping, making sheaves, threshing); through field preparation (ploughing and harrowing in teams of three or four, at 1.8 ha a day, sowing at 0.7 ha a day); hoeing and weeding (360–1,080 m^2 a day); to repairing channels and banks (1.6–3 m^3 a day) (Maekawa, 1990; Robson, 1999: 157–164). At the end of the accounting year, the work completed was compared with the work expected, and any deficit carried over to the following accounting period (Englund, 1991). Many of these theoretical labouring rates were still used to set word problems in Old Babylonian school mathematics, even though most had long fallen out of practical use (Robson, 1999: 93–110).

Building labour was managed in a similar way. One of the earliest extant building plans, and in some ways the most informative, is not a working document at all but a sculptural depiction of one (Figure 11). It is the focal point of an inscribed statue, one of a series of around twenty, depicting Gudea, who ruled a small state in southern Iraq at the end of the 22nd century BC. The inscription tells us that the city god Ningirsu had revealed to Gudea in a dream the layout of a new temple for him, the enclosure walls of which are outlined on the plan. It goes on to describe the construction of the temple, and how the statue was to be set up in its principal courtyard facing the statue of the god himself (Tallon, 1992: 41). Along the outer edge of the plan, the remains of a ruler are just visible. The ancient temple itself has been excavated, but sadly the statue was not recovered in situ; it was found with seven others in the ruins of a palace on the same site, built in the second century BC – placed there some two thousand years after its manufacture!

We also have less monumental house plans on clay tablets from the late third and early second millennia BC (Postgate, 1992: 91, 117; Robson, 1999: 148–152), which like the field plans give measurements and sometimes descriptions of the functions of the rooms. Contemporary administrative documents listing walls to be built and repairs to be made show that there were two

Figure 11 Statue of Gudea, city governor of Lagash, with a building plan and ruler on his lap *ca.* 2100 BC, southern Iraq; diorite; statue 93 × 46 × 62 cm, plan 26 × 16 cm (after Tallon, 1992: 42–43, figs. 12, 12a).

standard sizes of brick. Baked bricks, which look more like paving stones to us, but were used for prestige buildings and for flooring, measured $\frac{2}{3}$ cubit (*ca.* 33 cm) square; the cheaper sun-dried bricks, on the other hand, were $\frac{1}{2} \times \frac{1}{3}$ cubit (*ca.* 25 × 17 cm). Both were 5 fingers (*ca.* 8 cm) thick. Bricks, whatever their size, were counted in groups of 720. The number of 720s per unit volume, or brickage – 2.7 for square baked bricks, 7.2 for sun-dried bricks – was a useful constant in calculating materials for building work. Sometimes mortar was factored in as $\frac{1}{6}$ of the volume of a wall, in which case constants of 2.25 and 6 were used instead (Robson, 1999: 145–148). Standardised brick sizes were elaborated into a complex metrology in Old Babylonian school mathematics, with about a dozen types attested in many different sorts of word problems (Robson, 1999: 57–73). The theoretical brick sizes are very close to the measurements of ancient bricks recovered from Iraqi archaeological sites, where the square baked bricks are attested more or less continuously from the time of Gudea to the Persian period (sixth century BC) (Robson, 1999: 278–289).

EDUCATION

Not surprisingly, mathematics education is as old as mathematics itself. Even amongst the earliest cuneiform tablets from late fourth millennium Uruk we find exercises in bookkeeping and calculation which exhibit features we might expect from school work: they are anonymous, often incompetently executed, and result in conspicuously whole or round numbers (Englund, 1998: 106–110). Word problems, mathematical diagrams, and arithmetical tables are attested patchily from the mid-third millennium onwards (Friberg, 1987–90: 540–542), but our best and most abundant evidence for mathematics education comes from the Old Babylonian period. A few hundred cuneiform tablets contain between them over a thousand word problems, while the number of arithmetical tables surviving must run well into the thousands too. This is the subject matter of most modern surveys of 'Babylonian' mathematics (e.g. Neugebauer and Sachs, 1945; Friberg, 1987–90: 542–580; Høyrup, 1994; cf. Robson, 1996–) so my aim here is not particularly to summarise its contents but to explore how and why it was taught and learned. Most of the evidence we have comes from the eighteenth century cities of Nippur and Ur, between modern Baghdad and the Gulf (cf. Tinney, 1998).

Trainee scribes' first encounter with mathematical concepts was in the course of copying and learning by heart a standard list of Sumerian words, organised thematically and running to over 2000 entries. In the section on trees and wooden objects, for instance, the subsection on boats includes eight lines on boats of different capacities (1 *gur* $\approx$ 300 litres): '60-*gur* boat; 50-*gur* boat; 40-*gur* boat; 30-*gur* boat; 20-*gur* boat; 15-*gur* boat; 10-*gur* boat; 5-*gur* boat'. Similarly, the section on stone objects included a list of about thirty weights from 1 *gun* to 5 *gin* ($\approx$ 30 kg–4 g) (cf. Figure 12).[1]

They also learned weights and measures in a more structured fashion, writing out standard lists of capacity, weight, area, and length, in descending order of size, often with their SPVS equivalents (Friberg, 1987–90: 542–543). Equally, they copied a series comprising a division ('reciprocal') table followed by multiplication tables for forty sexagesimally regular numbers from 50 down to 2 (Friberg, 1987–90: 545–546). The procedure for all rote learning, whether of Sumerian words or arithmetical facts, was the same: first the teacher wrote out a model of 20–30 lines for the student to copy repeatedly on the same tablet, then the student progressed to writing out extracts on small tablets, and finally the whole series was written out on one, two, or three large tablets (Veldhuis in Tinney, 1998). Addition and subtraction facts, however, were never committed to memory in this way.

Figure 12 Bronze weight, inscribed in Akkadian 'Palace of Shalmaneser, king of Assyria, 5 royal *mana*'. One of a set of eight, *ca.* 725 BC. Nimrud, northern Iraq, 10.5 × 20 × 8 cm, 5.04 kg (after Curtis and Reade, 1995: no. 202).

Scribal students also had the opportunity to practice their arithmetical skills, making calculations or drawing geometrical diagrams on small round or roughly square tablets (Robson, 1999: 245–277). In many instances we can link those calculations with particular problem types which are set and given model solutions on other tablets, which we might think of as textbooks (cf. Figure 10; e.g. Fowler and Robson, 1998: 368–370). Those model solutions are essentially instructions in Akkadian, using as a paradigm a convenient set of numerical data which will produce an arithmetically simple and pleasing answer. Around half of such problems use real life scenarios such as the building, labouring, and inheritance contexts described above, but they should by no means be considered examples of narrowly 'practical' mathematics. Even when they use constants (brick sizes, labouring rates, etc.) known from administrative contexts, the problems themselves are clearly impractical: to find the length and width of a grain pile given their sum, for instance, or to find the combined work rate of three brick makers all working at different rates (Robson, 1999: 75, 221). In short, these scenarios are little more than window dressing for word problems involving mathematical techniques that working administrators and accountants were almost certainly never likely to need.

The remaining problems have traditionally been classed as 'algebra' and translated into arithmetised x–y symbolism. Jens Høyrup's groundbreaking study of Akkadian 'algebraic' terminology has shown, however, that the scribes themselves conceptualised unknowns much more concretely as lines, areas, and volumes (Høyrup, 1990a). These imaginary geometrical figures could then be manipulated, as described in the model solutions, until the magnitude of the unknown was found using techniques such as completing the square (cf. Fowler and Robson, 1998). We see then, that in the Old Babylonian period number had not yet completely shed its contextualised origins of 1500 years before: not only did it have magnitude but it still had dimension or measure. Just as administrative scribes recorded data and results in mixed metrological systems and used the SPVS only for intermediate calculations, even in the 'pure' setting of school mathematics the inherently metric (measured) properties of number were disregarded in favour of the SPVS solely for the duration of arithmetical operations.

Literacy, and literate numeracy, were professional skills in ancient Iraq, possessed with greater or lesser degrees of competence by a tiny percentage of the urban population. But if scribal education was no more than vocational training why, then, did Old Babylonian school mathematics far exceed the needs of working scribes, most of whom would have spent their lives as accountants or letter writers in bureaucratic institutions? One could ask the same of other aspects of their education: the long lists of rare and complex cuneiform signs, for instance, or the Sumerian literary epics. Some tentative answers may be found in the very subject matter that the students studied. A good many of the Sumerian proverbs (often found on the same tablets as arithmetical exercises: Robson, 1999: 246) and literary passages directly concerned or alluded to scribalism: the attributes of accomplished scribes were elaborated and extolled, while those of the incompetent were derided. High levels of literacy and numeracy were worthy of the most renowned of kings, endowed by Nisaba, goddess of scribal wisdom:

> Nisaba, the woman radiant with joy, the true woman, the scribe, the lady who knows everything, guides your fingers on the clay: she makes them put beautiful wedges on the tablets and adorns them with a golden stylus. Nisaba generously bestowed upon you the measuring rod, the surveyor's gleaming line, the yardstick, and the tablets which confer wisdom. (Praise poem of king Lipit-Eshtar (Lipit-Eshtar B), lines 18–24. Black *et al.*, 1998–: no. 2.5.5.2).

In short, the true scribe was not merely competent; he possessed divinely bestowed skills and wisdom which far exceeded the humdrum needs of his daily life yet were still somehow related to it (cf. Høyrup, 1990b: 67).

DIVINATION

The prediction of the future through the observation of ominous phenomena does not seem at first sight to be an activity rich in mathematical thinking. Extispicy, or divination by inspection of the livers of dead sheep, arose from the need to feed the gods, who in some sense inhabited beautiful statues of themselves housed in the temples of Mesopotamian cities. The gods fed, so the idea went, by smelling the offerings of food, drink, and incense made to them. These same gods, it was thought, decided the future of the world and recorded it all in cuneiform on the Tablet of Destinies. But they revealed their intentions in subtle ways, in particular – if the correct rituals were performed – in the entrails and especially in the livers of the sheep and goats sacrificed to them. If the expert diviners determined that the future the gods had in store was unfavourable, they could take measures to avert it. This was also a task for experts, and involved further prayers and rituals and sacrifices to the gods in order to persuade them to change their minds. A further liver divination would determine whether the procedure had been effective or not. The first clear evidence for liver divination comes from the end of the third millennium, when some years were named after high priests being chosen in this fashion. As well as these revealed omens, various happenings in the natural world and skies – so-called observed omens – could also be considered portentous.

Our richest sources of evidence are the compendia of omens, which are first attested in the early second millennium BC and reach their most elaborate and comprehensive form in eighth and seventh century Assyria (northern Iraq). By that time the omens had been collected into a series of around 100 tablets, divided into ten chapters. Its Akkadian name is *bârûtu* (seeing). The first nine chapters are organised according to the ominous organs of the body and work systematically through various features and defects of each. The omens operate by two sorts of general principle. First there are binary oppositions, such as right-left, up-down, large-small, and light-dark. Right was associated with a propitious omen, left with an unpropitious one. So while good health on the right side of the animals innards was auspicious, on the left it was inauspicious. Conversely, abnormalities were hoped for on the left, but not on the right, as in this extract from a divination prayer:

> Let the judges, the great gods, who sit on golden thrones, who eat at a table of lapis-lazuli, sit

> before you. Let them judge the case in justice and righteousness. Judge today the case of so-and-so, son of so-and-so. On the right of this lamb place a true verdict, and on the left of the lamb place a true verdict. [...] Let the back of the lung be sound to the right; let it be stunted to the left. Let the 'receptacle' of the lung be sound on the right; let it be split on the left (Starr, 1983: 37–38, lines 18–19, 30–31).

Secondly, various sorts of analogies were also used to make predictions. Punning played a prominent part, as did the association of specific features or dispositions of the internal organs with phenomena in the real world: mildew presaged ruin, berry-like excrescences portended warts, etc. Certain fortuitous markings had exact meanings too: the Weapon foretold war and death, the Foot suggested movement, and the Hole forecast death and disaster.

Particular features of the entrails could also be identified with individuals or institutions in the real world: on the liver the Palace Gate referred to the palace of course, and the Path meant the army on campaign. These associations were explicitly stated in the last of the ten chapters of the *bârûtu* omen series, which was called *multâbiltu*, or 'analysis'. Various attributes of the organs – length, thickness, massiveness, movement – were each linked with a general prediction – health, fame, power, happiness – and illustrated with an omen extracted from the first nine chapters (Starr, 1983: 6).

We see similar principles at work in collections of omens observed from real world phenomena, for instance the series concerning ominous births called in Akkadian *shumma izbu*, 'if a birth-anomaly'. This extract concerns the birth of kids which are a different colour to their mother goats:

> If a black goat gives birth to a yellow kid: that fold will be scattered; it will become waste; there will be anger from the gods.
> If a yellow goat gives birth to a black kid: that fold will be scattered; ...
> If a white goat gives birth to a black kid: destruction of the herd.
> If a black goat gives birth to a white kid: the fold of the man will scatter.
> If a red goat gives birth to a black kid: destruction of the herd.
> If a black goat gives birth to a red kid: destruction of the herd.
> (Leichty and von Soden, 1970: 175)

It is the birth of a black kid to a non-black goat that is unfavourable, as can be seen when the combinations of kids and goats of different colours are tabulated. Presumably other sorts of births are not ominous, for good or bad.

Goats/Kids	black	yellow	white	red
black		●	●	●
yellow	●			
white	●			
red	●			

But omens do not restrict themselves to a systematic, exhaustive enumeration of possible outcomes. Because, we remember, omens are signs from the gods, in theory almost anything is possible. So we find, for instance, goats giving birth to lions, wolves, dogs, and pigs. Similarly, the tablets of *shumma izbu* about other animals list equally impossible – better, improbable – birth events. Nevertheless, despite the lack of causal relationship between observation and prediction, we can detect some basic combinatoric and group theoretical concepts behind the development of divination.[2]

* * *

The chronological cut-off point for this survey roughly coincides with the end of the Assyrian empire in 612 BCE. A few decades after that point Iraq became little more than a province of the Persian empire (albeit a wealthy and important one), followed by periods of Alexandrian, Parthian, and Sassanian rule. It did not regain its political independence and cultural dominance until the rise of Islam in the sixth century CE. Nevertheless neither cuneiform nor mathematics died a sudden death (Friberg, 1993; Geller, 1997). On the contrary, both gained a new lease of life with the increasing cultic importance of celestial divination and the concomitant need to predict the movements of the heavenly bodies accurately. The detailed observations of heavenly bodies recorded in Babylon over the course of the first millennium BCE, the arithmetical schemes used to model their movements, and not least the SPVS in which those observations were recorded, were all crucial building blocks for Hipparchus and Ptolemy to lay the foundations of modern astronomy in turn-of-the-millennium Egypt (Toomer, 1988; Jones, 1993). Indeed, the SPVS (though not in cuneiform) remained the only universal and viable vehicle for astronomical and trigonometrical calculations until the coming of the Indian-Arabic decimal system that we use today. Old Babylonian-style cut-and-paste 'algebra' also heavily influenced the mathematical work of Diophantus and Hero in late Classical Egypt as well as early Islamic algebraists, although the direct links in this case are much more difficult to detect (Høyrup, 1994).

In summary, then, numerical and mathematical concepts were an integral part of the scribal worldview, running throughout ancient Iraqi literate culture (and non-literate culture too). The ancient Middle East witnessed not merely the 'infancy' of Western mathematical culture – though, as we have seen, it played a major role in the birth of astronomy – but also in a very real way housed the intellectual precursors of the marvellous flowering of Arabic mathematics in the early Middle Ages. Without this, much of the Classical mathematics we hold so dear would not have survived at all, and trigonometry, algebra, and algorithms – all of which have strong roots in the mathematics discussed here – might have looked very different, if they existed at all.

ACKNOWLEDGEMENTS

It is a pleasure to thank Helaine Selin, Niek Veldhuis and David Wengrow for their constructive contributions to the shape, tone, and content of this article.

NOTES

[1] The Assyrian royal *mana* of the early first millennium was double that of the traditional *mana*, which also continued to be used (Powell, 1987–90: 516).

[2] The famous Babylonian mathematical astronomy of the late first millennium BC also had its

roots in divination through observation of celestial phenomena (Koch-Westenholz, 1995; Rochberg, 1995; Veldhuis, 1999), but it is beyond the chronological scope of the article, and the competence of the author, to trace that development here.

BIBLIOGRAPHY AND FURTHER READING

Bahrani, Zainab. 'Conjuring Mesopotamia: imaginative geography and a world past.' In *Archaeology Under Fire: Nationalism, Politics and Heritage in the Eastern Mediterranean and Middle East*, Lynn Meskell, ed. London and New York: Routledge, 1998, pp. 159–174.

Black, Jeremy, Graham Cunningham, Eleanor Robson and Gábor Zólyomi, eds. *The Electronic Text Corpus of Sumerian Literature*. http://www-etcsl.orient.ox.ac.uk Oxford: University of Oxford, 1998–.

Caubet, Annie and Pouyssegur, Patrick. *The Ancient Near East*. Paris: Éditions Pierre Terrail, 1998.

Collon, Dominique. *First Impressions: Cylinder Seals in the Ancient Near East*. London: British Museum Press, 1987.

Collon, Dominique. *Ancient Near Eastern Art*. London: British Museum Press, 1995.

Cooke, Roger. *The History of Mathematics: a Brief Course*. New York: Wiley, 1997.

Curtis, J. E. and J. E. Reade, eds. *Art and Empire: Treasures from Assyria in the British Museum*. New York: Metropolitan Museum of Art, 1995.

Englund, Robert K. 'Administrative timekeeping in ancient Mesopotamia.' *Journal of the Economic and Social History of the Orient* 31: 121–185, 1988.

Englund, Robert K. 'Hard work – where will it get you? Labor management in Ur III Mesopotamia.' *Journal of Near Eastern Studies* 50: 255–280, 1991.

Englund, Robert K. 'Texts from the Late Uruk period.' In *Mesopotamien: Späturuk-Zeit und Frühdynastische Zeit*, Pascal Attinger and Markus Wäfler, eds. Göttingen: Vandenhoeck und Ruprecht, 1998, pp. 15–233.

Fowler, David and Eleanor Robson. 'Square root approximations in Old Babylonian mathematics: YBC 7289 in context.' *Historia Mathematica* 25: 366–378, 1998.

Friberg, Jöran. 'Mathematik.' In *RealLexikon der Assyriologie und vorderasiatische Archäologie* 7, Dietz O. Edzard, ed. Berlin and New York: Walter de Gruyter, 1987–90, pp. 531–585.

Friberg, Jöran. 'On the structure of cuneiform metrological table texts from the –1st millennium.' In *Die Rolle der Astronomie in Kulturen Mesopotamiens*, Hans. D. Galter, ed. Graz: RM Druck- und Verlagsgesellschaft, 1993, pp. 383–405.

Geller, M. 'The last wedge.' *Zeitschrift für Assyriologie* 87: 43–96, 1997.

Gibson, McG. 'Iraq since the Gulf War: the loss of archaeological context and the illegal trade in Mesopotamian antiquities.' *Culture Without Context* 1: 6–8, 1997.

Gibson, McG. and R. D. Biggs, eds. *The Organization of Power: Aspects of Bureaucracy in the Ancient Near East*. 2nd ed. Chicago: The Oriental Institute of the University of Chicago, 1991.

Goff, Beatrice Laura. *Symbols of Prehistoric Mesopotamia*. New Haven and London: Yale University Press, 1963.

Høyrup, Jens. 'Algebra and naive geometry. An investigation of some basic aspects of Old Babylonian mathematical thought.' *Altorientalische Forschungen* 17: 27–69; 262–354, 1990.

Høyrup, Jens. 'Sub-scientific mathematics: observations on a pre-modern phenomenon.' *History of Science* 28: 63–86, 1990.

Høyrup, Jens. *In Measure, Number and Weight. Studies in Mathematics and Culture*. New York: State University of New York Press, 1994.

Høyrup, Jens. 'Babylonian mathematics.' In *Companion Encyclopedia of the History and Philosophy of the Mathematical Sciences* 1, Ivor Grattan-Guinness, ed. London: Routledge, 1994, pp. 21–29.

Høyrup, Jens. 'Changing trends in the historiography of Mesopotamian mathematics: an insider's view.' *History of Science* 34: 1–32, 1996.

Jones, Alexander. 'The evidence for Babylonian arithmetical schemes in Greek astronomy.' In *Die Rolle der Astronomie in den Kulturen Mesopotamiens*, Hans. D. Galter, ed. Graz: RM Druck- und Verlagsgesellschaft, 1993, pp. 77–94.

Kilmer, Anne D. 'Sumerian and Akkadian names for designs and geometric shapes.' In *Investigating*

Artistic Environments in the Ancient Near East, Ann C. Gunter, ed. Washington, DC: Smithsonian Institution, 1990, pp. 83–91.

Koch-Westenholz, Ulla. *Mesopotamian Astrology. An Introduction to Babylonian and Assyrian Celestial Divination.* Copenhagen: Carsten Niebuhr Institute, 1995.

Leichty, E. and von Soden, W. *The Omen Series Shumma Izbu.* Locust Valley, New York: J. J. Augustin, 1970.

Leslie, Charles. 'Style tradition and change: an analysis of the earliest painted pottery from northern Iraq.' *Journal of Near Eastern Studies* 11: 57–66, 1952.

Liverani, M. 'The shape of Neo-Sumerian fields.' *Bulletin on Sumerian Agriculture* 5: 147–186, 1990.

Maekawa, K. 'Cultivation methods in the Ur III period.' *Bulletin on Sumerian Agriculture* 5: 115–145, 1990.

Neugebauer, Otto. *The Exact Sciences in Antiquity.* Princeton, New Jersey: Princeton University Press, 1952. Reprinted New York: Dover, 1969.

Neugebauer, Otto and Abraham J. Sachs. *Mathematical Cuneiform Texts.* New Haven: American Oriental Society and the American Schools of Oriental Research, 1945.

Nissen, Hans, Peter Damerow and Robert K. Englund. *Archaic Bookkeeping: Early Writing and Techniques of Administration in the Ancient Near East.* Chicago: University of Chicago Press, 1993.

Pingree, David. 'Hellenophilia versus the history of science.' *Isis* 83: 554–563, 1992.

Postgate, J. N. *Early Mesopotamia: Society and Economy at the Dawn of History.* London and New York: Routledge, 1992.

Powell, Marvin A. 'The antecedents of Old Babylonian place notation and the early history of Babylonian mathematics.' *Historia Mathematica* 3: 417–439, 1976.

Powell, Marvin A. 'Maße und Gewichte.' In *RealLexikon der Assyriologie und vorderasiatischen Archäologie* 7, Dietz O. Edzard, ed. Berlin and New York: Walter de Gruyter, 1987–90, pp. 457–530.

Reade, Julian. *Mesopotamia.* London: British Museum Press, 1991.

Roaf, Michael. *Cultural Atlas of Mesopotamia and the Ancient Near East.* Oxford: Facts on File, 1990.

Robson, Eleanor. 'Bibliography of Mesopotamian maths.' *http://it.stlawu.edu/~dmelvill/mesomath/erbiblio.html*, 1996–.

Robson, Eleanor. *Mesopotamian Mathematics: Technical Constants in Bureaucracy and Education, 2100–1600 BC.* Oxford: Clarendon Press, 1999.

Robson, Eleanor. 'Mesopotamian mathematics: some historical background.' In *Using History to Teach Mathematics: an International Perspective.* V. J. Katz, ed. Washington, DC: Mathematical Association of America, forthcoming 2000.

Rochberg, Francesca. 'Astronomy and calendars in ancient Mesopotamia.' In *Civilizations of the Ancient Near East*, vol. 3, Jack M. Sasson, ed. New York: Scribner's, 1995, pp. 1925–1940.

Roth, Martha. *Law Collections of Ancient Mesopotamia and Asia Minor.* Atlanta: Scholars Press, 1995.

Snell, Daniel. *Ledgers and Prices: Early Mesopotamian Merchant Accounts.* New Haven: Yale University Press, 1982.

Starr, Ivan. *The Rituals of the Diviner.* Malibu: Undena Publications, 1983.

Tallon, Françoise. 'Art and the ruler: Gudea of Lagash.' *Asian Art* 5: 31–51, 1992.

Tinney, Steve. 'Texts, tablets, and teaching. Scribal education in Nippur and Ur.' *Expedition* 40(2): 40–50, 1998.

Toomer, G. J. 'Hipparchus and Babylonian astronomy.' In *A Scientific Humanist: Studies in Memory of Abraham Sachs*, E. Leichty, M. de J. Ellis and P. Gerardi, eds. Philadelphia: The University Museum, 1988, pp. 353–362.

Van De Mieroop, Marc. *Cuneiform Texts and the Writing of History.* London and New York: Routledge, 1999.

Veldhuis, Niek. 'Reading the signs.' In *All Those Nations: Cultural Encounters Within and With the Near East*, H. L. J. Vanstiphout, ed. Groningen: Styx Publications, 1999, pp. 161–174.

Walker, C. B. F. *Cuneiform.* London: British Museum Press, 1987.

Washburn, D. K. and D. W. Crowe. *Symmetries of Culture: Theory and Practice of Plane Pattern Analysis.* Seattle: University of Washington Press, 1988.

JAMES RITTER

EGYPTIAN MATHEMATICS

SOURCES

Any examination of ancient Egyptian mathematics must start from a basic understanding of the problem of sources. Egypt presents the archeologist with a severe constraint with regard to written texts. Arable land – and with it the location of urban areas – now, and in the historic past, is restricted to the floodplain of the Nile delta (Lower Egypt) and a narrow band around the Nile river valley (Upper Egypt). The gradual raising of the water table level in the delta and the increasingly heavy population in the Nile valley has rendered substantial and extended digs in ancient urban areas virtually impossible. The vast majority of such endeavors have thus avoided the delta and, in the south, been situated in the desert areas, far from the ancient mud-brick settlements. They have concentrated almost exclusively on the stone-built cemeteries and temples, a tendency strengthened by the 'treasure-hunt' attitude of many early archeologists. Thus almost all our written sources come from the cemetery and funerary temple areas on the desert fringe.

Writing was, in Egypt as elsewhere in the ancient world, an essentially urban phenomenon.[1] This means that the source of most of our texts is funerary or religious, and for the great majority of other subjects, including mathematics, we are dependent on chance finds and indirect references. That is, we would have been had it not been for the existence of artificial villages, maintained by the State on the desert edge for the housing of the workers, priests and officials involved in the construction of royal tombs and the accomplishment of funerary rites. Such, for example, is the case of the Middle Kingdom (2000–1700 BC) village today called el-Lahun. Even the rapid and superficial excavation of this village uncovered a rich harvest of hundreds of administrative, legal, medical, and mathematical papyri, genres which are ordinarily missing from tombs and mortuary temples.

Even with such finds the number of documents available for a reconstruction of ancient Egyptian mathematics is limited. The mathematical texts proper consist of a handful of documents for the period which we shall be treating here, pharaonic Egypt prior to the Greco-Roman period (i.e., prior to 330 BC).

H. Selin (ed.), Mathematics Across Cultures: The History of Non-Western Mathematics, 115–136.

What we have are the remains of some seven papyri, of which only one is complete, a leather roll, a pair of wooden exercise tablets, and two ostraca, one a small inscribed limestone flake, the other an even smaller pottery sherd. Moreover, their time span is limited; all but the last two date to the Middle Kingdom and the ostraca are only a little later.[2]

WRITING AND METROLOGY

The end of the fourth millennium saw the birth of writing in Egypt and with it the first metrological signs. The earliest uses of writing in the Nile valley essentially reduce to three categories: titles and names of persons and places, names of commodities, and quantities of these last (Dreyer, 1999). If the analogy with the earliest writing in Mesopotamia is valid, then this is not the result of the 'luck of the spade' in uncovering texts but a reflection of the essential nature and scope of the earliest writing.[3] For writing in Egypt, as in Mesopotamia, was not created to provide a permanent record of the spoken language but rather to count and account for the production, distribution and consumption of commodities.[4] Numerical and related non-numerical information thus appear from the start, around 3200 BC, and, for some 500 years remain the sole attested use of the medium.[5]

It should be borne in mind that, from the beginning, there were two styles of writing, called since the Greek period, hieratic and hieroglyphic. Hieroglyphic writing, what one usually has in mind when thinking of Egyptian writing, was pictographic and employed until the end of traditional Egyptian culture for monumental inscriptions. The other form, hieratic, is equally old but cursive in nature, and was the standard form when writing with a brush and ink. Though most non-numerical signs had corresponding cursive and monumental forms, many numerical signs existed exclusively as abstract hieratic symbols; the rare attempts to transcribe them hieroglyphically for monumental inscriptions show how difficult a problem it was to invent hieroglyphs for these hieratic signs.

The numerical information that lies thus at the heart of early writing in Egypt is of a particular kind; it is metrological in nature, as befits its accounting role, with the value of the signs used dependent on the metrological system in which they are embedded.[6] The presence of a multiplicity of such systems – including those for counting discrete objects, for measuring lengths, areas, linen dimensions, grain capacities, and weights – and the limited repertory of signs used, implied a deal of polyvalence in the values of a given sign. Thus, for example, in the third millennium the same sign represents 10 units in the discrete counting system but 100 units in the grain capacity system. In all cases though the Egyptians used an additive system of notation, in which a unit was repeated as many times as necessary to express the value desired. [Editor's note: see Eleanor Robson's discussion of this same system in Iraqi mathematics.]

Examples of such variety are provided by some of the scenes engraved or painted on tomb walls in the Old Kingdom burials of high officials. Possibly the richest of these in terms of numbers of different systems is the representation of a market (Figure 1) from the joint tomb of Ni-ankh-khnum and Khnum-hetep, two high officials in the reign of the fifth-dynasty king Ni-user-ra (2450–2420), in the necropolis at Saqqara (Moussa & Altenmüller, 1977: pl. 24, fig. 10). Here we see a multitude of activities centering around the exchange of goods, with the hieroglyphic inscriptions reporting the (supposed) dialogues of the buyers and sellers; mentions are made of the discrete, linen, capacity and weight metrologies.

The other source for third-millennium metrological information is the administrative papyri of the Old Kingdom. Given the provenance of Egyptian material, it is inevitable that what we have – certainly a negligible fraction of what was actually produced – comes principally from the administrative centers attached to the funerary cults of dead kings; such have been found at the royal temple sites of Abusir and Saqqara.[7] In addition a few (still unpublished) non-funerary archive texts have been found in private tombs, such as those from Gebelein, Elephantine and Sharuna.[8] To these hieratic, 'working' texts should be added the hieroglyphic copies of hieratic administrative documents, transcribed on tomb walls as a special mark of favor by the pharaoh to one of his officials (Goedicke, 1967 and 1990).

The recent publication of many of these texts allows us, as mentioned above, to appreciate the multiplicity of metrological systems and the associated polyvalence of numerical signs. But the texts also permit us to put to rest certain long-standing hypotheses about the structure and origin of these signs, hypotheses which, as is often the case, have, since their initial introduction, hardened into widely-popularized 'facts'.

Such has been the fate of the 'Horus-eye fractions' and the supposed mythological origin of the lower sub-multiples (often called dimidiated fractions) of the grain capacity system with values of $\frac{1}{2}, \frac{1}{4}, \frac{1}{8}, \ldots, \frac{1}{64}$ of a basic unit called the *heqat*. The signs for these sub-multiples were, in this theory, supposed to represent the various parts of the eye of Horus, whose dismemberment and subsequent restoration plays an important part in Egyptian mythology. This account of things goes back to a brief note by Georg Möller (Möller, 1913) who based his tentatively advanced connection between the signs and the eye on late second-millennium hieroglyphic evidence. Accepted by Alan Gardiner in his influential *Egyptian Grammar* (Gardiner, 1957: 197), it has become a staple part of all introductions to Egyptian mathematics and metrology, another indication of the supposedly deeply religious nature of the Egyptians.[9] However, whatever the later Egyptian reinterpretation of the original hieratic signs may have been, the third-millennium papyri and information in the market-scene of Ni-ankh-khnum and Khnum-hetep in Figure 1 leave no doubt that the eye of Horus had nothing to do with the origins of the original hieratic signs (Posener-Kriéger, 1994). These correspond to no known hieroglyphic signs and, as mentioned earlier, very likely originally had none. With the exception of those used in the discrete system, the numerical hieratic signs were probably all non-representational from the outset.[10]

Another hoary chestnut is the purported high status of Egyptian scribes.

Figure 1 Market scene. Adapted from Ahmed M. Moussa and Hartwig Altenmüller, eds., *Des Grab Nianchchnum und Chnumhotep*, 1977. Reproduced with the kind permission of the Deutsches Archäologisches Institut, Cairo.

Figure 2 Scribes. Adapted from Ahmed M. Moussa and Hartwig Altenmüller, eds., *Des Grab Nianchchnum und Chnumhotep*, 1977. Reproduced with the permission of the Deutsches Archäologisches Institut, Cairo.

Here it is not new material but a careful use of long-available third-millennium evidence that helps us situate this group socially. Scenes of scribes are very common on Old Kingdom tomb walls, contextualized by their activities and relations to other employees of the landowners and high officials to whom these tombs belong. Copies of these scenes published in histories for a general readership almost always show the scribes in isolation, their reed pens behind their ears, their roll of papyrus or – more frequently – their wooden writing board in their hands. But that is never the complete picture; Figure 2 shows a typical full scene (like Figure 1, it comes from the tomb of Ni-ankh-khnum and Khnum-hetep, Moussa & Altenmüller, 1977: fig. 24). A scribe in the center of the scene registers on his wooden writing board the quantity of wheat and barley produced by each funerary domain in the possession of the owners of the tomb. This information is provided by the overseers of these properties, their arms clenched in a gesture of submission. In other versions of the same scene they are escorted by an armed policeman to ensure the accuracy of their rendering of accounts, and sometimes they are even shown being beaten. Here a second scribe gives the account to the steward of the estate, and this time it is the scribe who adopts the inferior's position-standing before his seated superior, he bends respectfully at the waist.

Scribes in Egypt, like their contemporaries in Mesopotamia, were accountants – calculators of work, rations, land and grain. Hard at work on their master's business – be he the State or a private landowner – they are shown with the external signs of their intermediate and subservient positions. They were not the movers and shakers of ancient society but they served those who were – well enough to be immortalized on their patrons' walls as the outward and visible sign of that concentration of power and wealth they labored so hard to measure and preserve.

THE MATHEMATICAL TEXTS

The rarity of our sources prevents us from grasping the manner in which calculations were actually carried out in third-millennium accounting. Starting with the advent at the end of the third millennium of that period of state centralization called the Middle Kingdom by modern Egyptologists, we have a new source – school mathematical textbooks, generally referred to in the literature simply as 'mathematical texts'.[11] That these papyri, our sole textual source for Egyptian mathematics, are, in fact, school texts is important for us, because it determines the range of questions that we can pose concerning Egyptian mathematics; they must be those for which answers may be reasonably sought in documents designed for the instruction of young scribes.

Neither based on religious ritual or mythological speculation, as some would have it, nor, as others claim, mere rules of thumb based on a trial-and-error empiricism, the mathematical texts show a sophisticated, pedagogic approach to the question of the organization and transmission of mathematical knowledge. One proof of this is the simple fact that the scribes successfully maintained for over three thousand years a complex, intellectually demanding economic and administrative structure; an analysis of the structure and functioning of these texts will bring out more clearly how they functioned in this role.

The only mathematical papyrus complete enough for a global analysis of its contents – the 'Rhind mathematical papyrus' (pRhind) – shows a clear organization which underscores its pedagogic nature.[12] It consists of 84 mathematical problems and solutions grouped by theme. The problem section of the recto of pRhind, for example, is organized in the following way:

- Equal division of different numbers of loaves of bread among 10 men (1–6)
- Addition and subtraction of fractions (7–23)
- Determining a number on the basis of its sum or difference with fractions of itself (24–34)
- Same problem for quantities of grain (35–38)
- Unequal division of loaves among men (39–40)
- Volumes of containers of different shapes (41–47)
- Areas of fields of different shapes (48–55)
- Linear dimensions of pyramids and cones(?) (56–60)

The verso contains a more miscellaneous collection of problems, including more on division of rations (63–66, 68) and the calculation of grain quantities in the production and exchange of bread and beer (69–78) and in the feeding of poultry and cattle (82–84).

Within each group the problems are arranged in subgroups by the methods used to solve the problem and, generally, by an increasing degree of complexity. Further, certain sections such as those on mensuration are careful to include subgroups of problems in which changes are rung on which measurements are to be taken as given and which are to be calculated. Examples of this are the first four problems posed on a page from pRhind (Figure 3):

Example of calculating a rectangular container of 10 in its length, 10 in its width and 10 in its height. What is then its capacity in grain?

A container into which have gone 75 hundred-quadruple-*heqat* of grain. It is how great by how great?

A container into which have gone 25 hundred-quadruple-*heqat* of grain. What are its dimensions?

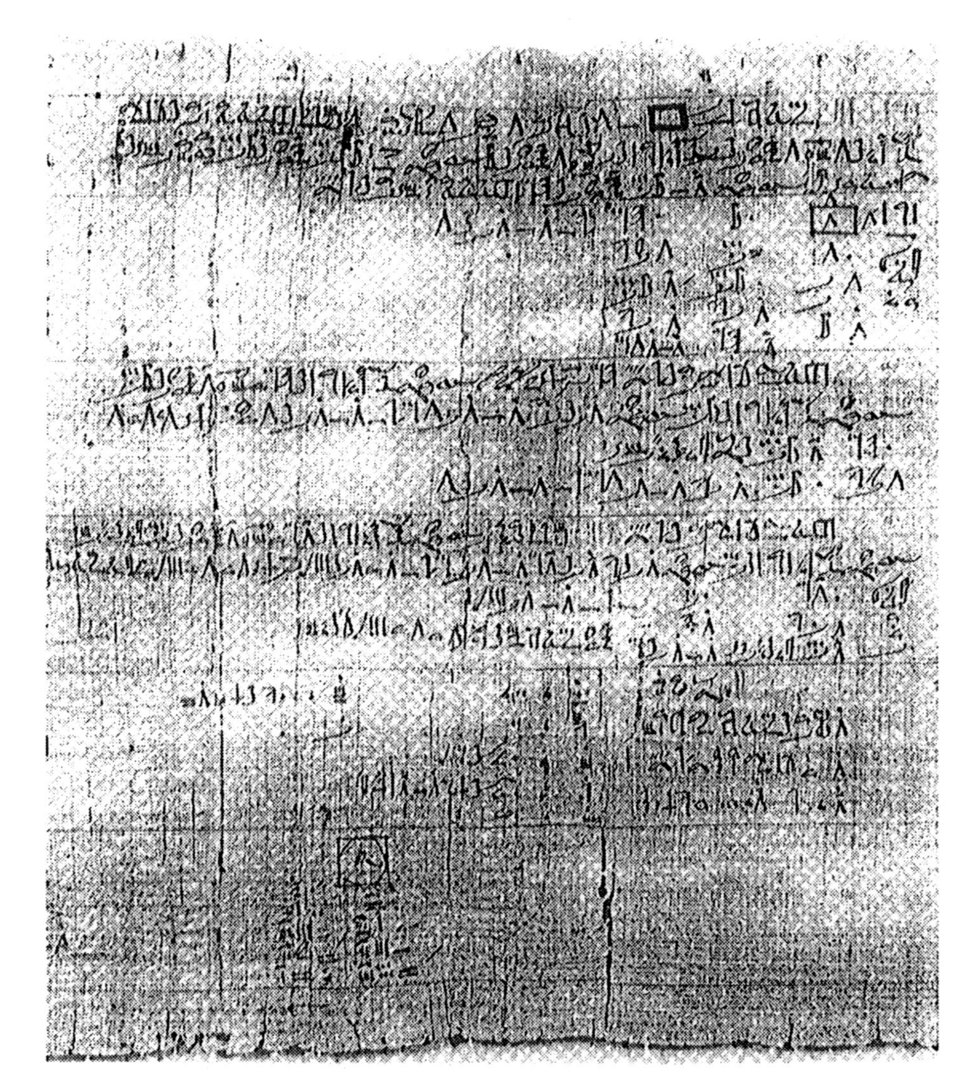

Figure 3 Details from the Rhind Papyrus.

If a scribe says to you: Let me know what $\frac{1}{10}$ becomes in a rectangular container and in a round container.

The last problem cited above asks for the conversion of the 'pure' fraction $\frac{1}{10}$ into metrological capacity values. This illustrates one of the most significant differences with third-millennium texts: the use of an *abstract* number system, i.e., one that is independent of any particular metrological system. This system is constructed by a combination of the discrete-system metrological signs for integers (given here in their hieroglyphic form):

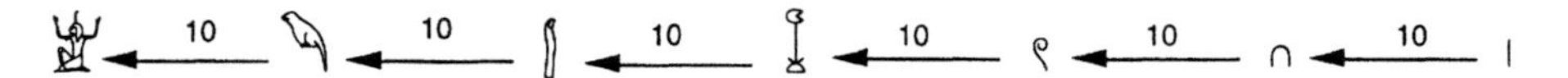

and the fractional signs used as the low-end of the linear measurement system, now attached not to the finger-length 𓂭 as was the case in their system of origin, but to 𓏺, the unit discrete element:

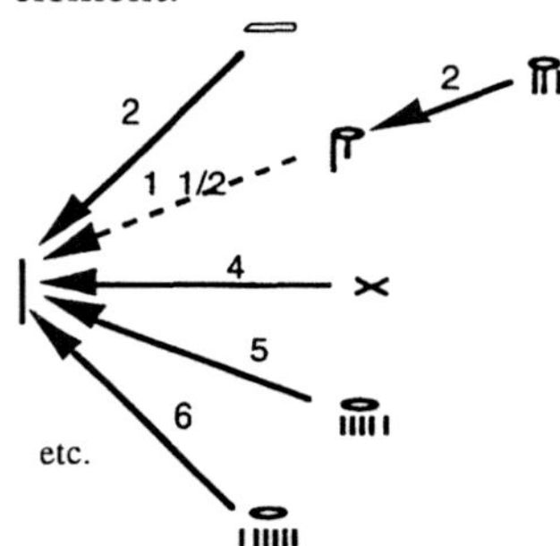

No metrological units of any kind are attached to the system and thus calculations need not constantly keep track of such. The elements of this system have a name in the mathematical papyri – *aha* – and the first three sections of the pRhind are devoted in fact to exercises in their manipulation.[13]

Another advantage of the abstract system is its capability to use, as intermediate results, mathematical objects fitting into no metrological system. Of course, once the solution is found it is necessary to convert it into the appropriate metrological system. Indeed, as was the case in contemporary Mesopotamia – and indeed in all societies prior to the advent of the metric system – a large part of mathematical training, in the Middle Kingdom and afterwards, involved learning how to transform the numerical data in a problem out of the metrological systems in which it was presented into the abstract system in which the calculations were performed and, finally, to transform the answer back into real metrological units.[14]

An example of how all this works is given by Problem 46 of the pRhind (Table 1), one of those illustrated in Figure 3 (fourth register from the top). The volume of a container is given in the standard Middle-Kingdom capacity unit of 'hundred-quadruple *heqat*' and the linear dimensions of the container are required.[15] The first step in the solution is the transformation of the given capacity into another capacity unit, the *khar*, 20 times smaller. The advantage of this is that the *khar* stands in a simple relationship to the linear measure of the cubit.[16] Since the problem presupposes that the length and breadth of the container are fixed at 10 cubits each, the height can be calculated by removing these two linear dimensions from the product by multiplying by $\frac{1}{10}$ and then again by $\frac{1}{10}$. What remains is the height – but not in cubits. It is the final multiplication by $\frac{2}{3}$ which yields the answer in cubits. If one were concerned with maintaining metrological units all the way through the calculation, it would have been necessary to invert the order and first convert *khar* into 'cubic cubits' (a unit which, in any case, never existed in the Egyptian system). Numerically, the calculations are almost always simpler in the order given and

Table 1 (pRhind 46)

A container into which have gone **25** hundred-quadruple-*heqat* of grain. What are its dimensions?
* You will calculate starting from 25, 20 times. It yields 500; this is its contents.
* You will calculate starting from 500. You will make its $\frac{1}{10}$: 50; its $\frac{1}{20}$: 25; $\frac{1}{10}$ of its $\frac{1}{10}$: 5; $\frac{2}{3}$ of $\frac{1}{10}$ of its $\frac{1}{10}$: $3\frac{1}{3}$.

Thus this container is 10 by 10 by $3\frac{1}{3}$.

Its working:

1	25
10	750
20	500

This is its contents.

1	500
$\frac{1}{10}$	50
$\frac{1}{10}$ of its $\frac{1}{10}$	1
$\frac{2}{3}$ of $\frac{1}{10}$ of its $\frac{1}{10}$	$3\frac{1}{3}$

This container yields 10 cubits by $3\frac{1}{3}$.

their order makes no difference in the abstract system where one is only concerned with *final* units.

We have already had occasion to mention details of some problem texts; the time has come to discuss this class of texts in a systematic way. They all present a similar structure:

- Problems are posed – and solved – in terms of specific or concrete numbers; not the area of a rectangular field but a field of such-and-such a length and width.[17]
- A solution is an algorithm, i.e., a set of prescribed steps, written in the second person; the pupil is directly addressed and instructed to follow a fixed sequence of operations.
- Generally, the solution is followed by a numerical verification (literally 'the procedure according to (its) form' or the like) of the correctness of the answer.

The pedagogical nature of these texts, with their invariable form and their carefully graded grouping, tells us nothing about the origins of the methods employed, any more than our present-day schoolbooks do. This, after all, is not their role. Neither the 'logic of discovery' nor the 'logic of justification' is visible here; it is a 'logic of transmission' which determines the content and the organization of the extant texts. This fact is sometimes forgotten, as for example in the numerous attempts to see in Problem 48 of the pRhind (bottom register of Figure 3) traces of a 'proof' of the Egyptian algorithm for determining the area of a circle. What the text shows is a square containing a square, another geometric figure inscribed within it and the dimension '9' written at the center. There is no text but the calculations which follow – the determination of the area of a square of side 9 and that of a circle of diameter 9 – show clearly the point of the exercise, the comparison of the areas of a circle and a square of the same 'size'. Thus the figure inscribed in the square is a simply a circle.

Attempts to see in it an inscribed polygon serving as a base for a rigorous derivation of an approximation to the area of a circle simply take no account either of the constraints on drawing small shapes in constricted areas on papyrus or of the nature of the source.

Beyond the structure of the problems there is the question of their mode of functioning. In fact, Egyptian mathematical problems work on three distinct levels.

The most general is that of the *strategy* employed to solve the problem. A common choice is that which is now called 'false position'. That is to say, a (false) solution is supposed, the result of calculation with this value is compared to a true value given in the statement of the problem, and the original choice of solution is corrected by the necessary factor. Problem 26 (Table 2) is a typical example of such an approach. A number is sought which, after its fourth is added to it, becomes 15. The trial number 4 is guessed (chosen for the commodity of calculating its $\frac{1}{4}$). Its fourth is calculated and added to it, the result is 5. Since this number is 3 times smaller than the given value of 15, the same factor of 3 is used to correct the initial value (4) of the solution, yielding the (correct) value of 12.

Table 2 (pRhind 26)

A quantity (*aha*); its $\frac{1}{4}$ is added to it. It becomes 15.

* Calculate starting from 4; you will make its $\frac{1}{4}$: 1.

[\1	4]
[\\$\frac{1}{4}$	1]

* [Add 4 to 1.]

Total: 5

* Calculate starting from 5, to find 15.

\1	5
\2	10

3 will be the result.

* Calculate starting from 3, 4 times

1	3
2	6
\4	12

12 will be the result.

The quantity: 12.

Its $\frac{1}{4}$: 3.
Total: 15.

[**The procedure as it occurs:**]

1	12
[$\frac{1}{2}$	6]
$\frac{1}{4}$	3

Total: 15

However similar problems may call for quite different solution algorithms, depending on the specific values of their data, and the apprentice scribe thus needed to know various methods and how to choose among them. Problem 34 of pRhind (Table 3) is one of a number of problems which appear in the same section as pRhind 26 but are solved by a completely different method. Here the sum of a quantity and its half and its quarter is 10 and now the choice of a trial answer is not quite so evident. The algorithm taught here, in place of that of 'false position', is the direct division of 10 by $1 + \frac{1}{2} + \frac{1}{4}$.

Whatever the choice of algorithm, the working-out of the solution is usually followed by a verification of the result in a final section entitled 'the procedure as it occurs' or, occasionally, 'trial'. These are not proofs in a contemporary sense, nor are they intended to be; they are numerical checks on the exactitude of the answer given, an assurance that the algorithm has been successfully applied.

The second level of structure is that of the *operations* necessary to carry out the solution algorithm: addition, subtraction, multiplication, division or root extraction, for example, though Egyptian operations also include others. Each operation corresponds to one step in the algorithm. In the example of pRhind 26 (Table 2) they are, in order: multiplication, addition, division, and multiplication, while pRhind 34 (Table 3) has only one step, division. Among these operations there are two, addition and subtraction, which are left implicit or simply combined with the preceding operation as in Step 2 of pRhind 26. The terminology for the other operations is fixed: multiplication of N by M is

Table 3 (pRhind 34)

A quantity; its $\frac{1}{2}$ and its $\frac{1}{4}$ are added to it. It becomes 10.

* [Calculate starting from $1\,\frac{1}{2}\,\frac{1}{4}$, to find 10.]

\1	$1\,\frac{1}{2}\,\frac{1}{4}$
2	$3\frac{1}{2}$
\4	7
\\$\frac{1}{7}$	$\frac{1}{4}$
$\frac{1}{4}\,\frac{1}{28}$	$\frac{1}{2}$
\\$\frac{1}{2}\,\frac{1}{14}$	1

Total: the quantity is $5\,\frac{1}{2}\,\frac{1}{7}\,\frac{1}{14}$.

Trial:

\1	$5\frac{1}{2}$	|	$\frac{1}{7}\,\frac{1}{14}$
\\$\frac{1}{2}$	$2\,\frac{1}{2}\,\frac{1}{4}$	|	$\frac{1}{14}\,\frac{1}{28}$
\\$\frac{1}{4}$	$1\,\frac{1}{4}\,\frac{1}{8}$	|	$\frac{1}{28}\,\frac{1}{56}$
Total	$9\,\frac{1}{2}\,\frac{1}{8}$		
Remainder	$\frac{1}{4}\,\frac{1}{8}$		

$\frac{1}{4}$ is 14.
$\frac{1}{8}$ is 7.
Total: 21

$\frac{1}{7}$	$\frac{1}{14}$	$\frac{1}{14}$	$\frac{1}{28}$	$\frac{1}{28}$	$\frac{1}{56}$
8	**4**	**4**	**2**	**2**	**1**

expressed by 'Calculate starting from *N*, *M* times', while division of *M* by *N* is 'Calculate starting from *N*, to find *M*'. These operations have a specific grammatical structure, always in the second person singular, using either the imperative or a specific verbal grammatical form.[18]

The third level is that of the *techniques* used to effectuate each operation and varies according to the specific values in play. Once again, the carrying-out of addition or subtraction is never explicitly shown, while multiplication and division are generally indicated in detail. The working of the operation can immediately follow the statement of the operation, as in pRhind 26, or can be placed in a special section at the end, called the 'working', as in pRhind 46. In all cases these techniques operate on a double column, one of which begins with 1, the other with the number one 'starts from', multiplicand or divisor. The appropriate techniques are applied uniformly to both columns. If your operation is multiplication, you add things in the first column until the other multiplicand is reached; if your operation is division, you add things in the second column until the sum is the dividend. The answer is then given by the sum of the corresponding entries in the other column.

To see how this works, let us look in detail at a certain number of examples. If we consider the working for the multiplication of 3 by 4 in the last step of pRhind 26 we find the following. The two columns start off with 1 and 3, the initialization of the calculation with the factor 3. Since this is a multiplication the scribe seeks the other factor 4, in the first column. This is obtained from the initial 1 by a series of doublings, one of the most frequent of Egyptian techniques. Having found 4 in the first column after two successive doublings, the scribe marks the row with a check mark '\' and finds the answer, 12, in the corresponding entry of the second column which he then records. Summing up, we have the following:[19]

Col. 1	*Col. 2*	*Technique*
1	3	*initialization*
2	6	*doubling*
\4	12	*doubling*

12 will be the result.

That the level of techniques is really independent of that of operations can be seen from the fact that a technique can be utilized in any operation where it is needed. Doubling, for example, is used to carry out not only the multiplication of step four of pRhind 26 but also the division of 15 by 3 in the preceding step of the same problem:

Col. 1	*Col. 2*	*Technique*
\1	5	*initialization*
\2	10	*doubling*

3 will be the result.

Since this is a division, the dividend 15 is sought in the second column, the rows whose sum is 15 (5 + 10) are checked, and the answer, 3, found by adding the checked entries (1 + 2) in the first column.

Doubling is a powerful technique – any two integers may be multiplied using it alone – and it is often presented as though it (and its inverse, halving) were the only techniques available. There are in fact quite a number of others taught in the mathematical papyri. Take, for example, the first step of pRhind 34, the multiplication of 25 by 20. The actual working given is:

Col. 1	*Col. 2*	*Technique*
1	25	*initialization*
10	250	*decupling*
\20	500	*doubling*

The other factor, 20, is reached not by four successive doublings, which would have yielded 1–4–8–16 in the first column, of which the 4 and the 16 would have been chosen (20 = 4 + 16), but in two steps only, by first multiplying by 10 (decupling) followed by a single doubling. Its inverse, dedecupling, exists also and is used, for instance, in the second and third rows of the calculation attached to the second operation in the same problem:

Col. 1	*Col. 2*	*Technique*
1	500	*initialization*
$\frac{1}{10}$	50	*dedecupling*
$\frac{1}{10}$ of its $\frac{1}{10}$	5	*dedecupling*

Multiplication by 10 in a decimal system such as that of the Egyptian abstract numbers is of course a conceptually simple technique and serves as a useful shortcut in many calculations. Of a different order of sophistication is the technique of inversion, illustrated by the fourth step of the division operation of pRhind 34:

Col. 1	*Col. 2*	*Technique*
1	$1\ \frac{1}{2}\ \frac{1}{4}$	*initialization*
2	$3\ \frac{1}{2}$	*doubling*
4	7	*doubling*
$\frac{1}{7}$	$\frac{1}{4}$	*inversion*

The passage from 4 and 7 to $\frac{1}{7}$ and $\frac{1}{4}$ is an example of the general technique which can be expressed as:

$$\begin{array}{cc} N & M \\ \frac{1}{M} & \frac{1}{N} \end{array}$$

for any two integers N and M, i.e., if M is N times a number then $\frac{1}{N}$ is $\frac{1}{M}$th of that same number.

FRACTIONS AND TABLES

This last calculation introduces another important aspect of Egyptian mathematics, the role of fractions. They arise frequently, a product of the techniques chosen to effectuate arithmetical operations. Halving, dedecupling, and, especially inversion will often produce fractions from an integer. With but one exception – the fraction $\frac{2}{3}$ – all fractions used in Egyptian mathematics were what we term today 'unit fractions', that is, of the form $\frac{1}{N}$. But it should be borne in mind that this was not an arbitrary restriction imposed by the Egyptians; from their point of view, each fraction was associated with the inverse of an integer and so was the fraction for that integer, one part of a whole divided into N parts.[20] The correct manipulation of fractions was an important – perhaps the most important – part of the mathematical training of young scribes.

The writing of these fractions was simple. A special sign, a small dot, was placed above the hieratic integer associated with the fraction; the fraction which we write $\frac{1}{N}$ was thus written N with a dot above (see Figures 3 and 4 for examples). In the rare hieroglyphic transcriptions made by the Egyptians themselves this was interpreted as the sign for a mouth, */r/*, though the ordinary hieratic sign for this was quite different. Sums of fractions which could not be simplified to a single unit fraction were necessarily kept as sums and written side by side in juxtaposition. Thus the Egyptian $1\frac{1}{2}\frac{1}{4}$ represented our $1 + \frac{1}{2} + \frac{1}{4} = 1\frac{3}{4}$.

No new operations are necessary for calculating with fractions; like integers, they can serve as factors in products, divisors and dividends in division. The techniques developed for integers – doubling, halving, decupling, dedecupling – continue to work correctly with fractions. But there is a difference. If halving poses no new problem for a fractional number – $\frac{1}{N}$ goes to $\frac{1}{2N}$ – the same can not be said for doubling. Let us look at the complete working for the division of pRhind 34, the division of 10 by $1\frac{1}{2}\frac{1}{4}$:

Col. 1	*Col. 2*	*Technique*
$\backslash 1$	$1\frac{1}{2}\frac{1}{4}$	*initialization*
2	$3\frac{1}{2}$	*doubling*
$\backslash 4$	7	*doubling*
$\backslash\frac{1}{7}$	$\frac{1}{4}$	*inversion*
$\frac{1}{4}\frac{1}{28}$	$\frac{1}{2}$	*doubling*
$\backslash\frac{1}{2}\frac{1}{14}$	1	*doubling*

Total: the quantity is $5\frac{1}{2}\frac{1}{7}\frac{1}{14}$.

We see that the use of doubling in the second, third and final entries poses no problem, but in the fifth step it is perhaps surprising. The second column there simply doubles $\frac{1}{4}$ to make $\frac{1}{2}$, but column 1 yields $\frac{1}{4}\frac{1}{28}$ for the double of $\frac{1}{7}$. The point is that since only unit fractions are admitted by the Egyptians in the writing of numbers, twice $\frac{1}{7}$ cannot be '$\frac{2}{7}$' even $\frac{1}{7}\frac{1}{7}$ is not permitted. Instead two or more distinct fractions must be found whose sum is twice that of the fraction to be doubled. Now $\frac{1}{4}+\frac{1}{28}$ is indeed equal to twice $\frac{1}{7}$, but how was this done by the scribe? How was the doubling of an 'odd' fraction in general effected?

We arrive here at an intrinsically difficult domain of calculation in Egyptian mathematics. There is no question of this being a reflection of the 'crude methods' or 'blind empiricism' of the Egyptian calculator, as a number of modern commentators have insisted. All systems of numeration – including our own – have such domains of difficulty; only their nature differs from one system to another.[21] The solution to the difficulty is, in general, always the same: one performs the calculations once and for all and commits them to writing in a form that can be used quickly and simply to find the result. That is, one constructs *tables.*

The doubling of an 'odd' fraction lies squarely in this domain of difficulty; consequently the Egyptian scribe used a table. In fact, we have his table; it constitutes a third of the recto of the pRhind and is placed at the very beginning of the papyrus. The process of doubling was so central to Egyptian calculations and the presence of fractions so ubiquitous that we even find a second, shorter copy of this table among the rare mathematical papyri, excavated with the fragments from el-Lahun (Figure 4).[22] A simplified translation of this last is given in

Figure 4 Illustration from the Kahun Papyrus UC 32159. Reproduced with the kind permission of the Petrie Museum of Egyptian Archaeology, London.

Table 4.[23] The table represents (as the more complete table in pRhind makes clear) the constitution of the number 2 from each of the odd integers N in turn (termed 'calling 2 out of N' in the pRhind table), which is equivalent to the doubling of the fraction $\frac{1}{N}$. For the doubling of any 'odd' fraction then the scribe has simply to refer to his table.

There are some new techniques that involve fractions in an essential way; we have already seen one, inversion. There is a further operation specific to fractions, an example of which we see at work in pRhind 46, in the last entry of the second step of the algorithm ($\frac{2}{3}$ of $5 = 3\frac{1}{3}$). This is the calculation of $\frac{2}{3}$ of a number and is performed directly.[24] If, as here, taking $\frac{2}{3}$ of an integer poses no particular problem for the scribe, calculating $\frac{2}{3}$ of a fraction, an operation which occurs in a number of problems, is more difficult; the scribe of pRhind had recourse to a table which he inserted at the beginning of the verso of the papyrus.

There is finally a pair of operations where the presence of fractions makes an important difference; simple addition and subtraction. If the addition of two integers was considered to need no explicit techniques, once fractions were involved the situation changes. The example of pRhind 34 shows clearly how addition of fractional numbers was done. In the verification that the division of 10 by $1\frac{1}{2}\frac{1}{4}$ does indeed yield $5\frac{1}{2}\frac{1}{7}\frac{1}{14}$, one multiplies this last number by $1\frac{1}{2}$ $\frac{1}{4}$ to see if 10 is obtained.

Col. 1		*Col. 2*		*Technique*
\1	$5\frac{1}{2}$	|	$\frac{1}{7}\frac{1}{14}$	*initialization*
\\$\frac{1}{2}$	$2\frac{1}{2}\frac{1}{4}$	|	$\frac{1}{14}\frac{1}{28}$	*halving*
\\$\frac{1}{4}$	$1\frac{1}{4}\frac{1}{8}$	|	$\frac{1}{28}\frac{1}{56}$	*halving*
Total	$9\frac{1}{2}\frac{1}{8}$			
Remainder	$\frac{1}{4}\frac{1}{8}$.			

The two successive halvings are sufficient to create the second factor, $1\frac{1}{2}\frac{1}{4}$, in the first column. Furthermore, the fact that halving is the only operation means that no recourse to tables is necessary in this part. To determine their sum the scribe separates each entry of the second column in two parts by a vertical

Table 4 (pUC 32159)

2	3	$\frac{2}{3}$		
	5	$\frac{1}{5}$	$\frac{1}{15}$	
	7	$\frac{1}{4}$	$\frac{1}{28}$	
	9	$\frac{1}{6}$	$\frac{1}{18}$	
	11	$\frac{1}{6}$	$\frac{1}{66}$	
	13	$\frac{1}{8}$	$\frac{1}{52}$	$\frac{1}{104}$
	15	$\frac{1}{10}$	$\frac{1}{30}$	
	17	$\frac{1}{12}$	$\frac{1}{51}$	$\frac{1}{68}$
	19	$\frac{1}{12}$	$\frac{1}{76}$	$\frac{1}{114}$
	21	$\frac{1}{14}$	$\frac{1}{42}$	

line: integer and dimidiated fractions on one side, all other fractions on the other. The sum of integers and dimidiated fractions being simple to calculate, the scribe sums just this part as his 'total' with the remainder necessary to complete 10 – in dimidiated fractions – labeled as a 'remainder'. The next question for the scribe is: Do the fractions on the other side of the vertical line sum to $\frac{1}{4}\frac{1}{8}$? The answer lies in the next part of the working. First the scribe chooses an integer for which the fractions which need to be summed will be integers or integers plus simple fractions; here his choice is 56.[25] Then his target fractions, $\frac{1}{4}$ and $\frac{1}{8}$, are written as $\frac{1}{4}$ of $56 = 14$ and $\frac{1}{8}$ of $56 = 7$. Their sum is 21; this is the meaning of the three lines:

$\frac{1}{4}$ is 14

$\frac{1}{8}$ is 7

Total 21

It only remains to verify that the fractions that need to be summed, expressed as parts of 56 add up to 21, and this is the purport of the final two lines of the problem. Precisely the same operation is carried out, with the equivalent parts out of 56 being written, in red ink, below the fraction in question:

$$\begin{matrix} \frac{1}{7} & \frac{1}{14} & \frac{1}{14} & \frac{1}{28} & \frac{1}{28} & \frac{1}{56} \\ \mathbf{8} & \mathbf{4} & \mathbf{4} & \mathbf{2} & \mathbf{2} & \mathbf{1} \end{matrix}$$

Their sum is 21, just like $\frac{1}{4}\frac{1}{8}$, and the verification is complete.

It should come as no surprise to learn that we also possess tables for the addition of fractions. A leather roll, once also part of the Rhind collection and now at the British Museum, carries in duplicate just such a table of sums of fractions (B.M. 10250; Glanville, 1927).

Given their central importance as the domain of intrinsic difficulty of Egyptian mathematics, it is natural to find fractions occurring precisely on the two Middle Kingdom mathematical texts which are not textbooks. They are wooden tablets on which there are exercises in the writing of proper names together with practice in the transformation between abstract fractions and capacity units (Cairo CG25367 and 25368; Daressy, 1901: 95–96 and pl. 52–53).[26] The small ostracon found in a fifteenth-century BC tomb near Thebes (Ostracon n° 153; Hayes, 1942: 29–30), one of the two examples of mathematical texts between those of the Middle Kingdom we have been looking at and the demotic mathematical texts of the Greco-Roman period,[27] is a school exercise. Found among other ostraca containing excerpts from the literary works that were a staple of Egyptian education, it is again a drill in the manipulation of fractions[28].

Tables and exercises in their use, as well as the solution algorithms of the problem texts, make up what we call Egyptian mathematics. The 'logic of transmission' at work here was the creation of a network of example-types, concrete problems whose solutions and articulation provided each student with a network of problem-resolution prototypes. Learning mathematics was learning how to select an algorithm by interpolating the new problems into the acquired network and then how to pick wisely the arithmetic techniques that would be necessary to carry out that algorithm for the specific numerical data of the new problem. If Egyptian mathematics is viewed not as a poor simulacrum of our proof-oriented mathematics but on its own ground, it will be seen to be a rational and practical response to the needs of Egyptian society.

Mathematics even provided a model for other domains of 'rational practice' in Egypt; a number of indices allow us to connect the corpus of Egyptian mathematical texts with other classes of documents from the Nile valley. The verbal form used in the algorithms is a comparatively rare one, seldom found outside mathematical, medical and a certain class of magical texts (Vernus, 1990: 64). The division of a domain into the two complementary classes of problem texts and tables exists for mathematics and medicine – and at least partially in the areas of law codes and calendrical and dream divination.[29] Whatever reaction this juxtaposition may inspire in us, it delimits a range of domains which for the ancient Egyptians constituted a privileged access to knowledge of the world around them.

In the place of mathematics in society, as in the algorithmic structure of the mathematical problem texts, the evidence from Egypt rejoins that from Mesopotamia, China and India, as well as from one of the mathematical traditions in Greece. This should in no way be construed as an argument for borrowing or for the 'diffusion' of mathematical knowledge from a unique center. The lack of documented intellectual contacts among these different cultures during their formative periods for mathematics militates against such a conclusion. More importantly, they differed in their choices of number system, of modes of writing, of arithmetical operations and of the techniques used to carry them out. But such widespread agreement is an argument for the efficacy of both the algorithmic approach and the transmission by networks of typical examples.

If mathematics was a model for privileged knowledge in Egypt, a model shared by almost all pre-modern cultures, why then the common attitude summed up in a recent handbook on the 'legacy of Egypt' which describes the supposed disarray of the Egyptian scribe in the face of elementary arithmetical calculations: 'we can see how hemmed in he was by his numerical system, his crude methods, and his concrete mode of thought.' (Harris, 1971: 40)? In fact, Egyptian mathematical texts had the misfortune to be rediscovered in the late nineteenth century, a period in which contemporary mathematical practice was much preoccupied by questions of proof and rigor, areas which were not part of Egyptian concerns, at least in the pedagogical domain. It was then a commonplace to see (one strand of) Greek mathematical thought as the unique precursor to any real activity in this domain.[30] At the beginning of the twenty-first century, other concerns, those of algorithms and effectivity, have regained center stage in large areas of mathematics and, with them, the possibility of seeing Egyptian mathematics with new eyes. We can perhaps more easily appreciate their mathematics for what it meant and how it was used by the Egyptians themselves, with their own needs and their own judgments, now that these have partially rejoined, even if temporarily, our own.

NOTES

[1] The dominance of funerary remains long raised doubts concerning the very existence of urban centers in Egypt (e.g. Wilson, 1951: 34). But see now Bietak, 1979 and Valbelle, 1990.

[2] It is true that the Rhind mathematical papyrus is later than the Middle Kingdom but it is a copy of a Middle Kingdom document.

[3] For an analysis of the nature of the archaic Mesopotamian texts, see Nissen *et al.*, 1994 and Ritter 1999.

[4] Studies which have seen in the birth of writing in Egypt a reflection of artistic, religious or pure ideological needs have not very seriously taken into account the nature of the texts themselves. Nor is this surprising given the almost total lack of interest within the Egyptological community for early texts. These lack after all the religious and politico-historical content which is the focus of interest of the majority of Egyptologists. It took, for example, nearly three quarters of a century after their discovery for the first publication of third-millennium papyri, those of the Abusir archive (Posener-Kriéger and de Cenival, 1968 and Posener-Kriéger, 1976)! The continuing publication of Old Kingdom papyrus archives by Posener-Kriéger, now interrupted by her death, along with the earlier pioneering work by Möller (1909), Sethe (1939), Schott (1951) and Kaplony (1963–1966), represented, until the resurgence of interest in the 1980s, practically the whole of synthetic publications on this subject.

[5] The earliest grammatically constructed phrases, i.e., records of a standard speech act, known to me are quotations of gods addressing King Djoser (third dynasty) in fragmentary bas-reliefs from a temple at Heliopolis. See Kahl *et al.*, 1995: 114–119.

[6] This, and other statements about third-millennium metrology and writing made here, will be discussed in my forthcoming book on the subject.

[7] The papyri were in the process of being published by Paule Posener-Kriéger up until her death. The only completely published archives are from the funerary temples of king Nefer-ir-ka-Ra (Posener-Kriéger and de Cenival, 1968 and Posener-Kriéger, 1976) and his wife Khenet-kaus (Posener-Kriéger, 1995) at Abusir. Still unpublished, except for scattered articles by Posener-Kriéger and others on points of detail, are the papyri from the temple of king Ra-neferef at Abusir and from disparate royal sources at Saqqara.

[8] See for now the general discussions in Posener-Kriéger, 1972 and Burkard and Fischer-Elfert, 1994.

[9] Note however that, alone, T. Eric Peet, in the second – and best – of the editions of the Rhind papyrus offered a healthy skepticism as to the significance of this for the origin of the signs (Peet, 1923a: 25–26).

[10] Here again the correspondence with the Mesopotamian case is striking, see Green and Nissen, 1987, Nissen *et al.*, 1994 and Ritter, 1999.

[11] The most influential or most recent surveys of Egyptian mathematics, aside from individual editions of texts, are Vogel, 1929; Neugebauer, 1934; Gillings, 1972; Couchoud, 1993. See also the forthcoming edition of the problem texts by Annette Imhausen.

[12] Properly Papyri 10057 and 10058 of the British Museum in London (plus fragments now in the Brooklyn Museum). For a detailed study, the edition of Peet (1923) remains the most trustworthy. For photographs of the newly-cleaned pRhind see Robins & Shute, 1987.

[13] Examples provided here are pRhind 26 (Table 2) and pRhind 34 (Table 3). Translations from the Egyptian in this article follow the following transcription rules:

Roman type represents black ink in the original; bold type, red ink;

Egyptian words are transcribed in italics;

Bullets (my addition) represent distinct steps in a solution algorithm.

Words added in square brackets are my additions, based on other sections of the papyrus.

[14] Of course this is the case even today for those countries like Great Britain or the United States which still use traditional metrological units. Indeed a great difference in teaching ancient mathematics to students in either type of culture today lies in their very different reactions to this fact. Students from 'metric' cultures find it incredible that such irrational time-wasting could ever have been tolerated; students from 'traditional' metrological cultures find it natural and obvious that this kind of work should constitute a good deal of arithmetical practice.

[15] This problem – finding the linear dimensions of a grain container of rectangular cross-section and of a given volume – is, of course, underdetermined. The preceding problems in the papyrus show that an additional constraint, that the length and width of the container are to be taken as 10 cubits each, is here to be understood.

[16] Indeed the *khar*, appearing only at the end of the Old Kingdom, was very likely introduced precisely in order to create a numerically simple link between capacity and length measures.

[17] The one case of a general algorithm (pRhind 61B) is discussed in Ritter, 1995a: 69–70.

[18] A good discussion of this form, called the *sedjem-kher-f* form in modern grammars of ancient Egyptian, is to be found in Vernus, 1990: 61–84.

[19] Words in italic, other than transcriptions of the Egyptian, represent my additions.

[20] For an interesting discussion of this point see Caveing, 1992. For the question of the origin of Egyptian fractions see Ritter, 1992 and for an analysis of fractions in general see Benoit, Chemla and Ritter, 1992.

[21] For a comparison of these areas of difficulty in the Egyptian and Babylonian systems of calculation see Ritter, 1995a.

[22] University College 32159. Published as papyrus 'Kahun IV.2' in Griffith, 1898: pl. VIII.

[23] The part that has been left out in this translation are those numbers which serve as a verification of the validity of the doubling. As an example, the full line which doubles $\frac{1}{5}$ actually reads: $5\ \frac{1}{3}\ 1\frac{2}{3}\ \frac{1}{15}\ \frac{1}{3}$, that is, '(2 times) $\frac{1}{5}$ is $\frac{1}{3}$ – ($\frac{1}{3}$ of 5 is) $1\frac{2}{3}$ – and $\frac{1}{15}$ – ($\frac{1}{15}$ of 5) is $\frac{1}{3}$ (– and $1\frac{2}{3}+\frac{1}{3}=2$)'. This is the structure of the pRhind table as well and is an indication of the importance that numerical verification played in Egyptian mathematics.

[24] Here we have the explanation for the placement of the $\frac{2}{3}$ operation (which corresponds to a change of capacity units) *after* the calculation of length and breadth; it is much simpler numerically to compute $\frac{2}{3}$ of 5 than of the 500 *khar* with which the calculation began. Since the system used is the abstract one, there is no need to keep track of intermediate units so long as the final ones are correct.

[25] Note that this is not a choice of the 'smallest common denominator' as we use it in summing fractions. Not only does such a concept have no meaning in the context of Egyptian fractions, but no necessity was felt for the resulting 'numerators' to be integers.

[26] Though published at the beginning of the twentieth century, their nature was not understood until the publication of T. Eric Peet, 1923.

[27] The problem texts have been published by Richard Parker in Parker, 1972 and 1975. A number of exercise ostraca from this period are to be found in Wångstedt, 1958: 70–71; Belli and Costa, 1981; Devauchelle, 1984: n° 3; and Bresciani *et al.*, 1983: n° 9, 17, 22, 30.

[28] The other ostracon, Turin N. 57170 (López, 1980:pl. 75), from the late New Kingdom period and found at Deir el-Medina, is too fragmentary to determine its exact nature, mathematics exercise or table.

[29] See the treatment of a similar grouping for Mesopotamia in Ritter, 1995b.

[30] For an analysis of such a truncated view of Greek and other mathematics, see Ritter and Vitrac, 1998. For the ways in which current attitudes towards mathematics determine judgments of past mathematics, see Goldstein 1995.

BIBLIOGRAPHY

Belli, Giula and Barbara Costa. 'Una tabellina aritmetica per uso elementare scritta in demotico.' *Egitto e Vicino Oriente* 4: 195–200, 1981.

Benoit, Paul, Karine Chemla and Jim Ritter, eds. *Histoire des fractions, fractions d'histoire* (Science Networks 10). Basel: Birkhäuser, 1992.

Bietak, Manfred. 'Urban archæology and the "town problem".' In *Egyptology and the Social Sciences*, Kent Weeks, ed. Cairo: American University in Cairo Press, 1979, pp. 97–144.

Bresciani, Edda, Sergio Pernigotti and Maria C. Betrò. *Ostraka demotici da Narmuti. I. (nn. 1–33).* Pisa: Giardini, 1983.

Burkard, Günter and Hans-Werner Fischer-Elfert. *Ägyptische Handschriften. Teil 4*, E. Lüddeckens, ed. Stuttgart: F. Steiner, 1994.

Caveing, Maurice. 'Le statut arithmétique du quantième égyptien.' In *Histoire des fractions, frac-*

tions d'histoire, Paul Benoit, Karine Chemla and Jim Ritter, eds. Basel: Birkhäuser, 1992, pp. 39–52.

Couchoud, Sylvia. *Mathématiques égyptiennes*. Paris: Le Léopard d'Or, 1993.

Daressy, Georges. *Ostraca (CGC 25001–25385)*. Cairo: Institut français d'archéologie orientale, 1901.

Devauchelle, Didier. 'Remarques sur les méthodes d'enseignement du démotique.' In *Grammata Demotika*, Heinz-J. Thiessen and Karl-T. Zauzich, eds. Würzburg: G. Zauzich Verlag, 1984, pp. 47–59.

Gardiner, Alan. *Egyptian Grammar* (3rd edition). London: Oxford University Press, 1957.

Gillings, Richard J. *Mathematics in the Time of the Pharaohs*, Cambridge, Massachusetts: MIT Press, 1972, reprinted New York: Dover, 1982.

Glanville, Stephen R. K. 'The mathematical leather roll in the British Museum.' *The Journal of Egyptian Archæology* 13: 232–239, 1927.

Goedicke, Hans. *Königliche Dokumente aus dem alten Reich* (Ägyptologische Abhandlungen 14). Wiesbaden: O. Harrassowitz, 1967.

Goedicke, Hans. *Die privaten Rechtsinschriften aus dem alten Reich* (Wiener Zeitschrift für die Kunde des Morgenlandes. Beiheft 5). Vienna: Verlag Notring, 1970.

Goldstein, Catherine. *Un théorème de Fermat et ses lecteurs*. St. Denis: Presses Universitaires de Vincennes, 1995.

Green, Margaret W. and Hans J. Nissen. *Zeichenliste der archaischen Texte aus Uruk* (ATU 2). Berlin: Gebr. Mann, 1987.

Griffith, Francis Llewellyn. *Hieratic Papyri from Kahun and Gurob* (2 vols.). London: B. Quaritch, 1898.

Harris, J. R. 'Mathematics and astronomy.' In *The Legacy of Egypt* (2nd edition), J. Harris, ed. Oxford: Clarendon Press, 1971, pp. 27–54.

Hayes, William C. *Ostraca and Name Stones from the Tomb of Sen-Mut*. New York: Metropolitan Museum of Art, 1942.

Kahl, Jochem, Nicole Kloth and Ursula Zimmermann. *Die Inschriften der 3. Dynastie. Eine Bestandaufnahme* (Ägyptologische Abhandlungen 56). Wiesbaden: O. Harrassowitz, 1995.

Kaplony, Peter. *Die Inschriften der ägyptischen Frühzeit* (Ägyptologische Abhandlungen 8; 3 vols.). Wiesbaden: O. Harrassowitz, 1963.

Kaplony, Peter. *Die Inschriften der ägyptischen Frühzeit. Supplement* (Ägyptologische Abhandlungen 9). Wiesbaden: O. Harrassowitz, 1964.

Kaplony, Peter. *Kleine Beiträge zu den Inschriften der ägyptischen Frühzeit* (Ägyptologische Abhandlungen 15). Wiesbaden: O. Harrassowitz, 1966.

Lopez, Jesús. *Ostraca ieratici: N. 57093–57319*. Milan: Istituto Editoriale Cisalpino-La Goliardica, 1980.

Möller, Georg. *Hieratische Paläographie. I. Bis zum Beginn der achtzehnten Dynastie*. Leipzig: J. C. Hinrichs, 1909.

Möller, Georg. 'Die Zeichen für die Bruchteile des Hohlmaßes und das Uzatauge.' *Zeitschrift für ägyptische Sprache und Altertumskunde* 48: 99–101, 1911.

Moussa, Ahmed M. and Hartwig Altenmüller. *Das Grab des Nianchchnum und Chnumhotep* (Archäologische Veröffentlichungen des Deutschen Archäologischen Instituts. Abteilung Kairo 21). Mainz: P. von Zabern, 1977.

Neugebauer, Otto. *Vorlesungen über Geschichte der antiken mathematischen Wissenschaften. I. Vorgriechische Mathematik*. Berlin: Springer Verlag, 1934. [An essentially unchanged 2nd edition was issued in 1969.]

Nissen, Hans, Peter Damerow and Robert K. Englund. *Archaic Bookkeeping. Writing and Techniques of Economic Administration in the Ancient Near East*. Chicago: University of Chicago Press, 1994.

Peet, T. Eric. *The Rhind Mathematical Papyrus*. London: University Press of Liverpool/Hodder & Stoughton, 1923a, reprinted Nendeln (Liechtenstein): Kraus Reprint, 1970.

Peet, T. Eric. 'Arithmetic in the Middle Kingdom.' *The Journal of Egyptian Archæology* 9: 91–95, 1923b.

Posener-Kriéger, Paule. 'Les papyrus de l'Ancien Empire.' In *Textes et langages de l'Égypte pharao-*

nique (Bibliothèque d'Étude 64; 3 vols.), Serge Sauneron, ed., vol. II. Cairo: Institut français d'archéologie orientale, 1972, pp. 25–35.

Posener-Kriéger, Paule. *Les Archives du temple funéraire de Néferirkarê-Kakaï* (Bibliothèque d'Étude 65; 2 vols.). Cairo: Institut français d'archéologie orientale, 1976.

Posener-Kriéger, Paule. 'Les mesures de grain dans les papyrus de Gébélein.' In *The Unbroken Reed. Studies in the Culture and Heritage of Ancient Egypt in Honour of A. F. Shore* (EES Occasional Publication 11), Christopher Eyre *et al.*, eds. London: Egypt Exploration Society, 1994, pp. 269–272.

Posener-Kriéger, Paule and Jean-Louis de Cenival. *The Abu Sir Papyri* (Hieratic Papyri in the British Museum. Fifth Series). London: British Museum, 1968.

Ritter, James. 'Metrology and the prehistory of fractions.' In *Histoire des fractions, fractions d'histoire*, Paul Benoit, Karine Chemla and Jim Ritter, eds. Basel: Birkhäuser, 1992, pp. 19–34.

Ritter, James. 'La médecine en – 2000 au Proche-Orient: une profession, une science?' In *Maladies, médecines et sociétés*, F.-O. Touati, ed., vol. II. Paris: L'Harmattan/Histoire au Présent, 1993, pp. 105–116.

Ritter, James. 'Measure for measure: mathematics in Egypt and Mesopotamia.' In *A History of Scientific Thought*, Michel Serres, ed. Oxford: Blackwell, 1995a, pp. 44–72.

Ritter, James. 'Babylon – 1800.' In *A History of Scientific Thought*, Michel Serres, ed. Oxford: Blackwell, 1995b, pp. 17–43.

Ritter, James. 'Metrology, writing and mathematics in Mesopotamia.' In *Calculi 1929–1999* (Prague Studies in the History of Science and Technology 3), Jaroslav Folta, ed. Prague: National Technical Museum, 1999, pp. 215–242.

Ritter, James and Bernard Vitrac. 'La pensée orientale et la pensée grecque.' In *L'Encyclopédie philosophique universelle. IV. Le Discours philosophique*, Jean-François Mattei, ed. Paris: Presses Universitaires de France, 1998, pp. 1233–1250.

Robins, Gay and Charles Shute. *The Rhind Mathematical Papyrus: An Ancient Egyptian Text.* London: British Museum, 1987.

Schott, Siegfried. *Hieroglyphen, Untersuchungen zum Ursprung der Schrift.* Mainz: Verlag der Akademie der Wissenschaften und Literatur, 1951.

Sethe, Kurt. *Vom Bilde zum Buchstaben. Die Entstehungsgeschichte der Schrift.* Leipzig: J. C. Hinrichs, 1939.

Struve, V. V. *Mathematische Papyrus des Staatlichen Museums der Schönen Künste in Moskau* (Quellen und Studien zur Geschichte der Mathematik A1). Berlin: Springer Verlag, 1930.

Valbelle, Dominique. 'Égypte pharaonique.' In *Naissance des cités*, Jean-Louis Huot, Jean-Paul Thalmann and Dominique Valbelle, eds. Paris, Nathan, 1990, pp. 255–322.

Vernus, Pascal. *Future at Issue. Tense, Mood and Aspect in Middle Egyptian: Studies in Syntax and Semantics* (Yale Egyptological Studies 4). New Haven: Yale University Press, 1990.

Vogel, Kurt. *Die Grundlagen der ägyptischen Arithmetik in ihrem Zusammenhang mit der 2/n-Tabelle.* Munich: Beckstein, 1929, reprinted Wiesbaden: Sändig, 1970.

Wångstedt, Sten V. 'Aus der Ostrakasammlung zu Uppsala. III.' *Orientalia Suecana* 7:70–77, 1958.

Wilson, John A. *The Burden of Egypt.* Chicago: University of Chicago Press, 1951.

JACQUES SESIANO

ISLAMIC MATHEMATICS

HERITAGE

Less than 150 years after Muḥammad's flight to Medina in 622 (which marked the beginning of the Islamic era), the vast territories which in the meantime had been conquered by the Arabs received a new capital: Baghdad, founded in 762 by Caliph al-Manṣūr. This was not just a political event, for among the institutions to be established in Baghdad was the 'house of wisdom' (*bayt al-ḥikma*), a kind of academy similar to the one created by the Greeks a millennium before in Alexandria which had remained the foremost scientific centre until late antiquity. The first task of the Baghdad Academy was to collect and synthesize the scientific knowledge available at that time.

For mathematics in particular, notable contributions came from three civilizations. The most ancient one was Mesopotamia. By the time of the disappearance of cuneiform writing, at the beginning of our era, the main mathematical and astronomical discoveries had reached Greece and been integrated into Greek science. Greek science itself was no longer alive but was still available through manuscripts, and translating and commenting on them became a primary task at the Baghdad Academy during the ninth century. Finally, Indian science was at this time very much alive and growing, and direct contact occurred, since Arabic chronicles of the time report a visit by Indian scholars to Baghdad during the early years of the Academy.

Mesopotamian heritage

From cuneiform texts in Akkadian going back to about 1800 BC, adapted from earlier Sumerian documents, it appears that by 2000 BC Sumerian mathematicians were able to solve numerically some problems and equations of the first two degrees with positive rational solutions. Later mathematical texts in cuneiform writing do not show any significant increase in mathematical knowledge, but considerable observation work was being carried out in astronomy. By the time of Hipparchus (second century BC), this information had reached Greece, and apparently continued to be used, since Ptolemy (*ca.* AD 150) mentions in section III.7 of his *Almagest* that from the time of Nabonassar

H. Selin (ed.), Mathematics Across Cultures: The History of Non-Western Mathematics, 137–165.

(747 BC) 'the ancient observations are, on the whole, preserved down to our time' (Toomer, 1984: 166). Both mathematical and astronomical computations used the sexagesimal base, which was already in use, together with the decimal one, around 3000 BC in Mesopotamia, even before the birth of cuneiform writing. When Mesopotamian astronomical records became available to the Greeks, they adopted, though for astronomical use only, the sexagesimal system, which thus remained alive and, through Greek transmission, was adopted by Indian and Islamic astronomers and has survived to the present day.

Indian heritage

The first major Indian mathematical and astronomical works date from the sixth and seventh centuries. At that time, the system used to designate numbers was already the positional one, with ten signs including the zero for the empty place.[1] Indian mathematicians had also developed the arithmetical operations adapted to that system, to the extent that numerical reckoning became an essential part of arithmetical and algebraic problems dealing with the practical needs of daily life and trade. In astronomy, some technical terms and the use of geometrical models to represent the planetary movements point to a Greek influence. However, unlike the Greek use of chords in plane trigonometry, the Indians used half-chords and thus introduced the sine and cosine. Since the Indian heritage arrived at an early date, Islamic science was from the outset marked by a knowledge of Indian arithmetic and astronomy, though in the latter case for a transitory period only.

Greek heritage

This limitation in time is a consequence of the rapidly growing influence of Greek science. A general feature of Islamic science is that having adopted the Greek scientific frame involving definitions, theorems and demonstrations, it came to have the same strength and precision as ancient research. More specific influences in mathematics gave rise to the use of geometrical demonstrations in algebra (as well as in the physical sciences), to the resolution of indeterminate equations of the second degree, to the study of number theoretical properties of integers, and to further inquiry into and extension of geometrical solutions of problems not solvable by means of ruler and compass alone – that is, problems leading algebraically to equations of a higher degree than the second. (The classical geometry of Euclid allows us, apart from performing the four arithmetical operations, the extraction of the square root only.)

Among the main works translated were, for mathematics, Euclid's *Elements of Geometry*, treatises by Archimedes on geometry and mechanics, half of Diophantus's *Arithmetica*, the largely accessible *Introduction to Arithmetic* by Nicomachus, the *Conics* of Apollonius, some treatises on spherics, and the main works of Ptolemy on astronomy and optics which also contain mathematical material. Several texts were translated more than once, or at least the initial translation was revised or commented on. This indicates the great care with which ancient Greek texts were treated, as do the efforts of some scholars anxious to search for better manuscripts when the available ones were defective (Apollonius (Toomer), 1990: xviii and 620–629). Indeed, Greek science was looked upon with the greatest admiration, to the point that a reference to 'the Ancients' (*al-qudamā, al-mutaqaddimūn*) had almost the weight of a formal proof.

The acquisition of the available Greek knowledge was completed during the ninth century. Islamic mathematics then entered its period of greatest productivity, which was to last into the fifteenth century, peaking from the tenth to the twelfth.

ARITHMETIC

Arithmetical reckoning

The first description of the Indian system of numerals and its use in arithmetical operations appears around 820 with the *Arithmetic* by Muḥammad al-Khwārizmī, a scholar of Persian descent living in Baghdad. Only two Latin versions, made in Spain during the twelfth century, survive today (Folkerts, 1997). Its purpose, as stated by the author in his introduction, is to explain 'the numbering of the Indians by means of nine signs'. The zero (which is generally not included among the digits) is denoted by a 'small circle'. After that follows the explanation of this new system of numeration, in which the value of the sign depends upon the place it occupies within the number. The author goes on to describe how to perform the operations of addition, subtraction, multiplication and division, as well as the particular cases of doubling and halving. After that comes square root extraction. All these operations are explained first for integers, then for proper fractions, and finally for sexagesimal fractions. This was to remain more or less the contents of most arithmetical textbooks in Arabic, but some books dropped the study of sexagesimal fractions (as being more relevant to astronomy) while others added the method for extracting cube roots.

A few words in use today bear testimony to the transmission of the Indian system to Europe by the Arabs. Thus, al-Khwārizmī's name was rendered in Latin as *Algorizmus*, through a misreading (the letters *kh* and *g* differ in Arabic merely by the position of a dot above or below the same sign), and because of the similarity between *t* and *z* in mediaeval writing, the transcription *Algoritmus* also occurred. But since in later times the true origin of the word was forgotten and a Greek one was conjectured, it came to be written as *algorithmus* and in the end to have a broader meaning than just that of reckoning with the Indian system of numerals. As to the word used in Sanskrit for zero, *shunya*, it was appropriately translated into Arabic as *ṣifr*, which means 'void'. In twelfth century mathematical works in Latin, *ṣifr* was transliterated both as *cifra* and *zefirum*. The first gave rise to the modern *cipher* and to *chiffre* and *Ziffer*, which in modern times came to characterize the whole system; the second, in an abbreviated Italian form, became *zero* (Menninger, 1934 and 1969).

The earliest arithmetic still extant in Arabic is that written by al-Uqlīdisī in Damascus in 952/53 (Saidan, 1973 and 1978). This is a sizeable compendium, in which the author also mentions other contemporary number systems (such as the ancient Greek one, still being used by the Byzantines). He also explains the changes the use of paper and ink made to computations with the Indian system (originally performed on the dust abacus). The fractions considered then were, unless sexagesimal, represented as the quotients of two integers; decimal fractions first occur in al-Uqlīdisī's treatise. He uses a mark placed over the last integral unit in order to indicate the separation from the subsequent, decimal part. However, although he does not miss any opportunity to insist on the quality and originality of his book in comparison with others', he does not claim for himself this, the most important novelty of his *Arithmetic*. We may infer from this that the invention is probably not his. In any case, it was apparently not widely in use, for decimal fractions rarely appear prior to the works of the Persian Ghiyāth al-Dīn al-Kāshī (fl. 1415), who explains how to use them and expresses in this way his approximation of π (see the section on geometry below). Their first, independent and isolated, appearance in Europe was in the *Compendion de lo abaco* of Fr. Pellos (1492), but they only came into real use a century later in the works of Stevin and Napier.

Root extraction

We mentioned above that arithmetical treatises usually taught how to extract the exact square, and sometimes also the exact cube root. For non square and non cubic integers, various approximation formulae were applied, such as

$$\sqrt[m]{N} = a + \frac{N - a^m}{(a+1)^m - a^m}$$

with $m = 2$ or 3 and $a^m < N < a^{m+1}$; thus a is the largest integral square or cube contained in N. This formula is also applicable to higher roots provided one has determined a.

The principle for the extraction of both square and cube roots is the same, and was extended in the Islamic countries to higher root extraction. Kāshī explains in his *Key of Arithmetic* how to extract the fifth root of 44240899506197 or, rather, how to find the largest possible integral fifth power a contained in it (Kāshī, 1977). The main steps of the procedure are as follows:

1. Divide the number, from the right side, into groups of five digits (according to the index of the root): 4424′08995′06197. Since there are three groups (it is irrelevant whether the last one is complete or not), the root will have three digits and thus be of the form $\alpha \cdot 10^2 + \beta \cdot 10 + \gamma$.
2. Let us first develop $(\alpha \cdot 10^2 + \beta \cdot 10 + \gamma)^5$. We shall find

$$10^{10} \cdot \alpha^5 + 10^9(5\alpha^4\beta + \alpha^3\beta^2) + 10^8(5\alpha^4\gamma + 2\alpha^3\beta\gamma + \alpha^2\beta^3)$$
$$+ 10^7(\alpha^3\gamma^2 + 3\alpha^2\beta^2\gamma) + 10^6(3\alpha^2\beta\gamma^2 + 2\alpha\beta^3\gamma + 5\alpha\beta^4)$$
$$+ 10^5(\alpha^2\gamma^3 + 3\alpha\beta^2\gamma^2 + \beta^5) + 10^4(2\alpha\beta\gamma^3 + 5\beta^4\gamma + \beta^3\gamma^2)$$
$$+ 10^3 \cdot \beta^2\gamma^3 + 10^2 \cdot 5\alpha\gamma^4 + 10 \cdot 5\beta\gamma^4 + \gamma^5.$$

3. Let us look for the greatest possible α with $\alpha^5 < 4424$. We find $\alpha^5 = 3125$ and shall thus replace α by 5, which gives

$$10^{10} \cdot 3125 + 10^9(3125\beta + 125\beta^2) + 10^8(3125\gamma + 250\beta\gamma + 25\beta^3)$$
$$+ 10^7(125\gamma^2 + 75\beta^2\gamma) + 10^6(75\beta\gamma^2 + 10\beta^3\gamma + 25\beta^4)$$
$$+ 10^5(25\gamma^3 + 15\beta^2\gamma^2 + \beta^5) + 10^4(10\beta\gamma^3 + 5\beta^4\gamma + \beta^3\gamma^2)$$
$$+ 10^3 \cdot \beta^2\gamma^3 + 10^2 \cdot 25\gamma^4 + 10 \cdot 5\beta\gamma^4 + \gamma^5.$$

Subtracting the numerical term $10^{10} \cdot 3125$ from the initial quantity leaves 1299′08995′06197.

4. Let us next determine β. We must surely have

$$31250000\beta + 1250000\beta^2 + 25000\beta^3 + 250\beta^4 + \beta^5 < 1299'08995.$$

The largest possible β is 3, giving 1056′95493. The remainder, with the group of the last five digits appended, becomes 242′13502′06197.

5. We must now choose γ so that the sum of the remaining terms,

$$39'45240'50000\gamma + 14887'70000\gamma^2 + 28'09000\gamma^3 + 2640\gamma^4 + \gamma^5,$$

is the closest (below) to the above quantity. Such is $\gamma = 6$, which produces 242′13502′06176. The remainder is therefore 21.

Using the above formula, we may thus give as the required approximation

$$536 + \frac{21}{414237740281}.$$

Remark: We have explained the extraction of α in an abbreviated modern way. Kāshī has instead a long algorithm represented in a table, where the results of all intermediate computations appear. But the principle of the reasoning is just the same.

Other arithmetical topics

Such an extension to higher root extraction supposes a knowledge of the coefficients in the development of $(a + b)^n$, known today as the Pascal or Tartaglia triangle since some of its properties were described by these two authors (neither of whom, by the way, claims its discovery for himself). Antiquity knew the developments for n equal to 2 and 3 (the latter in Diophantus). The development for any arbitrary power was certainly known around the year 1000, since we are informed by Samaw'al ibn Yaḥyā, two centuries later, that Abū Bakr al-Karajī (fl. 1010) described it in a (now lost) work. His excerpt from Karajī indeed shows complete familiarity with its construction (Ahmad and Rashed, 1972: 109–112). Starting from the first (right) column (Figure 1), Karajī constructs the following columns up to the fifth, then explains how this is used for the development of a binomial, and finally points out that the coefficients for any power 'till any chosen limit' can be

1	1	1	1	1	1	1	1	1	1	1	1
12	11	10	9	8	7	6	5	4	3	2	1
66	55	45	36	28	21	15	10	6	3	1	
220	165	120	84	56	35	20	10	4	1		
495	330	210	126	70	35	15	5	1			
792	462	252	126	56	21	6	1				
924	462	210	84	28	7	1					
792	330	120	36	8	1						
495	165	45	9	1							
220	55	10	1								
66	11	1									
12	1										
1											

Figure 1

determined in the way the first five columns were constructed: an element equals the sum of its neighbour and the one above it in the previous (right-hand) column. His figure also indicates at the top of the columns the power names (for which see the section on algebraic reckoning below). This triangle later appears in China (thirteenth century) and is common in sixteenth-century European writings.

The existence of the arithmetic triangle naturally leads to the question of whether combinatory problems were studied. In fact, they were already being approached in the eighth century, but in a grammatical context, for they dealt with the enumeration of letter combinations. The next trace is found in western Islamic mathematics, when in the thirteenth century the rule of deducing C_n^p from C_n^{p-1} by multiplying the latter by $\frac{n-p+1}{p}$ appears to have been known and thus also the possibility of computing the number of combinations of n things p at a time (Djebbar, 1997).

A topic studied from remotest times was that of summing the natural numbers and their powers, that is, computing

$$\sum_{k=1}^{n} k^t \quad t \text{ fixed.}$$

The sum of the natural numbers and of their squares (thus $t = 1$ & 2) was already known in Mesopotamia, that of the cubes in Greece. The next, more complicated, case

$$\sum_{1}^{n} k^4 = \left[\left(n^2 + \frac{n}{2}\right)(n+1)\right]\left[\left(\frac{n}{5} + \frac{1}{5}\right)n - \frac{2}{30}\right]$$

occurs in the tenth century, among other rules expounded by al-Qabīṣī (who does not claim the discovery for himself, as he does in some other cases) (Anbouba, 1982; Sesiano, 1987). No simple relation for the subsequent ones seems to have been known: this had to await the seventeenth century (for the next summations) and the early eighteenth (for Jakob Bernoulli's general solution). But an elegant relation is the one given by Ibn al-Haytham (965-ca. 1040; Suter, 1910/11) in his demonstrations of the cases $t = 3$ and $t = 4$, which can be generalized as follows: considering a rectangle with height $n + 1$ and base

$$\sum_{k=1}^{n} k^t$$

(Figure 2), we infer from the two expressions for its area that, in our symbolism,

$$(n+1) \sum_{k=1}^{n} k^t = \sum_{k=1}^{n} k^{t+1} + \sum_{r=1}^{n} \sum_{k=1}^{r} k^t.$$

ALGEBRA

Algebraic reckoning

(1) As with arithmetic, the history of algebra in the Arabic language begins with Khwārizmī. In his largely accessible, and probably not very original, *Short Account of Algebra* are already found what were to be the three main characteristics of mediaeval (Arabic, and later Latin) algebra (Rosen, 1831).

First, there is a complete absence of symbolism. Everything, including numbers, is written in words. Only a few designations, such as those for the powers of the unknown, are specific to algebra: 'thing' (*shay'*) is our x, 'amount' (*māl*) is x^2, 'cube' (*ka'b*) is x^3. The higher powers, found in later authors, are expressed, as were the Greek ones, by means of the words for x^2 and x^3: if the exponent is divisible by 3, thus $n = 3m$, the power x^n is expressed by *ka'b* repeated m times; if it is of the form $3m + 1$, twice *māl* is followed by $m - 1$ *ka'b*. If finally it is of the form $3m + 2$, one *māl* is followed by m *ka'b*. In an algebraic expression, the words for 'plus' and 'minus' are respectively *wa* (and) and *illā*

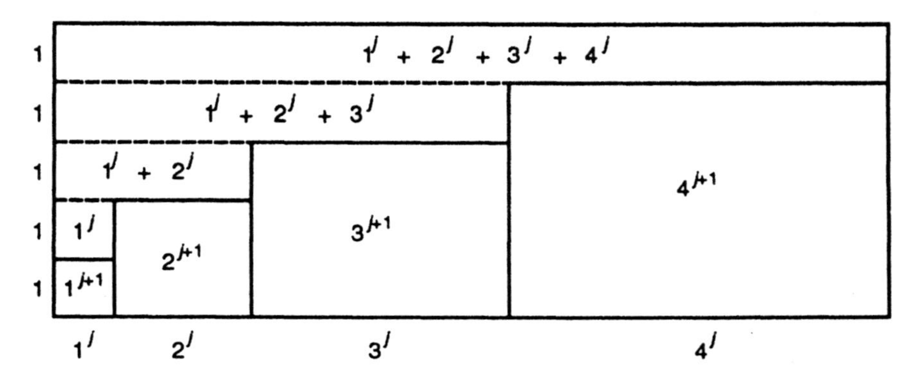

Figure 2

(minus). Since the terms of an expression are connected by *wa* anyway, *illā* is used to separate all the additive terms from all the subtractive ones in order to avoid the limit of the expression covered by a subtraction becoming undefined (Diophantus, who uses symbolism, juxtaposes the terms and likewise separates the two groups, in this case by the sign ⋔).[2] The use of higher powers found application not only in some problems which were reducible to the second degree but also in the development of algebraic reckoning. Thus Karajī, in his *Badīʿ fi 'l-ḥisāb* (Wonderful on Calculation), teaches the division of one polynom by another (Anbouba, 1964).

A second characteristic of mediaeval algebra is the recourse to geometrical figures to illustrate the rules of algebraic reckoning or the resolution formulae for equations. In that sense, algebra can be said to have not yet fully gained autonomy; the proof *more geometrico* was to remain for centuries the criterion of mathematical truth.

A third characteristic, which was, like the previous one, to last till late Renaissance times, is the reduction of the (then) solvable algebraic equations to six specific types with positive coefficients and at least one positive solution, namely the three 'simple' ones and the three 'compound' ones (*mufrada*, *muqtarana*):

$$ax^2 = bx$$

$$ax^2 = c$$

$$bx = c$$

$$ax^2 + bx = c$$

$$ax^2 = bx + c$$

$$ax^2 + c = bx.$$

(2) Once the equation of the problem has been set, it must be reduced to one of these types. (The form $ax^2 + bx + c = 0$ with a, b, c positive cannot occur since it does not have a positive solution.) In order to reach this final form, two operations are involved:

- *al-jabr*, 'the restoration', that is, the addition of the subtractive terms to both sides of the equation which is thus *restored* of its deficiency.
- *al-muqābala*, 'the opposition' or 'the compensation', that is, the removal of identical quantities from both sides.

To these two operations may be added a third one, *al-radd*, 'the reduction', namely the division of the whole equation by the coefficient of the highest power.

Let us consider, as an example, the fifth in Khwārizmī's set of problems. The question reads: *I have divided ten into two parts; then I have divided the first by the second and the second by the first, and (their sum) resulted in two dirhams and a sixth.*

Thus we have to solve

$$\begin{cases} u + v = 10 \\ \dfrac{u}{v} + \dfrac{v}{u} = 2 + \dfrac{1}{6} \end{cases}$$

Put $v = x$; since $u = 10 - x$, the second equation becomes

$$\frac{10 - x}{x} + \frac{x}{10 - x} = 2 + \frac{1}{6}$$

whence

$$100 + 2x^2 - 20x = (21 + \tfrac{2}{3})x - (2 + \tfrac{1}{6})x^2.$$

By *al-jabr*, this becomes

$$100 + (4 + \tfrac{1}{6})x^2 = (41 + \tfrac{2}{3})x.$$

Since there are no common terms to remove from both sides, we can omit the second operation and directly apply *al-radd*, thus finding

$$x^2 + 24 = 10x$$

with the (lesser) solution $5 - \sqrt{25 - 24} = 4$.

These transformations of an equation in order to obtain its reduced form are straightforward, and they were applied from the earliest times. Not surprisingly, they are already explained by Diophantus in his *Arithmetica* (introductions to Books I and IV[3]). In Arabic times, they had the specific names mentioned above, and *al-jabr* and *al-muqābala* were considered to be so characteristic of algebra that it was named after them. By the eleventh century, the science of *al-jabr* and *al-muqābala* had become simply the science of *al-jabr*, out of which the Latin translators and authors of the twelfth century made *algebra*.

(3) We chose the above example as an illustration because the division of a given number into two parts subject to a further condition was an extremely frequent type of problem. It appeared first in Mesopotamia, next in Greece, and then in Islamic countries, where, since the number chosen to be divided was usually ten, the name 'problems of the tens' (*masā'il al-'asharāt*) arose. Their general form

$$\begin{cases} u + v = 10 \\ f(u, v) = k \end{cases}$$

is not as banal as might seem at first. For, by choosing k within a certain range of values, or by taking as k some relation depending upon u and v, one may obtain irrational (or complex) solutions, or higher degree equations. Thus Bahā' al-Dīn al-'Āmilī (1547–1622) listed among problems considered then as 'impossible' the case of $k = u$, ending with the equation $u^3 + 100 = 8u^2 + 20u$ which, although it has three real roots, was not solvable (at least algebraically) in the Orient at that time (Nesselmann, 1843: I, 56; II, 57). Introduced by mediaeval Latin texts into Europe, this kind of problem remained widespread and is also directly connected with the resolution of the next two higher degree equations.

Examples of geometrical illustration

Although the justification of the resolution formulae by means of geometrical figures suggests some Greek influence, Khwārizmī does not mention the name of Euclid at all: his illustrations rely on an intuitive, visual geometry. But the second most important algebraist of the ninth century, the Egyptian Abū Kāmil (*ca.* 890), does explicitly refer to Euclid. Indeed, the main difference between Khwārizmī's book and Abū Kāmil's is that the second one is written specifically for mathematicians, that is to say, people trained in the study of Greek mathematics, chiefly Euclid's *Elements*. This has the advantage of reducing the explanations in the illustrations by simply referring to two theorems from Book II, as well as providing the means of actually constructing the solution using two theorems from Book V. Abū Kāmil does not perform this latter construction, but it is found in later treatises, sometimes together with the customary illustrations. This occurs for instance in an anonymous compilation which was written in 1004/5 and which, according to the author, was based on various sources.[4] The following are its illustrations for the compound equations; they do not refer to Euclid and thus follow the more accessible, traditional way.

(1) Case of $x^2 + px = q$ (Figure 3)

The only positive solution, as given in this and the other Arabic texts, corresponds to

$$x = \sqrt{(\tfrac{p}{2})^2 + q} - \tfrac{p}{2}.$$

Let the square ABCD represent x^2; let us extend AB by $BE = \frac{p}{2}$ and then complete the whole square AF, which then includes the squares AC and CF and the rectangles CG and CE. (In Greek and Arabic texts, rectangular figures are often designated by the letters at opposite angles.) From the construction,

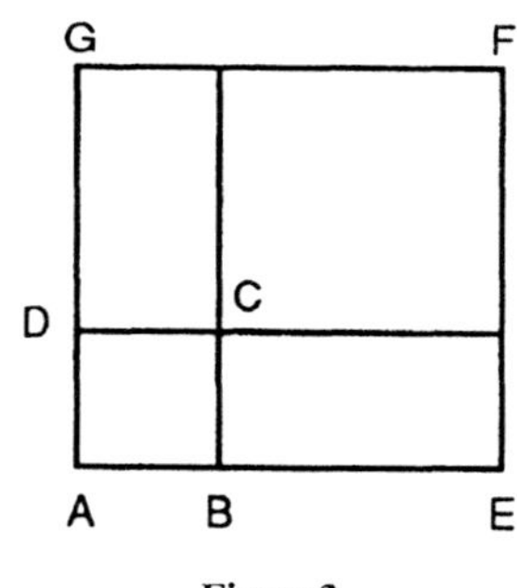

Figure 3

we know that

$$CE = CG = \tfrac{p}{2}x.$$

Consider now the figure formed by CG, CE and AC. According to the equation, it must equal q. Since $CF = (\frac{p}{2})^2$, the whole square AF is equal to $(\frac{p}{2})^2 + q$, but also, by construction, to $(x + \frac{p}{2})^2$. This illustrates the formula.

(2) Case of $x^2 = px + q$ (Figure 4)

The only positive solution is

$$x = \tfrac{p}{2} + \sqrt{(\tfrac{p}{2})^2 + q}.$$

Let ABCD represent x^2, EB be p, and let F be the middle of EB, so that

$$EF = FB = \tfrac{p}{2}.$$

Consider the completed figure, where $GD = GH = \frac{p}{2}$. Thus the rectangles GC and CF are each equal to $\frac{p}{2}x$, whence $DI + IB + 2 \cdot IC = px$. Since $IC = IJ$, using the equation we find that $GJ + AJ + JF = q$. Adding now to each side the square IJ, we obtain the relation $(x - \frac{p}{2})^2 = (\frac{p}{2})^2 + q$, from which the formula is deduced.

(3) Case of $x^2 + q = px$ (Figure 5)

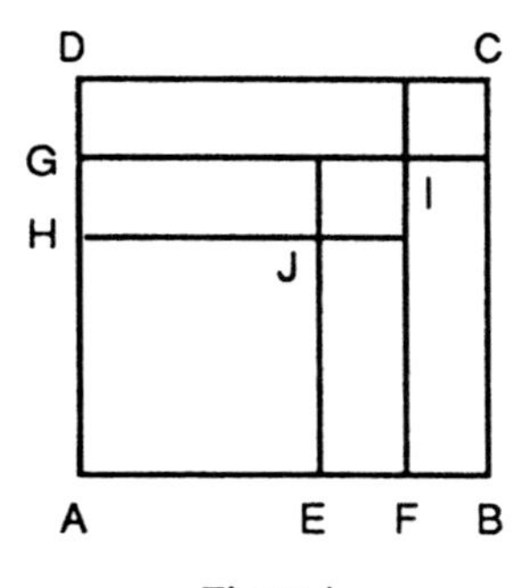

Figure 4

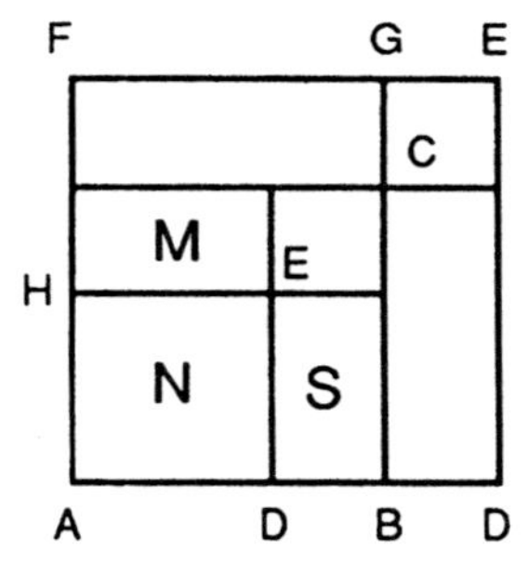

Figure 5

The formula is

$$x = \tfrac{p}{2} \pm \sqrt{(\tfrac{p}{2})^2 - q},$$

thus with two positive solutions (provided that the discriminant is positive). In the manuscript these two possibilities are represented in a single figure.

Let AB be $\frac{p}{2}$, thus AC $= (\frac{p}{2})^2$, and let AD represent the solution x, with either AD > AB or AD < AB, according to the two possible signs in the formula. We now complete the figure (keeping the same letters for the two solutions, as the manuscript does). Let us designate by N the smaller square AE and by M and S the (equal) rectangles adjacent to AE. In the case of the smaller solution,

$$\mathrm{AB} \cdot \mathrm{AD} = \mathrm{M} + \mathrm{N} \quad \text{and} \quad \mathrm{AB} \cdot \mathrm{AD} = \mathrm{N} + \mathrm{S} = \mathrm{AE} + \mathrm{S}.$$

In the case of the larger solution, represented by the whole square AE,

$$\mathrm{AB} \cdot \mathrm{AD} = \mathrm{M} + \mathrm{N} + \mathrm{S} + \mathrm{EC} + \mathrm{CD} \quad \text{and} \quad \mathrm{AB} \cdot \mathrm{AD} = \mathrm{AE} - \mathrm{CE} - \mathrm{CD}.$$

We find thus by addition that in both cases

$$2 \cdot \mathrm{AB} \cdot \mathrm{AD} = \mathrm{M} + \mathrm{N} + \mathrm{S} + \mathrm{AE}.$$

Now $2 \cdot \mathrm{AB} \cdot \mathrm{AD} = px$ and $\mathrm{AE} = x^2$, hence $\mathrm{M} + \mathrm{N} + \mathrm{S} = q$. Thus, considering the equality of the two squares EC and each of the two possibilities for AD,

$$(\tfrac{p}{2})^2 - q = (\tfrac{p}{2} - x)^2 = (x - \tfrac{p}{2})^2$$

which illustrates the formula.

Other algebraic topics

Khwārizmī's *Algebra* contained a small section on commercial mathematics, another one on geometry and a large section on inheritance problems. Abū Kāmil left practical geometry and estate sharing out. (He devoted a separate treatise to each of these topics.) His *Algebra* begins like that of his predecessor (but at a higher level) with algebraic theory (Book I) and numerous problems of application (Book II).[5] In Book IV, algebra is used for the determination of elements in regular polygons studied by Euclid where only quadratic equations can occur (see the section on regular polygons below). Here there are many irrational solutions, in the form of square or fourth roots, which explains why the previous part of the *Algebra* (Book III) deals with the resolution of quadratic equations with irrational solutions or irrational coefficients. This is the first instance of the systematic use and acceptance of such numbers, and even if the purpose – applying algebra to polygons – was restricted, it was significant: it was from that time on that irrationalities became more frequently accepted, at least in so far as they did not involve difficulties of expression due to the purely rhetorical form (see note 2). Book IV is followed by indeterminate problems of the second degree (see the section on number theory below), then by various problems often of a more recreational nature, such as cistern and pursuit problems. This also had a long-term effect: from then on, and up to modern times, most algebra books contained a section on such problems.

Abū Kāmil also wrote a separate treatise on systems of two linear indeterminate equations with positive integral solutions, of a kind already known from antiquity but which became most common in mediaeval times in such different parts of the world as Islamic countries, China and India (Suter, 1910/11; Tropfke, 1980). In these regions, the subject of such problems is mostly the purchase of a given number of birds with a given sum of money, whence the usual Arabic denomination of *masā'il al-ṭuyūr* (problems of the birds) for this kind of problem. Also given is the unit price for each kind of bird, and required is the number of birds of the ith kind. In our symbolism:

$$\begin{cases} \sum_{i=1}^{n} x_i = k \\ \sum_{i=1}^{n} p_i x_i = l \end{cases}$$

with k, l, p_i given.

In his introduction, Abū Kāmil expresses his admiration for such problems because of the quantity of solutions they may admit. Of his six examples, the first five (with three to five unknowns) have 1, 6, 96 (in fact: 98), 304, and 0 solutions. But the most remarkable one is the last, said by Abū Kāmil to have 2676 (in fact: 2678) solutions. One must buy 100 birds with 100 dirhams, namely ducks (2 dirhams each), pigeons ($\frac{1}{2}$ d.), doves ($\frac{1}{3}$ d.), larks ($\frac{1}{4}$ d.), and hens (1 d.). The system is thus:

$$\begin{cases} x_1 + x_2 + x_3 + x_4 + x_5 = 100 \\ 2x_1 + \frac{1}{2}x_2 + \frac{1}{3}x_3 + \frac{1}{4}x_4 + x_5 = 100 \end{cases}$$

Abū Kāmil begins by eliminating x_5 from the two equations:

$$100 - x_1 - x_2 - x_3 - x_4 = 100 - 2x_1 - \tfrac{1}{2}x_2 - \tfrac{1}{3}x_3 - \tfrac{1}{4}x_4,$$

whence

$$x_1 = \tfrac{1}{2}x_2 + \tfrac{2}{3}x_3 + \tfrac{3}{4}x_4. \qquad (*)$$

This leaves four unknowns. (He designates one by the usual *shay'*, two others by the coin names *dīnār* and *fals*, and the last by *khātam*, 'seal'.) Since $x_5 > 0$, for the admissible solutions must be strictly positive, we must have

$$x_1 + x_2 + x_3 + x_4 = \tfrac{3}{2}x_2 + \tfrac{5}{3}x_3 + \tfrac{7}{4}x_4 < 100.$$

From relation (*) we infer that x_1 will be integral if x_3 is divisible by 3, and if further either x_2 is odd and x_4 even but divisible by 2 only, or x_2 is even and x_4 divisible by 4. In the first case the possible values have the form

$$x_3 = 3, 6, \ldots; \quad x_2 = 1, 3, \ldots; \quad x_4 = 2, 6, \ldots;$$

and in the second case

$$x_3 = 3, 6, \ldots; \quad x_2 = 2, 4, \ldots; \quad x_4 = 4, 8, \ldots .$$

In order to obtain the solutions of the first case, Abū Kāmil fixes $x_3 = 3$, then takes $x_2 = 1$ and chooses x_4 as far as possible, does the same with $x_2 = 3$, and goes on with the next admissible values for x_2, thus obtaining 212 solutions. He then fixes $x_3 = 6$, but merely says that one will proceed as before. In the end, he adds, we shall have 1443 solutions corresponding to the values

$$x_2 = 1, 3, \ldots, 59; \quad x_3 = 3, 6, \ldots, 51; \quad x_4 = 2, 6, \ldots, 50.$$

There are two more solutions, for he has omitted those for $x_3 = 54$. For the second case, Abū Kāmil only mentions as an example the first solution, but it is clear from his text that he has reached the 1443 possible solutions.

Although such problems are common in later Islamic mathematics, no mathematician seems to have gone so far in the enumeration of all possible solutions. Traces of Abū Kāmil's problems are still found long after his death, for instance in the *Bāhir* (Ahmad and Rashed, 1972), which shows his study to have been widely spread.

By far the most influential of Abū Kāmil's works was his *Algebra*. In western Islam, it was the basis for the mathematical treatises on transactions (*muʿāmalāt*), which placed arithmetical reckoning and algebra in the context of commercial and daily life (and also provided some recreational problems). The best extant example of this trend in Spain is the Latin *Liber mahameleth* by Johannes Hispalensis (John of Seville), written around 1140 in Toledo in the Moorish tradition. In eastern Islam, Abū Kāmil's book was progressively superseded by later treatises, as a way was gradually found to deal with equations of the third degree (see the section on the cubic equation below).

GEOMETRY

Regular polygons

We have already mentioned Abū Kāmil's algebraic study of regular polygons. They were those of which Euclid had already taught the construction, namely the first regular polygons with a number of sides of the form $2^t \cdot 3$, $2^t \cdot 4$, $2^t \cdot 5$, $2^t \cdot 15$, t integer ≥ 0. (Since doubling the sides only requires halving each side and then erecting the perpendicular to determine the intersection with the circumscribed circle, only the construction of the polygons to 3, 4, 5, 15 sides is necessary.) Since they are exactly constructible by ruler and compass, the relations between the radius of the circumscribed (or inscribed) circle and the side of the polygon can always be algebraically reduced to the resolution of quadratic equations. Now, since the first two non-constructible regular polygons are those with 7 and 9 sides (both leading to a third-degree equation), they attracted the attention of scholars quite early on. Thus we know, through an Arabic version, of a heptagon construction by Archimedes.[6]

The method used by Euclid for constructing the pentagon (*Elements* IV.10–11), and later extended by Archimedes to operate the trisection of an angle, could be generalized in Islamic times to the construction of all polygons (except the easy ones with $n = 2^t$). The idea is the following: suppose we can construct an isosceles triangle with base x, equal sides a, angle at the summit

φ and equal angles at the base $k\varphi$, with k natural. Since $(2k+1)\,\varphi = 180°$, we have

$$\varphi = \frac{360°}{2(2k+1)}.$$

Thus, a circle having its centre at the triangle's summit and passing through the extremities of the base will appear to circumscribe a regular polygon of $2(2k+1)$ sides with length x and (linking every other angle) a regular polygon of $2k+1$ sides, from which any polygon with $2^t(2k+1)$ sides can be obtained.

Let us consider this generalized construction; this will also enable us to establish geometrically the relation between x and a and thus the corresponding algebraic equation.

• If $k = 1$, the triangle with its summit at the circle's centre is equilateral. Thus the side of the hexagon is simply the radius of the circle.
• If $k = 2$ (Figure 6, Euclid's construction), we wish to determine on the given straight line AC the segment DA $= x$, or CD $= a - x$. Since

$$\frac{AB}{BC} = \frac{BC}{CD}, \quad CD = \frac{BC^2}{AB} = \frac{x^2}{a}.$$

Hence

$$a - x = \tfrac{x^2}{a} \quad \text{and} \quad x^2 + ax = a^2,$$

with the solution $\frac{a}{2}(\sqrt{5} - 1)$. Since the equation is of the second degree, D can be determined by ruler and compass and, if AB is the radius, BC will be the side of the decagon. Note that $\frac{AB}{BC}$ is the golden section ratio, equal to 1.61803, ...

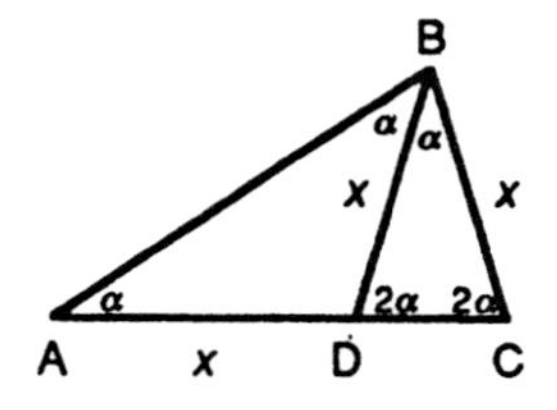

Figure 6

• Now let $k = 3$, and suppose the construction to be made. Draw perpendiculars BT, EU to CA (Figure 7). By similar triangles, we find as before CD $= \frac{x^2}{a}$ and therefore

$$DA = a - \tfrac{x^2}{a}; \quad \text{thus } UD = AU = \tfrac{a}{2} - \tfrac{x^2}{2a}.$$

By similar triangles also

$$\frac{AU}{AE} = \frac{AT}{AB}$$

whence

$$AB \cdot AU = AE \cdot AT = AE(AC - \tfrac{1}{2}CD)$$

that is

$$a(\tfrac{a}{2} - \tfrac{x^2}{2a}) = x(a - \tfrac{x^2}{2a})$$

or

$$a^3 - ax^2 = 2a^2x - x^3$$

and finally

$$x^3 + a^3 = ax^2 + 2a^2x.$$

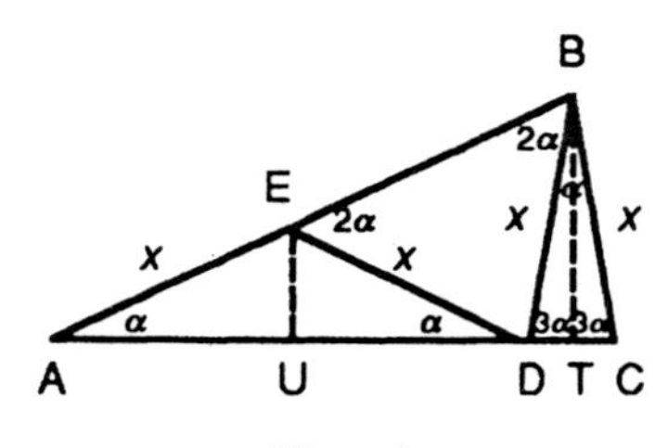

Figure 7

It appears now that the side of the heptagon, or that of the tetradecagon, cannot be constructed by ruler and compass; it can, however, be obtained using conic sections, well known to Islamic mathematicians by the translation of Apollonius's *Conics*. (Of the original eight Books, the first seven were translated into Arabic, whereas the first four only are still extant in Greek.) Thus, if a is given and, using conic sections, E is determined, x and D will be known. At the end of the tenth century mathematicians such as Abū'l-Jūd and Sijzī were able to construct this figure (Hogendijk, 1984).

The same construction underlies Archimedes' trisection of the angle, found in a text extant only in an Arabic version (Heiberg, 1910–15: II, 578). If BDC is the given angle, draw a circle around D. (The angle given is considered to be acute since the trisection of a right angle amounts to constructing an angle of 30°, which is half of the angle in the hexagon; see above, case $k=1$.) Then find the position of a point E on the circle in such a way that the extensions of the required straight line BE and of the given straight line CD meet at a point A, distant from E by the length of the radius. BAC is then a third of the given angle.

• Consider now $k=4$ (Figure 8). Besides $\mathrm{CD} = \frac{x^2}{a}$ and $\mathrm{AF} = x$, we have $\mathrm{BE} = x$, since the triangle BDE is equilateral. Thus

$$\mathrm{AE} = a - x \text{ and } \mathrm{AV} = (\tfrac{a}{2}) - (\tfrac{x}{2}).$$

Further, by similar triangles,

$$\frac{\mathrm{AV}}{\mathrm{AF}} = \frac{\mathrm{AT}}{\mathrm{AB}}.$$

Thus, as before,

$$\mathrm{AB} \cdot \mathrm{AV} = \mathrm{AF} \cdot \mathrm{AT} = \mathrm{AF}(\mathrm{AC} = -\tfrac{1}{2}\mathrm{CD})$$

that is

$$a(\tfrac{a}{2} - \tfrac{x}{2}) = x(a - \tfrac{x^2}{2a})$$

or

$$a^3 - a^2x = 2a^2x - x^3,$$

and finally

$$x^3 + a^3 = 3a^2x.$$

If E is found, x will be known. The determination of E was the subject of correspondence around the year 1000 between Abū'l-Jūd and Abū Rayḥān al-Bīrūnī, one of the most distinguished Persian scholars (Woepcke, 1851: 125–126).

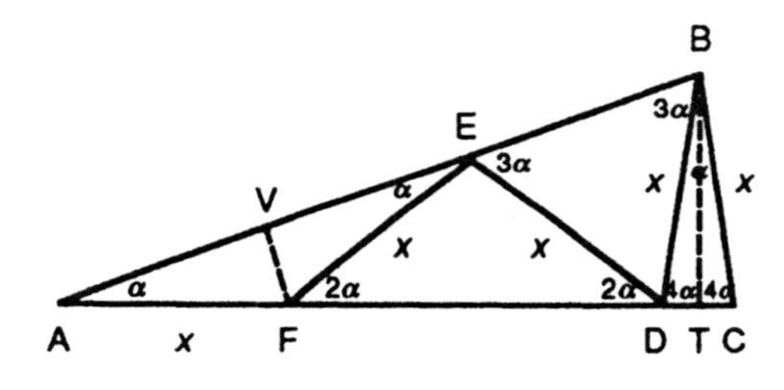

Figure 8

• Abū'l-Jūd claimed to have determined (supposedly in this same way) the side of the endecagon (Hogendijk, 1984: 263–264). He might have understood the principle, but he could hardly have reached the result by geometrical means. For, if $k = 5$ (Figure 9), we see that, as before, $\mathrm{CD} = \frac{x^2}{a}$ and so $\mathrm{AD} = a - (\frac{x^2}{a})$.

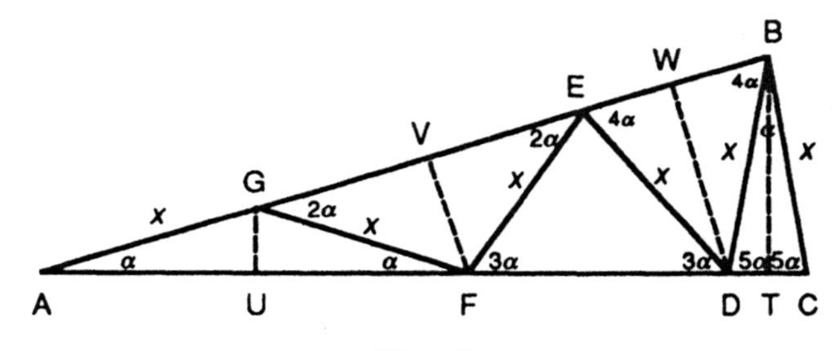

Figure 9

By similar triangles,

$$\frac{\mathrm{AW}}{\mathrm{AD}} = \frac{\mathrm{AV}}{\mathrm{AF}} = \frac{\mathrm{AU}}{\mathrm{AG}} = \frac{\mathrm{AT}}{\mathrm{AB}} = \frac{1}{a}\left(a - \frac{x^2}{2a}\right)$$

hence

$$\mathrm{AW} = \tfrac{1}{a}(a - \tfrac{x^2}{2a})(a - \tfrac{x^2}{a}) = a - \tfrac{3x^2}{2a} + \tfrac{x^4}{2a^3}$$

$$\mathrm{AU} = \tfrac{x}{a}(a - \tfrac{x^2}{2a})$$

and, since AF = 2AU,

$$AV = \frac{2x}{a^2}\left(a - \frac{x^2}{2a}\right)^2 = 2x - \frac{2x^3}{a^2} + \frac{x^5}{2a^4}.$$

Thus

$$GE = 2GV = 2(AV - AG) = 2x - \frac{4x^3}{a^2} + \frac{x^5}{a^4}$$

$$BW = \tfrac{1}{2}(AB - AE) = \tfrac{1}{2}(AB - AG - GE) = \frac{a}{2} - \frac{3x}{2} + \frac{2x^3}{a^2} - \frac{x^5}{2a^4}$$

whence

$$AW = AB - BW = \frac{a}{2} + \frac{3x}{2} - \frac{2x^3}{a^2} + \frac{x^5}{2a^4}$$

Equating this to the expression already found for AW, we have

$$a - \frac{3x^2}{2a} + \frac{x^4}{2a^3} = \frac{a}{2} + \frac{3x}{2} - \frac{2x^3}{a^2} + \frac{x^5}{2a^4}$$

that is,

$$x^5 + 3a^3x^2 + 3a^4x = ax^4 + 4a^2x^3 + a^5$$

which is of the fifth degree and beyond the mediaeval possibilities of resolution.

The cubic equation

A cubic equation with real coefficients has at least one, sometimes three, real roots. Considering, as with quadratic equations, only positive terms, there are fourteen cases in which the solution may be positive. Some isolated cases were already being considered in Greece, but all appear to have been identified around the year 1000. (The manuscript mentioned in note 4 lists them.) The geometric resolution of each was completed by the time of ʿUmar Khayyām (Omar Khayyam, *ca.* 1048–1130), for he solves them all in his *Algebra* by means of intersections of circles, parabolas and hyperbolas. These cases, with the type of solution and the curves used by ʿUmar Khayyām to construct the positive solution, are given below.

	Equation $(a, b, c > 0)$	*Solutions*	*Curves*
1	$x^3 = c$	$x_1 > 0;\ x_{2,3}\,\not\mathbb{C}$	P, P
2	$x^3 + bx = c$	$x_1 > 0;\ x_{2,3}\,\not\mathbb{C}$	C, P
3	$x^3 + c = bx$	$x_{1,2} > 0$ or $\not\mathbb{C}$; $x_3 < 0$	P, H
4	$x^3 = bx + c$	$x_1 > 0;\ x_{2,3} < 0$ or $\not\mathbb{C}$	P, H
5	$x^3 + ax^2 = c$	$x_1 > 0;\ x_{2,3} < 0$ or $\not\mathbb{C}$	P, H
6	$x^3 + c = ax^2$	$x_{1,2} > 0$ or $\not\mathbb{C}$; $x_3 < 0$	P, H
7	$x^3 = ax^2 + c$	$x_1 > 0;\ x_{2,3}\,\not\mathbb{C}$	P, H
8	$x^3 + ax^2 + bx = c$	$x_1 > 0;\ x_{2,3} < 0$ or $\not\mathbb{C}$	C, H
9	$x^3 + ax^2 + c = bx$	$x_{1,2} > 0$ or $\not\mathbb{C}$; $x_3 < 0$	H, H
10	$x^3 + bx + c = ax^2$	$x_{1,2} > 0$ or $\not\mathbb{C}$; $x_3 < 0$	C, H
11	$x^3 = ax^2 + bx + c$	$x_1 > 0;\ x_{2,3} < 0$ or $\not\mathbb{C}$	H, H

12	$x^3 + ax^2 = bx + c$	$x_1 > 0$; $x_{2,3} < 0$ or $\not{C}$	H, H
13	$x^3 + bx = ax^2 + c$	$x_1 > 0$; $x_{2,3} > 0$ or $\not{C}$	C, H
14	$x^3 + c = ax^2 + bx$	$x_{1,2} > 0$ or $\not{C}$; $x_3 < 0$	H, H

As we see, some cases always have one positive solution (1, 2, 4, 5, 7, 8, 11, 12); others have two if the curves intersect (3, 6, 9, 10, 14). Case 13 certainly has one and possibly three. ʿUmar Khayyām has in general recognized the possibility of two positive solutions but not this last isolated instance of three.

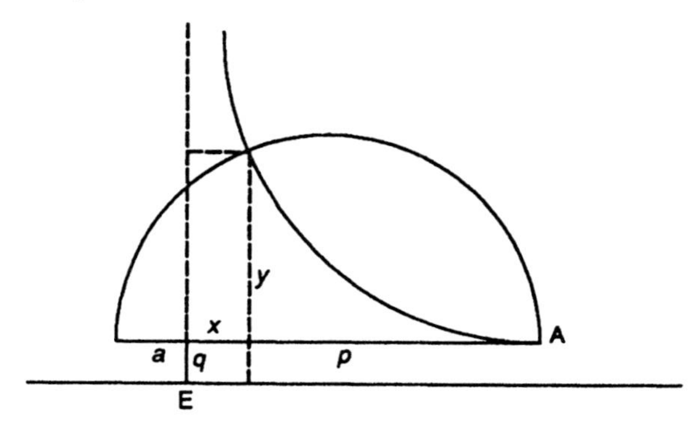

Figure 10

Let us, as an illustration, examine ʿUmar Khayyām's treatment of case 8, that is, $x^3 + ax^2 + bx = c$ (Woepcke, 1851: 46–47). Putting $q = \sqrt{b}$ and $p = \frac{c}{b}$, the equation takes the form $x^3 + ax^2 + q^2x = pq^2$. Let us draw (Figure 10) the segment q and, perpendicularily on either side of one of its extremities, a and p. Next, we draw a parallel to this straight line passing through the other extremity E. Finally, we draw a circle with $a + p$ taken as diameter, and an equilateral hyperbola having as asymptotes the straight line through q and the horizontal through E. Consider now their intersection, characterized by x, y in relation to E. From the property of the hyperbola, we know that

$$xy = pq;$$

subtracting qx from both sides, we have

$$x(y - q) = q(p - x)$$

or

$$\frac{y - q}{p - x} = \frac{q}{x}$$

or also

$$\frac{(y - q)^2}{(p - x)^2} = \frac{q^2}{x^2}.$$

But, according to the property of the circle, we have

$$(y - q)^2 = (x + a)(p - x).$$

Combining these two results, we find

$$\frac{x+a}{p-x} = \frac{q^2}{x^2}$$

whence

$$x^3 + ax^2 + q^2x = pq^2.$$

Thus, the segment x is the required solution. ʿUmar Khayyām mentions at the end that there is for this type neither distinction of cases nor impossibility. He means, as we would express it now, that there is *always* a real solution, which in addition is *positive*, so that no condition depending upon the coefficients is necessary.

Other geometrical topics

Conic sections also led to another type of research, that of determining the areas or volumes and centres of gravity of plane curves or solid bodies generated by a segment of parabola revolving about the axis. Such results, known to Archimedes, were discovered by Islamic mathematicians by other methods since those Archimedean writings dealing with these topics were not transmitted. For summing elements of surface, Thābit ibn Qurra had to divide the axis segment into unequal parts, a method also used later by Fermat.[7] Al-Ḥasan ibn al-Haytham, already mentioned in the section on arithmetic, solved the problem (known since mediaeval times as Alhazen's, from the Latin transcription of al-Ḥasan): given the positions of the eye, the object and a curved mirror (spherical, cylindrical or conical), find the points of reflection on the mirror.

The first three Euclidean postulates assert the possibility of drawing a straight line and a circle anywhere in infinite space; hence they also postulate the means of constructing them, namely ruler and compass. The fourth postulate asserts the invariability of right angles anywhere in space. Finally, the fifth is equivalent to requiring that through a given point one, and only one, straight line can be drawn which will be parallel to a given straight line. Of these postulates the fifth is the only one to have a hypothesis and a conclusion; for that simple reason, doubts about its necessity and attempts to prove it are seen from Greek times on. It was not until the second half of the nineteenth century that the necessity of its existence as a postulate (or of some equivalent statement) was finally demonstrated. Discussions by Islamic mathematicians about the Euclidean postulates involved such attempts, beginning in the ninth century with those of al-Jawharī and Thābit ibn Qurra. Most notable are those later on of Ibn al-Haytham and, in particular, ʿUmar Khayyām and Naṣīr al-Dīn al-Ṭūsī (1201–1274). By an approach similar to that taken in G. Saccheri's treatise of 1733, they deduced, by leaving out the fifth postulate, some simple propositions which are consistent with the first four postulates, or three of them only, and thus belong to what is called today non-Euclidean geometries (Jaouiche, 1986, 1988).

Spherical geometry was based on Menelaus's *Spherics* (and, in particular, its theorem III.1) and gave rise through Abū'l-Wafā' al-Būzjānī (940–997/8) to the law of sines for spherical triangles,

$$\frac{\sin a}{\sin \alpha} = \frac{\sin b}{\sin \beta} = \frac{\sin c}{\sin \gamma}$$

where a, b, c are the sides and α, β, γ the opposite angles (Carra de Vaux, 1892; Krause, 1936). Astronomical instruments, cartography, and also religious duties (determination of the *qibla*, the direction of Mecca), required representations of the celestial sphere or of the Earth, and many treatises deal with the study of various types of projection, including the stereographic one going back to Ptolemy (and Hipparchus). A new determination of the Earth's circumference was ordered by Caliph al-Ma'mūn (813–833) when it turned out that the one transmitted by the Greeks was much too long. (The Greek value had been converted to Roman miles in late antiquity, and it was thought in Islamic times that the Roman mile, approximately 1.5 km, was equivalent to the ancient mile used in the Near East, approximately 2 km.) Both celestial and terrestrial measurements required the use of precise instruments. The best known one is the astrolabe (*asṭurlāb*), considerably improved in Islamic times but, as its name indicates, of Greek origin.

Trigonometry, being linked to astronomy and geography, was mainly taught in treatises on astronomy or, for the purpose of local measurements, in those on geodesy. It can be said to have become an independent branch of mathematics with Ṭūsī, who devoted a separate treatise to it (Carathéodory, 1891). Islamic trigonometry had grown up out of two heritages. The influence of Indian trigonometry was predominant in the early ninth century, but, with the translation and study of Ptolemy's *Almagest*, diminished progressively. However, the Indian use of the half-chord of an arc instead of the whole chord (used by the Greeks) was mostly retained. To the Indian sine and cosine were added the tangent and cotangent as independent functions (Ḥabash al-Ḥāsib, fl. 830). A systematic exposition is that of Abū'l-Wafā', who also represented the tangent function on the tangent to the unit circle (whereas these functions appeared in India only in connection with the sundial) and established formulae such as $\sin(\alpha \pm \beta)$ in their modern form (Carra de Vaux, 1892). (The Greeks used the corresponding ones for chords.) Book IV of Bīrūnī's *Masudic Canon* is wholly devoted to trigonometry (Schoy, 1927; Bīrūnī, 1954–1956). By that time, trigonometry was also being used to solve some classical Greek problems leading to third degree equations, such as the trisection of an angle. But, again, the full development of trigonometry and its use to determine the six elements of any (plane or spherical) triangle on the basis of three known ones had to await Ṭūsī's treatise.

Finally, the determination of π was extended along the lines indicated by Archimedes: by computing the areas of a succession of regular inscribed and circumscribed polygons. Archimedes had started with a hexagon and doubled the sides until he obtained two polygons with $3 \cdot 2^5 = 96$ sides, which gave him the approximation

$$3 + \tfrac{10}{71} < \pi < 3 + \tfrac{10}{70}$$

whence the common mediaeval approximation $3 + \frac{1}{7}$. Kāshī went as far as computing the areas of a pair of regular polygons to $3 \cdot 2^{28}$ sides, thus determining π (or, rather, 2π, since he expresses it as the ratio of the circle's circumference to its area) to nine sexagesimal and sixteen decimal places (Luckey, 1953). (He computes the second from the first.) Another nice application of geometry, which appears in his *Key of Arithmetic* (Kāshī, 1977), was the computation he made of the domes of mosques (Dold-Samplonius, 1993).

NUMBER THEORY

Properties of some integers

Euclid's *Elements* (Books VII to IX) and Nicomachus's *Introduction to Arithmetic* account for the interest of Islamic mathematicians in the properties of integers, and, in particular, the relations between an integer and the sum of its divisors. Considering (in our symbolism) N to be a natural number and $s(N)$ the sum of the divisors of N other than N itself, the Greeks distinguished between 'abundant' numbers ($s(N) > N$), 'defective' numbers ($s(N) < N$) and 'perfect' numbers ($s(N) = N$). Euclid had demonstrated that $N = 2^{m-1}(2^m - 1)$ is perfect if $2^m - 1$ is prime; we know thanks to Euler that Euclid's formula indeed covers all even perfect numbers (possibly all perfect numbers, since it seems that odd perfect numbers do not exist). The Greeks had also found a pair of numbers, which were called 'amicable', where $s(N_1) = N_2$ and $s(N_2) = N_1$; that is, such that the sum of the divisors of one (without itself) equalled the other. This pair was $N_1 = 220$ and $N_2 = 284$. Most of this knowledge was developed quite early in Greece, partly already in the Pythagorean school (fifth century BC).

Here, two noticeable advances were made by Thābit ibn Qurra. He first gave a rule for easily finding defective and abundant numbers: considering the 'perfection' of $2^{m-1}p$, $p = 2^m - 1$, he showed that $2^{m-1}p$, with m given and p prime, is defective or abundant depending on whether $p > 2^m - 1$ or $p < 2^m - 1$ (Woepcke, 1852). In the same treatise, he also answered the question apparently left unsolved by the Greeks of finding amicable number pairs other than the smallest one. His rule, proved in true Euclidean manner, is: If $s = 3 \cdot 2^m - 1$, $t = 3 \cdot 2^{m-1} - 1$, $r = 9 \cdot 2^{2m-1} - 1$ ($m \neq 0, 1$) are prime, then $2^m \cdot s \cdot t$ and $2^m \cdot r$ are amicable. The severe restrictions on N_1 and N_2 (which must contain the same power of 2 as even factor and, respectively, one and two prime numbers at the first power as other factors) lead rapidly to large values while still making it possible to compute two more pairs quite easily, as did Descartes and Fermat in the seventeenth century after they rediscovered Thābit's rule.

An assertion by al-Baghdādī (*ca.* 1000) about the relation $s(N) = k$, with k an arbitrary natural number, may not have originated with him, since it occurs just as an incidental remark.[8] He asserts that $s(N) = k$ has no solution N for $k = 2$ and for $k = 5$. Now it is indeed believed that, of all the odd numbers, only 5 cannot be a sum of (proper) divisors; but if k is even, we know today that, besides 2, there are an infinite number of exceptions. How al-Baghdādī (or his source) arrived at these assertions may be reconstructed with some probability. For he also gives a simple rule whereby two (or more) numbers N_i with the same $s(N_i)$ can be found: Take any even number $2m$ and express it at least twice as the sum of two unequal primes:

$$2m = p_1 + p_2 = q_1 + q_2 = \dots;$$

then

$$N_1 = p_1 \cdot p_2, \quad N_2 = q_1 \cdot q_2, \dots,$$

will meet the requirement since

$$s(N_i) = p_1 + p_2 + 1 = q_1 + q_2 + 1 = \dots = 2m + 1.$$

Thus the resolution of $s(N_i) = 2m + 1$ can in a first approach be reduced to the question of expressing $2m$ as two or more sums of two unequal primes. Empirically, it will appear that two or more representations are always possible for the first even numbers, except for 4, 6, ..., 14, 38, and that the number of possibilities on average increases with the size of m. Considering now the exceptions, we find that one representation as the sum of two unequal primes is possible except for the first two, where these primes must be identical. From this we may draw two conclusions. The first is that, since the above rule would seem to allow us to solve $s(N) = 2m + 1$ for any $m \geq 4$, while for $m = 3$ and $m = 1$ obvious solutions are $N = 8$ and $N = 4$, respectively, only $s(N) = 5$ would have no solution. The second conclusion would have been easy to make, but apparently was not: that any even number larger than 2 is representable as the sum of two primes. Now this is the most famous still unproved number theoretical conjecture, suggested in 1742 by Christian Goldbach in a letter to Leonhard Euler.

Other topics

In the beginning of Book V of his *Algebra*, Abū Kāmil mentions that some of the indeterminate problems he will present were, in his time, 'the subject of discussion among mathematicians'. Now all the problems in question are of the second degree, and are solved by the very same kinds of methods as the ones found in Diophantus's *Arithmetica*, except that these problems are obviously not taken from Diophantus. There is no doubt that we have here the trace of a non-Diophantine Greek source. Book V also solves indeterminate equations of a type not found in Diophantus, such as general ones of the form $ax^2 + bx + c = \square$, or systems of such equations; but in some cases the solution found may be rational but not positive and the resolution therefore not valid according to ancient requirements (Sesiano, 1977; Diophantus, 1982: 81–82). Worth noting too is the classification by Karajī of the main methods and indeterminate equations found in Diophantus; his is the first known attempt to produce a synthesis of these ancient mathematical tools (Anbouba, 1964; Sesiano, 1977).

Another problem not found in Diophantus's *Arithmetica* is that of finding an integral x such that, with k a given integer,

$$x^2 \pm k = \square.$$

Al-Khāzin (fl. 930) and contemporaries studied the form the solution must take (Anbouba, 1979; Woepcke, 1860/1). A full treatise on this subject was later to be the *Liber Quadratorum* written in 1225 by Leonardo Fibonacci, who says he came across the problem at the court of Frederick II Hohenstaufen. But Arabic influence is obvious here, not only for the question itself but also for at least part of the solution. We also learn from the same sources of attempts to prove the impossibility of finding an integral or rational solution to $x^3 + y^3 = z^3$, the first case of the recently-proved Fermat conjecture, that $x^n + y^n = z^n$ is impossible to solve in integral or rational numbers for $n \geq 3$.

MAGIC SQUARES

One of the most impressive and original achievements in Islamic mathematics is the development of general methods for constructing magic squares. A magic square of order n is a square with n cells on its side, thus n^2 cells on the whole, in which different natural numbers must be arranged in such a way that the sum of each row, column and main diagonal is the same. When the n^2 first natural numbers are written in, as is usually the case, the constant sum amounts to $\frac{1}{2}n(n^2 + 1)$. Constructing such a magic square is possible for any $n \geq 3$. If, in addition, the squares left when the borders are successively removed also have this magic property, the square is called a *bordered* square (always possible for $n \geq 5$). If any pair of broken diagonals (that is, which lie on either side of, and parallel to, a main diagonal and together have n cells) shows the constant sum, the square is called *pandiagonal* (possible for $n \geq 4$ but not of the form $4k + 2$). Then there are *composite* squares: when the order n is a composite number, say $n = r \cdot s$ with $r, s \geq 3$, the main square can be divided into r^2 subsquares of order s. These subsquares, taken successively according to a magic arrangement for the order r, are then filled successively with sequences of s^2 consecutive numbers according to a magic arrangement for the order s, the result being then a magic square in which each subsquare is also magic. For general methods of construction, squares are usually divided into three categories: *odd-order* (or *odd*) squares are those with n odd, that is, $n = 2k + 1$ with k natural; *evenly-even* squares have their order n even and divisible by 4, thus $n = 4k$; *oddly-even* squares have an order n even but divisible by 2 only, whence $n = 4k + 2$.

Information about the beginning of Islamic research on magic squares is lacking. It may have been connected with the introduction of chess into Persia. Initially, the problem was a purely mathematical one; thus, the ancient Arabic designation for magic squares is *wafq al-aʿdād*, 'harmonious disposition of the numbers'. We know that treatises were written in the ninth century, but the earliest extant ones date from the tenth. One is by Abū'l-Wafā' and the other is a chapter in Book III of ʿAlī b. Aḥmad al-Anṭākī's (d. 987) *Commentary on Nicomachus's Arithmetic* (Sesiano, 1998 and forthcoming). It appears that, by that time, the science of magic squares was already established: bordered squares of any order could be constructed, and methods for the construction of a simple magic square of small order (up to $n = 6$) were known and could also be applied to composite squares. (Although methods for simple magic squares are easier to *apply* than methods for bordered ones, the latter are easier to *discover*.) The eleventh century saw the discovery of several ways to construct simple magic squares, in any event for odd and evenly-even ones. The much more difficult case of $n = 4k + 2$, which Ibn al-Haytham could solve only with k even, was probably not settled before the second half of the eleventh century. By that time, pandiagonal squares were being constructed for squares of evenly-even and of odd order, provided in the latter case that the order was not divisible by 3. (Little attention seems to have been paid, however, to the sum of the broken diagonals; such squares were considered of interest because their construction characteristically required the use of chess moves and because the initial cell, say the place of 1, could be chosen at random.) There were numerous treatises on magic squares in the twelfth century, and later developments tended to be limited to the refinement of existing methods. Since magic squares were increasingly associated with magic proper and used for divinatory purposes, many later texts merely picture squares and mention their attributes. Others do, however, keep the general theory alive even if their authors, unlike earlier writers, cannot resist adding various fanciful applications.

The link between magic squares and magic as such had to do with the association of each of the twenty-eight Arabic letters with a number (the units, the tens, the hundreds and one thousand). Thus it was sometimes possible to put a sequence of numbers corresponding to the letters of a proper name or the words of a sentence in, say, the first row and then complete the square so as to produce the magic property. But this involved a completely different kind of construction, which depended upon the order n and the values of the n given quantities. The problem is mathematically not easy, and led in the eleventh century to interesting constructions for the cases from $n = 3$ to $n = 8$. However, here again the subtle theory mostly was to end up in a set of practical recipes for the use of a wider readership.

This development was unfortunate for Europe, where by the late Middle Ages only two sets of simple magic squares associated with the seven known celestial bodies had been learned of through Arabic astrological and magic texts – whence the name – and without any indication as to their construction. (Whereas as early as the twelfth century some methods of construction had reached India and China, and later also Byzantium, as can be seen from the treatise on magic squares written *ca.* 1300 by Manuel Moschopoulos.) Thus the extent of Islamic research remained unknown for a very long time, since it has only recently been assessed and its importance recognized.

Of the two examples given, one (Figure 11) is found in al-Anṭākī's treatise; he explains a method for constructing bordered squares of odd order in which the numbers are separated according to parity. (Incidentally, this method is so refined as to make one wonder if the science of magic squares is not even older than we think.) The other example (Figure 12), that of a pandiagonal square constructed by chess moves, occurs in an anonymous treatise of the early

36	16	108	110	10	113	8	116	118	2	34
50	48	24	100	107	97	7	102	18	46	72
52	56	60	103	91	89	23	3	58	66	70
96	54	17	47	51	81	83	43	105	68	26
94	13	29	49	59	57	67	73	93	109	28
11	27	35	45	69	61	53	77	87	95	111
92	117	101	85	55	65	63	37	21	5	30
32	80	121	79	71	41	39	75	1	42	90
38	78	64	19	31	33	99	119	62	44	84
82	76	98	22	15	25	115	20	104	74	40
88	106	14	12	112	9	114	6	4	120	86

Figure 11

32	*39*	*60*	3	30	*37*	*58*	1
59	4	31	*40*	*57*	2	29	*38*
5	*62*	*33*	26	7	*64*	*35*	28
34	25	6	*61*	*36*	27	8	*63*
24	*47*	*52*	11	22	*45*	*50*	9
51	12	23	*48*	*49*	10	21	*46*
13	*54*	*41*	18	15	*56*	*43*	20
42	17	14	*53*	*44*	19	16	*55*

Figure 12

eleventh century (Sesiano, 1996). Starting with the numbers 1 and $\frac{n^2}{2}$ in the upper corner cells, the consecutive numbers (taken in ascending or descending order) are placed within each pair of rows with the knight's move and, near the square's border, the queen's move. The remaining half of the square is filled in by considering the subsquares of order 4 and writing in each empty cell the complement to $n^2 + 1$ of the number found in the cell diagonally distant of two cells (thus in the bishop's cell). This method can be used for the construction of a pandiagonal square of any order $n = 4k$.

NOTES

[1] Much has been written about this discovery, as a matter of fact a simple consequence of writing numbers on a dust board in columns with, at the top of each, the indication of the power of ten so

that in the column only the coefficient needed to appear. The introduction of a sign for the empty place, as used already in Greek astronomical tables, then made the use of symbols for the powers of ten altogether superfluous.

[2] Rhetorical algebra remains within the grasp of a modern reader. The expression of numbers in words originally had the advantage of avoiding copyists' errors and (in Arabic and Latin) of indicating their grammatical role, thus their function in the mathematical expression. The fundamental weakness of rhetorical algebra lies in its failure to indicate the length of an expression to which an operation is applied. We have seen how a possible ambiguity was avoided for subtraction. But there was no solution to the case of a succession of roots; the ambiguity already becomes apparent when two simple expressions such as $\sqrt{a+\sqrt{b}}$ and $\sqrt{a}+\sqrt{b}$ are expressed orally.

[3] Sizeable Greek works, thus also Arabic ones, are usually divided into Books (*βιβλία*, *maqālat*), that is, large chapters.

[4] Manuscript 5325 of the library Astan Qods in Mashhad.

[5] Abū Kāmil's *Algebra* is preserved in Arabic by a single, but very good manuscript (MS. Kara Mustafa Pa'mmsa 379, now Bayazıt 19046), reproduced in Abū Kāmil, 1986. A 14th-century Latin version, good in the mathematical part (less so for the more literary digressions) is published in Folkerts and Hogendijk, 1993: 215–252 & 315–452. A 15th-century Hebrew version (Books I–III) is found in Levey, 1966 [not reliable].

[6] More than two millennia after Euclid, Gauss (1777–1855) found and proved that the necessary and sufficient condition for an n-sided polygon to be constructible is that n have the form $2^t \cdot p_1 \cdot p_2, \ldots, p_k$ with t integer ≥ 0 and the p_i's different prime numbers of the form $2^m + 1$, of which five are known to exist (3, 5, 17, 257, 65537).

[7] The various treatises dealing with infinitesimal methods are found in Suter's works (see Bibliography, further reading, III).

[8] Saidan, 1985. There is a translation of this passage in Sesiano, 1991. A later occurrence of this rule is found in the work of the 17th-century Persian Yazdī; see Djafari, 1982.

BIBLIOGRAPHY

Abū Kāmil. *Kitāb al-Jabr wa l-muqābala*. Frankfurt: Institut für Geschichte der arabisch-islamischen Wissenschaften, 1986.

Ahmad, Salah and Roshdi Rashed, eds. *Al-Samaw'al: Al-Bāhir fi'l-jabr*. Damascus: Jāmi'a Dimashq, 1972.

Anbouba, Adel. *L'Algèbre al-Badī' d'al-Karagī*. Beyrouth: Université libanaise, 1964.

Anbouba, Adel. 'Un traité d'Abū Ja'far [al-Khazin] sur les triangles rectangles numériques.' *Journal for the History of Arabic Science* 3: 134–178, 1979.

Anbouba, Adel. 'Un mémoire d'al-Qabīṣī sur certaines sommations de nombres.' *Journal for the History of Arabic Science* 6: 181–208, 1982.

Apollonius. *Conics, Books V to VII* (2 vols.), Gerald Toomer, ed. New York: Springer-Verlag, 1990.

al-Bīrūnī. *Kitāb al-Qānūn al-Mas'ūdī* (Arabic). Hyderabad: Osmania University, 1954–1956.

Carathéodory, Alexandre. *Traité du quadrilatère attribué à Nassiruddin el-Toussy*. Constantinople: Osmanié, 1891.

Carra de Vaux, Bernard. 'L'Almageste d'Abû'lwéfa Albûzdjâni.' *Journal asiatique* ser. 8, vol. 19: 408–471, 1892.

Diophantus. *Arithmetica, Arabic translation attributed to Qusṭā ibn Lūqā*, Jacques Sesiano, ed. New York: Springer-Verlag, 1982.

Djafari Naini, Alireza. *Geschichte der Zahlentheorie in Orient*. Braunschweig: Klose, 1982.

Djebbar, Ahmed. 'Combinatorics in Islamic mathematics.' In *Encyclopaedia of the History of Science, Technology, and Medicine in Non-Western Cultures*, Helaine Selin, ed. Dordrecht: Kluwer, 1997, pp. 230–232.

Dold-Samplonius, Yvonne. 'The volume of domes in Arabic mathematics.' In *Vestigia Mathematica*, M. Folkerts and J.P. Hogendijk, eds. Amsterdam: Rodopi, 1993, pp. 93–106.

Folkerts, Menso, ed. *Die älteste lateinische Schrift über das indische Rechnen nach al-Ḫwārizmī*. Munich: Bayerische Akademie der Wissenschaften, 1997.

Folkerts, Menso and Jan Hogendijk, eds. *Vestigia Mathematica*. Amsterdam: Rodopi, 1993.

Heiberg, Johan, ed. *Archimedis opera omnia cum commentariis Eutocii* (3 vols.). Leipzig: Teubner, 1910–15, reprinted Stuttgart: Teubner, 1972.
Hogendijk, Jan. 'Greek and Arabic constructions of the regular heptagon.' *Archive for History of Exact Sciences* 30: 197–330, 1984.
Jaouiche, Khalil. *La théorie des parallèles en pays d'Islam.* Paris: Vrin, 1986. Relevant texts in Arabic in his *Naẓarīya al-mutawāziyāt fi 'l-handasa al-islāmīya.* Tunis: Bayt al-ḥikma, 1988.
al-Kāshī. *Miftāḥ al-ḥisāb*, Nādir al-Nābulsī, ed. Damascus: Jāmi'a Dimashq, 1977.
Krause, Max, ed. *Die Sphärik von Menelaos aus Alexandrien in der Verbesserung von Abū Naṣr b. 'Alī b. 'Irāq.* Berlin: Weidmannsche Buchhandlung, 1936.
Levey, Martin. *The Algebra of Abū Kāmil.* Madison: University of Wisconsin Press, 1966.
Luckey, Paul, ed. *Der Lehrbrief über den Kreisumfang (ar-Risāla al-muḥīṭīya) von Ğamšīd b. Mas'ūd al-Kāšī.* Berlin: Akademie-Verlag, 1953.
Menninger, Karl. *Zahlwort und Ziffer.* Breslau: Hirt, 1934, reprinted Göttingen: Vandenhoeck & Ruprecht, 1958; translated as *Number Words and Number Symbols*, Cambridge, Massachusetts: MIT Press, 1969.
Nesselmann, Georg. *Beha-eddin's Essenz der Rechenkunst* (2 Hefte). Berlin: G. Reimer, 1843.
Rosen, Frederick. *The Algebra of Mohammed ben Musa.* London: Oriental Translation Fund, 1831, reprinted Hildesheim: Olms, 1986.
Saidan, Ahmad. *Al-fuṣūl fi 'l-ḥisāb al-hindī li-(...)'l-Uqlīdisī.* Amman: Al-lajna al-urdunnīya li'l-ta'rīb wa'l-nashr wa'l-tarjama, 1973 [text]. *The Arithmetic of Al-Uqlīdisī.* Dordrecht: Reidel, 1978 [translation].
Saidan, Ahmad. *Abū Manṣūr al-Baghdādī: Al-Takmila fi 'l-ḥisāb.* Kuwait: Maṭba'at ḥukūmat al-Kuwayt, 1985.
Schoy, Carl. *Die trigonometrischen Lehren des persischen Astronomen (...) Bîrûnî, dargestellt nach al-Qânûn al-Mas'ûdî.* Hanover: Orient-Buchhandlung, 1927.
Sesiano, Jacques. 'Les méthodes d'analyse indéterminée chez Abū Kāmil.' *Centaurus* 21: 89–105, 1977.
Sesiano, Jacques. 'Le traitement des équations indéterminées dans le *Badī' fī 'l-ḥisāb* d'Abū Bakr al-Karajī.' *Archive for History of Exact Sciences* 17: 297–379, 1977.
Sesiano, Jacques. 'A treatise by al-Qabīṣī (Alchabitius) on arithmetical series.' *Annals of the New York Academy of Sciences* 500: 483–500, 1987.
Sesiano, Jacques. 'Two problems of number theory in Islamic times.' *Archive for History of Exact Sciences* 41: 235–238, 1991.
Sesiano, Jacques. *Un traité médiéval sur les carrés magiques.* Lausanne: Presses polytechniques et universitaires romandes, 1996.
Sesiano, Jacques. 'Le traité d'Abū 'l-Wafā' sur les carrés magiques.' *Zeitschrift für Geschichte der arabisch-islamischen Wissenschaften* 12: 121–244, 1998.
Sesiano, Jacques. 'Quadratus mirabilis.' In *New Perspectives on Science in Medieval Islam*, J. Hogendijk and A. Sabra, eds. Cambridge, Massachusetts: MIT Press, forthcoming.
Suter, Heinrich. 'Die Abhandlung über die Ausmessung des Paraboloides von el-asan b. el-asan b. el-Haitham.' *Bibliotheca Mathematica* ser. 3, vol. 11: 11–78, 1910/11.
Suter, Heinrich. 'Das Buch der Seltenheiten der Rechenkunst von Abū Kāmil el-Miṣrī. ' *Bibliotheca Mathematica* ser. 3, vol. 11: 100–120, 1910/11.
Toomer, Gerald. *Ptolemy's Almagest.* New York: Springer-Verlag, 1984.
Woepcke, Franz. *L'Algèbre d'Omar Alkhayyâmî.* Paris: Duprat, 1851.
Woepcke, Franz. 'Notice sur une théorie ajoutée par Thâbit ben Korrah à l'arithmétique spéculative des Grecs.' *Journal Asiatique* ser. 4, vol. 20: 420–429, 1852.
Woepcke, Franz. 'Recherches sur plusieurs ouvrages de Léonard de Pise, III.' *Atti dell'Accademia pontificia de' nuovi Lincei* 14: 211–227; 241–269; 301–324; 343–356, 1860/1.

FOR FURTHER READING

I. Bibliographical works

Brockelmann, Carl. *Geschichte der arabischen Litteratur* (5 vols.). Leiden: E. J. Brill, 1937– 49. [general].

Encyclopaedia of Islam (new edition), *Encyclopédie de l'Islam* (nouvelle édition), H Gibb, *et al.*, eds. Leiden: E. J. Brill, 1960–. [on scientists and topics; value unequal; for topics, see the headings in Arabic transcription, for example *manāẓir* for optics, *wafq* for magic squares].

Gillispie, Charles, ed. *Dictionary of Scientific Biography* (16 vols.). New York: Charles Scribner's Sons, 1970–1980.

ISIS Cumulative Bibliography, a bibliography of the history of science. London: Mansell, 1971–. [lists studies from 1913 on].

Krause, Max. *Stambuler Handschriften islamischer Mathematiker*. Quellen und Studien zur Geschichte der Mathematik, Astronomie und Physik, Abt. B, 3: 437–532, 1936.

Lexikon des Mittelalters, Robert Auty, *et al.*, eds. Munich/Zurich: Artemis, 1977–1999.

Matvievskaya, Galina and Rozenfeld, Boris. *Matematiki i astronomy musulmanskovo srednevekovia i ikh trudy (VIII-XVII vv.)* (3 vols.). Moscow: Nauka, 1983. [a complement to Sezgin].

Pearson, James D., comp. *Index Islamicus, a Catalogue of Articles on Islamic Subjects in Periodicals and Other Collective Publications*. Cambridge: W. Heffer and Sons, 1958–. [lists studies from 1906 on].

Selin, Helaine, ed. *Encyclopaedia of the History of Science, Technology, and Medicine in Non-Western Cultures*. Dordrecht: Kluwer, 1997.

Sezgin, Fuat. *Geschichte des arabischen Schrifttums*. Leiden: E. J. Brill, 1967–. [vol. V deals with mathematics before the mid-eleventh century and list extant studies and manuscripts].

II. General works

Anbouba, Adel. 'L'algèbre arabe aux IXe et Xe siècles. Aperçu général.' *Journal for the History of Arabic Science* 2: 66–100, 1978.

Berggren, John. *Episodes in the Mathematics of Medieval Islam*. New York: Springer-Verlag, 1986.

Braunmühl, Anton von. *Vorlesungen über Geschichte der Trigonometrie* (2 vols.). Leipzig: Teubner, 1900–1903.

Debarnot, Marie-Thérèse. *Al-Bīrūnī, Kitāb maqālīd 'ilm al-hay'a. La trigonométrie sphérique chez les Arabes de l'Est à la fin du Xe siècle*. Damas: Institut français, 1985.

Juschkewitsch, Adolf. *Geschichte der Mathematik im Mittelalter*. Leipzig: Teubner, 1964. [China, India, Islam, Europe]. French version of the section on Islam: Youschkevitch, Adolf. *Les mathématiques arabes (VIIIe–XVe siècles)*. Paris: Vrin, 1976.

Sesiano, Jacques. *Une introduction à l'histoire de l'algèbre*. Lausanne: Presses polytechniques et universitaires romandes, 1999. [selected topics from Mesopotamian to Renaissance algebra, samples of original texts].

Toomer, Gerald. 'Lost Greek mathematical works in Arabic translation.' *Mathematical Intelligencer* 6(2): 32–38, 1984.

Tropfke, Johannes. *Geschichte der Elementar-Mathematik*. Leipzig: de Gruyter, 1921–1924 (7 vols.); new edition of vols. 1–4, Leipzig: de Gruyter, 1930–1940; new edition of vols. 1–3 in one vol., Berlin: de Gruyter, 1980.

III. Fundamental studies by 19th and early 20th century scholars

Schoy, Carl: *Beiträge zur arabisch-islamischen Mathematik und Astronomie* (2 vols.). Frankfurt a. M.: Institut für Geschichte der arabisch-islamischen Wissenschaften, 1988.

Suter, Heinrich. *Beiträge zur Geschichte der Mathematik und Astronomie im Islam* (2 vols.). Frankfurt a. M.: Institut für Geschichte der arabisch-islamischen Wissenschaften, 1986.

Wiedemann, Eilhard. *Aufsätze zur arabischen Wissenschaftsgeschichte* (2 vols.). Hildesheim: Olms, 1970.

Wiedemann, Eilhard. *Gesammelte Schriften zur arabischen Wissenschaftsgeschichte* (3 vols.). Frankfurt a. M.: Institut für Geschichte der arabisch-islamischen Wissenschaften, 1984. [does not include the studies reprinted in 1970].

Woepcke, Franz. *Etudes sur les mathématiques arabo-islamiques* (2 vols.). Frankfurt a. M.: Institut für Geschichte der arabisch-islamischen Wissenschaften, 1986.

Y. TZVI LANGERMANN AND SHAI SIMONSON

THE HEBREW MATHEMATICAL TRADITION

PART ONE (Y. Tzvi Langermann)

Numerical Speculation and Number Theory. Geometry

PART TWO (Shai Simonson)

*The Evolution of Algebra**

In Part One of this essay, we look in detail at the arithmetic and numerology of Abraham Ibn Ezra, and we continue with a broad description of Hebrew contributions in geometry. In Part Two, we shift the focus to algebra and its evolution from a geometric to a combinatorial subject, as seen in the works of Levi ben Gershon and Abraham Ibn Ezra.

NUMERICAL SPECULATION AND NUMBER THEORY

Abraham Ibn Ezra (1092–1167) is the central figure in the Hebrew mathematical tradition. It was he who wrote what is probably the most influential Hebrew treatise on arithmetic, *Sefer ha-Mispar* (*Book of the Number*); *inter alia*, that book seems to have played an important role in the transmission of the Hindu-Arabic numerals to the west. In addition, it was he who, by means of some enticing, cryptic remarks in his biblical commentaries, suggested a deep mathematical interpretation for some fundamental religious concepts. His commentaries were very widely read; they engendered a flock of supercommentaries whose authors did their best to spell out in full the ideas which Ibn Ezra only hinted at. Ibn Ezra also wrote *Sefer ha-Eḥad* (*Book of the One*), an arithmological[1] monograph, and *Sefer ha-Shem* (*Book of the Name*), a reflection upon the divine names in which some mathematical themes are developed as well.

Ibn Ezra's most significant musings are found in a few excursuses from his verse by verse explication of the Torah. Let us look at one of these, which is found in his remarks to Exodus 33:21. The properties of the tetragrammaton YHWH are discussed, and the numerical values of the letters of the Hebrew

* Supported by Grant STS-9872041 from the National Science Foundation.

H. Selin (ed.), Mathematics Across Cultures: The History of Non-Western Mathematics, 167–188.

alphabet ($Y = 10$; $W = 6$; $H = 5$) are the basis for Ibn Ezra's observations. We shall translate the text phrase by phrase, elaborating within square brackets upon the extremely terse wording; we readily acknowledge our debt to some supercommentaries written in the fourteenth century.

Said Abraham, the author. I have already commented upon the name that is written but not pronounced. It is a proper noun [shem ha-esem, literally, 'the name of identity'], and its identity is the Glory. When you add up all of the letters, it amounts to 72. [The sum here is: Y + YH + YHW + YHWH = 10 + 15 + 21 + 26 = 72.] This why our sages said that it is the ineffable name. [In fact, it seems that the notion of a name composed of 72 letters is post-Talmudic.] If you add the square of the first to the square of the middle, in truth it is like the number of the Name. [The squares on the first number, 1, and on the middle of the first ten digits, i.e. 5, add up to 26, which is also the sum of the numerical values of YHWH.] Likewise the conjunctions of the five planets. [There are 26 possible combinations of the five planets, taken five, four, three, or two at a time.] So also, if you add up the letters which a person pronounces in half of the Name, they are equal to the number of the Name. [If one pronounces the first letters, *yod* – written YWD, and numerically equivalent to 20, and *hē*, written HA and numerically equivalent to 6, then they too add up to 26.] If you take the sum of the squares of the even numbers of the first grade, it is equal to the collective sum of the half of the name. [$2^2 + 4^2 + 6^2 + 8^2 = 120 = 1 + 2 + 3 + \ldots + 15$; 15 is the value of YH, which is half of the Name.] If you multiply half of the Name by the other half, the result is the [sum of the] squares of the odd numbers [of the first grade]. [$YH \cdot WH = 15 \cdot 11 = 165 = 3^2 + 5^2 + 7^2 + 9^2$.] If you subtract the square on the first from the square on the [first] two, the result is the cube on the second letter. [The square on the first letter, $Y = 10$, is 100; and the square on the first two letters, $YH = 15$, is 225; their difference is 125, the cube on the second letter, $H = 5$.] And if you subtract the square on the first two from the square on the first three, the result is the cube on the third letter. [The square on the first three letters, $YHW = 21$, is 441; and $441 - 225 = 216 = 6^3$.] This Glorious Name is the One which is self-standing. It does not require anything before it. For if you look at it from the aspect of counting, where it is the beginning of the All, and all counting derives from ones, [then] it is the one which is the All.

It is important to define as precisely as we can the type of 'number mysticism' which Ibn Ezra practices. He is drawing upon the neopythagorean tradition, in particular, the work of Nicomachus of Gerasa (second century), whose widely repercussive *Introduction to Arithmetic* devotes considerable attention to series and their sums. As we have just seen, this is just the type of mathematical insight which appeals to Ibn Ezra. Moreover, 'the All' is an important neoplatonic concept; and neoplatonic cosmology often went hand in hand with neopythagorean numerical speculations.

In particular, it is crucial to distinguish between Ibn Ezra's enterprise and the very popular hermeneutical exercise known as *gematriah.* The latter find deep secrets in the numerical values of Hebrew words. Ibn Ezra unequivocally distances himself from the practice, for example, in his explication of Genesis 14:14, which narrates Abraham's mobilizing his 318 interns in order to rescue his nephew Lot. In his commentary to the verse, Ibn Ezra takes the bold step of rejecting a Talmudic homily (*Nedarim* 32a) which asserts that the 318 interns refer to just one person, Abraham's faithful slave Eliezer, since, in gematriah, ALY'ZR = 318. He states emphatically, 'Scripture does not speak in gematriah.'

According to the practitioners of gematriah, secrets are contained in the words and letters of the Hebrew alphabet; arithmetic is a tool for ferreting out these secrets. In Ibn Ezra's view, the matter is reversed: it is number which contains the secrets, and the Hebrew spellings reflect the deep meaning which these numbers have. Thus, for example, the divine names YHWH and AHYH (pronounced *eheyeh*, it is the 'I am that I am' of Exodus 3:14) are spelled with the letters which stand for the numbers 1, 5, 6, and 10, precisely because these numbers have tremendous mathematical import. As Ibn Ezra explains in the third chapter of his *Sefer ha-Shem*, the number one is 'the cause of all number, but is not itself a number.' Ten resembles one, since it is both the end of the units and the beginning of the tens. Moreover, 5, 6, and 10 all share this property in common: all powers of these three numbers will maintain in the final digit the same final digit which they have in the first power. (So, e.g. $5^2 = 25$, $5^3 = 125$, and so forth). It seems to me that the key concept here is cyclicity or replication, an idea which resonates well with the Pythagorean concept that numbers are the building blocks of creation. The other properties of these letters, their shapes, and their application in Hebrew grammar, all ramify from their peculiar mathematical characteristics in the base ten system.

Ibn Ezra was interested primarily in numbers rather than shapes; he specialized in arithmetic, not geometry. He was expert in astronomy and, accordingly, he was well versed in geometry, as every astronomer must be. He seems to have regarded the circle as a sacred form. The Hebrew letter *yod*, which stands for the number ten, has the shape of a semi-circle, 'to indicate the whole [circle]. The reason is, that it encircles the All' (*Sefer ha-Shem*, ch. 3). In his shorter commentary to Exodus 26:37, he depicts the candelabrum in the Tabernacle as semicircular; this unusual explication must also be traced to the meaning which he finds in that geometric form. In one passage which Ibn Ezra included both in his commentary to the Torah (*ad* Exodus 3:14) and, with some elaboration, in his *Sefer ha-Shem* (Chapter 6), he tied together the special properties of the circle and the number ten. As usual, his remarks are laconic and cryptic, which amplified the interest which they generated. Isaac Israeli, the eminent fourteenth-century astronomer, and Mordecai Comteano, the fifteenth century Byzantine polymath, both spelled out the theorem which Ibn Ezra only hinted at. With the aid of simple modern algebra, the purpose of Ibn Ezra's construction is readily perceived.

The passage reads as follows (in the commentary to Exodus): 'If we make the diameter of the circle equal to this number [ten], and draw a chord in its third, then the number of the isosceles triangle will be equal to the number of the perimeter, and so also the rectangle within the circle.' In Figure 1, the diameter AB is assigned the number 10. Chord CD cuts off one-third of the circle, as does chord EF. Ibn Ezra asserts that the area or 'number' of triangle ACD = rectangle EFDC. This is easily proven. Right triangles ECI and JIA are equal, as are right triangles FDK and JKA. The trapezium CIKD is common to triangle ACD and rectangle EFDC. Adding to the trapezium triangles ECI and FDK produces the rectangle, and adding their equal, i.e. triangles JIA and JKA, produces the triangle ACD. Hence triangle ACD = rectangle EFDC.

This 'number', in turn, is said to be equal to the perimeter of the circle, provided, we recall, that the length of the diameter has been given as 10. To

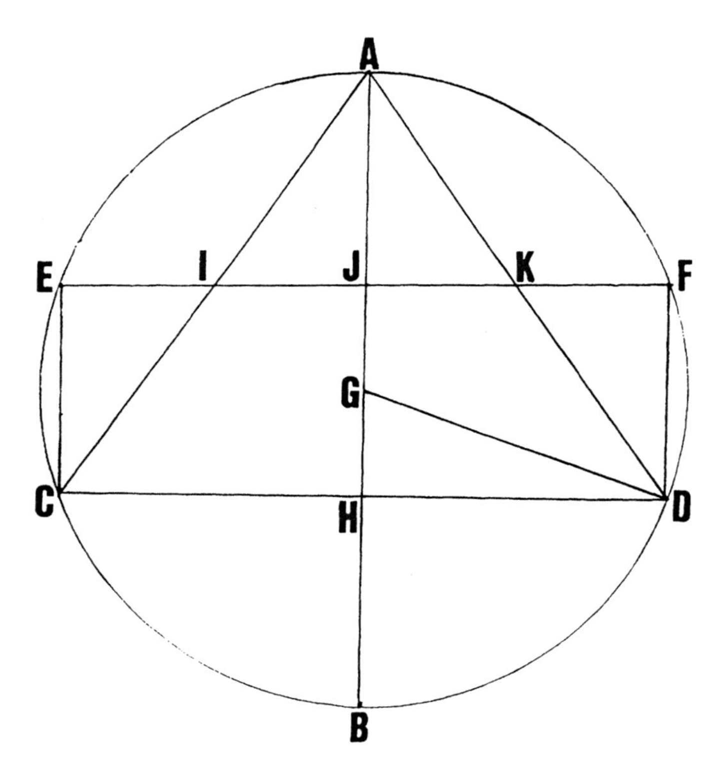

Figure 1 Ibn Ezra's construction.

be sure, the numbers of the triangle and rectangle are in square units, whereas the perimeter is a linear measure, but Ibn Ezra is interested only in the number, not in what it measures. Now if we call the diameter d, then $\mathrm{GH} = \frac{d}{6}$ and $\mathrm{GD} = \frac{d}{2}$. Applying the Pythagorean theorem, we find that $HD = \frac{(d\sqrt{2})}{3}$, so $CD = \frac{(2d\sqrt{2})}{3}$, and the area of rectangle $\mathrm{EFDC} = \frac{(2d^2\sqrt{2})}{9}$. The perimeter is πd. Since, as emphasized above, we are interested only in the numbers, we set the perimeter equal to the area of the rectangle; after dividing through by d, we are left with $\pi = \frac{(2d\sqrt{2})}{9}$. Simple computation (on a pocket calculator) shows that $\frac{(2\sqrt{2})}{9}$ is approximately equal to 0.314. Hence, only when $d = 10$ will the rectangle yield a fairly good approximation to π. This, we believe, is the property which Ibn Ezra wished to demonstrate.

Like other pre-modern mathematicians, Ibn Ezra regarded neither zero nor one as numbers. Zero serves only as a place-marker in decimal notation, and one is the primal source from which numbers are generated. Thus the first true number is two. As he states in his *Sefer ha-Eḥad*, 'two is the beginning of all number.' Nonetheless, here and there in his mathematical writings, Ibn Ezra treats both zero and one just as he would any other number. We have already seen that he groups together the zero in 10 along with 5 and 6 when noting that all three are preserved in successive powers of their respective numbers.

As for one: among the first properties listed in *Sefer ha-Eḥad* are the facts that one is 'square root and cube root, square and cube.' However, at the beginning of the seventh chapter of his *Sefer ha-Mispar*, he offers a prosaic explanation for one being equal to its own square, namely, that the squares of fractions are less than their roots, and the squares of whole numbers are greater. On the other hand, it is noteworthy that Ibn Ezra – in contravention of the arithmological tradition – does not include a chapter on the number ten in his *Sefer ha-Eḥad*. True, *Sefer ha-Eḥad* is concerned almost exclusively with mathematics and, as Ibn Ezra states at the end of the chapter of the one, 'computation with the ten resembles that of the one.' In other words, with the number ten, we have simply moved into the next grade or power of numbers. Nonetheless, astrological, cosmic, and other meanings are adduced in *Sefer ha-Eḥad* (especially in connection with the number 4), and moreover, the arithmological tradition – and Ibn Ezra's own religious musings, as we have seen above – attaches considerable significance to the number 10. Ibn Ezra's decision to omit this number from his arithmology demonstrates, at the very least, that he took the conformity of the one to the ten very seriously indeed.

GEOMETRY

The branch of mathematics in which Jews applied themselves most diligently was geometry. There are a number of possible explanations for this. First, geometry was intimately connected with logic, especially logical demonstration. In the medieval period, algebraic rules were often merely illustrated by way of numerical example, but geometrical propositions – including those that are the equivalent of such basic algebraic formulae as the binomial expansion – were proven rigorously. Indeed, geometrical demonstration in some ways represented to thinkers of the period the pinnacle of scientific certainty, an ideal which other branches of learning could strive to emulate but, in general, not attain.

Closely connected to this is the fact that many of the basic concepts of medieval natural philosophy, including some notable notions which were of crucial significance for theology, were debated within a geometrical context. The very intricate (and as yet unstudied) discussions in Levi ben Gershon's commentaries on Averroes' *Physics* (who, in turn, is commenting upon Aristotle) are part of this tradition. Theologians generally agreed that the deity was the infinite, unbounded, noncorporeal mover of a finite and corporeal universe. Notions such as infinity, motion, and even corporeality were analyzed geometrically. Hebrew theological tracts very often contain geometrical proofs connected to some of these notions.

Finally, we should take note of the close connection between geometry and astronomy. The Hebrew tradition in astronomy was very strong, and astronomers used geometrical models to describe the planetary motions. A sound education in geometry was requisite for any serious study of astronomy.

The corpus of Hebrew geometrical writings consists almost exclusively of translations. At first this may appear a bit strange: why are there not a few works on arithmetic, and even some on algebra, but almost none in geometry?

The simple answer is that Hebrew scholars, like those throughout the West and a good part of the East, relied upon the great synthesis of Euclid. The *Elements* were translated from the Arabic into Hebrew, along with the commentaries of al-Fārābī and Ibn al-Haytham and a number of glosses which explained differences between the two Arabic translations (by Thābit and al-Hajjāj). A smaller number of Arabic writings connected to the *Elements* (including the Arabic translation of the *Elements* itself) were transcribed into the Hebrew alphabet but not translated. Euclid's short optical treatises were also available in Hebrew, as were some writings of Archimedes and Eutocius.

There exist as well a few precious codices containing unique Hebrew translations of geometrical tracts – in some cases, the only surviving copy of a thread in geometrical thinking which occupied Hellenistic and Arabic as well as Jewish mathematicians. These manuscripts are a blessing for historians of mathematics, and they also tell us something about the cultural priorities of the medieval Jewish communities. The most important codex of this type is ms Oxford-Bodley d. 4 (Neubauer and Cowley 2773). Much like a modern university professor, worried that one shoddy piece of work could destroy a reputation established by years of hard work, the translator, Qalonymos ben Qalonymos (d. after 1329) appended a personal note begging the indulgence of his readers, and informing them that he was forced to work quickly and from a defective manuscript, so eager was his patron to obtain a Hebrew copy. Mss. Hamburg, Levi 113, and Milan, Ambrosiana 97/1, are two other noteworthy codices.

A few distinct topics generated particular interest, and around them there evolved a distinct Hebrew tradition. Perhaps the most important of these is the corpus of demonstrations of the property of an asymptote to a curve. Maimonides, in his *Guide of the Perplexed*, part 1, chapter 73, noted the asymptotes to the hyperbola whose properties are demonstrated by Apollonius in his *Conics*. Maimonides studied the *Conics* in Arabic. However, that text was never translated into Hebrew, and Hebrew readers of the *Guide* required an independent demonstration of this particular property. In fact, the notion that two lines could continuously approach each other without ever meeting stimulated thinkers in other cultures as well. In addition to Apollonius' discussion (for those with access to this *Conics* and the necessary background in order to understand it), Nichomedes' proof utilizing the conchoid was studied, and some original proofs based on Euclid were expounded. However, it seems that, in the wake of Maimonides' statement, Jewish mathematicians applied themselves to this problem with particular vigor. At least six distinct proofs are presented in the Hebrew literature (Lévy, 1989). Although the passage from the *Guide* was the immediate stimulus, interest in this issue finds its place in larger movements in cultural history; these have been explored by Freudenthal (1988).

Another interesting and hitherto unexplored Hebrew tradition is comprised of the texts on spherical geometry (often called sphaerics for short). Mastery of this branch of geometry was a necessary prerequisite for serious students of astronomy. However, the codicological evidence – by which I mean the existence of numerous codices exhibiting similar features and transmitted independently of astronomical writings – suggests that there was considerable interest in sphaerics from the point of view of pure mathematics. At the core of this tradition are two Hellenistic texts, treatises on the sphere written by Menelaos and Theodosios. The Hebrew versions were – as was nearly always the case[2] – translations from the Arabic. Short tracts by Jābir ibn Aflah and Thābit ibn Qurra on transversals are often included. Occasionally some other relevant materials are covered; for example, the Milan codex cited above contains some lengthy and unstudied commentaries.

What gives this tradition its coherence and uniqueness are the extensive glosses that are found in the margins of most manuscripts. Most of these are initialed; the signatures have not been deciphered. Many are signed by the Hebrew letters *dalet* and *tav*, which may mean *divrei Thābit* (the words of Thābit) but also *divrei talmid* (the words of a student).

Hebrew scholars were not on the whole as excited by the quadrature of the circle as were their Latin counterparts. There exists but a single Hebrew monograph on the subject, written by a certain 'Alfonso' who, recent scholarship suggests, is the fourteenth century apostate and philosopher, Abner of Burgos. Unfortunately, his treatise, *Meyashsher 'Aqov* (Straightening out the Crooked) survives in only one, incomplete copy. Alfonso clearly realized that squaring the circle was impossible within the framework of conventional, Euclidean geometry; he set as his goal the description of some sort of 'higher reality' where this would be possible. The surviving fragments establish the philosophical and mathematical foundations upon which, presumably, this higher reality is constructed. The subtlety and erudition displayed in this treatise are in a class by themselves, and it is a pity that the complete treatise is not extant.

By way of Aristotle, Jewish scholars were acquainted with one of the earliest Greek attempts at a solution, namely Hippocrates' quadrature of the lunule (a crescent shaped figure constructed from two circles). According to Aristotle, Hippocrates thought that his successful squaring of the lunule indicated that the entire circle could be squared, and the Stagirite made use of this as an example of faulty reasoning. About half a dozen Hebrew manuscripts contain various reconstructions of Hippocrates' proof, usually as glosses to Aristotle or his commentators.

THE EVOLUTION OF ALGEBRA

When we say algebra, we mean the study of solutions of equations, and polynomial identities. In ancient times, this was intrinsically tied up with geometric notions. The connection of algebra with geometry is prominent in Greek mathematics, and continues in the geometric proofs of the algebraic results of Islamic mathematicians. Eventually, this dependence disappeared, and the full power of algebra came to bear with symbolic notation and combinatorial identities.

When one looks at the history of algebra by concentrating on results, it seems to have made little progress for over 3000 years and then taken a sudden leap. After all, the solutions to linear and quadratic equations were well known to the Babylonians, and the solution to the general cubic equation did not come until the Renaissance. Throughout these 3000 years, the Greeks, Indians, Chinese, Muslims, Hebrews and Christians seem to have done no more than present their own version of solutions to linear and quadratic equations, which were well known to the Babylonians. Diophantus, Bhaskara, Jia Xian, al-Khwārizmī, Levi ben Gershon and Leonardo of Pisa, all provide examples of this.

However, the history of algebra is subtler than that, with small changes accumulating slowly. There were many concepts which needed to mature before the breakthrough in algebra was ready to occur. Independence from geometry, a comfortable symbolic notation, a library of polynomial identities, and the tool of proof by induction, were all stepping stones in the history of algebra. Hence what at first seems like new presentations of the same old stuff needs to be considered more carefully.

To this end, to appreciate this particular view, it is useful to focus upon the liberation of algebra from its geometric roots and its subsequent connection with combinatorics. Our intention here is to present the algebra of Levi ben Gershon and Abraham ben Meir ibn Ezra, and to argue that although their results were perhaps already well known, their approach represents an evolution of algebra from a geometric subject to a combinatorial subject.

LEVI BEN GERSHON AND *MAASEH HOSHEV*

Levi ben Gershon (1288–1344), rabbi, philosopher, astronomer, scientist, biblical commentator and mathematician, was born in Provence in the south of France and lived there all his life. Through his writings, he distinguished himself as one of the great medieval scientists and a major philosopher. He wrote more than a dozen books of commentary on the Old Testament, a major philosophical work, *MilHamot Adonai* (Wars of God), a book on logic, four treatises on mathematics, and a variety of other scientific and philosophical commentaries. *MilHamot Adonai* has a section on trigonometry and a long section on astronomy, including the invention of the Jacob's Staff, a device to measure angles between heavenly bodies which was used for centuries by European sailors for navigation, a discussion of the *camera obscura*, and original theories on the motion of the moon and planets. A complete bibliography on Levi and his work can be found in Menachem Kellner's list (Kellner, 1992).

Levi was highly regarded by the Christian community as a scientist and mathematician. They referred to him as Leo Hebraus, Leo de Balneolis or Maestro Leon. However, despite his originality and reputation, much of Levi's scientific and mathematical work did not heavily influence his successors. It is not clear to what extent he played a role in the transmission of Hellenistic mathematics from the Arab world to Western Europe.

Levi wrote four major mathematical works.

1. *Maaseh Hoshev*, Levi's first and largest mathematics book, is extant in two editions, the first completed in 1321 and the second in 1322 (Simonson, in

press). It is known best for its early illustrations of proofs by mathematical induction (Rabinovitch, 1970).

2. A commentary on Euclid was completed in the early 1320s, with ready access to Hebrew translations. Included is an attempt to prove the fifth postulate.
3. *De Sinibus, Chordis et Arcubus*, a treatise on trigonometry in *MilHamot Adonai*, was completed in 1342 and dedicated to the Pope.
4. *De Numeris Harmonicis*, completed in 1343, was commissioned by Phillip de Vitry, Bishop of Meaux, and immediately translated from Hebrew into Latin. Philip was a musicologist interested in numbers of the form $2^n 3^m$, called harmonic numbers. Levi proves that the only pairs of harmonic numbers that differ by one are (1, 2), (2, 3), (3, 4) and (8, 9). The book is relatively short and the original Hebrew is lost.

Maaseh Hoshev, The Art of Calculation, is Levi's first book. The title is a play on words for theory and practice, *Maaseh* corresponding to practice and *Hoshev* corresponding to theory. Levi writes in his introduction to the book:

> It is only with great difficulty that one can master the art of calculation, without knowledge of the underlying theory. However, with the knowledge of the underlying theory, mastery is easy ... and since this book deals with the practice and the theory, we call it Maaseh Hoshev.

Maaseh Hoshev is a major work in two parts. Part one is a collection of 68 theorems and proofs in Euclidean style about arithmetic, algebra and combinatorics. Part two contains algorithms for calculation and is subdivided into six sections: (a) Addition and Subtraction, (b) Multiplication, (c) Sums, (d) Combinatorics, (e) Division, Square Roots, and Cube Roots, and (f) Ratios and Proportions. A large collection of problems appears at the end of Section (f) in part two. Not including this collection of problems, parts one and two are about the same size, and the problems are about half the size of each part. The text in part two and the problem section often refer back to the theorems of part one. Levi lists Euclid's *Arithmetic*, books 7–9, as prerequisite reading.

SQUARE ROOT EXTRACTION FROM LEVI BEN GERSHON'S *MAASEH HOSHEV*

Before we look at Levi's method, let us review the ancient Babylonian method which is intuitive and serves as a starting point in order to understand Levi's method more easily. This method may be familiar to many readers. Given N, it works by guessing some number as a first approximation of the square root of N, and then improving the guess repeatedly and converging to the correct answer.

Let a_i be the i-th approximation and let $a_1 = 1$ (any number is just as good). Then

$$a_{n+1} = \frac{(a_n + \frac{N}{a_n})}{2}.$$

For example, the first three approximations for $N = 2$ are:

$$a_1 = 1, \qquad a_2 = 1.5, \qquad a_3 = 1.416666$$

The Babylonian method dates from 1800 BCE. The Babylonians left no proof or explanation of their ideas. The method can be thought of both algebraically and geometrically. Nowadays, perhaps due to all the drilling students have in algebra, the idea is often presented with a commonsense algebraic approach. Algebraically, if you guess that the square root of N is *Old*, and *Old* is smaller than the actual square root, then $\frac{N}{Old}$ will be bigger than the actual square root. Then their average,

$$\frac{(Old + \frac{N}{Old})}{2}$$

is a better guess than Old. This is a simple and straightforward explanation, but with a geometric view, numbers are distances and products are areas.

Geometrically, constructing the square root of N can be done by constructing a square whose area is equal to N. This can be accomplished by constructing a rectangle with area N, and repeatedly reconstructing the rectangle until it is close to a square. We start by setting the smaller side of the rectangle to *Old*, and the larger side to $\frac{N}{Old}$. Now we take half the excess of $\frac{N}{Old}$ over *Old* and move it from the bigger side to the smaller side, thereby 'squaring' the rectangle; please see Figure 2. The new square has side

$$\frac{(Old + \frac{N}{Old})}{2}.$$

However, the area of this square is larger than the area of the rectangle by the amount of the small missing square in the bottom right corner of the figure. Hence our new guess is smaller than the actual square root, but larger than our old guess. Our new guess becomes the side of a new rectangle, and we repeat the process. As we do, the small missing square gets smaller and smaller, and hence the side of the squared rectangle gets closer and closer to the actual square root.

LEVI'S METHOD FOR SQUARE ROOT EXTRACTION

Levi's idea is not new. Some readers may recognize it as the method taught to them in school. It is seen in geometric form in Chinese sources that predate him by a few centuries. Indeed, it is natural to think of the algorithm geometrically, but Levi does not. He provides no figures or references to any relevant geometric notion. It is noteworthy that his focus is simply on the algebraic identity $(a + b)^2 = a^2 + b^2 + 2ab$.

Levi's method for extracting square roots of perfect squares from Part 2 Section (e) of *Maaseh Hoshev* is shown below. An explanation of the algorithm in plain English appears afterward with some hints and analysis to help the reader unravel Levi's style. Be forewarned that Levi's writing is difficult. He makes no use of equations, and very little symbolic notation. Even the simplest

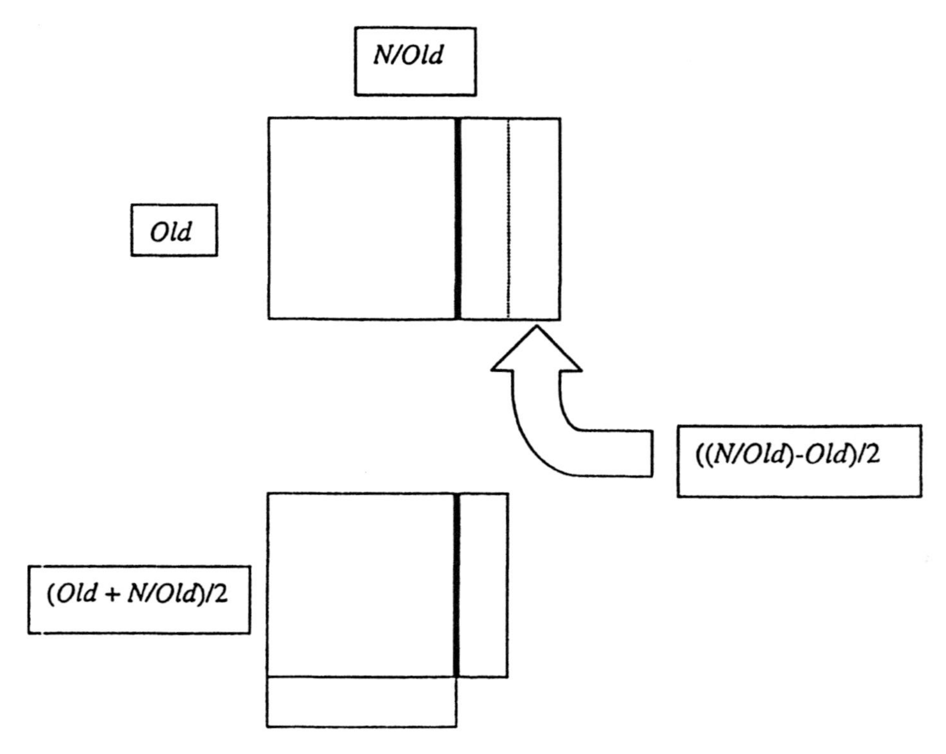

Figure 2 Geometric view of the Babylonian method for square root extraction.

relationships and theorems are stated in longwinded prose. It may be best to read this section lightly and return later for a more careful reading.

> *In order to extract the square root of a perfect square, write the number in a row according to its levels, and examine the last level to see whether it is odd. If it isn't odd, then bring down the digit from the level before it, so that the last digit is from an odd level.*
> *Afterward, find the square closest to this number, but still smaller than it, and write the root of this square down in the root row, beneath the previous row, in the middle level between the first and last levels. And this is what we call the 'result' row. And the square of this root is subtracted from the top row, and what remains is divided by twice the root. But be careful that after the division, there still remains an amount about as big as the square of the root that resulted from the division. Then write this result in the root row, in the level whose distance from the level that you divided is equal to the distance from that level to the first level.*
> *Now take this result and multiply it by twice the root that exists, and by itself. This result is subtracted from the top row. And so you should do, until there is not one thing left in the top row.*
> *For example, if you wished to find the square root of 82646281, since the last level is the eighth, bring down the one before it to give 82. And 81 is the square*

closest to this, and its root is 9. Write 9 down in the root row in the fourth level which is half way between the seventh and the first. And the square of 9 in the fourth, is 81 in the seventh. Subtract this from 82, and 1 in the seventh is left over. And this cannot be divided by twice 9, which is the current root, so bring down the one before it to give 16, and still we cannot divide by twice 9, so bring down the one before it to get 164. Divide this by twice the current root, which is 18, and the result is 9 which is the resulting root. And write this in the root row in the fourth level after the level of 164. Multiply it by twice the current root and itself, and subtract the result from the top row, and 18181 is left. Divide this by twice the current root, which on the previous side is 18 and 2 tenths, and the result is approximately 1 in the first level. And we write this in the root row in the first level. Multiply by twice the current root and itself, and subtract the result from the top row, and there is nothing left in the top row. And so the square root of the given number is 9 thousand and 91, and that is what was requested.
And if you wish, you can check this by multiplying the root row by itself and getting the top row. And this works because of what I have already explained, that when one number is added to another, the square of their sum is equal to the squares of the two numbers and twice the product of one with the other.

			1	8	1	8	1
	1	6	4	6	2	8	1
8	2	6	4	6	2	8	1
				9	0	9	1

For those readers trying to untangle Levi's presentation, there are two things to note. First, Levi is not completely consistent with his reference to the 'top row', nor does his example show a situation where one must 'be careful that after the division, there still remains an amount about as big as the square of the root that resulted from the division', but there is still enough detail here to extract the algorithm. Second, when Levi speaks of writing a digit in a particular 'level', he is referring to the column for that number. For example, 6 in the 4th level is his way of saying 6000.

Levi's idea is to construct the square root one digit at a time, always making sure that the square of the current approximation remains less than the original number N. If the current approximation is a, then we need to account for $N - a^2$ by adding an appropriate b to the approximation. To this end, he reminds us that $(a + b)^2 = a^2 + b^2 + 2ab$, hence adding b accounts for $2ab + b^2$ of the required $N - a^2$. Given N and a, Levi computes b by dividing $N - a^2$ by $2a$. This gives a good guess, since $2ab + b^2 = 2ab$, when b is small in comparison to a.

Here is an example in his style. Let us try 3224990521. Here the last level is the tenth level, so we look at 32 rather than 3. The largest square smaller than 32 is 25, and its root is 5, so we write 5 in the middle column (under the second 9), i.e. 50,000. Then we subtract 25 from 32 and get 7 which is not divisible by 2*5, so we look at 72, which is divisible by 2*5, and the result is about 7 in the fourth column, i.e. 7000. So we take 7 and multiply by $2*50+7$, giving us 749, which is bigger than 724. Whoops! We forgot that one must 'be careful that after the division, there still remains an amount about as big as the square of the root that resulted from the division'. In this case, the square of the root is 49 and there is only 24 remaining after the division, i.e. $724-700$. So we take 6 and multiply by $2*50+6$. This gives 636, which we can subtract from 724, giving 88. So our current root is 56,000. Now we bring down another digit and divide 889 by $2*56$, and we get approximately 7 in the third column (i.e. 700), which we multiply by $2*560+7$ to get 7889, which we subtract from 8899 to get 1010. The current root now is 56,700. We bring down another digit, giving 10100, so that we can divide by $2*567$ and get approximately 8 in the second column (i.e. 80). We multiply 8 by $2*5670+8$ to get 90,784. Note that the 10100 is the prefix of 10,100,521, while 90,784 is the prefix of 9,078,400. So we subtract 90784 from 101005 to get 10221. The current root is now 56,780. Then we bring down another digit to get 102212 so that we can divide by $2*5678$, and we get approximately 9 in the first column (i.e. 9). Then we multiply 9 by $2*56780+9$ to get 1022121, and we subtract this from 1022121 which leaves us with nothing in the top row. So our final answer is 56789.

The following modern presentation of Levi's method is easier to follow but is not faithful to Levi's style or language. To calculate the square root of a perfect square N according to Levi's algorithm:

1. Pair up the numbers from right to left, and count the pairs. This leaves either a pair of digits at the left end or a single digit. Let P be the number of pairs, not counting the last digit or last pair of digits, and let M be the (one or two digit) number which remains on the left end. (For example, given 2415732, $P=3$ and $M=2$; given 2389397496, $P=4$ and $M=23$.)
2. Find the largest perfect square smaller than M, and let its square root followed by P 0's be the first approximation. Let A be this approximation.
3. Let $R=N-A^2$.
4. Divide R by $2A$, and let B be the result excluding all but the most significant digit. (Be very careful though, that $2AB+B^2$ is not larger than R. If so, then subtract one from the most significant digit of B and try again).
5. Let $R=R-(2AB+B^2)$. Let $A=A+B$.
6. If $R=0$ then the answer is A and we are done, otherwise go back to step 4.

Here is a final example. Let $N=16384$.

1. $P=2$ and $M=1$.
2. $A=100$.
3. $R=16384-10000=6384$.
4. $B=6384/200\approx 30$, but $2(100)(30)+(30)(30)$ is larger than 6384, so $B=20$.
5. $R=6384-2(100)(20)+(20)(20)=1984$ and $A=100+20=120$.
6. R is not 0 so go back to step 4.

4. $B=1984/240\approx 8$, so $B=8$.
5. $R=1984-2(120)(8)+(8)(8)=0$, and $A=120+8=128$.
6. $R=0$ so the answer is 128.

Levi's method, at first glance, seems very different from the Babylonian method. There are two main differences. First, Levi always keeps his approximations smaller than the actual square roots and improves the approximations by adding small amounts, while the Babylonian method (except perhaps for the first guess) keeps the approximations greater than the actual square root, subtracting small amounts at each iteration. Second, Levi's method converges from below at a steady rate of one digit per iteration, while the Babylonian method converges from above at an accelerating rate.

For example: Let $N = 16{,}384$. Then Levi's approximations will be 100, then 120, and finally 128. The Babylonian method will have the following approximations: 1, 8192.5, 4097.2499, 2050.6243, 1029.307, 522.6123, 276.9813, 168.0667, 132.7759, 128.0859, etc. It is interesting to compare the two methods with respect to ease of calculation and speed of convergence.

However, the similarity between Levi's method and the Babylonian method can be seen with some analysis and careful notation. If A is the current approximation of the square root, then the new approximation according to the Babylonian method is:

$$\frac{(A + \frac{N}{A})}{2} = A + \frac{(N - A^2)}{2A}$$

and according to Levi it is:

$$A + \left\{\frac{(N - A^2)}{2A}\right\},$$

where the symbols { and } mean that the enclosed division excludes all but the most significant digit, and that this digit is occasionally decreased when necessary, as described in step 4 above.

CUBE ROOT EXTRACTION ALGORITHM FROM LEVI BEN GERSHON'S *MAASEH HOSHEV*

In the same section, Levi also discusses a method for extracting cube roots. He does this with a similar algorithm centering around the identity $(a + b)^3 = a^3 + 3ab(a + b) + b^3$ (Theorem 62 from Part 1 of *Maaseh Hoshev*), in place of the identity $(a + b)^2 = a^2 + b^2 + 2ab$ (Theorem 6 from Part 1 of *Maaseh Hoshev*).

Given N, Levi constructs the cube root as he did for the square root, digit by digit, in order from the most significant to the least significant. Let a be the cube root constructed so far. Levi says to let the next digit be

$$\left\lfloor \frac{(N - a^3)}{(3a(a + 1))} \right\rfloor.$$

This approximation is based on the fact that $N \approx a^3 + 3ab(a + b)$, when $b \ll a$. However, it sometimes makes b too large. Levi warns to be careful that there is enough leftover to account for the b^3, and that the ratio of $N - a^3 - 3ab(a + 1) - b^3$ to $N - a^3$ is bigger than the ratio of $b - 1$ to $a + b$.

If the conditions are not met, he instructs us to subtract one from b, and try again to meet the conditions.

The first condition is similar to his warning in the square root algorithm and ensures that the current cube root is not too large due to the extra b^3. The strange second condition is similar. It is needed to account for the fact that $3a(a+1)$ is smaller than $3a(a+b)$. After b is added to the current cube root, Levi wants

$$N > (a+b)^3 = a^3 + 3a^2b + 3ab^2 + b^3 = (a+b)^3 = a^3 + 3ab(a+b) + b^3.$$

He gives the identity

$$3ab(a+b) = 3ab(b-1) + 3ab(a+1),$$

and so equivalently he wants $N - a^3 - 3ab(a+1) - b^3 > 3ab(b-1)$. This last condition is the same as his required second condition when

$$N - a^3 \approx 3ab(a+b).$$

To prime the pump, Levi gets the first approximation of the cube root of N by examining a left prefix of N of at most three digits. (Note that Levi writes his numbers in left to right order from most significant to least significant digit, as we do, despite the fact that the flow of the Hebrew around it is going right to left. A note of interest to Hebrew readers is that Levi uses the first nine Hebrew letters to stand for the digits 1–9, and does not use the standard Hebrew numbering scheme unless the numbers are smaller than 1000.)

To decide whether the prefix is one, two or three digits, he numbers the digits of N from right to left, in increasing order starting with 1. The prefix desired is the shortest prefix that ends in a digit numbered $3x+1$. He instructs us to find the largest cube smaller than this prefix and place its cube root in column $x+1$, as the first digit.

Levi uses the example 654321.

He considers the prefix 654, and therefore places 8 as the first digit of the cube root in column 2. Hence the first approximation of the cube root is $a =$ 80. He then subtracts 512000 from 654321 to get 142321. He approximates the next digit by dividing 142321 by (3)(80)(81)(3), and ignoring the remainder to get 7. Then he checks his two conditions. For the first condition he calculates $142321 - 7(3)(80)(81) - 7^3 = 5894$, and notes that it is greater than zero. Then he notes however, that $\frac{5894}{142321}$ is not greater than $\frac{6}{87}$. Hence he instructs us to try 6 instead of 7. Repeating the checks of the two conditions, he finds that $142321 - 6(3)(80)(81) - 6^3 = 25465$, and that $\frac{25465}{142321} > \frac{5}{86}$. Then he goes ahead and subtracts $6(3)(80)(86) + 63$ from 142321 to get 18265. As an aside he notes that due to the identity $3ab(a+b) = 3ab(b-1) + 3ab(a+1)$, an easier way to calculate 18265 is simply to subtract 3(80)(5)(6) from 25465. At this point Levi notes that it is impossible to get any closer to the cube root with integers because adding one to the original cube root of 80 adds at least $3(80)(81) =$ 19440, which is already more than what we have currently left over, that is 18265.

$$\begin{array}{l} 86 \\ 654321 \\ \underline{512} \\ 142321 \\ \underline{124056} \\ 18265 \end{array}$$

Levi notes that it may be cumbersome to do the calculations required to make sure that his two conditions hold and thereby ensure that the next digit is not too high. To this end, he gives an alternate *easier* method, guaranteed never to generate digits that are too high. That is, he guarantees that at any stage of the algorithm, $(a+b)^3 < N$, where a is the current cube root, b is the new amount added, and N is the original number. Note that this new method, in contrast to the previous one, does not guarantee that the digit will be the largest possible. The previous method guarantees a lower bound on the new digit, and this method ensures an upper bound.

Let the last digit of a be in the ith column from the right. He instructs us to let the next digit equal the current leftover divided by $\frac{1}{10}$ of $((a+10^{i-1})^3 - a^3)$. In the previous example the current leftover is 142321, $i = 2$, and $\frac{1}{10}$ of $(90^3 - 80^3) = 21700$, therefore the next digit equals $\frac{142321}{21700} \approx 6$. In general, this shortcut sets the next digit equal to the integer part of:

$$\frac{(N - a^3)}{(3a^2 10^{i-2} + 3a10^{2i-3} + 10^{3i-4})}$$

Since $b = 10^{i-2}$ times this digit, this implies that

$$N = a^3 + 3ba^2 + \ 3ba10^{i-1} + b10^{2i-2}.$$

Note that $(a+b)^3 < N$ when $b < 10^{i-1}$. But b is always smaller than 10^{i-1}, since it begins with the $(i-2)$th digit, hence the next digit calculated with this method is never too high.

Levi does not claim that this new easier method always gives the largest possible digit. In fact, it does not. For example, the cube root of 6859 is calculated with 10 as the starting approximation. This leaves over 5859, and according to this method, $\frac{(20^3 - 10^3)}{10} = 700$, and so the second digit equals $\frac{5859}{700} \approx 8$, rather than the correct value of 9. Hence Levi simply provides two methods, one that gives an upper bound on the next digit, and one that gives a lower bound, letting the reader choose.

Levi's algebraic fluency is indicated by his manipulation of the various polynomial identities which form the basis of this cube root algorithm. It is distinct from the geometric approach of his predecessors. We see this same algebraic focus in the problem section and in part one of *Maaseh Hoshev.*

A critical edition of all 21 problems in this section, including details on the two different editions of *Maaseh Hoshev* can be found in 'The Missing Problems of Gersonides, A Critical Edition' (Simonson, in press). Here we present a problem on certain simultaneous linear equations that shows Levi's algebraic sophistication.

Given A, B and C, find X, Y and Z, such that

$$X + \frac{(Y+Z)}{A} = Y + \frac{(X+Z)}{B} = Z + \frac{(X+Y)}{C}.$$

He gives the solution as:

$$X = C + (B - A) + (A - 2)BC,\ Y = X + 2(B - A)(C - 1) \text{ and}$$
$$Z = Y + 2(A - 1)(C\text{-}B).$$

Note that there are in fact an infinite number of solutions to this problem. Levi understands this and explains that these three values are the 'basic' solution. He continues to explain that the presence of an additional constraint, like the value of any of X, Y or Z, or the sum of any two, etc., determines a unique solution. This unique solution is determined by setting up the appropriate proportions with the basic solution. His proof is long and tedious, and is based on theorems 44–52 from Part 1 of *Maaseh Hoshev*.

44. $ab + a = (b + 1)a$, and $ab + b = (a + 1)b$.
45. Given $a < b < c$, then $c(b - a) + a(c - b) = b(c - a)$.
46. Given $2 < a < b$, then $2(a - 2)(b - 1) + b + (a - 2) + (b - a) = 2(a - 1)(b - 1)$.
47. Given $a < b$, then $ba + (b - a) = (a - 1)(b - 1) + b + (b - 1)$.
48. Given $2 < a < b$, then $2(b - 1)(a - 2) + b + (a - 2) + (b - a) = (a - 1)b + (a - 2)(b - 1) + (b - a)$.
49. Given $a < b < c$, and $d = a - 2$, then $2(c - 1)(b - a) + cd + (c - 1)d + c + (b - a) + (c - b) = 2(b - 1)(c - 1)$.
50. Given $a < b < c$, and $d = a - 2$, then $2(c - 1)(b - a) + (cb)d + (c - b)(a - 1) + c + (b - a) = (b - 1)(c - 1)a$.
51. Given $a < b < c$, then $(c - 1)(b - a) + c + (b - a) = c(b - a + 1)$.
52. Given $a < b < c$, then $(c - 1)(b - a) + (a - 1)(c - b) + c + (b - a) = b(c - a + 1)$.

In the first part of *Maaseh Hoshev*, Levi prepares the reader for the algorithms in Part 2 with a collection of theorems. Many of these show Levi's algebraic bag of tricks. Some are shown in the last section and are used in the solution to Problem 21. Other theorems include formulae for sums of consecutive integers, squares and cubes. In the last case, he makes elegant use of mathematical induction (Rabinovitch, 1970). He also has a complete proof of some fundamental combinatorial identities. A list of these theorems with brief commentary in parentheses appears below. Of course, Levi has no notation for these theorems and so they are all heavily worded. For example, Theorem 27 looks like this:

> When you add consecutive numbers starting with one, and the number of numbers is odd, the result is equal to the product of the middle number among them, times the last number.

And Theorem 65 looks like this:

> When there is a given number of different elements and the number of permutations of a second given number of these elements selected from the first given number, and smaller than it, changing in the order and in their elements, is the third given number, then the number of permutations of the successor to the second number from these elements is the product of the third given number by the excess of the first number over the second number.

ALGEBRAIC IDENTITIES

26. $1 + 2 + \ldots + n$, where n is even, is equal to $(\frac{n}{2})(n + 1)$. (Literally, half the number of terms times the number of terms plus 1. Proof works from

outside in, in pairs showing that each pair sums to $n+1$, and there are $\frac{n}{2}$ pairs).

27. $1+2+\ldots+n$, where n is odd, is equal to $\frac{(n+1)}{2}n$. (Literally the middle term times the number of terms. Proof works from inside out, in pairs showing that each pair sums to twice the middle term.)
28. $1+2+\ldots+n$, where n is odd, is equal to $(\frac{n}{2})(n+1)$. (Literally, half the last term times the number after the last term. Proof uses proportions, algebra-like idea and 21.)
29. $1+3+5+\ldots+(2n-1)=n^2$. (Literally, the square of the middle term. Proves it first for an odd number of terms, then an even number.)
30. $(1+2+\ldots+n)+(1+2+\ldots+n+(n+1))=(n+1)^2$.
31. $2(1+2+\ldots+n)=n^2+n$. (Proof uses 30.)
    [Corollary: The sum $1+2+\ldots+n=(\frac{n^2}{2})+(n/2)$.]
32. $1+(1+2)+(1+2+3)+\ldots+(1+2+\ldots+n)=2^2+4^2+6^2+\ldots+n^2$, n even; and $1^2+3^2+5^2+\ldots+n^2$, n odd. (Proof uses 30.)
33. $(1+2+3+\ldots+n)+(2+3+4+\ldots+n)+\ldots+n$
    $=1^2+2^2+3^2+\ldots+n^2$.
    (Proof uses a counting argument.)
34. $(1+2+3+\ldots+n)+(2+3+4+\ldots+n)+\ldots+n+1+(1+2)+(1+2+3)+\ldots+(1+2+\ldots+(n-1))=n(1+2+3+\ldots+n)$.
    (Proof uses a counting argument.)
35. $(n+1)^2+n^2-(n+1+n)=2n^2$.
36. $(1+2+3+\ldots+n)+(2+3+4+\ldots+n)+\ldots+n-(1+2+3+\ldots+n)$
    $=2(2^2+4^2+6^2+\ldots+(n-1)^2)$, $n-1$ even; and
    $2(1^2+3^2+5^2+\ldots+(n-1)^2)$, $n-1$ odd. (Proof uses 33 and 35.)
37. $n(1+2+3+\ldots+(n+1))=3(1^2+3^2+5^2+\ldots+n^2)$, n odd; and $3(2^2+4^2+6^2+\ldots+n^2)$, n even. (Proof uses 32, 34 and 36.)
38. $(n-(\frac{1}{3})(n-1))(1+2+3+\ldots+n)=1^2+2^2+3^2+\ldots+n^2$.
    (Proof uses 32, 33, 34 and 37.)
39. $\frac{n^2-n}{2}=(1+2+\ldots+(n-1))$. (Proof uses 30.)
40. $\frac{n^2-n}{2}+n=(1+2+\ldots+n)$. (Proof uses 30.)
41. $(1+2+3+\ldots+n)^2=n^3+(1+2+3+\ldots+(n-1))^2$.
    (Proof uses 30 and 6.)
42. $(1+2+3+\ldots+n)^2=1^3+2^3+3^3+\ldots+n^3$.
    (Proof by induction using 41.)
43. Let $m=1+2+3+\ldots+n$, then $1^3+2^3+3^3+\ldots+n^3$
    $=1+3+5+\ldots+(2m-1)$.

COMBINATORIAL IDENTITIES

63. $P_{n+1}=(n+1)P_n$, where P_n is the number of different ways to order n elements. (Corollary: $P_n=n!$)

64. $P_{n,2}=n(n-1)$, where $P_{n,m}$ is the number of ways to order m elements out of n.

65. $P_{n,m+1} = P_{n,m}(n-m)$.
(Corollary: $P_{n,m} = n!/(n-m)! = n(n-1)(n-2)\ldots(n-(m+1))$.)

66. $P_{n,m} = C_{n,m}P_m$, where $C_{n,m}$ is the number of ways to choose m elements out of n without regard to order.

67. $C_{n,m} = \frac{P_{n,m}}{P_m}$.

68. $C_{n,n-m} = C_{n,m}$.

ABRAHAM BEN MEIR IBN EZRA'S *SEFER HAMISPAR*

Abraham ben Meir Ibn Ezra (1090–1167), mathematician, philosopher, grammarian and biblical commentator, lived in Spain two hundred years before Levi. Ibn Ezra does not prove his results in a rigorous Euclidean fashion like Levi, nor does he show any explicit use of mathematical induction; however he does present some of the same knowledge of algebraic identities. He was not as original a mathematical thinker as Levi, but more of a careful organizer who presented the known mathematics of his day. Although he was not the mathematician that Levi was, he too showed early signs of an algebraic tradition in the process of extrication from geometric notions. He exhibits certain combinatorial theorems in his works on astrology while discussing the conjunctions of planets (Katz, 1993). However, his only purely mathematical work is *Sefer Hamispar*, in many ways a primitive version of Part 2 of *Maaseh Hoshev*.

Sefer Hamispar, 1146, is a book on arithmetic and calculation in seven sections with a short introduction explaining the decimal system and the use of 0 as a place keeper. Note that both Ibn Ezra and Levi use the decimal system for integers but base 60 for fractions, as was common in that day. The seven sections are: Multiplication, Division, Sums, Differences, Fractions, Ratios, and Square roots.

Ibn Ezra has no algorithm for cube root extraction but does discuss a method for square roots based implicitly on the identity $(a+b)^2 = a^2 + b^2 + 2ab$, and the related approximation

$$\sqrt{(x^2 \pm A)} = \frac{x \pm A}{2}.$$

This identity and approximation were known to the Babylonians and certainly the Greeks and have very simple geometric interpretations, similar to those discussed earlier. Ibn Ezra, however, makes no reference to the geometric viewpoint.

Unlike Levi who provided careful proofs, general examples and special cases, Ibn Ezra's style is to explain his ideas through numerous examples, relying on quantity to substitute for rigor. Here are two of his examples for calculating the closest integer square root of a given integer.

$$\sqrt{7500} = \sqrt{(8100-600)} \approx 90 - \frac{600}{180} \approx 86.$$

$$\sqrt{600000} = \sqrt{(640000-40000)} \approx 800 - \frac{40000}{1600} \approx 74.$$

He checks his calculations by squaring the answers, implicitly using the identity $(a-b)^2 = a^2 + b^2 - 2ab$. For example, $(800-26)^2$ gives

$$640000 - 51600 + 26^2 = 587724.$$

Through many examples he shows how to use the identities

$$ab = \sqrt{(a^2b^2)}, \quad \frac{a}{b} = \sqrt{(\tfrac{a^2}{b^2})}, \quad (\tfrac{a}{b})^2\, b^2 = a^2, \quad \text{and} \quad (a \pm b)^2 = a^2 + b^2 \pm 2ab$$

in order to break larger problems up into smaller ones. For example, he shows how to calculate 19^2 if you know 7^2, by calculating $(\frac{19}{7})^2 = (2+\frac{5}{7})^2 = 2^2 + 2(2)(\frac{5}{7}) + (\frac{5}{7})^2 = 4 + \frac{20}{7} + \frac{25}{49} = 7 + \frac{2}{7} + \frac{4}{49}$, and then multiplying by 7^2 to get $343 + 14 + 4 = 361$.

In Section 3 on sums, he explicitly states the formula for the sum of the first n integers in two different ways similar to Levi's Theorems 26–28. He adds what he claims is his own formula, which is the same as Levi's Theorem 31. He also states the formula for the sum of the first n squares, Levi's Theorem 38, and he adds the well-known formula for summing geometric series.

He concludes the section on square roots with a nontrivial identity, showing that indeed he had a sophisticated sense of algebra. This identity is shown by examples of special cases followed by the general case, albeit no proof is given. Furthermore, this identity can not be found in Levi's work.

He starts with the identity $2(a^2+b^2) - (a-b)^2 = (a+b)^2$. Then he generalizes this to $3(a^2+b^2+c^2) - (a-b)^2 - (b-c)^2 - (a-c)^2 = (a+b+c)^2$. He then states the case for four and finally states the general theorem:

> The square of the sum of any number of integers equals the sum of the squares, times the number of integers in your sum, minus the squares of the differences between all pairs of integers in your sum.

The first special case, $2(a^2+b^2) - (a-b)^2 = (a+b)^2$, is easily thought of geometrically. One way is the figure below:

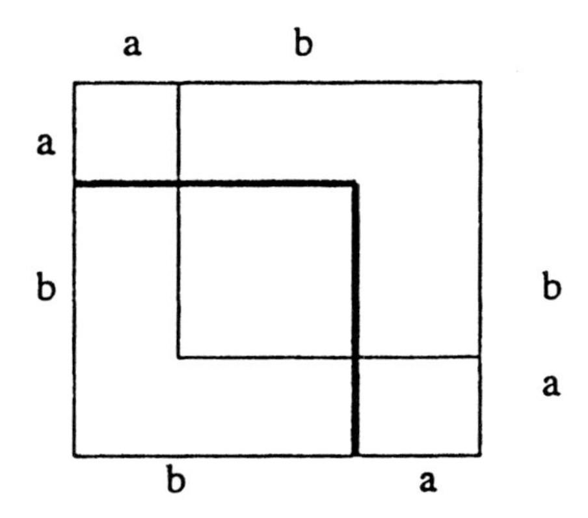

Figure 3 A nontrivial identity.

It is not as easy to imagine how this could be completely generalized geometrically. However it is easy to prove this theorem by induction, using just the simple identities available to Ibn Ezra. In the equations that follow, we assume that the sums over all i and j are from 1 to n. We wish to prove by induction on n that:

$$(*) \qquad n \sum_{i=1}^{n} a_i^2 = \left(\sum_{i=1}^{n} a_i \right)^2 + \sum_{1 \leq i < j \leq n} (a_i - a_j)^2$$

Note that from $(a \pm b)^2 = a^2 + b^2 \pm 2ab$, one can derive:

$$(**) \qquad (n+1)a_{n+1}^2 + \sum_{i=1}^{n} a_i^2 = \sum_{i=1}^{n} (a_i - a_{n+1})^2 + a_{n+1}^2 + 2a_{n+1} \sum_{i=1}^{n} a_i$$

Adding both sides of (**) to both sides of (*), gives us the $(n+1)$st case of the (*) theorem, namely

$$(n+1) \sum_{i=1}^{n+1} a_i^2 = \left(\sum_{i=1}^{n+1} a_i \right)^2 + \sum_{1 \leq i < j \leq n+1} (a_i - a_j)^2.$$

Ibn Ezra concludes by explaining that the number of these differences equals the sum of the consecutive numbers from 1 up to and including one less than number of integers in your sum. That is, he is well aware of the identity:

$$C(n, 2) = 1 + 2 + 3 + \ldots + (n-1).$$

* * *

The work of Medieval Hebrew mathematicians shows the progression of algebra from a geometric subject to a symbolic one. Through algorithms, creative puzzles, inductive arguments and algebraic identities, their work shows an emphasis on the combinatorial aspects of algebra.

NOTES

1 Arithmology is a term coined, I believe, by Armand Delatte, one of the premier historians of Pythagoreanism; it refers to the class of writings dealing with the mathematical properties of the first ten numbers and their significance for religion and philosophy. Note that Ibn Ezra's monograph deals only with the first nine numbers.

2 There are a few instances of direct Hebrew translations from the Greek, for example, a passage from Pappus (Langermann, 1996: 52).

BIBLIOGRAPHY

Chemla, Karine and Pahaut, Serge, 'Remarques sur les ouvrages mathématiques de Gersonide.' In *Studies on Gersonides – A 14th Century Jewish Philosopher-Scientist*, Gad Freudenthal, ed. Leiden: Brill, 1992, pp. 149–191.

Freudenthal, Gad. 'Maimonides' "Guide of the Perplexed" and the transmission of the mathematical tract "On Two Asymptotic Lines" in the Arabic, Latin and Hebrew Medieval traditions.' *Vivarium* 26: 113–140, 1988.

Goldstein, B. R. 'Scientific traditions in Late Medieval Jewish communities.' In *Les Juifs au regard de l'histoire. Mélanges en l'honneur de Bernhard Blumenkranz*, G. Dahan, ed. Paris: Picard, 1985, pp. 235–247.

Katz, Victor J. *A History of Mathematics: An Introduction*. New York: Harper Collins, 1993.

Katz, Victor J. 'Combinatorics and induction in Medieval Hebrew and Islamic mathematics.' In *Vita Mathematica: Historical Research and Integration with Teaching*, Ron Calinger, ed. Washington, DC: Mathematical Association of America, 1996, pp. 99–107.

Kellner, Menachem. 'An annotated list of writings by and about R. Levi ben Gershon. In *Studies on Gersonides – A 14th Century Jewish Philosopher-Scientist*, Gad Freudenthal, ed. Leiden: Brill, 1992, pp. 367–414.

Lange, Gerson. *Sefer Maassei Chosheb-Die Praxis des Rechners, Ein hebraeisch arithmetisches Werk des Levi ben Gerschom aus dem Jahre 1321*, Frankfurt am Main: Louis Golde, 1909.

Langermann, Y. T. and Hogendijk, J. P. 'A hitherto unknown Hellenistic treatise on the regular polyhedra.' *Historia Mathematica* 11: 325–326, 1984.

Langermann, Y. T. 'The mathematical writings of Maimonides.' *Jewish Quarterly Review* 75: 57–65, 1984.

Langermann, Y. T. '"Sefer Uqlidis" by an anonymous author' [Hebrew]. *Kiryat Sefer* 54: 635, 1984.

Langermann, Y. T. 'The scientific writings of Mordekhai Finzi.' *Italia: Studie ricerce sulla storia, la cultura e la letteratura degli ebrei d'Italia* 7: 7–44, 1988.

Langermann, Y. T. 'Medieval Hebrew texts on the quadrature of the lune.' *Historia Mathematica* 23: 31–53, 1996.

Levey, Martin. 'The Encyclopaedia of Abraham Savasorda: a departure in mathematical methodology.' *Isis* 43: 257–264, 1952.

Levey, Martin. 'Abraham Savasorda and his algorism: a study in early European logistic.' *Osiris* 11: 50–64, 1954.

'Levi ben Gershon.' In *Encyclopedia Judaica*, vol. 11. Jerusalem: Keter Publishing House, 1971, pp. 91–94.

Lévy, Tony. 'L'étude des sections coniques dans la tradition médiéval hébraïque: ses relations avec les traditions arabe et latine.' *Revue de l'Histoire des Sciences* 42: 193–239, 1989.

Lévy, Tony. 'Le Chapitre I, 73 du *Guide des Égarés* et la tradition mathématique hébraïque au moyen âge: un commentaire inédit de Salomon b. Isaac.' *Revue des Études juives* 147: 307–336, 1989.

Lévy, Tony. 'Gersonide, commentateur d'Euclide: Traduction annotée de ses gloses sur les Eléments.' In *Studies on Gersonides, a Fourteenth Century Jewish Philosopher-Scientist*, Gad Freudenthal, ed. Leiden: Brill, 1992, pp. 83–147.

Lévy, Tony. 'Hebrew mathematics in the Middle Ages: an assessment.' In *Tradition, Transmission, Transformation*, F. J. Ragep and S. Ragep, eds. Leiden: Brill, 1996, pp. 71–88.

Lévy, Tony. 'The establishment of the mathematical bookshelf of the Medieval Hebrew scholar: translations and translators.' *Science in Context* 10(3): 431–451, 1997.

Müller, Ernst. *Abraham Ibn Esra. Buch der Einheit.* Berlin: Welt-Verlag, 1920.

Rabinovitch, Nahum L. 'Rabbi Levi ben Gershon and the origins of mathematical induction.' *Archive for History of Exact Sciences* 6: 237–248, 1970.

Silverberg, Moshe. *Sefer Hamispar of Rabbi Abraham ibn Ezra.* Frankfurt am Main: Y. Kaufmann, 1895.

Simonson, Shai. 'The missing problems of Gersonides, a critical edition Parts I and II.' *Historia Mathematica*, forthcoming.

Simonson, Shai. 'Mathematical gems of Levi ben Gershon.' *Mathematics Teacher*, forthcoming.

Simonson, Shai. *The Mathematics of Levi ben Gershon, the Ralbag*.' Jerusalem: Bar-Ilan University Press, forthcoming, 2000.

Steinschneider, M. *Die hebraeishe Uebersetzungen des Mittelalters und die Juden als Dometscher.* Berlin: Olms, 1893.

Steinschneider, M. *1893–1901, Die Mathematik bei den Juden.* Reprint, Hildesheim: G. Olms, 1964.

Sternberg, Shlomo, ed. *Studies in Hebrew Astronomy and Mathematics by Solomon Gandz.* New York: Ktav, 1970.

THOMAS E. GILSDORF

INCA MATHEMATICS

This paper is about the South American indigenous group commonly called the Incas and their mathematics. The first term we must clarify is 'Inca', by which we refer to a collection of many groups who had a common government, religion and language, but were of distinct cultural origins. When we speak of the 'Inca Empire', we refer to the territory controlled by the Inca from about 1400 to 1560 AD, though we will see that there were other groups before and during those years that had significant influence on the Incas and their mathematics. The first Incas started near Cuzco in present-day Peru and persistently moved on neighboring groups until they formed an enormous empire that included part or all of Peru, Ecuador, Bolivia, Chile, Argentina, and southern Colombia.

We will first discuss several environmental and cultural factors that may have influenced the development of Inca mathematics, in order to give a cultural perspective and to emphasize the interdependence of mathematics and non-mathematical factors.

GEOGRAPHY, WEATHER AND OTHER DIRECT INFLUENCES ON MATHEMATICS

A first consideration in the mathematics of the Incas and of other groups in the region is that of geography and climate. The Inca territory included regions of coastal desert, high rugged Andes, the Lake Titicaca inter-mountain area, seacoast along the Pacific, and jungle on the eastern edge of the Andes as they descend into the vast Amazon basin. It is worthwhile to add that the coastal desert conditions are quite extreme in parts and that the Andes range contains many areas that are rugged enough to be accessible only by foot.

In terms of weather, there are numerous unstable climate and geological patterns such as earthquakes, droughts, floods and the corresponding effects of periodic occurrences of el Niño.[1] These geographical aspects indicate several needs for mathematics. For example, all successful groups in the dry regions had to have some kind of effective water control in the form of irrigation and aqueducts. Next, those in the high altitudes had to have some form of flexible

H. Selin (ed.), *Mathematics Across Cultures: The History of Non-Western Mathematics*, 189–203.

mountain agriculture, such as terrace farming. Moreover, in many parts of the Inca territory, the people had to construct bridges to cross deep canyons and difficult mountain areas. In effect, civil and agricultural engineering were crucial elements of survival. Finally, as in almost any cultural group, knowledge of astronomy was important in predicting planting and harvesting seasons, approximate weather changes, and general time keeping. Mathematics is also necessary for these activities.

Another factor relevant to the development of mathematics in the Inca region is that of economy. As we have mentioned, the Inca empire consisted of many groups that eventually were absorbed into the Inca system. Starting with very early organized groups and extending even to the present, trade has been an important factor in the societies in the Inca region. In addition, the Inca economy included an extensive taxation system. Later, we will see that this taxation, among other activities, could be recorded on the Incas' *quipus*.[2]

RITUAL, DECORATION, WEAVING AND OTHER FACTORS

Although the previous section describes some uses for mathematics that are environmental and economic, there are other influences on mathematics that are equally powerful in terms of shaping the development and appreciation of mathematics in a society. In fact, such connections between mathematics and culture are part of the rather new field of ethnomathematics. The term has two distinct meanings currently in use. D'Ambrosio (1990), who coined the term, takes it to mean a general anthropology of mathematical thought and practice. Ascher (1991), on the other hand, has defined it as the study of mathematical concepts in small-scale or indigenous cultures. See the articles by Eglash, D'Ambrosio, and Wood in this volume. Ascher (1991), Closs (1986), Pacheco (1998), Urton (1997), and Zaslavsky (1973) also provide good places to start in the study of ethnomathematics. In the case of the Inca, we mention here briefly some connections between mathematics, ritual and decoration.

Throughout the city of Cuzco there are remains of some of the 328 markers, called *huacas*, along imaginary lines called *ceques* (Moseley, 1992: 78–79). These huacas not only had ritual and social significance, they had connections with astronomy and mathematics. In particular, 328 corresponds closely to the sidereal year, and the Inca used several of the ceque lines as part of astronomical observations such as the June solstice. Furthermore, the ceques and huacas were crucial components of the Incan religious ceremonies and in fact the huacas are more accurately described as sacred shrines.

In terms of decoration, the works of Ascher and Ascher (1981), Ascher (1991), Moseley (1992) and Paternosto (1996) contain excellent descriptions of Inca decoration and mathematical concepts such as symmetry. Let us discuss this point about decoration further. It is worthwhile to pause for a moment to remind ourselves that in the following description, we are taking a Western point of view in our analysis, and that this point of view is not necessarily the same as the perception used by the original makers of such decorations.

Suppose we make a pattern by repeating the letter X, as follows: XXXXXXX,

Figure 1 Inca strip pattern. Reprinted with permission from Ascher, Marcia, *Ethnomathematics: a Multicultural View of Mathematical Ideas.* Pacific Grove, California: Brooks/Cole Publishing, 1991, Figure 6.4a, p. 162. Copyright CRC Press, Boca Raton, Florida.

on something like a horizontal strip of a decoration. We notice that this pattern remains unchanged if we reflect across a vertical line, glide it left or right, reflect it across a horizontal line through the middle of the pattern, or even if we rotate it by 180 degrees (using the middle of one X about which to rotate). Thus, this pattern has several types of symmetry. On the other hand, if we use the letter P, we obtain PPPPPPPPPPP. This pattern has a glide symmetry like with X, but has symmetry neither with respect to vertical nor horizontal lines, nor rotation of 180 degrees. In general, collections of symmetries of geometric objects can be described by the mathematical structure known as a group. The interested reader may find more details of such symmetries from an ethnomathematical point of view in chapter six of Ascher (1991) and mathematical details of symmetry groups in chapters 27 and 28 of Gallian (1998). For our purposes, we examine a few patterns specific to the Incas.

Figure 1 shows some examples of Inca patterns. We can see that the second and third patterns are much like our XXXXXX example above, being symmetric with respect to the same kinds of changes. The first pattern shows symmetry with respect to reflections across a vertical or horizontal line (recalling that the reflection across a horizontal line supposes that the line passes through the middle of the figure) and with respect to glides. Yet, the figure is not symmetric with respect to rotation, because it would then appear upside down. Finally, we remark that color can also affect symmetry. Indeed, if we were to color in the right two points of the star shape in the second pattern, then it would no longer be symmetric with respect to vertical reflection.

A FEW RELEVANT GROUPS

There were many groups that formed what we are calling the Incas, and we would like to mention just a few of those groups. These groups had their own mathematical concepts and practices, and as the Incas absorbed these groups into their system, they almost certainly made use of some of these concepts.[3]

Of the many competing cultural groups in what became Inca territory, we can first mention the Moche (also called the Chimor empire or the Chimú). The Moche group was the largest Inca rival, and they inhabited the northern

Peruvian coast during the Early Intermediate period to the Late Intermediate period (about AD 0 to 1470). Their capital was at Chan Chan, and the Moche practiced agriculture in several of the river valleys that descend from the Andes to the Pacific, including the Moche River. The Incas did not overcome them until 1470.

Another group is the Huari of the central Andean highlands of Peru during the Early Intermediate and Middle Horizon periods (about AD 600 to 1000). The Huari engaged in advanced irrigated terrace farming and had quite mathematical artwork. The Aymara kingdoms of the Lake Titicaca region of the Late Intermediate period (about AD 1000 to 1400) consisted of several groups, e.g., the Colla and Lupaqa. Also, the Aymara had their own language group, and some dialects of Aymara are still spoken today. They also had an extensive terrace agriculture and domesticated llamas and alpacas.

The last of this sampling of groups is the Nazca of the southern Peruvian coast of the Early Intermediate period (about BC 400 to AD 500). The Nazca were quite mathematically sophisticated, as is partially seen in the lines and figures they drew in the high Andean desert. It is still not clear what the lines represent, but there are connections between the lines and both the astronomy and ritual of the Nazca.[4]

THE INCA EMPIRE

Tahuantinsuyu, 'Land of the Four Quarters', is the word the Inca used to describe their own territory, and we would like to describe some pertinent features of the Inca empire here. We will see that there are ramifications of our discussion here regarding their mathematics.

The Incas did not use a writing system as we know it. We will see that the *quipu* represents a system of information keeping that could well have served in place of writing. Meanwhile, however, it is also the case that the Inca groups made use of oral tradition, whereby information, history and social practices are passed along via oral descriptions.[5]

There are problems with accurate information on the history of the Incas. Because the only studies of Inca culture have taken place since the Spanish conquest in the 1500s, information is either profoundly culturally biased, as in the case of most Spanish chroniclers, or is a study of a group that has profoundly changed in nature, as is the case in studying present-day descendants. Thus, accurate information is difficult. For our purposes, two sources considered to be relatively accurate are those of Pedro Cieza de León and Felipe Guamán Poma de Ayala.[6] Guamán Poma is particularly known for the many drawings of Inca culture that he made, a few of which are presented in Figure 3.

How could the Incas manage a large territory that covered such challenging terrain? They did, and a primary factor in their management was their remarkable road and communication system, as depicted in Figure 1. Information was conveyed to all of the four corners of Tahuantinsuyu by runners who passed the information from station to station. Although similar to the Pony Express, the difficult changes in geography of the territory made human runners the most efficient and fastest form of communication.

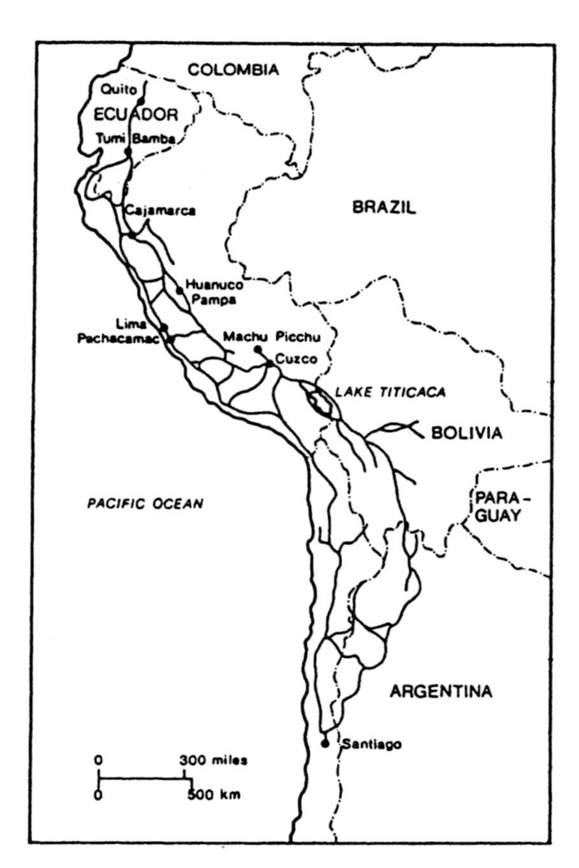

Figure 2 The Inca Road System. From Michael E. Moseley, *The Incas and their Ancestors*, New York: Norton, 1992. Used with their kind permission.

Another feature of the Incas worth mentioning is that as they expanded into regions of other groups, they allowed those groups a certain amount of local control. There are two advantages to this attitude. One is that the subjugated groups would not be as likely to reject Inca rule. The other, relevant to our theme of mathematics, is that by allowing the conquered groups to retain some original culture, the Incas could use and improve mathematical and engineering ideas of those groups, such as astronomy and agriculture. At least some of the mathematics of the Incas was probably diffused from other groups.

INCA NUMBER WORDS

In a study of Inca mathematics, there are two distinct aspects that we must consider: number words and number symbols (representation). These two concepts are not the same. The first things we must clarify have to do with Quechua, the language used by the Inca. Until the 1960s, it was commonly thought that Quechua originated and spread with Inca expansion. Quechua in fact originated in northern central Peru and later split into essentially two branches (Mannheim, 1985; Mannheim, 1991; Stark, 1985; Urton, 1997 and Weber, 1989). For our intentions, we take as our model Weber's description of Huallaga (Huánuco) Quechua, which is often referred to as *Quechua I* or *Quechua B*. As Weber indicates (p. 1), Huallaga Quechua seems to have suffered fewer changes than some other Quechua dialects, which implies that our discussion here of number words should have some accuracy to it. The Incas counted with a decimal system. In the next section on the quipu, we will see that by clever knot placement, this system was even positional.

A basic list of number words from Huallaga Quechua follows:

huk	1	*qanchis*	7
ishkay	2	*pusaq*	8
kimsa	3	*isqon*	9
chuska	4	*chunka*	10
pichqa	5	*pachak*	100
soqta	6	*waranqa*	1000.

The format of more complex number words is: *[multiplier] {nucleus} (adder)*.

The nucleus is always a power of ten. Some examples below demonstrate the above notation, namely, multipliers – [ ], nuclei – { } and adders – ( ).

1. *isqon pachak*: [9] {100} $= 9 \times 10 = 900$.
2. *qanchis chunka pichqa*: [7] {10} + (5) = 75.
3. 347,002 = [[3] {100} ([4] {10} (7))] {1000} (2).
   kimsa pachak chuska chunka qanchis waranqa ishkay.

THE QUIPU (*KHIPU*)

The Incas counted with a decimal system. They had a large territory to control and utilized, for example, a complicated system of taxation. The Incas kept track of this information using a carefully planned record keeping device formed by knotted strings, called a *quipu* (also *khipu*). A quipu serves as a type of database, not a counting device. Blank quipus were first formed, then filled with information in the form of knots on the cords. A quipu consists of several types of cords. There is always a *main* cord, to which other, *pendant*, cords are tied. There can be cords tied to the pendants, called *subsidiary* cords, and there can be *top* cords and *dangle-end* cords. Quipus can have just a few cords, or up to thousands. Some of the cords, such as top cords, often have totals of values of other cords as their values. Marcia and Robert Ascher undertook an exhaustive study of quipus. They provide details on 215 quipus (1978 and 1988) and information on their cultural and mathematical aspects (1981 and Ascher, 1983). In Chapter three, the Aschers explain to us how the carefully planned stone walls of Inca architecture and the geometric style of Inca pottery decorations indicate a cultural tendency toward order and precision. In particular, in Section 12 of that chapter, we see that the quipus, with their orderly designs and their careful arrangement of strings and knots, convey to us an image of a methodical and precise society. In addition, the Aschers point out to us that this orderliness is illustrated by the Incas' practical choice of cloth strings for the quipus, as opposed to wood or stone. The practicality comes from the Incas' desire for a portable record-keeping device[7]. We will discuss the mathematical aspects of quipus in this and the next section.

On quipus there are three types of knots: *Simple* knots representing powers of ten, *long* knots with several loops representing digits between 2 and 9, and *figure eight* knots representing the number one. The spacing of knots relative to other knots indicates value with respect to other knots, so in this way the Inca system is positional. In addition, the color of the strings is important. Different colors can represent different types of data. This organization using color is comparable to using mathematical notation. For example, instead of denoting real number variables by x and vector-valued variables by $\mathbf{x}$, we could just as well distinguish those quantities by the color of the variable.

Here are some examples of basic concepts of quipus.

A person selling jewelry has 12 bracelets, 31 pairs of earrings and 110 small pins. The quipu to record these quantities would have four pendant cords tied to the main cord. Those cords could be described using colors red, blue, green and purple as follows (see Figure 4):

Bracelets: a red cord, with one simple knot for 10 (in the proper place to indicate 10) and a long knot with two twists for 2.

Earrings: a blue cord, with three simple knots for 30, then a figure eight knot for 1.

Pins: a green cord, with a simple knot in the hundreds place and another simple knot in the tens place.

Total: a purple cord with knots to indicate the total: one simple knot in the hundreds place, five simple knots in the tens place and one long knot with three twists. Note how the positions of the knots play a role in the values.

Apart from using a positional numbering system, the Inca knew the concept of zero (Ascher, 1986: 271 and Ascher, 1983). On one hand, by leaving space on strings where there is no data present, the quipus indicate that the Inca utilized the concept of a vacant position that contributes to the value of a number. For example, with a knot in the hundreds place and seven loops for digits, we distinguish between 107 and 117 by observing whether there is a knot in the tens place. In addition, they knew that 'nothing' by itself can be considered a number. To illustrate, consider the following situation (from M. Ascher, personal communication): To keep a record of shoes in several families, we could make a quipu in which a red cord denotes the father's shoes, a white cord denotes the mother's and a blue cord denotes the children's shoes. If the cords are grouped by family, then a cord group with an empty blue cord implies that the children in that family do not have shoes, while a missing blue cord in a cord group implies that the family in question has no children.

Figure 3 Four drawings of quipus. Reprinted from *Nueva Corónica y Buen Gobierno* by Felipe Guaman Poma. Original manuscript at Det Kongelige Bibliotek, Copenhagen, Denmark. Used with the kind permission of the library.

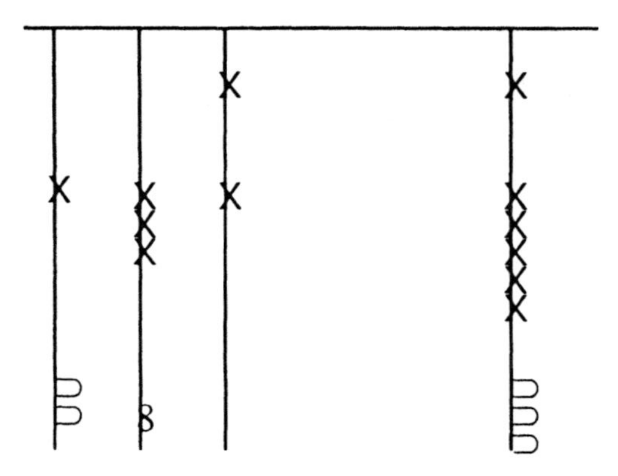

Figure 4 The quipu of Example 4.

MORE COMPLICATED QUIPUS

The previous example indicates a rather basic version of a quipu. The cord and knot structure allows for unlimited levels of complication. In fact, in most of the references where there are discussions of quipus, there are also discussions of a 'quipu maker' or *quipucamayoc*. The quipucamayocs were experts at constructing and interpreting quipus. They had to have significant mathematical knowledge and had to be able to explain the information on complicated quipus to the Inca royalty. Here we have an instance of mathematicians having high social status.

Quipus can be made more complex in several ways. One way is to group colors, while another is to associate groups of cords by spacing. Combinations of spatial grouping and color coding are possible. The following example illustrates one possibility, four colors repeated three times.

In this case we can express the values on the cords as: Each cord is an element of the ith group, jth color, which we may write symbolically as q_{ij}, $i = 1, 2, 3$, $j = 1, 2, 3, 4$.

With more colors or grouped into more sets of color groups, the quipu can be organized on more levels. For example, with sixty cords having five sets of twelve, and each of those sets organized into three groups of four colors, we may indicate the cords as: q_{ijk}, $i = 1, \ldots, 5$, $j = 1, 2, 3$, $k = 1, \ldots, 4$.

Another way to arrange a quipu utilizes a mathematical concept known as

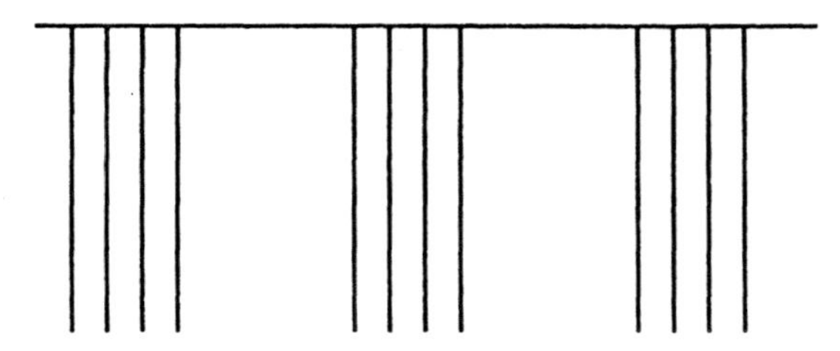

Figure 5 A quipu with four colors repeated three times.

a tree structure. In this case, levels of information with increasing complexity can be obtained by attaching more levels of subsidiaries of pendants to other subsidiaries, pendants, and continuing this process to as many levels as desired. We illustrate with an example.

> Suppose we are considering the hierarchy of a government system that is organized by districts. Assume each district has the same structure given by the tree structure depicted in Figure 6. Thus, each district has an executive position E, and two management positions, M_1 and M_2, with M_1 having four subordinates and M_2 having three subordinates. Next, suppose we would like to log the number of hours worked in seven districts over a span of five years. How would we express this information on a quipu? First, we note that there will be seven trees repeated five times for a total of thirty-five trees. We can let t_{ij} denote each tree, where $i = 1, \ldots, 7$, and $j = 1, \ldots, 5$. We can group the cords in five groups of seven. Now, on each t_{ij}, how do we arrange the cords? By the structure of the tree given, each tree t_{ij} would have one pendant for position E, and two subsidiaries for M_1 and M_2. On the M_1 subsidiary there would be four more subsidiaries for the subordinates and on the M_2 subsidiary there would be two more subsidiaries.

Combinations of patterns and more levels of hierarchies are possible.

The above example can be easily generalized to show that non-numerical data can be stored on quipus. Such information was indeed stored by the Incas. In particular, in chapter 4 of Ascher and Ascher (1981), we see that evidence exists to indicate that *quipucamayocs* held a social status somewhere between that of royalty and common people. Moreover, they worked within the system of Inca governmental bureaucracy. From this, we can conclude that the quipucamayocs stored information about populations, supplies, taxes and other specific data on their quipus. The storing of large amounts of data on quipus could certainly have served as a form of writing, and such a use would at least partially explain the lack of evidence of more familiar forms of writing. From a mathematical point of view there is a question of how mathematically sophisticated the Inca were. Ascher and Ascher (1981) provide descriptions of many quipus from both a cultural and a mathematical viewpoint.

Our last example describes a quipu which is simple in construction but contains striking mathematical properties (Ascher and Ascher, 1981: 149–151).

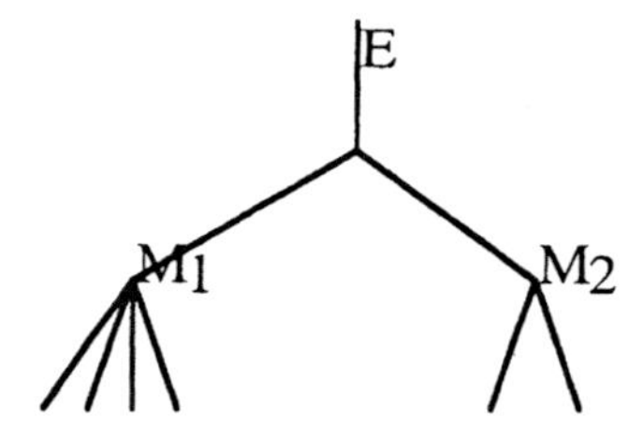

Figure 6 A tree structure for a government system.

The intention of the example is to show that genuine mathematical development was taking place in Inca society.

> Two rather small quipus tied together, one with seven pendants and three subsidiaries, one with three pendants only, for a total of thirteen values. As with the examples above, we denote by q_{ij} the values of cord ij. On a subset of pendants, we arrange the values in tabular form, as in the matrix in Figure 7.

$$\begin{bmatrix} q_{11} & q_{12} & q_{13} \\ q_{21} & q_{22} & q_{23} \\ q_{31} & q_{32} & q_{33} \end{bmatrix}$$

Figure 7 The values of nine cords of the quipu of Example 6. Each entry represents the value of a pendant cord.

Among other properties, the following relations hold:

$$\text{(a)}\quad \frac{q_{11}q_{12}q_{13}}{q_{31}q_{32}q_{33}} = \frac{q_{31}q_{32}q_{33}}{q_{21}q_{22}q_{23}}, \qquad \text{and} \qquad \text{(b)}\quad q_{31}q_{32}q_{33} = q_{11}q_{22}q_{33}.$$

Furthermore, the fractions $\frac{11}{14}$, $\frac{7}{8}$, $\frac{34}{33}$, appear in many places. If we denote

$$B = \tfrac{7}{8}, \quad C = \tfrac{34}{33}, \quad p = q_{12}, \quad \text{and} \quad r = q_{13},$$

then the matrix above can be expressed as in Figure 8:

$$\begin{bmatrix} CB^2 & p & r \\ C^2B^6p & BCp & C^2B^2r \\ CB^4p & Cp & CB^2r \end{bmatrix}$$

Figure 8 The pendant values of Figure 7, relabeled to show the relationships between them.

YUPANA

The last example in the previous section, along with other quipu descriptions in places like Ascher and Ascher (1981), indicate that the Inca performed rather complicated computations. As indicated above, the quipu was a record keeping device and not used for calculating. In fact, the knots on known quipus are tight, implying that the values on them are fixed. This leads us to ask how they made the computations. The answer to this question is still not understood. One possibility arises in one of the drawings of quipus made by Guaman Poma shown in Figure 3 D.

The rectangular grid of solid and unfilled dots appears to be a kind of counting board; however, no explanation of the grid is given. Certainly there would be reason to believe that the grid has mathematical meaning because of its appearance with a quipu. A curious aspect of the grid is that each row contains eleven dots, something that obviously does not coincide with a decimal counting system. In addition, because Guaman Poma's work occurred after the Spanish conquest of Peru, the grid does not necessarily represent a device of the Inca.

On the other hand, proposals of mathematical uses of the grid have been given, with the term *yupana* (Quechua for the verb to count) used to denote it. Several such descriptions can be found in Mackey *et al.* (1990), Wassén (1990), Higuera (1994), and Burns Glynn (1981 and 1990). For example, in Mackey *et al.*, we find an explanation of the grids as kinds of abaci, as we can see in Figure 9. The bottom row would represent, from left to right, units of ones, fives, fifteens and thirty. The second row from the bottom would represent units of tens, counted by fives, respectively. The next row would count by hundreds, then the row after that by thousands, and finally, in the top row, by tens of thousands. An example is the following: If in the bottom row there are four filled places in the ones box and two in the fifteens box, then the total for that row would be two times fifteen plus four, equaling thirty-four. Note that simply filling or emptying a particular dot could easily change values on such a grid. The mathematical methods for counting that are presented make logical sense, and there are other possibilities, such as filling the eleventh dot as a place holders for powers of ten.

Figure 10 shows photos of three-level rectangular Inca stone figures that resemble grids in some ways, that could also have served as abaci for counting. Radicati (1990: 219–234) proposes three possible uses of the stone figures: as counting devices, as architectural models, or as games. He does not propose

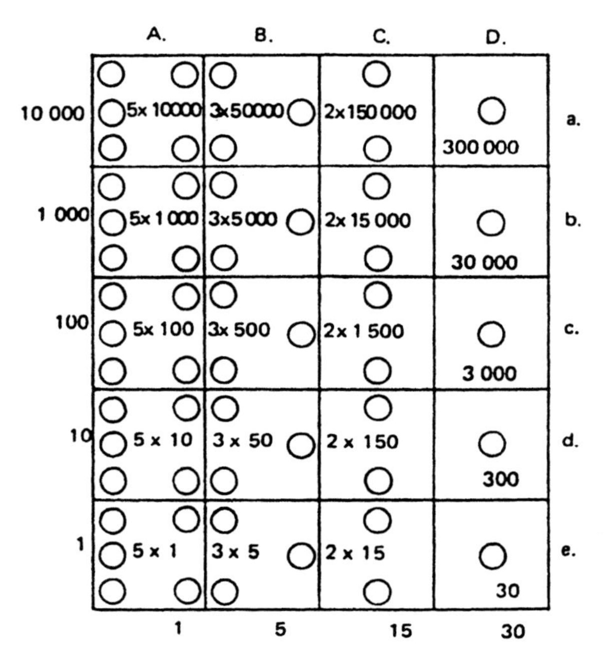

Figure 9 Possible computation scheme for an Inca dot pattern. From Carol Mackey *et al.*, editors, *Quipu y Yupana*. Lima: CONCYTEC, 1990, p. 214.

Figure 10 Stone figures, possibly used for counting. From Carol Mackey *et al.*, editors, *Quipu y Yupana*. Lima: CONCYTEC, 1990, p. 221.

many specific rules for them, but instead discusses anthropological evidence that does or does not seem to support the various possible applications.

Descriptions of observations of Incas counting on these rectangular forms are given, but here we encounter the problem of accuracy. Moreover, the works of Burns Glynn (1981, 1990) also contain interpretations of symbols that appear in weaving patterns as possible hieroglyphs. A conclusive explanation of the grids, rectangular forms, and weaving patterns still has not been given, although the works cited here point toward possible patterns of counting or symbols. More investigation could lead to new information on this.

ACKNOWLEDGEMENTS

The author would like to express his gratitude to several people who made useful comments about the paper and/or about the topic of Inca mathematics. They are: Elisa Alcántara G.; Professors José Vargas, Edgar Vera, Alejandro Ortiz F., Roxana López Cruz, Clara Lucía Higuera Acevedo, Oscar Pacheco, and Gary Urton. I am especially thankful to Professor Marcia Ascher, whose comments and suggestions significantly improved the quality of this paper. Finally, I wish to thank Helaine Selin for her patience with my many questions.

NOTES

[1] See pages 85–86 of Aveni, 1990, or Chapter 1 of Hadingham, 1987, for example.

[2] A good reference to the ideas of trade and interaction of the various groups can be found in Moseley, 1992.

[3] A quite complete general reference on the various groups and their interactions with the Inca is Moseley, 1992.

[4] We would also like to mention in passing the importance of Maria Reiche, a German-born mathematics teacher whose work in the 1940s was crucial to the preservation and study of the Nazca culture and lines. See Aveni, 1990, for a complete discussion of the Nazca, and also Hadingham, 1987, where there is information on the life of Maria Reiche.

[5] See Schneider, 1994 for a general description of oral tradition. Also, an interesting description of Inca culture and literature can be seen in Lara, 1960.

[6] In Ascher and Ascher, 1981: 3, the reader may find a detailed description of original works and translations of Cieza de León's work, and in the bibliography we have listed a reference to the works of Guamán Poma (1936).

[7] See also Section 2 of Chapter 4 of Ascher and Ascher, 1981.

BIBLIOGRAPHY

Alcina Franch, José and Josefina Palop Martínez. *Los Incas, el Reino del Sol*. Madrid: Ediciones Anaya, 1988.

Ascher, Marcia. *Ethnomathematics: A Multicultural View of Mathematical Ideas*. Pacific Grove, California: Brooks/Cole, 1991.

Ascher, Marcia. 'Mathematical ideas of the Incas.' In *Native American Mathematics*, Michael P. Closs, ed. Austin: University of Texas Press, 1986, pp. 261–289.

Ascher, Marcia. 'The logical-numerical system of Inca quipus.' *Annals of the History of Computing* 5(3): 268–278, 1983.

Ascher, Marcia and Robert Ascher. *Code of the Quipu: A Study in Media, Mathematics, and Culture*. Ann Arbor: University of Michigan Press, 1981. (Reprinted as *Mathematics of the Incas: Code of the Quipu*, New York: Dover Publications, 1997.)

Ascher, Marcia and Robert Ascher. *Code of the Quipu: Databook*. Ann Arbor: University of Michigan Press, 1978 and *Code of the Quipu: Databook II*. Ithaca, Ascher and Ascher, 1988 (on microfiche at Cornell University Archives, Ithaca, New York).

Aveni, Anthony, ed. *The Lines of the Nazca*. Philadelphia: The American Philosophical Society, 1990.

Buechler, Hans and Judith-Maria Buechler. *The Bolivian Aymara*. New York: Holt, Rinehart and Winston, 1971.

Burns Glynn, William. *Legado de los Amautas*. Lima: Editora Ital Peru, 1990.

Burns Glynn, William. *La Escritura de los Incas.* Boletín de Lima, No. 12, 13, 14. Lima: Editora Los Pinos, 1981.

Closs, Michael P., ed. *Native American Mathematics.* Austin: University of Texas Press, 1986.

D'Ambrosio, U. *Etnomatematica.* Sao Paulo: Editora Atica, 1990.

Gallian, Joseph. *Contemporary Abstract Algebra.* 4th ed. Boston: Houghton-Mifflin, 1998.

Guaman Poma de Ayala, Felipe. *El Primer Nueva Corónica y Buen Gobierno* (1614), Paris: Université de Paris Istitut d'Ethnologie, 23, 1936. (Original manuscript at Det Kongelige Bibliotek, Copenhagen, Denmark.)

Hadingham, Evan. *Lines to the Mountain Gods; Nazca and the Mysteries of Peru.* New York: Random House, 1987.

Higuera Acevedo, Clara L. 'La yupana Incaica: Elemento histórico como instrumento pedagógico.' In *Proceedings of the Meeting of the International Study Group on Relations between History and Pedagogy of Mathematics*, Sergio Nobre, ed. Brazil: UNESP, 1994, pp. 77–89.

Lara, Jesus. *La Literatura de los Quechuas.* Cochabamba, Bolivia: Editorial Canelas, 1960.

Mackey, Carol and Hugo Pereyra, Carlos Radicati, Humberto Rodriguez, Humberto and Oscar Valverde, eds. *Quipu y Yupana, Colección de Escritos.* Lima: CONCYTEC, 1990.

Mannheim, Bruce. *The Language of the Inka Since the European Invasion.* Austin: University of Texas Press, 1991.

Mannheim, Bruce. 'Southern Peruvian Quechua.' In *South American Indian Languages, Retrospect and Prospect*, Harriet E. Manelis and Louisa S. Stark, eds. Austin: University of Texas Press, 1985, pp. 481–515.

Moseley, Michael E. *The Incas and Their Ancestors; The Archeology of Peru.* London: Thames and Hudson, 1992.

Myers, Sarah. *Language Shift Among Migrants to Lima, Peru.* Chicago: University of Chicago Department of Geography, Research Paper no. 147, 1973.

Pacheco, Oscar. *Ethnogeometría.* Santa Cruz (Bolivia): Editorial CEPDI BOLIVIA, 1997.

Paternosto, César. *The Stone and the Thread; Andean Roots of Abstract Art*, trans. Esther Allen. Austin: University of Texas Press, 1996.

Radicati di Primeglio, Carlos. 'Tableros de escaques en el antiguo Perú.' In *Quipu y Yupana, Colección de Escritos*, Carol Mackey *et al.*, eds. Lima: CONCYTEC, 1990, pp. 219–234.

Schneider, Mary Jane. *North Dakota Indians, An Introduction.* Dubuque, Iowa: Kendall/Hunt, 1994.

Stark, Louisa R. 'Ecuadorian highland Quechua: history and current status.' *In South American Indian Languages, Retrospect and Prospect*, Harriet E. Manelis and Louisa S. Stark, eds. Austin: University of Texas Press, 1985, pp. 443–479.

Stark, Louisa R. 'The Quechua language in Bolivia.' In *South American Indian Languages, Retrospect and Prospect*, Harriet E. Manelis and Louisa S. Stark, eds. Austin: University of Texas Press, 1985, pp. 516–545.

Urton, Gary. *The Social Life of Numbers, A Quechua Ontology of Numbers and Philosophy of Arithmetic.* Austin: University of Texas Press, 1997.

Wassén, Henry. 'El antiguo abaco peruano según el manuscrito de Guaman Poma.' In *Quipu y Yupana, Colección de Escritos*, Carol Mackey *et al.*, eds. Lima: CONCYTEC, 1990, pp. 205–218.

Weber, John David. *A Grammar of Huallaga (Huánuco) Quechua.* Berkeley: University of California Press, 1989.

Zaslavsky, Claudia. *Africa Counts: Number and Pattern in African Culture.* Chicago: Lawrence Hill Books, 1973.

MICHAEL P. CLOSS

MESOAMERICAN MATHEMATICS[1]

OLMEC

The Olmec world has a special place in the sequence of cultures that developed in the region that has come to be known as Mesoamerica (Figure 1). It flourished along Mexico's Gulf Coast between 1200 and 400 BC. Because it exhibited early achievements in art, politics, religion, and economics that appear to be ancestral to later developments in Mesoamerica, it has earned the reputation of being a kind of 'mother culture' of all later civilizations in the region. The same culture is now known to have deep roots in the Pacific littoral of Chiapas state in Mexico and neighbouring Guatemala as well as in the hills of Guerrero state in Mexico. Thus, the Olmec heartland may be more extensive than once believed.

Evidence of Olmec writing is meager. Two short texts, of three signs each, and a handful of possible glyphic notations in iconographic context are known. These all date between 1100 and 400 BC. Two descendent script traditions flanked the Olmec homeland: Oaxacan and Southeastern.

ZAPOTEC

The geographical locus of the Oaxacan script lies mainly in the central valleys of Oaxaca state in Mexico. Beginning around 600 BC, an identifiable culture arose in the region. Since this area has been occupied by Zapotec speakers for millennia and the Oaxacan script has a long history in the area, it is not surprising that scholars have concluded that the Zapotec language underlies the script. The largest urban centre in the region developed at Monte Albán. It reached a peak population between 600 and 800 AD. By the turn of the millennium at 1000, it had entered a state of decline and depopulation.

It is in Zapotec writing that we find the earliest examples of the use of a sacred calendar of 260 days known as the Sacred Round. It is formed by the permutation of the sequence of numbers from 1 to 13 with a sequence of 20 day names. This calendar was of such fundamental importance to the Zapotecs that they often named their children after the date in the 260-day calendar on which they were born, a practice also followed elsewhere in Mesoamerica.

H. Selin (ed.), Mathematics Across Cultures: The History of Non-Western Mathematics, 205–238.

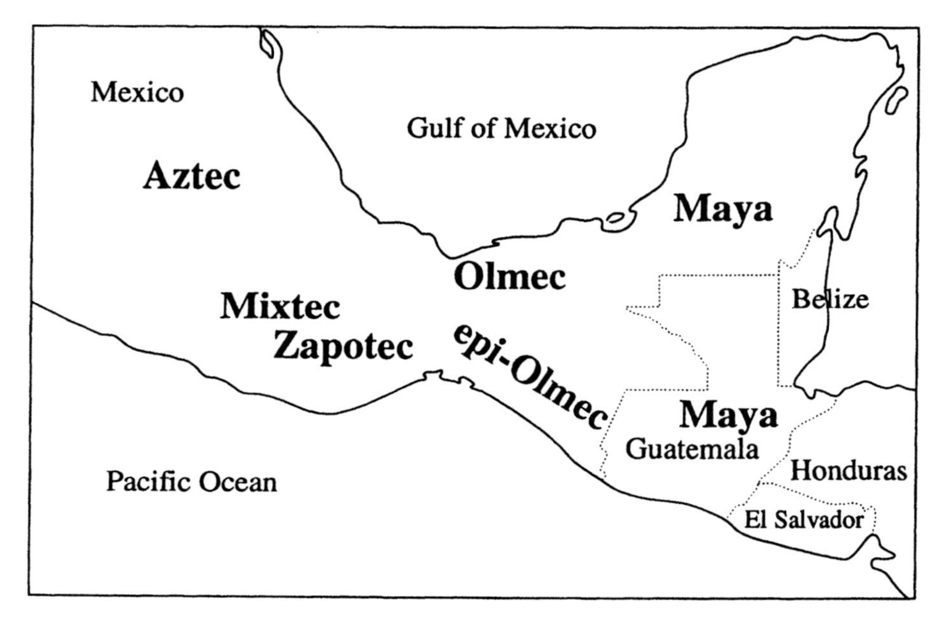

Figure 1 Map of Mesoamerica showing approximate locations of the cultures discussed in this article. [Drawing by M. P. Closs.]

Thus, in texts of the period, a 260-day calendar notation may well be the proper name of an historical individual or the name of a deity rather than a reference to a date in the Sacred Round.

The common numerals among the Zapotec were composed of bars with value 5 and dots with value 1. Since number systems in the region (and in fact throughout Mesoamerica) were vigesimal, that is base 20, numerals for digits from 1 to 19 were required. The digits were represented by a simple tally of bars and dots yielding the desired value. The most common usage of numerals is found in recording dates of the Sacred Round or calendrical names derived from such dates. For these purposes, only the numerals from 1 to 13 were required.

The oldest datable monument in Mesoamerica exhibiting numeral usage is Monument 3 from San José Mogote about 15 kilometres north of Monte Albán. The contemporaneous date of the monument is around 600 BC as determined by the secure stratigraphic context in which it was excavated. It is the earliest dated example carrying a calendrical notation in Mesoamerica. The carved stone, illustrated in Figure 2, shows a naked sacrificial victim with eyes closed and a stream of blood flowing from his open chest after the removal of his heart. The name of the victim is written in a brief glyphic caption below the left shin. This consists of the day sign 'Eye' (glyph L; corresponding to the 16th day, Vulture, in the Aztec day list – after Urcid-Serrano, 1992) above an elaborate numeral 1 informing us that the (calendrical) name of the victim is 1 Eye (1 Vulture).

One of the elite individuals portrayed in a narrow relief on the South Platform at Monte Albán with his accompanying name glyph is illustrated in Figure 3. In this instance, the glyphic name has been deciphered to yield a combined calendrical and personal name similar to those recorded in *lienzos* and colonial legal documents in which names are written in Zapotec using the Latin alphabetic script (Urcid-Serrano, 1992: 288). The name has been read phonetically as **Yoho-Neza-Pe-Loo,** literally 'Temple-Road-Nine-Monkey' and yields the name *Yohoneza Peloo*, that is Walker/Traveler 9 Monkey.

Other examples of Sacred Round dates or calendrical names using bar and dot numerals are shown in Figure 4a–b. Occasionally, in the earliest texts, the Zapotec used fingers to represent single units as indicated in Figure 4c–d, a usage that has not been documented in cases where the coefficients are greater than 2.

The earliest examples of a system of chronological reckoning known as the year bearer system are found in Zapotec writing. It is attested in texts dating from around 300 BC. The system was used to name 365-day years after the 360th day, the so-called year bearer, in the previous 365-day year. The existence of such a system implies the existence of a 365-day solar calendar, sometimes referred to as the Vague Year. This type of calendar is also ubiquitous in Mesoamerica and consists of eighteen 'months' of 20 days with a residual period of 5 days.

The interlocking of the 260- and 365-day calendars meant that only four of the 20 day names could serve as year bearers. In turn, since the day dates

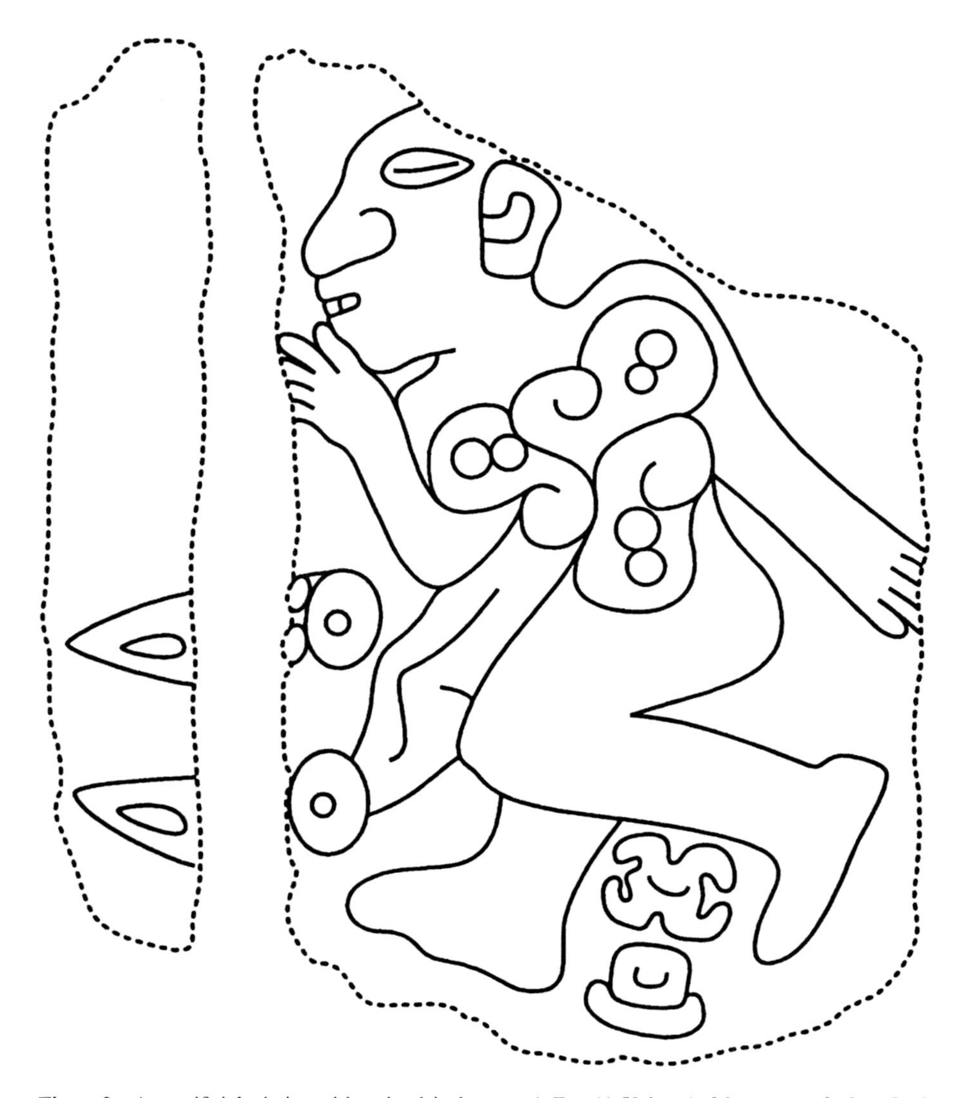

Figure 2 A sacrificial victim with calendrical name 1 Eye (1 Vulture), Monument 3, San José Mogote. [Drawing by M. P. Closs after Marcus and Flannery, 1996: fig. 137.]

admitted only coefficients from 1 to 13, the system generated a cycle of 52 named years. This cycle of 52 Vague Years spans a total of 18,980 days and is commonly referred as the Calendar Round. It is the period in which paired dates from the Sacred Round and Vague Year begin to repeat. In Zapotec writing, the Sacred Round dates that serve as year bearers are marked with a characteristic glyphic prefix. Examples of year bearers are depicted in Figure 4e–f.

There are Zapotec glyphs which sometimes carry numerical coefficients but whose meaning remains opaque. One of them, catalogued as glyph I, has

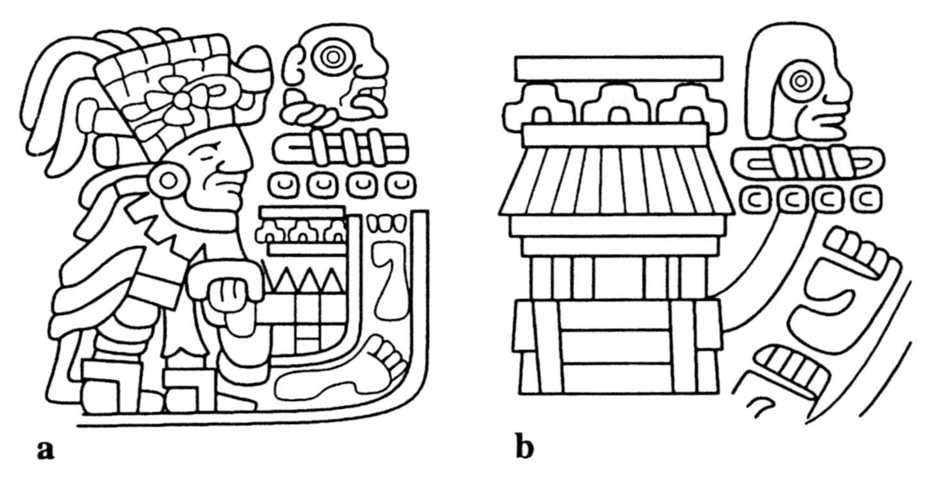

Figure 3 (a) Portrait and name glyph of Yohoneza Peloo (Walker/Traveler 9 Monkey), Monument 9, South Platform, Monté Albán. (b) A second instance of the name glyph of Yohoneza Peloo, Monument 9, South Platform, Monté Albán. [Drawing by M. P. Closs after Marcus and Flannery, 1996: fig. 261.]

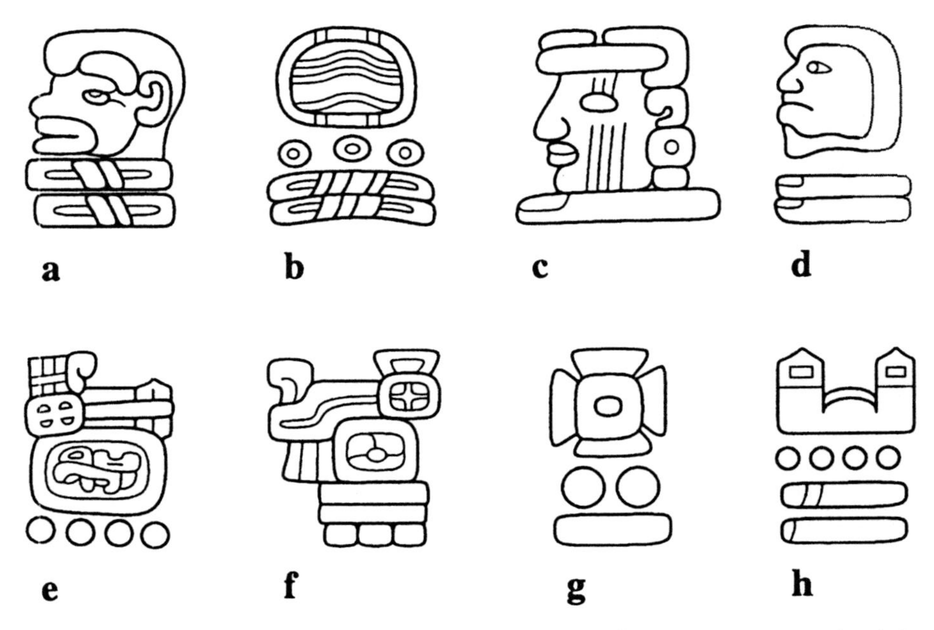

Figure 4 (a) 10 Monkey, Frieze, Lambityeco; (b) 13 Water, effigy in the Museo Regional de Oaxaca; (c) 1 Xipe (1 Flint), Monument 140, Building L, Monte Albán; (d) 2 Lord (2 Flower), Monument 142, Building L, Monte Albán; (e) year 4 Lightning (4 Wind), Monument 140, Building L, Monte Albán; (f) year 13 Movement, Monument 2a, South Platform (corners), Monte Albán; (g) 7 glyph I, Monument 16a, Zaachila; (h) 14 glyph W, Monument 142, Building L, Monte Albán. [Drawing by M. P. Closs after Urcid-Serrano, 1992.]

coefficients in about one third of its appearances. An example of glyph I with a coefficient of 7 is shown in Figure 4g. In attested epigraphic contexts, glyph I never has a coefficient exceeding 13.

The only Zapotec glyph that takes a coefficient of more than 13 is catalogued as glyph W. Its meaning is still under discussion. An example of glyph W with a coefficient of 14 is illustrated in Figure 4h.

EPI-OLMEC

In Southeastern Mesoamerica, a distinctive script tradition arose about 150 BC and continued for several hundred years. This tradition has been referred to as epi-Olmec because the archaeological cultures of the area descended from that of the Olmecs. The script may also be descended from an Olmec hieroglyphic system but too little of that has been recovered to confirm or disprove the connection. Justeson and Kaufman (1993) have partially deciphered the epi-Olmec script and identified it as an early form of Zoquean, the ancestor of four languages still spoken today in former Olmec areas of Veracruz, Tabasco, Oaxaca, and Chiapas. Since speakers of other languages in these areas were intrusive, and because there is a correlation of cultural artifacts and practices widely diffused throughout Mesoamerica with a related vocabulary derived from early Zoquean, scholars have concluded that the Olmecs spoke the same language.

It is in the epi-Olmec texts that we find the first use of a remarkable system of absolute chronology known as the Long Count. In this system the chronological distance of a given Calendar Round date [a date listed in both the 260-day and 365-day calendars] from a fixed but unrecorded base date far in the past is recorded. The time intervals are measured by a composite chronological count consisting of a vigesimal count of *tuns* (periods of 360 days) and distinct counts of *winals* (20 day periods) and *k'ins* (days). We find two vigesimal multiples of the tun in the script: the *k'atun* (20 tuns) and the *baktun* (400 tuns). [The terms for these time periods are anachronistic in being Mayan, the script in which the Long Count was first recognized.]

The chronological count is a system of metrology whose relationships are conveniently expressed in the following factor diagram.

$$\text{baktun} \xleftarrow{20} \text{k'atun} \xleftarrow{20} \text{tun} \xleftarrow{18} \text{winal} \xleftarrow{20} \text{k'in}$$

The base date, or zero point of the absolute chronology, is the Calendar Round date 4 Ahau 8 Cumku (in the Classic Maya system) that fell in the year we designate as 3114 BC. It is a mythological date falling more than 3000 years before the earliest epi-Olmec inscriptions and more than 1500 years before the Olmec awakening.

A typical Long Count record, taken from Stela 1 at La Mojarra in the state of Veracruz, Mexico, is illustrated in Figure 5a. It begins with a characteristic sign, called an introducing glyph, and continues with an account of the chronological interval separating the unrecorded base date from the contemporaneous date of the text. The epi-Olmec scribes recorded this time span using bar and

Figure 5 Long Count records: (a) 8.5.16.9.7, 5 Manik 15 Pop, Stela 1, La Mojarra; (b) 7.16.6.16.18, 6 Etz'nab [1 Uo], Stela C, Tres Zapotes; (c) (7.16.) 3.2.13, 6 Ben [16 Xul], Stela 2, Chiapa de Corzo. [Drawing by M. P. Closs: (a) after Winfield Capitaine, 1988; (b) and (c) after Coe, 1976.]

dot numerals embedded in a place value system in which counts were written vertically: the number of baktuns in the highest position, the number of k'atuns in the next position, and the number of tuns, winals, and k'ins in successively lower positions. Thus the Long Count record in question displays a count of 8 baktuns, 5 k'atuns, 16 tuns, 9 winals, and 7 k'ins, conventionally transcribed as 8.5.16.9.7. It informs us that

$$8(144{,}000) + 5(7200) + 16(360) + 9(20) + 7(1) = 1{,}193{,}947$$

days (approximately 3268.85 years) have elapsed since the mythological base date of the calendar. Computation shows us that it must lead to a calendar round date of 5 Manik 15 Pop (in the Classic Maya system), corresponding to the Gregorian date July 13, AD 143 (following the widely accepted 584285 calendar correlation formula). The 260-day calendar date of 5 Manik is indicated at the bottom of the Long Count record. In this instance the day sign coefficient of 5 is missing, due to erosion on the monument, but has been restored in the drawing. The date 15 Pop in the 365-day calendar is recorded immediately below the introducing glyph above a conventional sign for the tun of 360 days.

The oldest example of a complete Long Count record in Mesoamerica is found on Stela C from Tres Zapotes, Veracruz (Figure 5b). It records a count of 7.16.6.16.18 beginning with the customary introducing glyph and leading to the date 6 Etz'nab 1 Uo (September 2, 32 BC). The oldest monument employing this system is Stela 2 from Chiapa de Corzo (Figure 5c). The Long Count record is fragmentary but can be reconstructed to yield (7.16.)3.2.13 leading to 6 Ben 16 Xul (December 8, 36 BC). It is noteworthy that these epi-Olmec monuments are more than 300 years older than the oldest known Maya monuments using the same system of absolute chronology.

Stela 1 at La Mojarra contains two Long Count records and several chronological records of a type that are known as distance numbers. The latter are counts that lead from a given date, usually fixed in the absolute chronology, to later dates in the inscription. For example, in various places in the text we have references to counts of 13 tuns (Figure 6a), 4 winals (Figure 6b), 9 days (Figure 6c), and 13 days (Figure 6d). There are also counts of single days represented without any numerical coefficient (Figure 6e). In one instance, there is a count of 1 tun in which the numeral is represented by a finger (Figure 6f). There also appears to be a count of 23 days in which the value of 20 is represented by a distinctive sign (Figure 6g). The usage of distance numbers, including a distinctive sign for 20, anticipates later Maya practice.

The same monument also contains an example of a numeral being used in a proper name, that of a deity called Ten Sky (Figure 6h). This deity resurfaces as a god of Venus among the later Maya.

MAYA

Maya writing appears to derive from a Mixe Zoquean script attested at the highland Guatemalan site of Kaminaljuyú, a script closely related to that of the epi-Olmec. Maya writing conventions also closely resemble those of the epi-Olmec. In addition, the Maya employed bar and dot numerals and the same calendrical and chronological structures that are found in epi-Olmec writing.

The Dresden Codex, an ancient Maya book, has a few examples of numerals being used to enumerate objects. These are found in the cycle of New Year's ceremonies outlined in the codex where counts of nodules of copal incense are specified as offerings for the celebrations (Figure 7a). More frequently, numerals feature in the names of deities and historical figures. Thus, for example, the

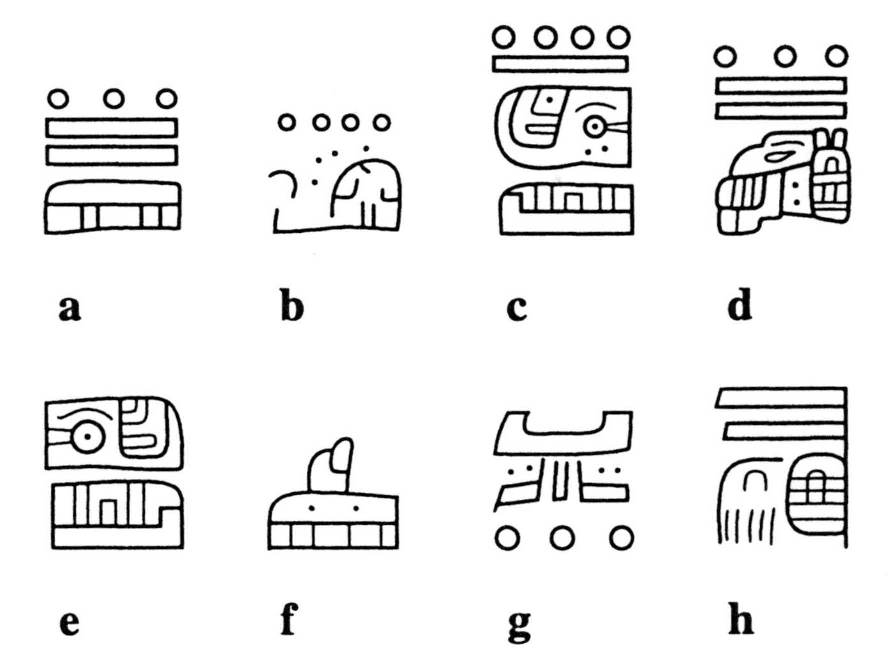

Figure 6 Glyphs from Stela 1, La Mojarra: (a) 13 tuns, I1; (b) 4 winals, N1–2; (c) 9 days, R11–13; (d) 13 days, T19–20; (e) [one] day, G5–6; (f) 1 tun, T8; (g) 23 [days], T15; (h) 10 Sky, P21–22. [Drawing by M. P. Closs after Winfield Capitaine, 1988.]

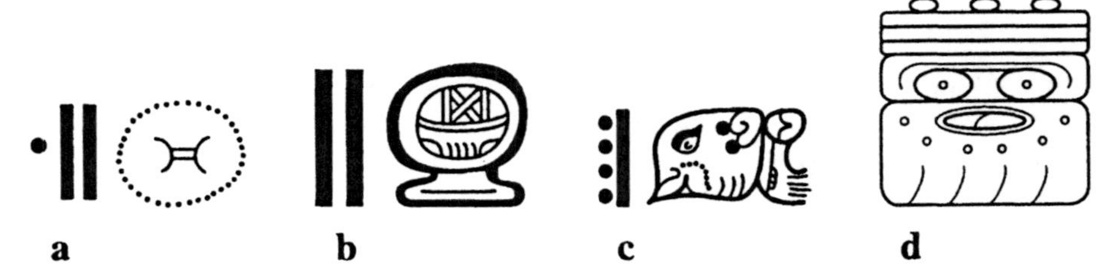

Figure 7 (a) An offering of eleven nodules of copal incense (pom), Dresden Codex 27; (b) the name of the Venus god Ten Sky (Lahun Chan), Dresden Codex 47b; (c) the name of the god Nine Strides (Bolon Yocte), Dresden Codex 60b; (d) the name of a ruler of Naranjo, 18-Ub (Waxaklahunub), Stela 32, Naranjo. [Drawing by M. P. Closs.]

name glyphs of the Venus deity Ten Sky, Lahun Chan, (Figure 7b), the deity Nine Strides, Bolon Yocte, (Figure 7c), and an ancient ruler of Naranjo known as 18-Ub, Waxaklahunub, (Figure 7d), all incorporate numerical coefficients. However, the greatest usage of numbers by far occurs in the calendrical and chronological records of the Maya.

If an event had to be anchored absolutely in time, Maya scribes engaged the Long Count chronology. In the Maya texts, these chronological statements are usually found at the beginning of texts and so are often referred to as Initial Series. If an inscription discusses several events, it is usually only the first which is anchored by an Initial Series. The date of the next event is then linked to that of the first by a distance number. Later events are similarly linked to those events which precede them in the text. It was also the Maya custom to identify an event in time by a Calendar Round date consisting of a specification of the given day in both the 260-day Sacred Round and the 365-day Vague Year.

On Maya monuments, Initial Series typically include period glyphs indicating the time periods involved. For example, the Long Count record in Figure 8a opens with a standard introductory glyph. It has a variable central element that depends on the month position of the terminal date. The introductory glyph is followed by the Initial Series 9.13.17.12.10 in which the bar and dot numerals 9, 13, 17, 12, and 10 are prefixed to signs that represent the periods of the baktun, k'atun, tun, winal, and k'in, respectively. (Note: Maya texts are usually read in paired columns from top to bottom. The crescent shapes in the coefficients of 17 and 12 are for aesthetic balance only and do not have numerical value.) The Initial Series leads to the Calendar Round date 8 Oc 13 Yax (August 27, AD 709). Distance numbers are treated in a similar way but there is a notable exception: a sign for the k'in period is often suppressed as may be seen in Figure 8b, which records a count of 2.3.5.10. It may also be observed, as happens here, that the order of the periods in distance numbers is generally opposite to that in Initial Series.

By contrast, in the Maya codices, Long Count records lack the introductory glyph and both Initial Series and distance numbers routinely use place value notation. Figure 8c shows an Initial Series of 9.16.4.10.8 leading to 12 Lamat [1 Muan] (November 12, AD 755) while Figure 8d shows a distance number of 3.5.

In a remarkable addition to the corpus of numerals, Maya scribes used portrait heads and even entire anthropomorphic figures to represent numbers. These elaborate head variant and full-figure numerals are singular in their beauty and are unrivaled in any other script. The Maya identified each of the numbers from 1 to 13 with a particular deity and this made it possible to substitute the appropriate deity head for the corresponding number. Thus, for example, the head for 1 is that of a young earth goddess recognizable by a long lock of hair running in front of the ear, a bob of hair or ornament on the forehead, and an IL sign on the cheek. Figure 9a depicts a count of 1 k'atun and employs the head of this goddess as the numeral 1.

Head variants for the numbers from 14 to 19 were obtained by combining the death head skull for 10 with the characteristic features of the numbers from 4 to 9. An interesting combination of the heads for 10 (skull) and 6 (hafted axe in eye) and a monkey variant of the k'in sign are used to depict a count of 16 k'ins in Figure 9b.

An example of a full-figure variant is shown in Figure 9c. In this instance, the Maya scribe chose to portray the number 9 by a human figure with jaguar patches on the arm and leg and a jaguar ear since the number 9 is a jaguar deity. In addition, another characteristic element often found with the head variant of the number 9, a *yax* sign meaning 'new, green', appears as a headdress

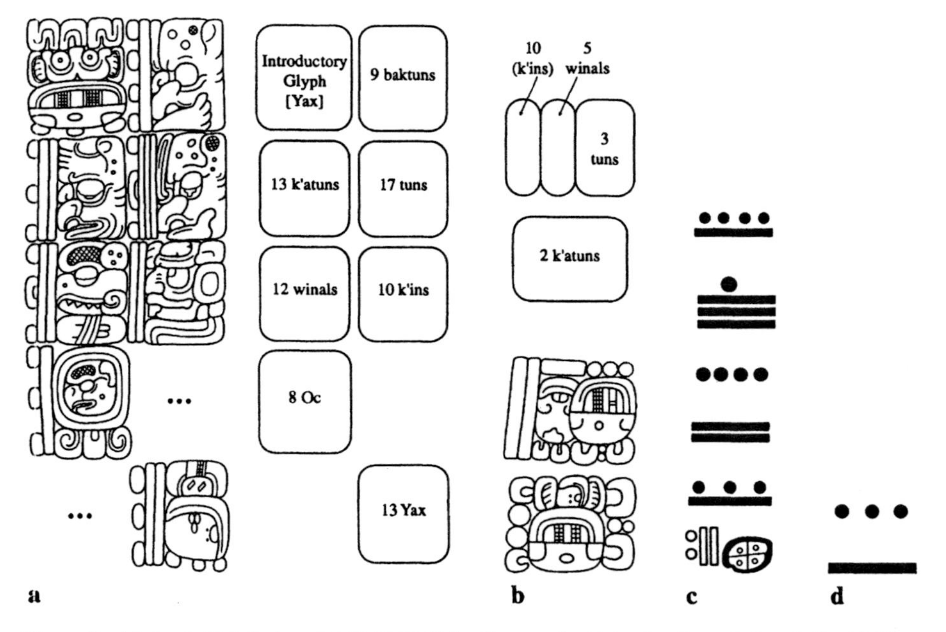

Figure 8 (a) An Initial Series of 9.13.17.12.10 leading to the Calendar Round date 8 Oc 13 Yax, Lintels 29 and 30, Yaxchilan; (b) a distance number of 2.3.5.10 linking two events on Lintel 30, Yaxchilan; (c) an Initial Series of 9.16.4.10.8 leading to the Sacred Round date 12 Lamat, Dresden Codex 52a; (d) a distance number of 3.5, Dresden Codex, 73b. [Drawing by M. P. Closs: (a) and (b) after Graham, 1979.]

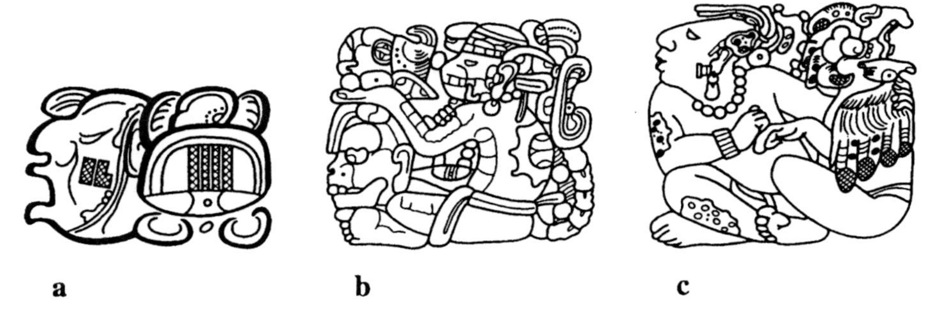

Figure 9 (a) 1 k'atun, Tablet of the 96 Glyphs, Palenque; (b) 6 + 10 = 16 k'ins, Lintel 48, Yaxchilan; (c) 9 baktuns, Palace Tablet, Palenque. [Drawing by M. P. Closs: (a) after a drawing of Merle Greene Robertson; (b) after Graham, 1979; (c) after a drawing of Linda Schele.]

on the human numeral. The combination of the anthropomorphic 9 holding a bird variant of the baktun period then indicates a count of 9 baktuns.

Despite their complexity, the head variant and full-figure variant numerals are functionally equivalent to the basic bar and dot numerals in all contexts. The Maya scribes sometimes also used a finger to represent the number 1 and employed other numeral variants (Closs, 1986: fig. 11.21).

In addition to signs for the numbers from 1 to 19, the Maya also had a sign for 20. It was most often used in lunar counts to give the age of the moon or to specify that the length of the lunation was of 29 or 30 days. Figure 10a illustrates a length of lunation count of 20 + 9 = 29 days. The symbol for 20 was also used in distance numbers. Figure 10b shows a relatively rare instance

Figure 10 (a) A length of lunation count of 20 + 9 = 29 days, Lintel 47, Yaxchilan; (b) a distance number of 2 tuns and 20 + 16 = 36 days, Stela 22, Tikal; (c) a distance number of 20 + 13 = 33 days, Dresden Codex 18c. [Drawing by M. P. Closs.]

of such usage in the inscriptions while Figure 10c illustrates one of many such examples in the codices.

There is reason to credit the Maya with the first invention of a zero symbol. It is absent in the surviving epi-Olmec texts but is very common in the Maya inscriptions. Zeros are found in many chronological counts in the Dresden Codex where they occur in positional contexts just as other numerals. Most Maya glyphs come in several variants and the same is true of the zero sign. The zeros in the codices are identifiable as shells (Figure 11a) and are always painted red. In most cases, the zero shells are stylized and simplified (Figure 11b). In the inscriptions, the most common form of the zero is shaped somewhat like a three quarter portion of a Maltese cross (Figure 11c).

There has been some argument about whether the Maya zero symbols represent a zero count or a count completed (Thompson, 1971: 137–138). In the case of the common variants in the codices and inscriptions, the arithmetical environment in which they are engaged is such that a value of zero is necessary. There are three classic proofs of the zero hypothesis. (i) In the Venus Table of the Dresden Codex there is a distance number of 8 days that is rendered in

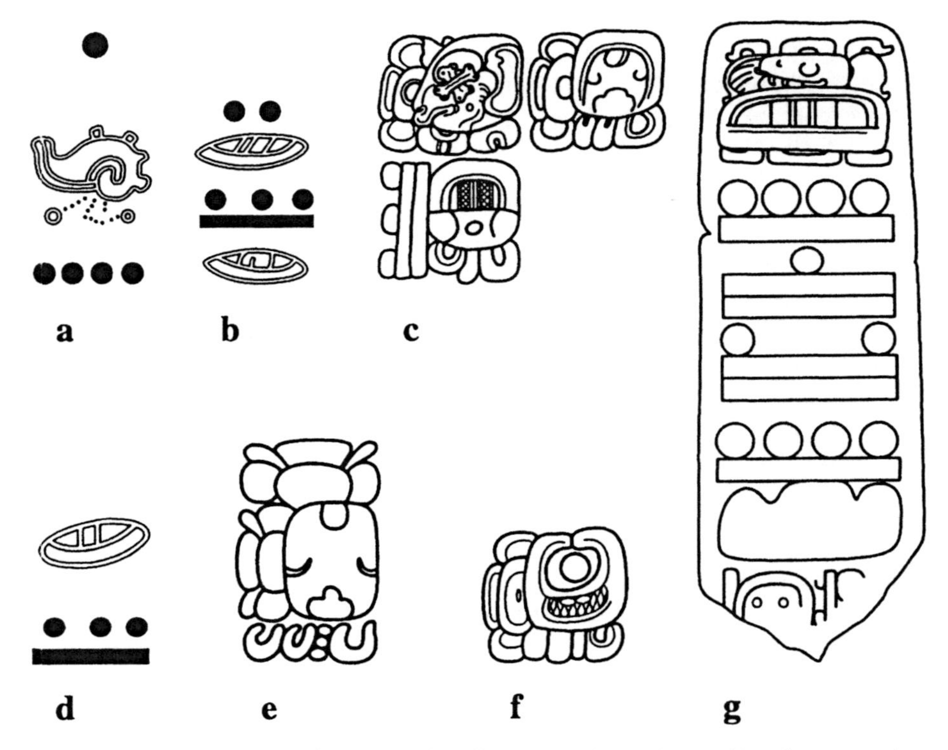

Figure 11 Maya zeros: (a) a distance number of 1.0.4, Dresden Codex 64; (b) a distance number of 2.0.8.0, Dresden Codex 64; (c) a distance number of 13.0.0, Palace Tablet, Palenque; (d) a distance number of 0.8, Dresden Codex 48; (e) a distance number of 0.0, Monument 11 (Stela K), Quirigua; (f) a distance number of $0 + 20 = 20$ days, Tablet of the Cross, Palenque; (g) an Initial Series of 9.11.12.9.0 incorporating positional notation and a zero, Stela 1, Pestac. [Drawing by M. P. Closs.]

positional notation as 0.8 (Figure 11d). The superfluous count of zero winals is included to preserve the uniformity of the text since the other numbers in the same register all occupy two positions. (ii) On Stela K from Quirigua there is a distance number of 0 days represented as a count of 0.0 (Figure 11e). In this instance, the numerically unnecessary distance number emphasizes that the following event occurs on the same date as the previous event. (iii) The Tablet of the Cross from Palenque records a distance number of 20 days in the form of a zero sign prefixed to the special symbol for 20 (Figure 11f). From analogous cases in which the sign for 20 is used, the whole can only mean $0 + 20 = 20$ days.

Despite the clear evidence that the common zero symbols do have a sense of zero, there are also less common 'zero variants' which are used in more restricted contexts. It is possible that these variants carry a sense of a count completed in addition to the usual meaning of a zero.

The oldest Maya text exhibiting a zero sign within the place-value system is Stela 1 at Pestac (Figure 11g). The Initial Series on the monument, 9.11.12.9.0, leads to 1 Ahau 8 Cumku and securely dates it to February 8, AD 665. (The monument has an erroneous Sacred Round coefficient.) In addition, zero signs are found within Initial Series having period glyphs at much earlier dates. The oldest such monument is the poorly preserved Stela 19 from Uaxactun. It carries an Initial Series of 8.16.0.0.0 dating it to February 3, AD 357. The oldest Maya monument with an Initial Series and period glyphs but without a zero sign is Stela 29 from Tikal dating to AD 292.

The Maya had a number of ways of referring to the base date of the absolute chronology. On Monument 3 (Stela C) from Quirigua (Figure 12), a scribe referenced the base date by the formula 13.0.0.0.0, 4 Ahau 8 Cumku (August 13, 3113 BC).

It may be noted that the common zero appears in the winal position while a variant zero appears in the k'atun, tun, and k'in positions.

Some scholars have argued that the Initial Series in this formula is to be understood as a count of 13 baktuns from an earlier chronological base (Thompson, 1971: 149). It has also been argued that the Long Count chronology is not absolute but is embedded within a nested sequence of ever larger periods. As a result, it is claimed that the Long Count is only apparently linear over an extended time interval but in reality is cyclical (Freidel *et al.*, 1993: 63). I think these notions are incorrect and that we are dealing with a simple Maya convention for recording the zero point of the chronology.

In a telling example (Figure 13), a scribe from Yaxchilan referenced a historical Calendar Round date of 3 Muluc 17 Mac by the exotic Long Count record

13.13.13.13.13.13.13.13.9.15.13.6.9, 3 Muluc 17 Mac (October 21, AD 744)

rather than by a standard Initial Series. Nevertheless, the date which is referenced is very clearly the same as that which would be determined by a standard Initial Series of 9.15.13.6.9. The 13s attached to the string of higher periods do not have a numerical significance.

This idea is reinforced by yet another exotic Long Count record on Stela 1

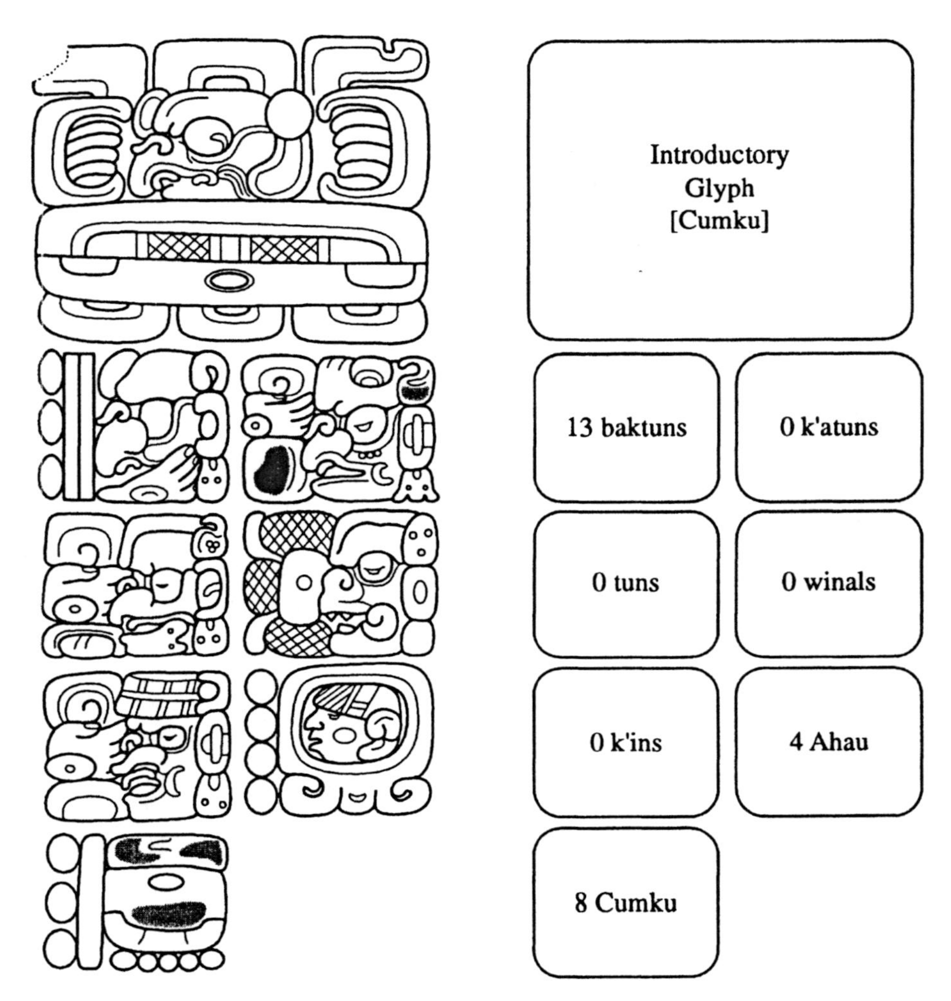

Figure 12 The base date of the Long Count expressed by an Initial Series of 13.0.0.0.0 leading to 4 Ahau 8 Cumku, Monument 3 (Stela C), Quirigua. [Drawing by M. P. Closs after Maudslay, 1889–1902.]

at Coba. In this instance the scribe referenced the chronological base date of 4 Ahau 8 Cumku in the following way:

13.13.13.13.13.13.13.13.13.13.13.13.13.13.13.13.13.13.13.13.0.0.0.0 4 Ahau 8 Cumku

In this and other related formulas the function of the 13s is to assert that the placement in the chronology is absolute whether we are referring to a period of 20^2 tuns in a typical Long Count record or to the much longer periods of 20^{10} tuns on Hieroglyphic Stairway 2 at Yaxchilan or 20^{21} tuns on Stela 1 at Coba.

The misunderstanding of these expressions has created other misconceptions concerning the nature of the chronological count. Indeed, it has been argued

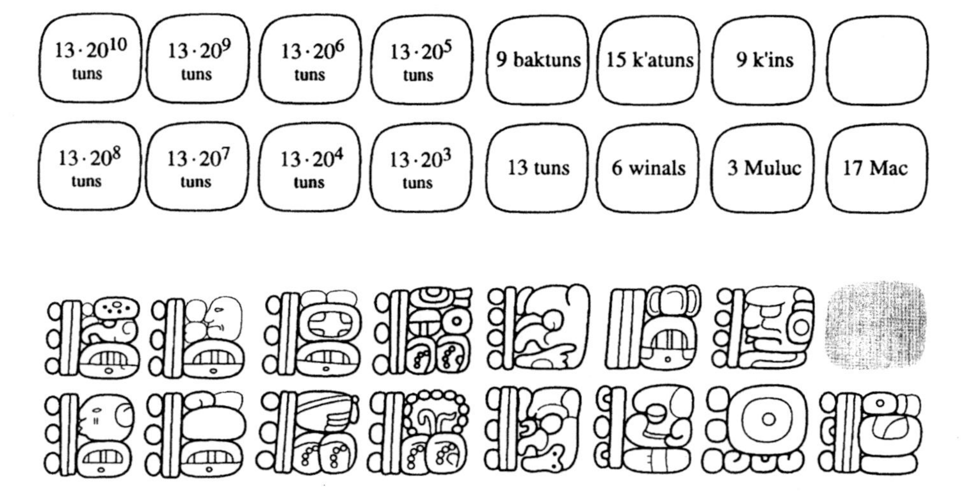

Figure 13 An exotic Initial Series of 13.13.13.13.13.13.13.13.9.15.13.6.9 leading to 3 Muluc 17 Mac, Hieroglyphic Stairway 2, Yaxchilan. [Drawing by M. P. Closs after Graham, 1982.]

that the chronological periods higher than the baktun are no longer vigesimally based and that when a count of 13 is completed in such a period then the next higher period increases by 1 (Freidel *et al.*, 1993: 63). There is little or no evidence to sustain that notion since the Maya, like other Mesoamerican peoples, counted vigesimally. It is only in calendrical calculations with the Sacred Round that residue arithmetic base 13 was applied. This likely explains why the Maya attached 13s to the higher chronological periods in these exotic records. It is the simplest way in which the 260-day Sacred Round is automatically preserved in eternity without any error or defect in contemporary calculations. The fact that the 365-day Vague Year is not preserved was of lesser significance.

That the count of tuns in the chronological count is strictly vigesimal is demonstrated by calculations involving the higher periods and is in agreement with the traditional idea that the pictun, the period immediately above the baktun, is 8000 tuns. In fact, the Dresden Codex contains several calculations in which distance numbers larger than usual are involved. Figure 14 shows two such calculations involving distance numbers located within the coils of a serpent in upright position, on the open jaws of which sits a peccary wearing an elaborate headdress. One of the numbers is recorded in black and the other in red to avoid confusion. The computation is to add the distance number to the Calendar Round date at the top and thereby reach the Calendar Round date at the bottom. It can be verified that the scribe did his work correctly and that it is in keeping with the vigesimal nature of the tun count.

A minimal Initial Series of 8 days is recorded in the Eclipse Table of the Dresden Codex. This table commensurates lunar and solar eclipse cycles with the Sacred Round. Its functional base is a date 12 Lamat in historical time. However, as was their wont (and there are many examples of similar calculations), the Maya astronomers linked this historical date to a mythological like-in-kind date as close to the base date of the absolute chronology as possible. This mythological base is a date 12 Lamat that occurs exactly 8 days after the zero point at 4 Ahau 8 Cumku. The codex scribe has referenced the Long Count position in a text that runs: [4] Ahau, 8 Cumku, 12 Lamat, 8 k'ins, **ti-ba** (Figure 15a). It says that from the chronological base at 4 Ahau 8 Cumku we advance to 12 Lamat (the base date of the Eclipse Table) – a total distance of 8 days. The final glyph is read phonetically as *ti'ba*, a term that appears in the *Diccionario Maya Cordemex* with the meaning 'in person' or 'by itself'. It refers to the fact that the '8 days' stands 'in person' or 'by itself' as the Initial Series. It is significant that the scribe denoted the Initial Series by a simple record of 8 days and not by 13.0.0.0.8 as many modern scholars have done. It demonstrates that the day after 4 Ahau 8 Cumku was simply 1 in the Long Count chronology.

In their inscriptions the Maya tracked the age of the moon and also counted lunations as either 29-day or 30-day months since they did not employ fractions. They grouped lunations using varying proportions of 29-day and 30-day months in order to keep their moon age records in agreement with astronomical reality when doing long range calculations into the past or future. From their

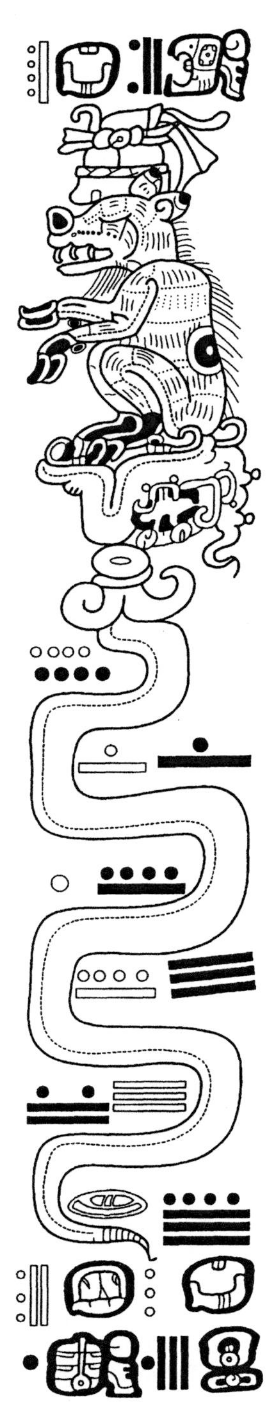

Figure 14 A double calculation involving large distance numbers: (i) 9 Kan 12 Kayab + 4.6.1.9.15.0 [in red] = 3 Kan 17 Uo; (ii) 9 Kan 12 Kayab + 4.6.9.15.12.19 [in black] = 13 Akbal 1 Kankin, Dresden Codex 62. [Drawing by M. P. Closs.]

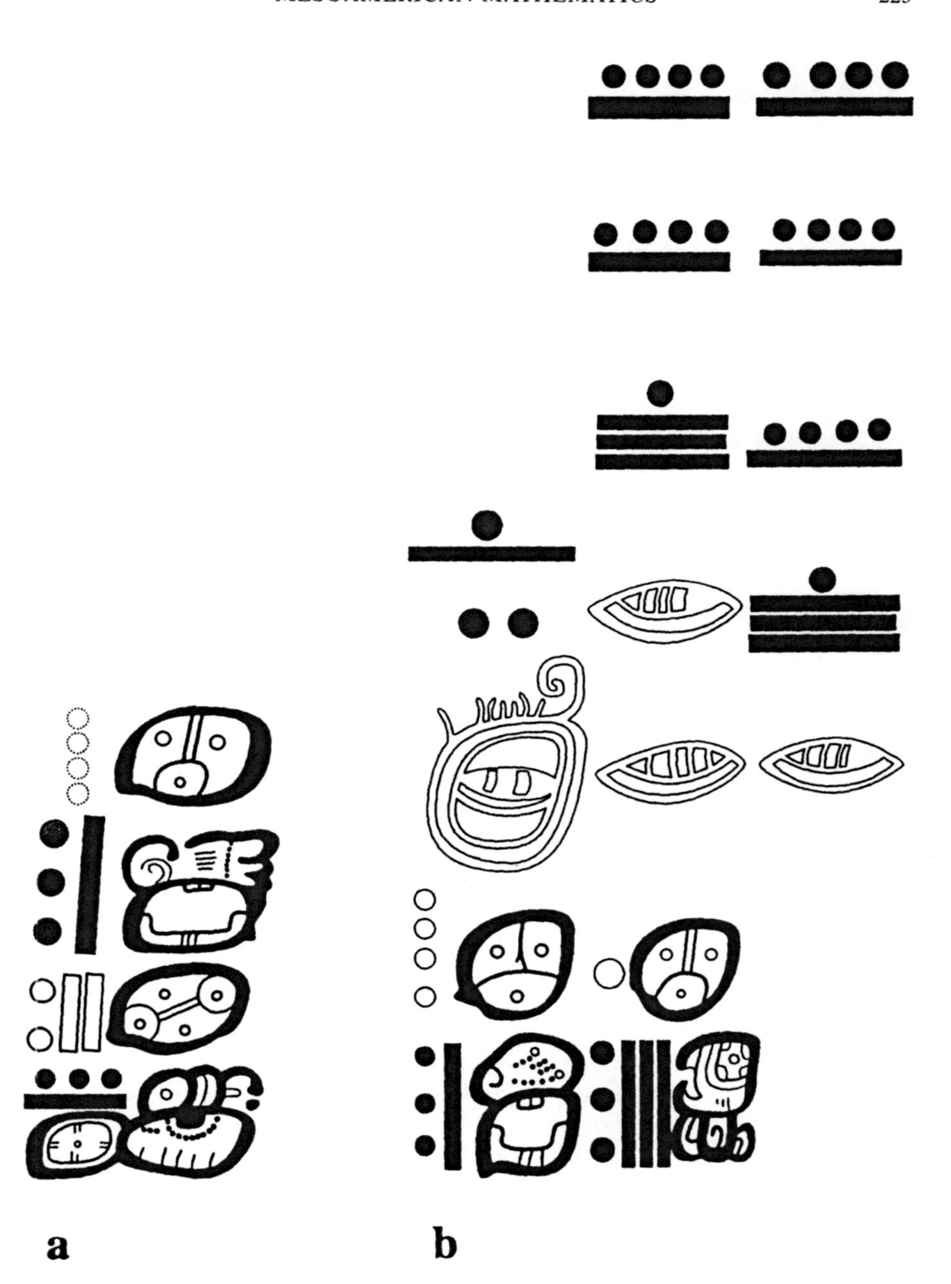

Figure 15 (a) An Initial Series of 8 days in the Eclipse Table, Dresden Codex 51A; (b) a ring number calculation anchoring the Venus Table, Dresden Codex 24:

– 6.2.0 before 4 Ahau 8 Cumku at 1 Ahau 18 Kayab (mythological base)

+ 9.9.16.0.0 (companion number)

9.9.9.16.0 at 1 Ahau 18 Kayab (historical base).

[Drawing by M. P. Closs.]

calculations it is known that the Maya employed different formulas. The Eclipse Table of the Dresden Codex is based on 46 multiples of the 260-day calendar, that is $46 \times 260 = 11{,}960$ days, which is commensurated with 405 lunations. In fact, the relationship is very accurate since 405 moons are approximately 11,959.89 days. This demonstrates how the Maya could be precise without the necessity of fractions in their integer-based arithmetic.

The Maya also recorded dates of mythological events occurring before the base date of their chronology by counting backwards from the zero point. Negative Initial Series of this type are distinguished in the codices by a notation that has led them to be called 'ring numbers'. The finest example of a ring number that functioned as a negative Initial Series and the calculation in which it figured occurs in the Venus Table of the Dresden Codex. In this instance, the historical base date of the table is a Calendar Round date of 1 Ahau 18 Kayab. The Maya astronomers linked this to a like-in-kind date as close to the base date of the absolute chronology as possible. That date fell 6 tuns, 2 winals and 0 k'ins before 4 Ahau 8 Cumku. In order to anchor the table at this mythological base date the scribe recorded a negative Initial Series (ring number) of 6.2.0. He then recorded a companion number to link the mythological base to the historical base of the table which he also anchored by an Initial Series (Figure 15b).

In analyzing the motions of Venus, the Dresden Codex uses a commensuration of the 365-day calendar with a mean Venus calendar of 584 days based on the formula

$$8 \times 365 = 5 \times 584 = 2920.$$

At the end of this period, Venus and the sun once again occupy the same relative positions in the sky as they did at the beginning of the period. It has long been known that the companion number in Figure 15b, recorded in Maya style as 9.9.16.0.0, is divisible by 260, 365, and 584, all key calendrical and astronomical periods. As a result it is very useful as a calculation factor relating the 260-day sacred calendar, the 365-day solar calendar, and the 584-day Venus calendar. Another consequence of this divisibility is that its prime factorization, $(2^5)(3^2)(5)(13)(73)$, consists of relatively small prime numbers. The factorization is not coincidental and gives additional emphasis to the importance of residue arithmetic in scribal calculations.

There is an interesting example of a mythological date on the Tablet of the Cross at Palenque which is anchored by an Initial Series of 12.19.13.4.0. The mathematics of the text make it clear that this is equivalent to a negative Initial Series or ring number of 6.14.0. The date marks the birth of an ancestral mother goddess on 8 Ahau 18 Zec. Because of the form of the Initial Series, it has sometimes been regarded as a count of 12.19.13.4.0 from a previous chronological base. Since such presumed bases are never explicitly mentioned in the Maya inscriptions, the proposition is a dubious one. Moreover, the intent in selecting a mythological date is to locate it as closely as possible to the base date of 4 Ahau 8 Cumku subject to constraints that have been imposed in the historical period. This is easy to see in the case of the two astronomical examples discussed previously. In the present instance, one of those constraints is certainly the fact that Pacal, the greatest of the Palenque kings, was also born on 8 Ahau. The other constraints which went into determining this specific date are less evident. Nevertheless, the essential link is to the date 4 Ahau 8 Cumku. The Palenque scribe expressed this relationship using a subtractive notation rather than ring number notation:

$$-6.4.0 = 13.0.0.0.0 - 6.14.0 = 12.19.13.4.0.$$

The fact that dates were linked to the base date of the chronology by counting either backwards or forwards from it underlines the absolute nature of the Long Count chronology.

In many inscriptions, scribes used a briefer format to engage the Long Count. It was common to anchor dates by linking them to major stations in the absolute chronology such as the end of a baktun, the end of a k'atun, or the end of a 5 tun, 10 tun, 13 tun, or 15 tun period within a k'atun. Figure 16 depicts several examples of these notations.

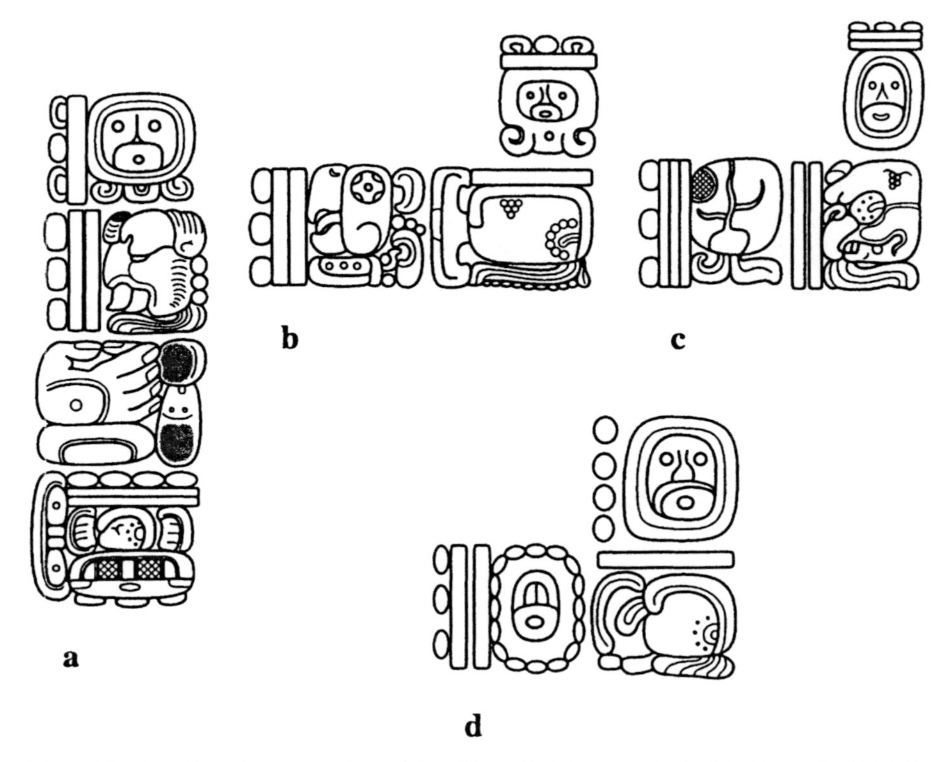

Figure 16 Period ending expressions: (a) 6 Ahau 13 Muan, end of 14th k'atun [9.14.0.0.0] (December 5, AD 711), Stela 3, Piedras Negras, Guatemala; (b) 6 Ahau 13 Kayab, 5 tuns, [9.17.5.0.0] (December 29, AD 775), Monument 3 (Stela C), Quirigua, Guatemala; (c) 13 Ahau 18 Kankin, 10 tuns [9.10.0.0.0] (December 6, AD 642), Tablet of the Sun, Palenque, Chiapas, Mexico; (d) 4 Ahau 13 Mol, 5 tuns lacking [9.11.15.0.0] (July 28, AD 667), Lintel 2, Piedras Negras, Guatemala. [Drawing by M. P. Closs.]

An alternative system of anchoring dates in time, known as the Short Count, was employed in the Yucatan during the Maya Classic and is also found in colonial times in the Books of Chilam Balam (post-Conquest Maya books employing the Latin alphabet). It was not an absolute chronology like the Long Count and was only accurate within a period of 13 k'atuns (approximately 256 years). Nevertheless, it was less ambiguous than the year bearer system which repeated itself in a period of slightly less than 52 years. In the Short Count events were anchored by reference to the name of the k'atun (the Ahau date on which the k'atun ended) as well as to the numbered tun in the k'atun in which the date fell. Examples of such statements from the site of Xcalumkin are illustrated in Figure 17. The first anchors an event in the 2nd tun of K'atun 2 Ahau and the second anchors a different event in the 13th tun of K'atun 2 Ahau. In this instance, given the context, it is possible to determine the specific K'atun 2 Ahau in question as the one that ended on the date 2 Ahau 13 Zec having Long Count position 9.16.0.0.0 [May 9, AD 751]. Without the supplementary contextual data it would not be possible to fix the date in the Long Count precisely since other possibilities such as 9.3.0.0.0 [2 Ahau 18 Muan; January 30, AD 495] or 10.9.0.0.0 [2 Ahau 13 Mac; August 15, AD 1007] also satisfy the Short Count criteria.

The visibility of mathematics in ancient Maya culture goes beyond the use of number and calculation in the Maya texts. It also appears in the iconography. It is helpful to have some exposure to this in order to have a better sense of the place of mathematics in the life of ancient Mesoamerica.

There are pictures in the Maya codices in which deities are shown in the act of writing. Some of these are further embellished with an arithmetical scroll coming out of the god's mouth indicating that he is engaged in a mathematical activity. One such example (Figure 18) shows the rain god Chak sitting in a pool or large cistern, holding a paint brush in one hand and an ink pot in the other. Out of his mouth pours a stream of numbers: 1, 7, 2, 2, 12, 13, 3. The remaining iconography and the associated glyphic text also reveal an unusual interest in dates and chronological intervals not often seen in the almanacs of the Madrid Codex.

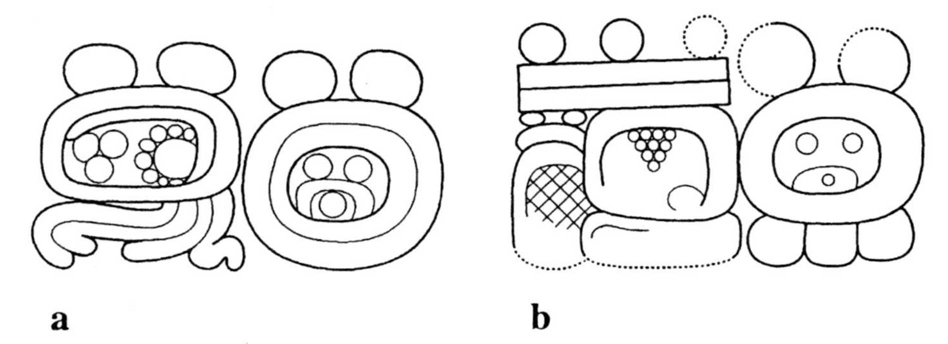

Figure 17 The Short Count: (a) [in the] second tun [of Ka'tun] 2 Ahau, Panel 5, Xcalumkin; (b) [in the] thirteenth tun [of K'atun] 2 Ahau, Panel 2, Xcalumkin. [Drawing by M. P. Closs.]

Figure 18 The Maya rain god Chak with a number scroll pouring out of his mouth, Madrid Codex 73. [Drawing by M. P. Closs.]

On Classic Maya pottery there are also many scenes of scribes engaged in mathematical activity. Perhaps the most notable example is a vessel, roughly dated to around AD 750, depicting a classroom scene in which a god of scribes is conducting a mathematics lecture (Figure 19). He is shown seated with a codex in front of him, with brush pen in his left hand, and a speech scroll issuing from his mouth saying, '11, 13, 12, 9, 8, 7'. Two students, at least one of whom is paying rapt attention, are seated in front facing the teacher.

Other pots exhibit scribes with number scrolls emanating from their armpits which distinguishes them from scribes represented without such scrolls. Thus on one ceramic vessel, also from around AD 750, two scribes seated on the ground are shown with pens in hand busily writing in codices (Figure 20). Only the second of the scribes is marked with a number scroll. I have provided evidence elsewhere that such scribes are mathematical specialists (Closs, 1994).

That mathematical scribes should be singled out in this way is not unusual.

Figure 19 A classroom scene in which a scribe god is giving a mathematics lecture to two apprentices. [Drawing by M. P. Closs after Kerr, 1989: 67.]

In fact, Diego de Landa, the second bishop appointed to serve in the Yucatan shortly after the Spanish conquest, wrote an account of the Maya derived from information provided by native scribes. In it, he refers to the mathematical techniques of the Maya scribes as 'the computation of the k'atuns' and he tells us that this 'was the science to which they gave the most credit, and that which they valued most and not all the priests knew how to describe it' (Tozzer, 1941: 168). Consequently, we know that not all scribes were mathematical specialists and that those who were had a greater prestige. They were recognized as a specialized subgroup of the scribal class. In this regard, the situation at the time of the conquest seems to have been no different than it was during the Classic period.

There are indications that the scribal curriculum was divided into two principal categories, one pertaining to arts and letters and the other to mathematical sciences. This is suggested by the vase with the mathematics lecture (Figure 19). Indeed, it also has a classroom scene on the other side which specifically refers to teaching in general. It once again shows the scribe god in front of two apprentices but in this case the glyphic scroll coming out of the mouth contains a verb 'to lecture' without any reference to mathematics. In addition, there is another ceramic vessel (Closs, 1997: 649), from the same general period, showing two patron gods of scribes, one seated behind the other. Of these two deities, one is bedecked with the mathematical scroll and the other is not. The pairing is not unlike that seen in the vessel of the two Maya scribes illustrated above.

From Landa and other early sources we can identify those portions of the scribal curriculum which pertained to the mathematical sciences. This would include astronomy, chronology and calendrics, divination and prophecy, and genealogy.

Figure 20 Two Maya scribes, the second of whom has a numerical scroll emanating from the armpit, writing in codices. [Drawing by M. P. Closs after Robicsek and Hales, 1981: Vessel 71.]

MIXTEC

The Mixtec occupied a region of southern Mexico comprised of the western third of the present-day state of Oaxaca, the eastern section of the state of Guerrero, and the southern part of the state of Puebla. There are several pre-Conquest historical documents from the region that provide information on Mixtec history beginning about AD 750. In these documents we also find some evidence of the use of numbers by the peoples of that area.

In the Mixtec codices a simple tally of dots was used to represent the numbers from 1 to 13. For numbers greater than 5, dots were often grouped by fives but such groups were not replaced by bars as in the earlier systems. In some manuscripts, the numeral dots are always represented as a pair of concentric circles; in others, only the numerals 1 and 2 are occasionally represented in this way.

The day signs in the Mixtec and Valley of Mexico (Aztec) manuscripts are depicted in the same way. Calendrical names are assigned to children according to the day in the 260-day calendar on which a person is born. They are represented in exactly the same way as the corresponding day date. In the historical manuscripts individuals are named by their calendrical name and also by a second name, referred to as a personal name or nickname.

The year bearer system was heavily used by the Mixtec for naming years. They did not use the same set of year bearers as the Zapotec. A similar system was known by the Maya in a variant form but was not exploited for chronological purposes in their glyphic texts. Because of the structure of the 260-day and 365-day calendars only four day names could function as year bearers. In the Mixtec calendar (and also in the later Aztec calendar) these are the days Reed, Flint, House, and Rabbit. The presence of a year date in the Mixtec histories is signalled by an interlaced 'A–O' sign accompanied by the appropriate year bearer.

Figure 21 illustrates a scene from the Codex Nuttall and demonstrates the interplay of calendrical names, dates and year signs in a Mixtec manuscript. The upper register portrays lord 8 Deer 'Tiger Claw' and lady 13 Serpent 'Flowered Serpent' in a palace. 8 Deer's calendrical name and his personal name appear below his figure. He is also depicted wearing a deer headdress. 13 Serpent's name appears in two parts: her calendrical name is in front of her figure and her personal name constitutes her headdress. 13 Serpent is offering 8 Deer a bowl of chocolate, symbolic of marriage. The wedding took place in the year 13 Reed (AD 1031), shown by the year sign attached to the year bearer 13 Reed, on the day 12 Serpent.

The lower register mentions the two sons of the couple. The first son (reading from right to left) 4 Dog 'Tame Coyote' was born on the day 4 Dog in the year 7 Rabbit (AD 1058). The second son 4 Alligator 'Serpent Fire Ball' was born two years later on the day 4 Alligator in the year 9 Flint (AD 1060).

AZTEC

The Aztec of central Mexico gained prominence once they settled in their capital of Tenochtitlan, present-day Mexico City. Their appearance is relatively

Figure 21 The Mixtec lord 8 Deer Tiger Claw and his family, Codex Nuttall 26c. [Drawing by M. P. Closs after Nuttall, 1975.]

late in Mesoamerican history but they rapidly created an expansionist empire. There are many pre-Conquest Aztec monuments carved in stone and several Colonial period Aztec manuscripts that carry pictographic texts with associated dates, names, and places.

The artifacts indicate that the Aztec employed the same notation for the 260-day calendar dates as did the Mixtec. Figure 22a, shows a sequence of

a

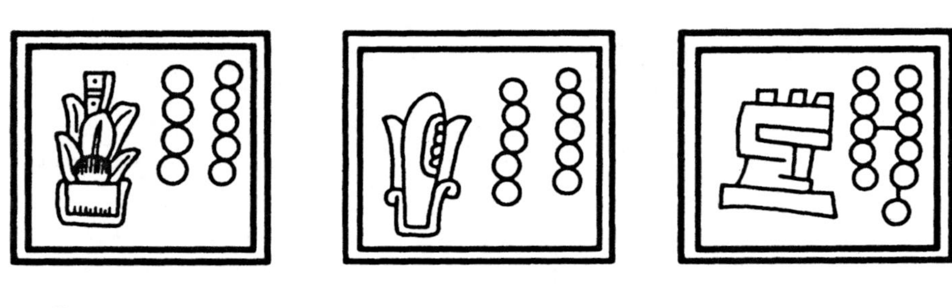

b

Figure 22 (a) Three successive dates in the 260-day calendar: 8 Dog, 9 Monkey, and 10 Grass, Codex Telleriano-Remensis 13r; (b) three successive years: 9 Reed (AD 1475), 10 Flint (AD 1476), and 11 House (AD 1477), Codex Telleriano-Remensis 37r,v. [Drawing by M. P. Closs.]

three successive Sacred Round dates from an Aztec manuscript. Dates such as these also served as calendrical names since it was the custom that among the names a child was given was that of the day on which he or she was born.

Many of the Aztec deities also had calendrical names. Thus, for example, the god Tonacatecuhtli, 'Lord of our Sustenance', who is said to have created the world, is also known as 7 Flower. His wife Tonacacihuatl, 'Lady of Our Sustenance', a maize goddess, is also called 7 Serpent. And the god of the sun, Tonatiuh, was sometimes called 4 Movement after his feast day and the expected day on which the world would end with earthquakes and an eclipse.

The Aztec also employed the year bearer system. The year was named after a day within it but there is some disagreement among authors on exactly which day that was. In the Aztec manuscripts, the 360th day of the 365 day calendar was used and the name of the year bearer was the Sacred Round date of that day. Like the Mixtec, the Aztec year bearers were Reed, Flint, House, and Rabbit. However, for some unknown reason, the correlation of the year bearers differs by 12 years in the two systems. It is also known that other year bearers were used in other times and places.

In the Aztec manuscripts, placing the year bearer inside a rectangular blue cartouche indicated a particular year. Figure 22b, illustrates a sequence of three successive years.

Since each year bearer carries a numeral coefficient from one to thirteen, there are 52 year bearers in total. The resulting cycle of 52 years is the Calendar Round cycle described earlier. It combines 73 cycles of the 260-day calendar with 52 cycles of the 365-day calendar. A cycle of 52-years was sometimes represented as a bundle of canes that was tied up and named after the last year of the cycle. Figure 23 illustrates a stone model of such a 52-year cycle named after the year 2 Reed.

For calendrical purposes, a simple tally of dots running from one to thirteen was sufficient. However, it was not adequate for representing larger counts required in the enumeration of years or tribute items. For these purposes, the Aztec employed additional symbols (Figure 24). There were signs for the basic vigesimal units of 20 (a flag), 400 (hair or a growth of garden herbs), and 8000 (a bag of cacao beans). In the latter two cases, the symbols are representative of the literal meanings of the numerical terms for 400, *tzontli*, and 8000, *xiquipilli*. To represent a count of objects an economical tally of the basic units was used.

Figure 23 A stone bundle representing a 52-year cycle ending in the year 2 Reed. [Drawing by M. P. Closs.]

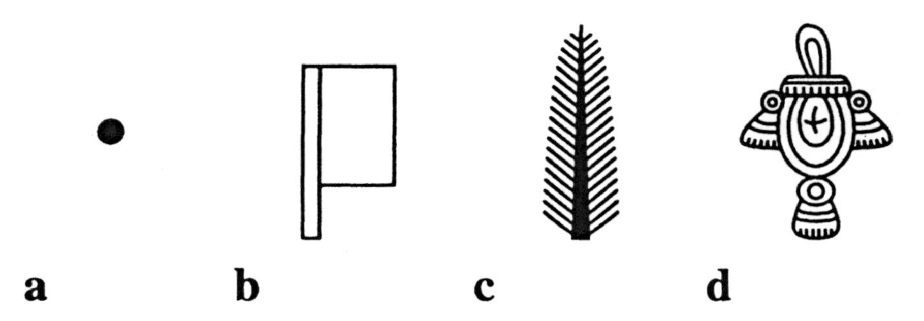

Figure 24 Aztec numerals: (a) 1; (b) 20; (c) 400; (d) 8000. [Drawing by M. P. Closs.]

Years of 365 days were represented pictorially by a circular symbol of turquoise coloured blue. One section of the Mendocino Codex gives the different stages in the education that a boy or girl receives from its parents over the years. The tasks which children must learn are depicted and their daily food ration is given by a tally of tortillas. Thus, for example, when boys and girls are thirteen years old, indicated by a count of 13 blue disks, their daily ration is two tortillas (Figure 25a). At this age the boys are going out in a canoe to cut reeds for domestic use. The girl is taught to grind corn, make tortillas and cook. Facing her is the grinding-stone and below that the hearth (three stones). On the earth is the *comalli* on which the maize cakes were baked and next to it a pot containing limewater.

There is similar record of this type that refers to those who have reached their senior years. It notes that when a man or woman has achieved the age of $(3 \times 20) + 10 = 70$ years (Figure 25b) they are entitled to get drunk, a privilege denied to those who are younger.

The Vatican Codex contains other counts of years which measure far larger

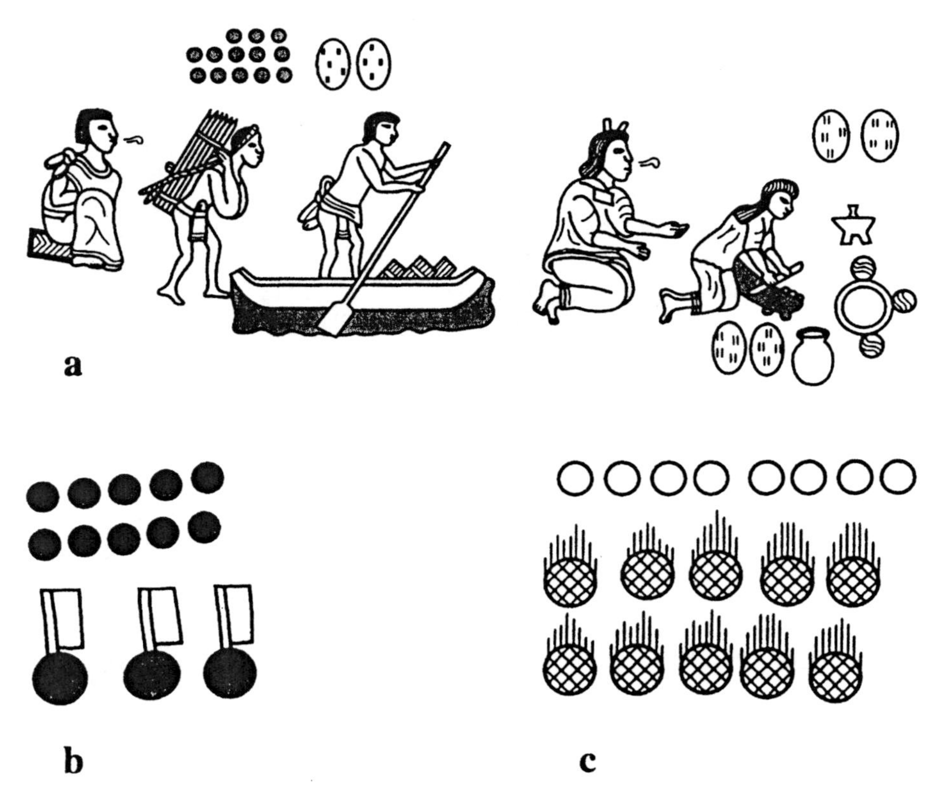

Figure 25 Aztec chronological counts: (a) age of 13 years, food ration 2 tortillas, Codex Mendoza 60; (b) age of $(3 \times 20) + 10 = 70$ years, Codex Mendoza 71; (c) count of $(10 \times 400) + 8 = 4008$ years, Vatican Codex 7. [Drawing by M. P. Closs; (c) after Thomas, 1900, fig. 38.]

intervals. In one reference to the years of the second age of the world there is a count of

$$(10 \times 400) + 8 = 4008$$

years (Figure 25c). The usual blue disk marking a year is shown with hair (*tzontli*), a term that also means 400, to create a symbol for 400 years.

Some of the Aztec codices contain lists in which the quantities of various tribute items to be received from conquered towns are indicated by numerals. For example, in one such list the Matrícula de Tributos demands 20 military costumes and 20 shields of a particular type to be paid by a certain group of tributary towns every 80 days (Figure 26a). The same list demands many other items including 400 women's blouses (Figure 26b) and 400 fine ponchos or blankets with black stripes and a length of 2 brazas (Figure 26c), a braza being about 1.5 metres. In the latter case, two fingers indicate the double length.

One of the largest numerals recorded in the Aztec manuscripts appears in a text referring to the dedication of the Main Temple of Tenochtitlan in 1487. The completion of the two shrines at the summit of the structure marks the culmination of a renovation project begun in 1483. The inauguration of the Main Temple was commemorated by a dedicatory New Fire and by numerous human sacrifices. The victims are represented by three white sacrificial figures and numerals indicating the numbers slain – in this instance,

$$(2 \times 8000) + (10 \times 400) = 20{,}000 \text{ (Figure 26d).}$$

Other early sources report a much greater number of sacrifices on this occasion, as high as 80,000, so the present record is a conservative one. Figure 26d includes one of the sacrificial figures depicted in a characteristic generic form. He is identified as Mazatecuhtli (Deer Lord) by the head of a deer wearing the diadem of a ruler. His home city of Xiuhcoac (Place of the Turquoise Snake) is indicated by a blue snake below the nametag.

A novel system of numerical notation was used in Aztec land documents. It was devised for the specific purpose of recording the measurements of perimeters and areas of land holdings in at least two locations in the Valley of Mexico. The basic numerals were a vertical tally stroke, a bundle of five strokes linked at the top, a large dot, and a corn glyph (*cintli*). The values of the measurements depended on the position of these numerals on maps of a given land holding.

An example from the Códice de Santa Maria Asunción, dating from around 1545, is discussed in detail elsewhere (Harvey and Williams, 1980, 1986; Closs, 1997). Here another version of the same system from a late 16th century Texcocan locality is considered (Figure 27). In this instance, the field on the left has its sides marked using lines and dots, a line equal to one linear unit and a dot equal to 20 linear units. Other symbols on the sides refer to fractions of a linear unit. The recorded lengths, going counter-clockwise beginning with the left side, are 61, 42, 60+, and 47+.

The field on the left is joined to a field on the right with a dotted line. The field on the right is the same field as on the left but is now inscribed with notations giving the area. The units marked on the outside of the field are

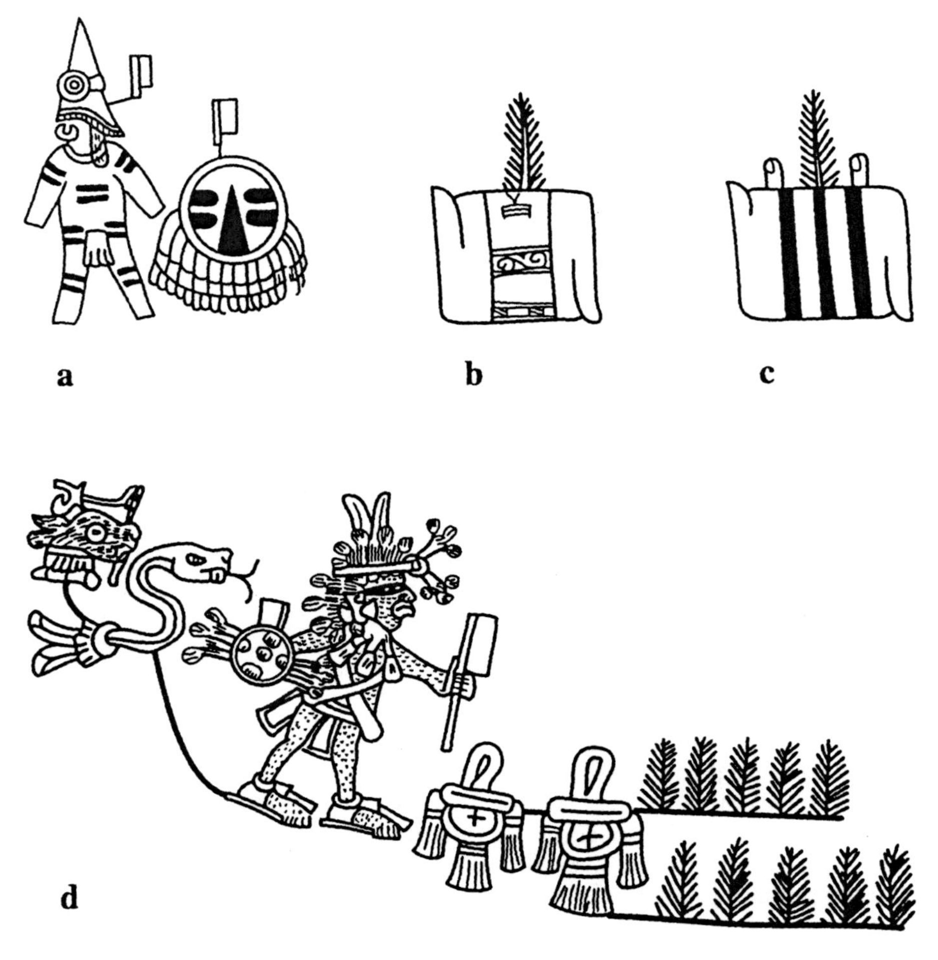

Figure 26 Aztec tribute items: (a) 20 military costumes with 20 shields, Matrícula de Tributos 17; (b) 400 women's blouses, Matrícula de Tributos 17; (c) 400 fine striped ponchos or blankets with a length of 2 brazas, Matrícula de Tributos 17. (d) The numeral recording the $(2 \times 8000) + (10 \times 400) = 20{,}000$ human sacrifices offered at the dedication of the Main Temple at Tenochtitlan in 1487, Codex Telleriano-Remensis 39r. [Drawing by M. P. Closs.]

basic units of area; those marked inside the field are similar units multiplied by 20. Thus the area of the given field is represented as

$$20 \times ((6 \times 20) + 12) + 110 = 2650 \text{ square units.}$$

Although the area of a quadrilateral cannot be computed from its perimeter, if it is approximately rectangular then it may be estimated by the product of the average of the opposite sides. Thus the area of the given field is roughly (ignoring fractions whose values are unknown) $60.5 \times 44.5 = 2692$ square units.

The problem of how the Aztec determined the area of fields in these land

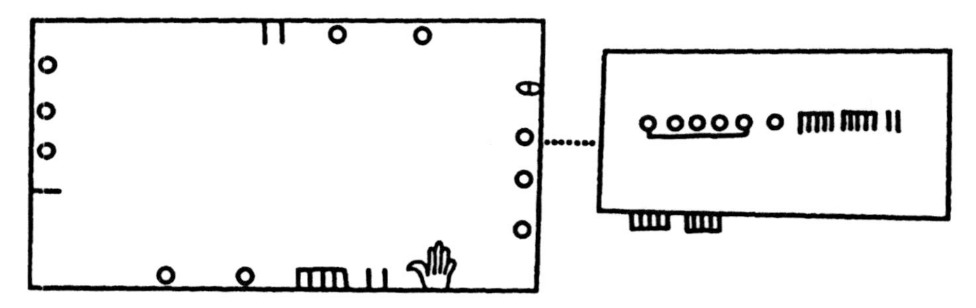

Figure 27 Paired fields, one giving the perimeter and the other the area, from a late 16th century document in the vicinity of Texcoco. [Drawing by M. P. Closs.]

documents has not been resolved. Many of the results given in the manuscripts appear to be close to modern estimates; several are less so.

NOTES

[1] This article does not contain a detailed description of Mesoamerican calendar systems nor does it describe techniques for doing calculations in these systems. Such details are available in a variety of sources including Broda de Casas, 1969; Closs, 1986; Payne and Closs, 1986; Thompson, 1971. It may also be noted that a variety of computer programs have been developed to effect such calculations. Several of these are available on the Internet. A program developed by the author is now available in an interactive format online at:
<http://www.mpiwg-berlin.mpg.de/staff/kantel/mayacalc/mayacalc.html>.

Broda de Casas, Johanna. 'The Mexican calendar as compared to other Mesoamerican systems.' *Acta Ethnologica et Linguistica* 15, 1969.

Closs, Michael P. 'The mathematical notation of the ancient Maya.' In *Native American Mathematics*, Michael P. Closs, ed. Austin: University of Texas Press, 1989, pp. 291–369.

Payne, Stanley E. and Michael P. Closs. A survey of Aztec numbers and their uses. In *Native American Mathematics*, Michael P. Closs, ed. Austin: University of Texas Press, 1989, pp. 213–235.

Thompson, J. Eric S. *Maya Hieroglyphic Writing: An Introduction.* Norman: University of Oklahoma Press, 1971.

BIBLIOGRAPHY

Closs, Michael P. 'The mathematical notation of the ancient Maya.' In *Native American Mathematics*, Michael P. Closs, ed. Austin: University of Texas Press, 1986, pp. 291–369.

Closs, Michael P. 'Mathematicians and mathematical education in ancient Maya society.' In *Selected Lectures from the 7th International Congress on Mathematical Education*, David F. Robitaille, David H. Wheeler and Carolyn Kieran, eds. 7th International Congress on Mathematical Education (Université Laval, Québec, Canada). Sainte-Foy: Les Presses de l'Université Laval, 1994, pp. 77–88.

Closs, Michael P. 'Mathematics of the Aztec people.' In *Encyclopaedia of the History of Science, Technology, and Medicine in Non-Western Cultures*, Helaine Selin, ed. Dordrecht, The Netherlands: Kluwer Academic Publishers, 1997, pp. 622–625.

Closs, Michael P. 'Mathematics of the Maya.' In *Encyclopaedia of the History of Science, Technology, and Medicine in Non-Western Cultures*, Helaine Selin, ed. Dordrecht, The Netherlands: Kluwer Academic Publishers, 1997, pp. 646–651.

Coe, Michael D. 'Early steps in the evolution of Maya writing.' In *Origins of Religious Art and*

Iconography in Preclassic Mesoamerica, H. B. Nicolson, ed. Los Angeles: UCLA Latin America Center Publications, 1976: 109–122.

Freidel, David, Linda Schele and Joy Parker. *Maya Cosmos: Three Thousand Years on the Shaman's Path*. New York: Morrow, 1993.

Graham, Ian. *Corpus of Maya Hieroglyphic Inscriptions, Vol. 3, Part 2*. Cambridge, Massachusetts: Peabody Museum of Archaeology and Ethnology, Harvard University, 1979.

Graham, Ian. *Corpus of Maya Hieroglyphic Inscriptions, Vol. 3, Part 3*. Cambridge, Massachusetts: Peabody Museum of Archaeology and Ethnology, Harvard University, 1982.

Harvey, Herbert R. and Barbara J. Williams. 'Aztec arithmetic: positional notation and area calculation.' *Science* 210: 499–505, 1980.

Harvey, Herbert R. and Barbara J. Williams. 'Decipherment and some implications of Aztec numeral glyphs.' In *Native American Mathematics*, Michael P. Closs, ed. Austin: University of Texas Press, 1986, pp. 237–259.

Justeson, John S. 'The origin of writing systems: Preclassic Mesoamerica'. *World Archaeology* 17(3): 437–458, 1986.

Justeson, John S. and Terrence Kaufman. 'A decipherment of epi-Olmec hieroglyphic writing.' *Science* 259: 1703–1711, 1993.

Kerr, Justin. *The Maya Vase Book: A Corpus of Rollout Photographs of Maya Vases*, Volume 1. New York: Kerr Associates, 1989.

Marcus, Joyce and Kent V. Flannery. *Zapotec Civilization: How Urban Society Evolved in Mexico's Oaxaca Valley*. London: Thames and Hudson, 1996.

Maudslay, Alfred P. *Archaeology. Biologia Centrali-Americana* (5 vols.). London: R.H. Porter and Dulau and Co., 1889–1902.

Nuttall, Zelia, ed. *The Codex Nuttall, A Picture Manuscript From Ancient Mexico*. New York: Dover Publications, 1975.

Payne, Stanley E. and Michael P. Closs. 'A survey of Aztec numbers and their uses.' In *Native American Mathematics*, Michael P. Closs, ed. Austin: University of Texas Press, 1986, pp. 213–235.

Quiñones Keber, Eloise. *Codex Telleriano-Remensis: Ritual, Divination, and History in a Pictorial Aztec Manuscript*. Austin: University of Texas Press, 1995.

Robicsek, Francis and Donald M. Hales. *The Maya Book of the Dead: The Ceramic Codex*. Charlottesville, Virginia: The University of Virginia Art Museum, 1981 (distributed by the University of Oklahoma Press).

Ross, Kurt. *Codex Mendoza, Aztec Manuscript*. Fribourg: Productions Liber, 1978.

Thomas, Cyrus. *Numeral Systems of Mexico and Central America*. Smithsonian Institution, Bureau of American Ethnology, 19th Annual Report, Part 2, 1900, pp. 853–955.

Thompson, J. Eric S. *Maya Hieroglyphic Writing: An Introduction*. Norman: University of Oklahoma Press, 1971.

Tozzer, Alfred M. *Landa's Relacion de las Cosas de Yucatan*. (A translation, edited with notes). Papers of the Peabody Museum, Vol. 18. Cambridge, Massachusetts: Harvard University, 1941.

Urcid-Serrano, Javier. *Zapotec Hieroglyphic Writing*. PhD Dissertation, Yale University, 1992.

Winfield Capitaine, Fernando. *La Estela 1 de La Mojarra, Veracruz, México*. Research Reports on Ancient Maya Writing, 16, Washington, DC: Center for Maya Research, 1988.

DANIEL CLARK OREY

THE ETHNOMATHEMATICS OF THE SIOUX TIPI AND CONE

The people inhabiting the plains region of North America were able to construct sophisticated communities, trade routes and communications systems that reached throughout pre-Columbian North and Mesoamerica. The Great Plains, called a sea of grass by early European explorers, is an area encompassing an enormous geographic region stretching east of the Mississippi River to the Rocky Mountains, and from modern day Texas to the sub-arctic region of Southern Canada. When examining the history and cultures of this region, one quickly becomes aware of a completely different way of thinking. This mode contrasts radically with our westernized one, which is dominated by squares and boxes – we live in square rooms and buildings on square blocks; we measure things using square feet or meters. When examining Native American culture, art, and mathematics one comes to see a different pattern. In their language, customs, architecture, religion, and ceremony, it is the circle that is dominant. For the Plains peoples, the circle was imbued with tremendous power and symbolism; it was the symbol of their link with the universe. A look at the literature pertaining to the Plains peoples uncovers a number of archetypal symbols, including the pipe, the buffalo, and the tipi. All of these held a definite power associated with the time, space, people and age they commanded. But what becomes obvious, when looking further, is the unique interconnectedness of it all through the metaphor of the circle. For the purposes of this chapter, we will limit our study to the mathematics of the tipi.

What is it about the tipi[1] that contains mathematics? Is it the obvious employment of the triangle, cone, circle, and oval? Is it the use of patterns and a definite function in the way that the structure was erected? Or is it the manner in which the fabric was laid out, cut and stitched together, and decorated to form a half circle before being hoisted into place? Though space does not allow us to examine all of these in detail, the reader will be introduced to a number of these questions and solutions. In so doing we will first begin with a discussion of Native American mathematics.

H. Selin (ed.), Mathematics Across Cultures: The History of Non-Western Mathematics, 239–252.

WHAT IS NATIVE AMERICAN MATHEMATICS?

Native American mathematics is part of the entirety of the mathematics generated by humanity throughout recorded as well as unrecorded history. Some characteristics of ethnomathematics that are important are:

1. It is limited in techniques, because it is based on limited sources. On the other hand, its creative component is high, because it is free of formal rules and follows open criteria.
2. It is specific, because it is limited in context, although it is less limited than *ad hoc* knowledge of the universal character of mathematics that pretends to be free of context.
3. It operates using metaphors and symbol systems that are related psychometrically, while mathematics operates with symbols that are condensed from rational form (D'Ambrosio, 1998: 26).

The diverse peoples of this region flourished and lived in relative harmony with their environment and neighbors for many thousands of years. Until the introduction of imported diseases and invading populations from the east created a disastrous decline in their fortunes, the early North Americans created a worldview that has come to influence the rest of humanity in numerous and profound ways (Weatherford, 1988). In adapting to the environmental realities of this region, the Plains peoples operated within a unique paradigm that allowed them to observe and predict the cycle of the seasons, the movement of the buffalo and the stars, their navigation on the 'sea of grass', and the construction and organization of their villages and towns. They made use of methods and accompanying traditions that allowed them to construct a variety of home styles and to govern their communities (Nabakov and Easton, 1989). Unlike western mathematics, this mathematics was intimately tied to a spiritual view of the world. One can best understand the differences in the paradigm by contrasting the two views (Stevenson, 1996; Hall, 1976):

Traditional Aboriginal values	*Western values*
Individual, extended family and group concerns	Individual and immediate family concerns
Cooperation	Competition
Holistic view of nature	Homocentric view of nature
Partnership with nature	Exploitation of nature
Renewable resource economy	Nonrenewable resource economy
Sharing by all of land and resources	Private ownership of land and resources
Sharing and wealth distribution	Saving and wealth accumulation
Nonmaterialistic orientation	Materialistic orientation
Age and wisdom valued	Youth and beauty valued
Polychronic time[2]	Monochronic time[3]

The unity of all things was expressed in the circle. The circle is the perfect figure, because each part is an equal distance from the center. Everything moves and lives in a circle. We come from the earth, live on the products of the earth, and return to the earth. Old Indians tell us that the tipis were made in a circle to express this cycle of living. The unity of all things is also expressed in a circle. The winds blow in a great circle. The seasons of the year are like a circle in that they always come back again and again to where they started (Bryde, 1971: 80). The Oglala Sioux, Black Elk said,

> Everything the Power of the World does is done in a circle. The sky is round, and I have heard that the earth is round like a ball, and so are all the stars. The wind in its greatest powers whirls. Birds make their nests in circles, for theirs is the same religion as ours. The sun comes forth and goes down again in a circle. The moon does the same, and both are round. Even the seasons form a great circle in their changing, and always come back where they were. The life of a man is a circle from childhood to childhood, and so it is in everything where power moves (John G. Neihardt, *Black Elk Speaks*. New York: William Morrow and Co., 1932, pp. 164–165; quoted in Lombardi and Lombardi, 1982: 18).

A MATHEMATICAL STUDY OF THE TIPI

Spatial geometry is inherent in the shape of the tipi. The spatial geometry of the tipi itself was used to symbolize the universe in which the Plains peoples lived. This lightweight and easily transportable home was also part of a larger cosmology, of which the fireplace and altar points were the center:

> Even the home was symbolic – a church, a place of worship – as well as a mere place to eat and sleep. The floor of the tipi represented the earth – the Mother. The lodge-cover was the sky above – the Father. The poles linked mankind with the heavens. The little altar behind the fireplace showed the relationship of man to the spiritual forces surrounding him. ... the poles represent the four directions, and the lifting pole at the rear is the one that 'holds up the sky.' The Indian usually placed his altar in the west, where the sun set, thus keeping the fire alive all through the night in the place where the sun disappeared (Laubin and Laubin, 1989: 241).

Numerous sacred dances incorporated the circle in the use of the hoop, by dancing or moving around the center of the space in which many ceremonies took place. The tipi, medicine wheel, sun dance structures, and sweathouses all used the circle and oval, each structure serving as a gentle reminder to users of their place in the larger circle of life. Village and camp centers were also laid out in a circle fashion, reminding the villagers of the four sacred points of the compass, the winds, the four seasons, and the universe in which they were full partners (Laubin and Laubin, 1989; Nabakov and Easton, 1989).

Often the tipi is confused with the wigwam. Both words mean the same thing, and refer to a dwelling. But it is the word *tipi* from the Sioux language that refers to a conical skin tent or dwelling common among the prairie peoples. *Wigwam* refers to the dome-shaped oval shelters, thatched with bark or reed mats, used by people of the eastern woodlands of North America such as the Potawatomi, Chippewa, and Kickapoo peoples (Nabakov and Easton, 1989). Laubin and Laubin (1989: 15) used the word tipi to 'designate the Plains type of shelter and "wigwam" to designate the bark or mat covered Woodland dwelling even though both types are now covered with canvas.'

The tipi is the final evolution of the conical tent, used extensively in northern and colder regions of the world. A conical shaped[4] structure easily sheds wind, rain, and snow. At the same time the addition of smoke flaps allowed users to make adjustments for a good smoke draw independent of the direction of the wind.

However, it is important to know that the 'fully developed plains tipi' was not a true cone.

> Its steeper, rear side braced the tilted structure against the prevailing westerly winds, and its eastern doorway faced the rising sun. Nor was its plane (base) a true circle; the wider end of its egg-shaped base lay in the rear (Nabakov and Easton, 1989: 150).

Tipis were in use long before the Europeans arrived in North America. Apparently the earlier forms of the tipi were small (around 12 to 14 feet) to facilitate travel when there were only dogs to transport the poles. With the reintroduction of the horse in the 16th century, it became easier to obtain and transport longer poles and more buffalo hides for construction of larger structures. Both George Catlin (1841) and Alfred Jacob Miller (1951) showed tipis from the 12-foot to 18-foot range. Tentsmiths (1998) found that the 16-foot size was the easiest to transport and still have room for two or three adults. The larger tipis from the reservation period probably came into being because it was no longer necessary to move them as often. It seems that early Anglo-American use of the tipi was limited to an occasional stay while interacting with a tribe for trade or marriage.

PRINCIPAL ATTRIBUTES OF A TIPI

True tipis do not form a perfect cone. They lean slightly to the back (west) to allow the smoke hole to be directly over the centrally located fire pit. The base or 'footprint' is more of a wide egg than a true circle, so the measure is often the average of its diameter.

The tipi is made up of three basic components: a pole arrangement, forming a foundation of either three or four poles, the cover which stretched over the outside of the poles, and a liner (*ozan*) suspended from the inside of the poles, surrounding the inside of the structure. Various tribal styles of this sort were apparent throughout the Plains region of North America (Laubin and Laubin, 1989). The differences were primarily in the arrangement of the poles. These were the components determining the cut of the hides or canvas coverings and the shape of the base, in turn determining the size of the overall structure.

An open fire pit in the center of the tipi is another attribute, with its accompanying smoke vent over the center (but not the top of the cone) of the construction. If the tipi were a true cone, the smoke hole would be at the center, but this proved difficult for a number of important reasons. A vent directly centered over the cone would be too large to be effective. The early Native Americans solved this problem by gently moving the cone towards the back (west) – away from the door aperture, thus allowing for the point where the poles meet to be relatively free of the smoke hole (Laubin and Laubin, 1989). This minor geometric modification allowed for a far smaller aperture and for the addition of smoke flaps that could be adjusted depending on wind and other climatic factors. The crossing of the poles – the center point of the cone – is then at the uppermost part of the smoke hole instead of the middle, so 'that it is possible to close the hole entirely with the use of flaps or ears' (p. 17). In smaller tipis with fewer poles, this is not considered a problem, but in larger tipis, this is extremely critical. Pole placement is essential to the correct setting up of the tipi.

> A number of observers have reported the placement of the poles upon the ground, but have neglected to mention the far more essential information of how they are placed in the crotches of the foundation poles. The order of placement in the four-pole type is practically the reverse of that in the three-pole type, but the solution of the problem is the same in both. The majority of the poles are grouped in the front crotch of the foundation – in the smoke hole and away from the cover at the back of the tipi – making it possible to fit the cover tightly and smoothly instead of bunching it around an ungainly and bulky mass of poles (Laubin and Laubin, 1989: 16–17).

Judging by the number of tribes using the tripod foundation, the majority of tipi dwellers realized that the tripod or three-pole foundation provided the best solution to the problem (Nabakov and Eaton, 1989). The three-pole tipi is stronger than the four-pole tipi, because the quadripodal foundation necessitates a number of outside ropes whereas the tripodal foundation is more firm (Laubin and Laubin, 1989). A more in depth look into this from a mathematical point of view follows in the next section.

> It seems ... that the four-pole people were mainly those who live in the Northwest, in or close to the mountains, which would be one explanation why they retained this method. The winds in these regions are not as strong as farther out in the open prairies (Laubin and Laubin, 1989: 214).

TRIPODAL VS. QUADRIPODAL FOUNDATIONS OF THE TIPI

Mathematics can explain why a tripod is more flexible than a quadripodal or four-legged structure. Imagine three points, A, B, and C, that are not collinear. There are an infinite number of planes that pass through points A and B that contain the straight line AB. Only one of these planes can pass through point C. Therefore we can say that these three points determine a plane, if and only if they are non-collinear. This signifies that these 'three non-collinear points exist on one plane and that these three collinear points determine only one plane' (Machado, 1988). For example, a well known math problem from Brazil, states[5]

> It is common to find tables with 4 (four) legs exactly supported on a flat floor. However, to balance them you might need to place a chock under one of the legs if you want to make them more stable. You can explain this using a geometrical argument. Why doesn't this happen with a 3-legged table?
>
> Reply: This can be explained using the postulate for the determination of a plane. 'Given any three non-collinear points, there is only one plane on which these same three points exist.' In the 4-legged table, there is the possibility that one of the legs would not belong to the same plane. A table that has 3 legs, therefore, is always balanced (Bucchi, 1994).

The structure of the tipi appears to be perfectly adapted for the harsh environment in which it was used. It had the advantage of providing a stable structure and was lightweight and portable. At the same time it withstood the prevailing winds and extremely variable weather of this region. Let us look at this information mathematically.

The base formed by the tripod is a triangle we shall call: $\triangle ABC$ (seen above as N for north, S for south and D for first door pole).

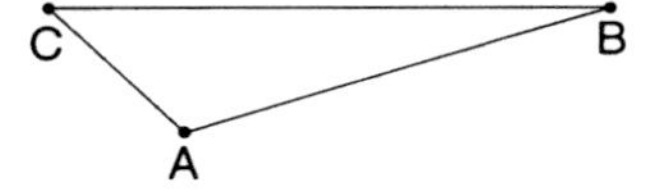

One can then determine the midpoints of each of the sides of $\triangle ABC$, as points M, N and P.

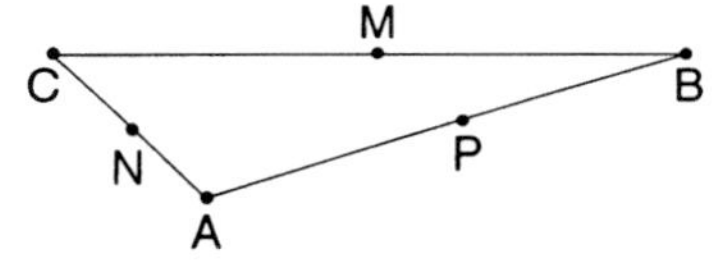

The reader can match each vertex of $\triangle ABC$ to the mid-point of each opposing side, which then gives us the following straight lines: AM, BN, and CP.

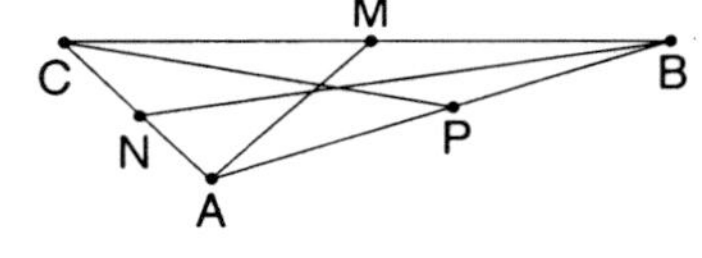

These straight lines form the medians[6] of each line segment. One can see that the crossing medians form only one point. This point is called its centroid. Archimedes demonstrated that medians of a triangle meet at the balance point or center of gravity which he called the centroid. Native Americans place their fire and altar at this point in the tipi. Cartographers call this point the geographic center.

Also essential is the size of the tripod in order to construct the right sized tipi in relation to its outer covering. Poles are measured using the cover itself as reference (Laubin, 1989). The tipi skin is folded in half and the poles are laid together before they are tied to form the tri or quadripod frame, which forms the foundational base for the structure. Once this length is determined, it can be marked or tied at that point.

DETERMINING THE HEIGHT OF THE TIPI

The tripod is raised to check the correct angles between poles visually and to adjust the height. Corrections are made until the desirable distance between the tipi height and distance between poles is achieved. There exist at least two 'experimental' processes that were probably used to check the accuracy of the work: In one, a type of pendulum is constructed by tying a rock to a rope. The pendulum is positioned at the meeting point of the poles. To check the best height and the best pitch between poles, we have to observe the point where the pendulum falls vertically to the soil. In the other, one pole is placed vertically to the ground in relation to the meeting point of the poles. If the pole is vertical in relation to the ground and the meeting point of the poles, this also verifies the best height and angle for the tipi.

Mathematically this situation can be explained where the pendulum and poles function as the height of the tipi (straight-line path) that falls perpendicularly in the center of the triangle forming the tripod on the ground. When this straight line is perpendicular to the ground, the point is determined using the centroid (the break-even point, center of gravity, and center of mass) of the triangle. Thus the Native Americans found that this equally distributed the area in the space outlined by the tripod.

This type of foundation, using the centroid of the triangle determined by the tripod and tipi base, equally distributes the area delimited on the ground. This form also allows the operating load of the structure to be distributed equally, forming a stable structure.

GEOMETRY OF THE BASE

Geometry can be used to explore how the dwellers determined the center of the circular base of the cone. This can be done using mathematical modeling and the existing triangle formed by the tripod.

As mentioned earlier, tipis do not constitute a true or perfect cone. The conic shape allows for efficient diffusion of light, which easily and efficiently covers all of the internal living space. The tipi scaffolding itself is constructed by placing the poles around the foundation. The same sequence depends upon the type (three or four pole) of the foundation, thus assisting in the determination of the base area. For the purpose of this study, the tipi base is considered to be a circle, though by the time the poles form the base, it is actually a slightly oval or egg shape.

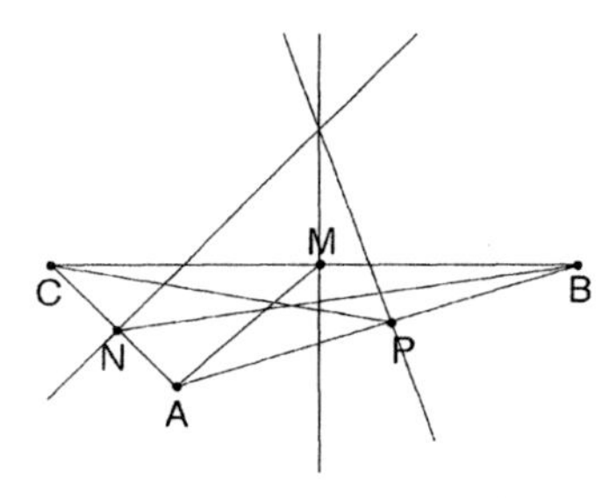

Using a triangle formed by the tripod, the vertices can then be determined. Mathematically, the poles must be placed as if they were points on this circumference; they therefore determine the external circumference of the tipi base. By tracing each one of the sides of the triangle, using straight lines passing perpendicularly through the midpoint of each side of the triangle, it can be shown that these straight lines are perpendicular bisectors.[7] When we look at the figure, we can verify that the perpendicular bisectors of a triangle are intercepted (they are crossed or cut) at the same point that is an equal distance from the vertex of the triangle. This point is called the circumcenter.[8]

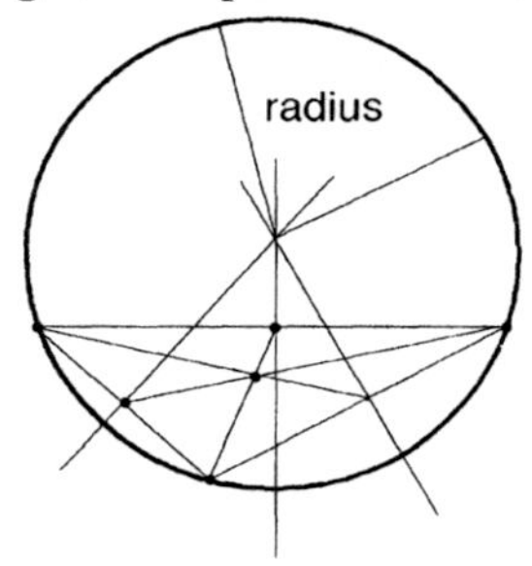

The center of the tipi holds a definite power and holiness. It is more than just necessity or aesthetics that went into finding the center of the Sioux home. The effusion of soft light into the interior space created a mood and a calming sensation. Like the great European cathedrals or the architecture created by the Aztecs, Incas and Mayas, the Plains peoples created an elegant as well as dramatic effect through the use of conic geometry, space, and light.

AREA OF THE CIRCULAR SECTION AND LATERAL AREA OF THE COVER

The tipi area cover is similar to an area of a circular section, as in this representation:

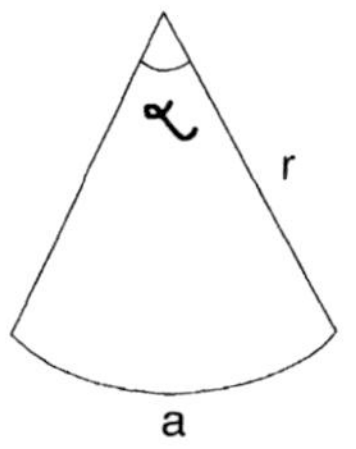

Area	Length
$\pi.r^2$	$2.\pi.r$
Area of the circular section	a

Area of the circular section $x2\pi.r = \pi.r^2.a$
Area of the circular section $= \pi.r^2.\frac{a}{2.\pi.r}$
Area of the circular section $= r.\frac{a}{2}$

The area of the tipi cover is equal to the lateral area of a similar cone. The lateral area is a circular section of the radius 'g' corresponding to the generator of the cone and forming a complementary arc '$2\pi r$'. Therefore a complementary arc of the cover and of the same complementary (perimeter) of the circle forms the base of the tipi.

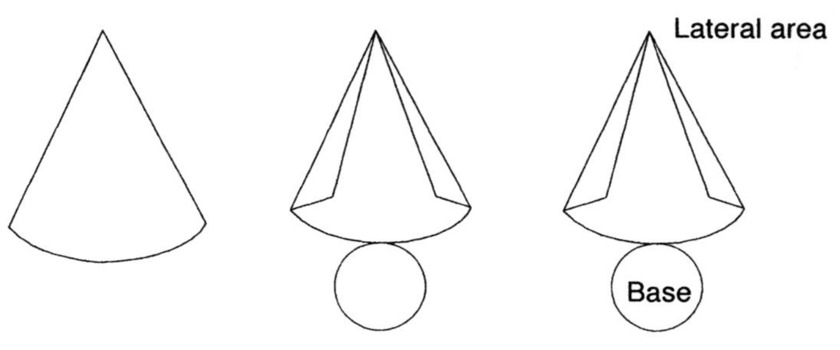

$$\text{Lateral area} = \frac{ra}{2} = \frac{2\pi rg}{2} = \pi rg \text{ where } a = 2.\pi.r \text{ and } r = g.$$

How do we determine the existing relation between the radius of the base of the tipi and the radius of the lateral area of the cover mathematically?

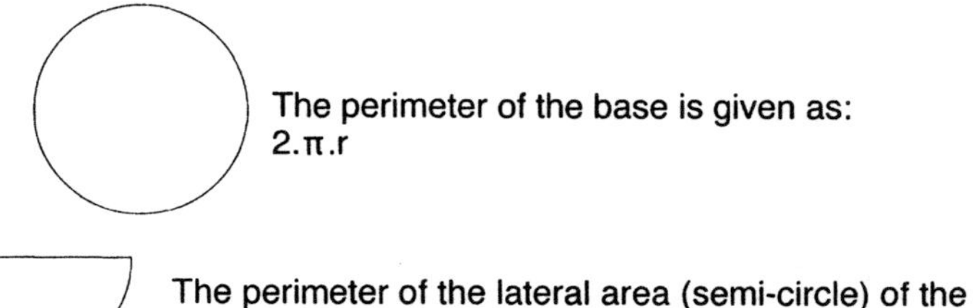

The formula below is used to explain the perimeter of the lateral area (semi-circle) above. This is used to find the area of the semi-circle and is then divided by the area of circle (2).

$$\frac{2\pi R}{2}$$

Let us call r the radius of the base of the tipi and R the radius of the lateral area of the cover. Therefore the perimeter of the base forming the tipi is equal to the semi-perimeter of the lateral area of the cover, and is therefore a semi-circle. We have

$$2\pi r = \frac{2\pi R}{2}, \qquad 2\pi r = \pi R, \qquad \frac{R}{r} = 2.$$

The radius of the lateral area of the cover is the double of the radius of the tipi base.

* * *

As demonstrated in this essay, architecture and mathematics are interrelated. The paradigm that diverse cultures use or work within evolves out of unique interactions between their language, culture and environment. The tipi used by the Sioux evolved from a perspective that did not use the idea of square, with the possible exception as its employment in decorative motifs. Instead, the idea of a 'sacred circle' was evident in everything that was done, including how the people lived, were housed, organized their structures and communities, worshiped, even entered a structure or space.

When we study the evidence left to us of the mathematics used by the Plains cultures, the reality of that same mathematics as used in this unique cultural context, brought about by ethnomathematical techniques and the tools of mathematical modeling, allows us to see a truly different reality. These methodologies included the development of strategies such as the sharing of evidence, the justification of results, the demonstration of creativity, the development of personal initiative, confidence, and the capacity for facing challenges. These strategies are a blend of aboriginal and western values which are vitally important for modern life. They also are useful in allowing us to capture a glimpse of the reality that once existed in the minds and souls of the people who originally inhabited the great 'sea of grass'.

Modern mathematics is international in character, as witnessed by concepts transmitted, studied and developed in all parts of the world (Closs, 1986). In the unique case of the Plains peoples, symbolic descriptions of 'universal' mathematical concepts is presented in 'notation independent of language and culture' (Closs, 1986: 1). The unintended consequence of this has been the almost complete loss of evidence of indigenous thinking and problem solving traditions related to this period of time and place. Modern mathematics is a relatively recent phenomenon and represents a continuation of mathematical developments originating in Europe and the Middle East. Mathematical historians have traditionally limited their sources to those that lead to modern mathematics. They have paid very little attention to the mathematics developed by peoples not directly contributing to it, or by cultures that were conquered or colonized during the European expansion.

> The intellectual activity known as 'mathematics' was developed in certain areas of the world in which a tradition of writing had already been established. However, most cultures did not, and still do not have such a tradition. If we state that the mathematical skills or the systems of numeration and calculus in these cultures are rudimentary, we make a judgement from the wrong perspective. Indeed, we use the exception – cultures with a written tradition, especially a western intellectual tradition – as a measure for all others (Gnerre, 1996: 71).

This most certainly is a problem with the exploration of the geometry found in the tipi. However, what can be learned from the mathematics exemplified by the tipi can be summarized in six points:

1. Circle-based mathematics is evident. Further exploration of this area will reveal what it must have been like to measure and judge the world using primarily the circle. This is especially interesting as the original peoples of the Americas made minimal, if any use at all, of the wheel for transportation.
2. An early form of democracy was employed and may have been directly influenced or inspired by the sacred geometry of the dwellings the people used and constructed. The opposite is most surely feasible: that the dwellings and architecture may have been employed as an outgrowth of early forms of this democracy.
3. Non-written forms of literacy were employed, as most certainly the tipi was also used to convey information. Not unlike the cathedrals of Europe, many tipis displayed important information, using symbolic representations about the use of the structure and its inhabitants.
4. The adaptation of either a 3- or 4-pole foundation strongly suggests that knowledge of the environment, and the tipi, was kept and passed down from generation to generation using some form of literacy. This was most certainly done by women. The tipi was kept, maintained, and constructed by women in their respective societies.
5. The use of the semi-egg shape vs. a true circular base demonstrates the acquisition of a certain physical knowledge as linked to the world of nature. We cannot really know if the originators of the tipi observed the phenomenon in nature (egg) or came to it by themselves by experience and experimentation. I am intrigued by the obvious symbolic connections between motherhood, fertility, and the shape of the tipi base as it rests gently on the 'Earth Mother'.
6. The space inside the tipi economically increased light and eased communication in a manner that lifted the spirit. The conic shape of the structure encouraged anyone who entered to look to the sky. The materials used to construct the tipi allowed filtered light to enter the structure softly. These links most certainly employed geometry.

Standing as we are at the beginning of a new millennium, many people sense a change around them. It is my hope that this change may come to include an expanded awareness of the relationship between science and culture – with an inclusive awareness that the perspectives of all peoples, cultures, and experiences are of value and deserve an equal voice.

What has intrigued me as I continue to study mathematics from diverse contexts is the ability of many peoples to create a world where all members of a society became competent in the day to day tasks that needed to be done. A lack of access and equity seems to be a modern phenomenon (Eisler, 1988). I am convinced that we have much to learn from the paradigms found in indigenous or aboriginal cultures. Despite our technological and scientific accomplishments, we have managed to alienate and disenfranchise major proportions of our very own communities (Mander, 1991). This does not seem as evident, and was most certainly was not as severe, in the case of the Plains Indian people. Almost everyone dwelt together with a sense of participation and lived with that same sense of being worthy members of their communities. All people deserve the right to learn to use the tools that allow them to reach their full potential. This is just one of the important lessons left to us by the original inhabitants of the Americas.

> One expects then that in this evolutionary phase of the species, with all its arrogance, envy and great power ... will instead yield to a place of respect for diverse peoples, in solidarity, in contribution to a preservation of a common patrimony (D'Ambrosio, 1997).[9]

Ethnomathematics might be characterized as a tool used to act in the world. The mathematics used in the Plains Indian culture, in their unique context, studied using ethnomathematical techniques and the tools of mathematical modeling, allows us to see a different reality and gives us insight into science done in a different way.

ACKNOWLEDGEMENTS

I wish to thank my dear friend, colleague and research partner Milton Rosa of Amparo, Brazil, whose assistance in looking at numerous drafts of this work has been more helpful than words.

NOTES

[1] The word tipi is often spelled teepee and refers to the Dakota word; the current literature on the subject prefers the spelling tipi.

[2] Polychronic time operates with several things happening at once, stressing involvement and interaction of and between people, and the completion of transactions.

[3] Monochronic time emphasizes schedules, segmentation, and promptness over and above completion and human interaction.

[4] Though not a tipi by any stretch of the imagination, my favorite conic edifice is the 50 story metropolitan cathedral of Maringá, Paraná, Brazil.

[5] (UNICAMP) É comum encontrarmos mesas com 4 (Quatro) pernas que, mesmo apoiadas em um piso plano, balançam e nos obrigam a colocar um calço em uma das pernas se a quisermos firme. EXPLIQUE, usando argumentos de geometria, porque isso não acontece com uma mesa de 3(Três) pernas? Resposta: Isto se explica pelo postulado da determinação de um plano'Três pontos não colineares determinam um plano'. Na mesa com quatro pernas, há a possibilidade de a extremidade de umas das pernas não pertencer ao mesmo plano determinado pelas outras três pernas, por isso, algumas vezes, elas balançam ... Matemática (Segundo Grau) – Paulo Bucchi – Volume Único – São Paulo – Editora Moderna – 1994.

[6] The median is the straight line connecting the midpoint of the opposing side of the triangle and its vertex.

[7] The perpendicular bisector of a straight line is the straight line perpendicular to its midpoint. The perpendicular bisector is also a geometric point of the plan formed for the points that are equidistant (they are to the equal angle) of the extremities of a line.

[8] The circumcenter is the point of intersection of the perpendicular bisectors of the sides of a triangle. The circumcenter is the center of the circumscribed circumference to the triangle and that passes though its vertices.

[9] Espera-se então que nesta fase da evolução da espécie, a arrogância, a inveja e a prepotência – historicamente vistas como distorções capitais no relacionamento do indivíduo com sua totalidade

interior e igualmente com a realidade exterior – cedam lugar ao respeito pelo diferente, à solidariedade, à colaboração na preservação do patrimônio comum.

BIBLIOGRAPHY

Araujo, M. A. 'Matemática: Acerto de contas.' *Educação* (197): 16–19, 1997.

Benatti, L. 'Disciplina liga matemática ao cotidiano.' São Paulo: *Folha de São Paulo*, August 4, 1997.

Bucchi, P. *Matemática (Segundo Grau) Volume Único*. São Paulo: Editora Moderna, 1994.

Borba, M. 'Ethnomathematics and education.' *For The Learning of Mathematics* 10(1): 39–43, 1990.

Caraway, C. *Plains Indian Designs*. Owings Mills, Maryland: Stemmer House Publications, 1984.

Catlin, G. *Letters and Notes on the Manners, Customs, and Conditions of the North American Indians, Written During Eight Years of Travels 1832–39*. New York: Wiley and Putnam, 1841.

D'Ambrosio, U. *Transdisciplinaridade*. São Paulo: Editora Palas Athena, 1997.

D'Ambrosio, U. *Educação matemática: da teoria à prática*. São Paulo: Papirus Editora, 1996.

D'Ambrosio, U. *Ethnomathematics: the Art or Technique of Explaining and Knowing*. International Study Group on Ethnomathematics, Las Cruces, New Mexico, 1998.

D'Ambrosio, U. 'Ethnomathematics and its place in the history and pedagogy of mathematics.' *For the Learning of Mathematics* 5(1): 44–48, 1985a.

D'Ambrosio, U. *Socio-Cultural Bases for Mathematics Education*. Campinas: UNICAMP, 1985b.

D'Ambrosio, U. *Intercultural Transmission of Mathematical Knowledge: Effects on Mathematical Education*. Campinas: UNICAMP, 1984.

Dolce, O. and Pompeo, J. N. *Fundamentos da Matemática Elementar-geometria plana*. São Paulo: Editora Atual, 1993.

Eisler, R. *The Chalice and the Blade*. San Francisco: Harper, 1988.

Freire, P. *Pedagogia da Autonomia: Saberes Necessários à Prática Educativa*. São Paulo: Editora Paz e Terra, 1997.

Freire, P. *Na Escola que Fazemos: Uma Refexão Interdisciplinar em Educação Popular*. Petrópolis: Editora Vozes, 1987.

Gilmore, M. R. *Prairie Smoke*. New York: AMS Press, Inc., 1966.

Hall, E. T. *Beyond Culture*. New York: Doubleday, 1976.

Hassrick, R. B. *The George Catlin Book of American Indians*. New York: Promontory Press, 1977.

James, C. *Catch the Whisper of the Wind: Inspirational Stories from Native Americans*. Deerfield Beach, Florida: Heath Communications, 1995.

Laubin, R. and Laubin, G. *The Indian Tipi: its History, Construction and Use*. 2nd ed. Norman: University of Oklahoma Press, 1977.

Lombardi, F. G. and Lombardi, G. S. *Circle without End*. Happy Camp, CA: Naturegraph Publishers, 1989.

Lopes, E. T. and Kanagae, C. F. *Desenho Geométrico. Texto e Atividades*. São Paulo: Editora Scipione, 1995.

Mander, J. *In the Absence of the Sacred: The Failure of Technology and the Survival of the Indian Nations*. San Francisco: Sierra Club Books, 1991.

Machado, A. S. *Matemática – Temas e Metas (Segundo Grau) – Áreas e Volumes*. São Paulo: Atual Editora, 1988.

Miller, A. J. *The West of Alfred Jacob Miller (1837)*. Norman: University of Oklahoma Press, 1951.

Nabakov, P. and Easton, R. *Native American Architecture*. Oxford: Oxford University Press, 1989.

Orey, D. 'Mathematics for the 21st century.' *Teaching Children Mathematics* 5(4): 241, 1998.

Orey, D. 'Ethnomathematical perspectives on the NCTM Standards.' *International Study Group on Ethnomathematics Newsletter* 5(1): 5–7, 1989.

Orey, D. 'Logo goes Guatemalan: an ethnographic study.' *The Computing Teacher* 12(1): 46–47, 1984.

Orey, D. 'Mayan math.' *The Oregon Mathematics Teacher*, 1982, pp. 6–7.

Powell, A. B. and Frankenstein, M., eds. *Ethnomathematics: Challenging Eurocentrism in Mathematics Education*. Albany, New York: State University of New York Press, 1997.

Regents of the University of California. *Multicultural Education and the American Indian*. Los Angeles: American Indian Studies Center, UCLA, 1979.

Rosa, M. *Matemática: Seqüências e Progressões*. São Paulo: Editora Érica, 1998.

Stevenson, M. G. 'Indigenous knowledge in environmental assessment.' *Arctic* 49(3): 256–264, 1996.

Tentsmiths. *http://www.tenstsmiths.com/page34.htm*, 1998.

Weatherford, J. *Indian Givers: How the Indians of the Americas Transformed the World*. New York: Random House, 1988.

Williamson, R. A. *Living the Sky: The Cosmos of the American Indian*. Boston: Houghton-Mifflin, 1984.

WALTER S. SIZER

TRADITIONAL MATHEMATICS IN PACIFIC CULTURES

The three traditional regions of the Pacific are Polynesia, Melanesia, and Micronesia. Polynesia lies in the east, and is roughly contained in the triangle with corners in New Zealand, Hawaii, and Easter Island. The northwest boundary bends somewhat, excluding Fiji but including Tuvalu (the former Ellice Islands) and Kapingamarangi. The main parts of Polynesia, besides those places already mentioned, include Tonga, Samoa, French Polynesia (the Austral Islands, the Cook Islands, the Society Islands, the Tuamotu Archipelago), and the Marquesas Islands. Melanesia starts in Papua/New Guinea and extends east and southeast to Polynesia. Included in Melanesia are the Bismarck Archipelago (the Admiralty Islands, New Britain, New Ireland), the islands of the Torres Straight, the Solomon Islands, New Caledonia, the Loyalty Islands, Vanuatu (the New Hebrides), Santa Cruz, the Banks Islands, and Fiji. Micronesia lies to the north of Melanesia and includes the Marshall Islands, the Marianas (including Guam), Kiribati (the Gilbert Islands), and the Caroline Islands (including Yap, Chuuk or Truk, Palau, and Pohnpei). There is some disagreement among writers as to exactly where to draw the boundaries for the regions, and among the people living near the boundaries one can frequently observe traits of more than one region.

People first arrived in these regions about 50,000 years ago. Because of the ice age it would have been possible to reach New Guinea from Asia without losing sight of land. Ancestors of the present Melanesians came to that region about 4000 to 6000 years ago, Micronesia was first inhabited starting about 3500 years ago, and Polynesia began to be settled about 3000 years ago. Full settlement of the islands was not completed for about 2000 years. Especially for Polynesia there is no good evidence for contact with non-Pacific peoples from the time of settlement until European sailors arrived, generally in the 1700s. The same is true for the other regions after about 200 BC, except that European contacts came earlier, from 1521 on (when Magellan's expedition stopped briefly in Guam).

In a work like this one is tempted to include the nearby islands of Indonesia (including Borneo and the Celebes) and the Philippines. Some of the people in these regions were in much closer contact with Asia, but others kept to

H. Selin (ed.), Mathematics Across Cultures: The History of Non-Western Mathematics, 253–287.

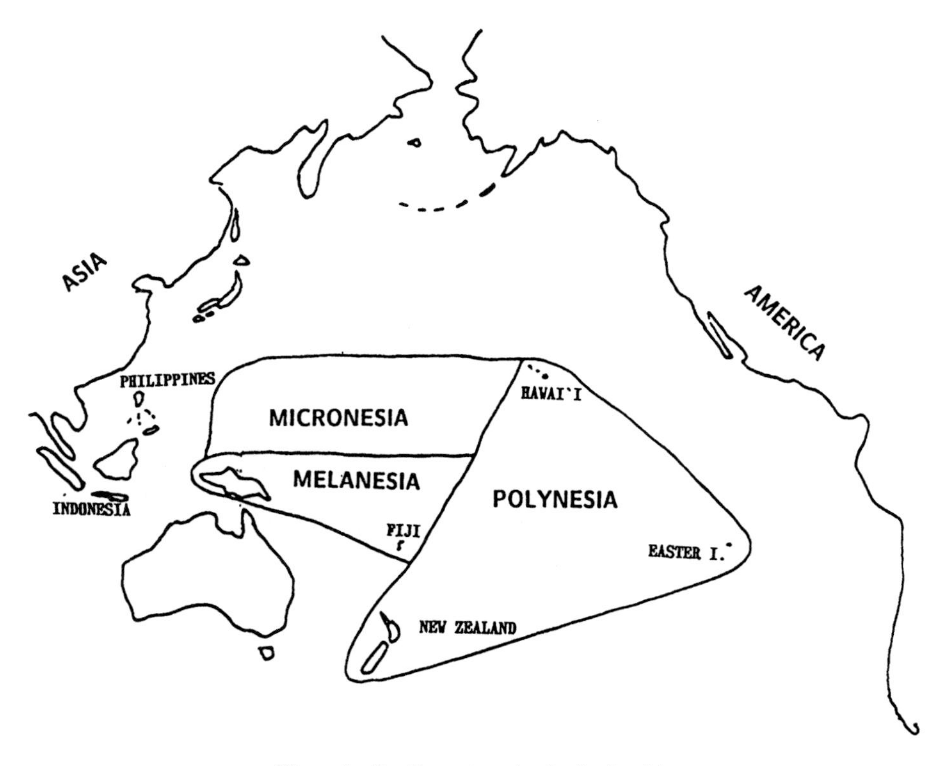

Figure 1 Pacific regions (author's sketch).

traditional ways and seem to exhibit little Asian influence. I have included some accounts from these regions where I think Asian influence was not a factor, because I have found them to be interesting examples of how traditional cultures have developed mathematical concepts.

Besides giving the geographic scope of this work, I should also outline its mathematical content. Especially in a setting like this care should be taken to explain why topics are chosen for inclusion. Here the bias of Western mathematics appears: all the topics chosen – numeration, geometry, games, relations – relate to subjects studied by present day Western mathematicians. The common thread which unites them all under the heading of mathematics is the way logical thinking is applied to the various situations to allow the practitioner to analyze them and to reach conclusions.

NUMERATION

Pacific cultures traditionally indicated numbers using primarily spoken words. Numbers were originally neither written words nor written numerals. Still, number systems were frequently quite well developed and displayed discernible structure as (usually) base five, base ten, or base twenty systems.

Gilbertese numbers: a pure base ten system (Micronesia)

The residents of Kiribati (the Gilbert Islands) have a number system which is linguistically a pure base ten system. Words for numbers one through ten show no connections. Words for multiples of ten through ninety show a base derived from the corresponding unit word and a suffix (*bwi*) indicating that the root is multiplied by ten. Units are added on with the connective *ma* (but with hundreds but no tens, the *ma* is absent). A new word is introduced for one hundred, and then the same pattern is followed again, and likewise for thousands. A table of some values follows.

Table 1 Gilbertese number words

1	*teuana*	10	*tebwina*	100	*tebubua*	1000	*tenga*
2	*uoua*	20	*uabwi*	200	*uabubua*	2000	*uanga*
3	*tenua*	30	*tenibwi*	300	*tenibubua*		
4	*aua*	40	*abwi*	400	*abubua*		
5	*nimaua*	50	*nimabwi*	500	*nimabubua*		
6	*onoua*	60	*onobwi*	600	*onobubua*		
7	*itiua*	70	*itibwi*	700	*itibubua*		
8	*wanua*	80	*wanibwi*	800	*wanibubua*		
9	*ruaiua*	90	*ruabwi*	900	*ruabubua*		
37	*tenibwi ma itiua*			782	*itibubua wanibwi ma uoua*		

In Gilbertese ten thousand is *tebwina tenga* and one hundred thousand is *tebubua tenga*. One million is *teuana te mirion*, an evident variant of the English (Bibina, 1974).

Other bases in Polynesia

Across Polynesia there is considerable agreement among the different languages for the number words from one to nine (Turner, 1884: 374–375). There is more disparity in the words for ten, but still many similarities, and from ten through nineteen they show a base ten structure. At twenty some languages introduce a totally new word, suggesting a base twenty system. In the Maori language (New Zealand) there are two ways to express twenty, *rua tekau* which means two tens and *hokotahi*, a totally new word. For odd multiples of ten there is just one way to express the number, as a multiple of ten. But for even multiples of ten there are two variants, as a multiple of ten and as a multiple of twenty. One hundred has two variants, *kotahi rau*, a new expression corresponding to a base ten system, and *hokorima* or five twenties, corresponding to a base twenty terminology. Although there are these vestiges of a base twenty system in some number terminology, simple words for higher powers of ten but not for higher powers of twenty show that it evolved into a regular base ten system. This is evident from the H M Ngata online English-Maori dictionary.

Traditional Hawaiian number words showed a much different system. Numbers to forty were constructed as shown in Table 2 (Clark, 1839).

From forty on the system operated for a time as a base forty system; instead of a new word for one hundred, this number was expressed as two forties plus

Table 2 Traditional Hawaiian number words

1	*akahi*	11	*umikumamakahi*
2	*alua*	12	*umikumamalua*
3	*akolu*	13	*umikumamakolu*
4	*aha*	20	*iwakalua*
5	*alima*	21	*iwakaluakumamakahi*
6	*aono*	22	*iwakaluakumamalua*
7	*ahiku*	30	*kanakolu*
8	*awalu*	31	*kanakolukumamakali*
9	*aiwa*	40	*kanaha*
10	*umi*		

twenty (*elua kanaha me ka iwakalua*). The next new word was introduced for four hundred (ten forties): *lau*.

Further new words came at four thousand (*mano*), forty thousand (*kini*) and four hundred thousand (*lehu*). Thus the traditional Hawaiian number system showed a mixed base ten, base forty structure. (Hawaiian usage has changed since the early 1800s. Modern Hawaiian usage does have a simple word for hundred – *haneli* – and has stopped using the old words for four-, forty-, and four hundred thousand).

Maenge number words: base twenty, subbase five (Melanesia)

The Maenge people of New Britain Island in Melanesia had a base five system for numbers up to ten, a base ten system up to twenty and a base twenty system to go higher. As they did not have the need to count beyond four hundred, however, this means that words for all numbers were constructed from the words for one to five, ten and twenty. The following table of number words and examples shows how this system worked (Panoff, 1970).

Table 3 Maenge number words

1	*kena*	10	*tangulelu*
2	*lua*	11	*tangulelu va kena*
3	*mologi*	19	*tangulelu va lima va tugulu*
4	*tugulu*	20	*giaukaena*
5	*lima*	40	*giaukaenu lua*
6	*lima va kena*	100	*giaukaenu lima*
7	*lima va lua*		
399	*giaukaenu tangulelu va giaukaenu lima va giaukaenu tugulu va tangulelu va lima va tugulu*		

Four hundred, the limit of the traditional number system, was *giaukaenu giaukaenu*, or, if it was clear that four hundred and not forty was intended, *giaukaenu lua*.

Other systems in Melanesia

A thorough account of number words used in Papua/New Guinea and nearby Melanesia is given in the seventeen volume work *Counting Systems of Papua*

New Guinea by Glendon A. Lean (1991). For other parts of Melanesia one source is R. H. Codrington's *The Melanesian Languages* (1885). Unfortunately, Codrington gives less information about number words than one frequently wants. Most of the number systems recorded have base five, base ten, or base twenty.

An example of a base five system (according to Codrington) is the Sesake system in use on Three Hills Island in Vanuatu. This structure is displayed in the words for numbers through twenty, but as no words are given for numbers beyond twenty one cannot be certain that a pure base five system was intended. It would seem that the traditional system had little occasion to go beyond twenty.

Table 4 Sesake number words

1	*sekai*	5	*lima*	9	*le veti*
2	*dua*	6	*la tesa*	10	*dua lima*
3	*dolu*	7	*la dua*	20	*dualima dua*
4	*pati*	8	*la dolu*		

The Kewa of central New Guinea have a system which is primarily a base four system, as an examination of their number words shows (Franklin, 1962). The *mala* in the expression for sixteen is another Kewa word for four, and thus sixteen is expressed as four fours and not with a simpler expression for a power of the base. The expression for twenty (five fours, using an alternate word for five) also does not fit a base four pattern. These numbers, however, are at the upper bounds of this version of their number system.

Table 5 Kewa number words

1	*pameda*	5	*kode*	9	*kilapona kode (kilapona pameda)*
2	*lapo*	6	*kodelapo*	10	*kilapona kodelapo*
3	*repo*	7	*koderepo*	11	*kilapona koderepo*
4	*ki*	8	*kilapo*	12	*ki repo*
16	*ki mala*	17	*ki malana kode*	20	*kisu*

Extent of number systems

In looking at accounts of number words and systems, one frequently sees comments that a tribe did not count beyond twenty in one case, or in another case four hundred. The general comment is that number words were always developed to count as high as was necessary, whether this limit was less than a hundred or less than a thousand or more than a thousand. However some groups developed words for numbers well beyond the limits of practical counting. Some of the words for larger numbers are noted in Table 6.

An interesting account showing a knowledge of how number systems work to give larger and larger numbers is given by William Mariner, an English seaman who lived for several years in the early 1800s on the island of Tonga (Mariner, 1816). Mariner was fifteen years old when the ship he was working

Table 6 Words for larger numbers

Gilbertese (Kiribati, Micronesia)	1000	*tenga*
	10000	*tebwina tenga*
	100000	*tebubua tenga*
Marquesas (Polynesia)	1000	*mano*
Tahiti (Polynesia)	1000	*mano*
	10000	*manotini*
	100000	*rehu*
	1000000	*'iu*
Samoa (Polynesia)	1000	*afe*
	10000	*mano*
	100000	*'ilu*
Tonga (Polynesia)	1000	*afe*
	10000	*mano*
	100000	*kilu*
Tuvalu (Polynesia)	1000	*afe*
	10000	*sefulu afe*
	100000	*selau afe*
Fiji (Melanesia)	1000	*udolu*
	10000	*oba*
	100000	*vatu lon*
	1000000	*vetelei*
Motu (Banks Islands, Melanesia)	1000	*tar*
	10000	*gerebu*
	100000	*domaga*

on put in for repairs on Tonga. A skirmish with the natives left the rest of the crew dead, and Mariner was adopted for a while by the family of a chief. He lived there for four years, when he was rescued by another British ship. On his eventual return to England he wrote of his experiences in Tonga, and in his work pointed out a misconception in a report of an earlier visitor to Tonga. This visitor had desired to learn something of the natives' system of numeration, and presumably by pressing for words for higher and higher powers of ten was able to record words for powers of ten up to a quadrillion. Looking at the recorded words, Mariner noted they were correct up to one hundred thousand, the limit of their legitimate number vocabulary at the time. After one hundred thousand the natives had filled in with other words, sometimes quite vulgar, or nonsense syllables. The natives knew what was expected, however, and were happy to provide their European questioner with responses!

Using the body to count

Many societies in Melanesia had standardized ways of counting on the body. They took the natural tendency to count on fingers, or fingers and toes, and extended it using other parts of the body to get to higher totals. Lean and others give many examples of this practice; the correspondence used by the Kewa people as recounted by the Franklins is described in Table 7.

Table 7 Kewa body counting

1	(left) little finger	14	upper upper arm
2, 3, 4	successive fingers	15	shoulder
5	thumb	16	shoulder bone
6	heel of palm	17	neck muscle
7	palm	18	neck
8	wrist	19	jaw
9	forearm	20	ear
10	large arm bone	21	cheek
11	small arm bone	22	eye
12	above elbow	23	inner corner of eye
13	lower upper arm	24	between eyes
		25–47	reverse order, using right side of body

Other standardized counting procedures using parts of the body among tribes of New Guinea went to 47 (Wiru and Ialibu), 37 (Kutubu), 27 (Telefomin), 23 (Durmut Mandobo and Ayom), 15 (Pole), and 14 (Huli and Duna).

Fractions

Older accounts of Pacific languages in general do not give constructions for naming fractions. Some writers will offer a translation of the English word one half, but often the native word best translates back into English as a part. The usual practice seems to have been to change the units used in measuring: instead of saying one third of an arm span, one would express the same distance in terms of handspans. In cases where the languages have been modernized expressions have been introduced for fractions.

Recording numbers

Physical objects are frequently used as an aid to counting or as a means of recording a tally. An instance where objects are used to help count is given in the traditions of the Keraki people of Papua. Here the objects to be counted were the *taitu* (an edible tuber) harvested from one person's field. The goal was not necessarily to come up with a definite total, but to determine that the harvest was sufficient for the coming year. The target goal was 1296 *taitu*. The counting had about it something of a ritual, with expert counters brought in for the ceremony. From the pile of harvested tubers two men would each remove three and put them down together, while a tally keeper would count 'one'. The process would be repeated for counts of 'two', 'three', 'four', and 'five'. With the sixth load of six the tally keeper threw a *taitu* to another counter, and the procedure continued as before. The goal was reached when the third counter got six *taitu*, indicating that 1296 had been counted. A similar way of counting tubers is practiced by the natives of Saa, in Melanesia, except that the count is done base ten and a yam is set aside to represent each hundred.

Other objects were commonly used to assist in counting and to give a record of the total. Stones, shells, and twigs or leaf stems were often used for these purposes. In Florida (Melanesia) one method of counting guests at a feast involved collecting a small object from each, thus getting a record of the count. Another way to record a count was to carve notches in a stick; this technique was used in the Philippines and in Melanesia.

On the islands of the Torres Straight tallies were kept by tying small sticks onto a string, one stick for each object being counted. Sticks were frequently crafted with a groove for the string so the sticks would not easily become separated from the tally string. Counts were preserved in this way to keep track of sea turtles killed in open waters, of kingfish speared, and of dugong killed. One example of such a tally contained ninety-two sticks.

Longer lasting records of counts are preserved in knots tied into a string. This technique is used in the Solomon Islands and among the Kayan tribe of Borneo to keep track of the number of days until an event, for example, with one knot cut off as each day passes. It was also reported among some Philippine people as a way to keep track of loans. One account from Hawaii gives an unusual report of keeping records on string:

> The tax gatherers, though they can neither read nor write, keep very exact accounts of all the articles, of all kinds, collected from the inhabitants throughout the island. This is done principally by one man, and the register is nothing more than a line of cordage from four to five hundred fathoms in length. Distinct portions of this are allotted to the various districts, which are known one from another by knots, loops, and tufts, of different shapes, sizes, and colors. Each taxpayer in the district has his part in this string, and the number of dollars, hogs, dogs, pieces of sandal-wood, quantity of taro &c., at which he is rated, is well defined by means of marks, of the above kinds, most ingeniously diversified (Montgomery, 1832, vol. 2, p.71).

GEOMETRY

Most cultures display considerable geometric understanding through their manufactured products, whether these are craftwork, buildings, weapons, utensils or decorations. The geometry displayed shows understanding of basic shapes such as angles, straight lines, circles and other curves, and rectangles and other polygons. It also shows understanding of concepts such as symmetry, directions and magnitude, and relative position. The range of topics is so wide that in a summary like this whole topics must be omitted.

Straight lines

Straight lines are evident in the design of many buildings, in the layout of farming plots, in boats, and in the crafting of spears and other weapons. In the first two settings above, precision is not particularly important, yet frequently buildings or plots were laid out by stretching string or vine tight between two points to assure proper alignment. In this context, a description by Hortense

Powdermaker of construction methods in New Ireland (Melanesia) is of interest (1933):

> Putting the posts in the ground in a straight line is a skilled job and not every man can do it. There were ten men in a village of two hundred and thirty people who were skilled in this. If a man is notable to do it he gets one of these skilled men to mark off the ground in a straight line with a vine.

With spears or other weapons it is very important to fashion the weapon with a straight shaft for reasons of accuracy. A straight spear can be thrown with much greater accuracy and force than a spear which is crooked. Accounts from Truk (Micronesia) and Papua/New Guinea describe how artisans heat and bend the wooden shafts to straighten them, checking by sighting along the shaft to see that it is actually straight. Pictures of natives in Papua/New Guinea holding perfectly straight six- or eight-foot spears are striking (Figure 2).

Hose and McDougall (1912) give a very curious account of straightening by craftsmen of the Kayan tribe of Borneo. Here the artisans were concerned with making blowguns for hunting. The design involved hollowing out a long straight shaft, which was then heated and formed to the desired pattern, so

Figure 2 New Guinea warriors. Note the straight spears. From Gardner, Robert, and Karl G. Heider. *Gardens of War: Life and Death in the New Guinea Stone Age*. New York: Random House, 1968.

that sighting through the bore from one end one could see exactly half the circular opening at the other end. Thus the tube would have a slight bend to it, so that when a spear point was lashed at the other end and the tube held horizontally the weight of the spear point would complete the job of straightening it out.

Circles

Circles are common design motifs. Some groups favor circular plans for their houses, and circles occur frequently as parts of decorative designs. The laying out of the design for circular houses shows at least two different understandings of the form. The most common way of obtaining a circular shape on the ground for a house is to stretch a line from a central point and trace at a fixed distance in a circle. A contrasting understanding of the concept of a circle is displayed in the way the Dugum Dani of Papua lay out the outline of their houses (Heider, 1970). They start out with a loop of heavy vine which they lay on the ground and adjust until the form obtained is everywhere evenly rounded, thus getting a circle. Although the final shape is the same, the method of obtaining it and the understanding behind the method are different.

Other basic shapes

In addition to straight lines and circles one sees a variety of other standard shapes. Usual shapes for houses include rectangular, circular, hexagonal (long parallel sides with a point on either end), and a shape with parallel straight sides and semicircular ends. In decorative designs one sees triangles, diamonds, meanders, zig-zags and star shapes, as well as crescents, ovals, spirals, and curved shapes which defy attempts at naming. The shapes used in a particular situation might depend on the object involved, but a person would typically have a wide repertoire of shapes to choose from.

Parallel lines and parallel constructions

The notion of parallel lines is a common one and is usually understood as lines which are everywhere equidistant. In some instances in the Pacific care was taken to obtain parallels, and in similar circumstances in other cultures deliberate deviations from the parallel were made.

Many Pacific islanders navigate the ocean with canoes. A single-hulled canoe – long, narrow, and high – would be very unstable, so either a double hulled canoe or a canoe with a small outrigger (float) on one side was used. In Fiji, where double-hulled canoes were the norm, careful measurements were made to have the two hulls' center lines everywhere equidistant. According to canoe builder Deve Tonganivalu (Haddon, 1936), 'Much care was taken to assure the accurate spacing of the beams; if they were badly fitted or if the spacing were wider at one end than the other the canoe would be a slow sailer.'

One would expect the same care to be taken in aligning outriggers. Indeed in reports from Mokil in the Caroline Islands this is the case. Careful measurements were made to make the outrigger ride parallel to the canoe hull, with the center of the outrigger directly opposite the center of the main hull.

However, in other parts of the Pacific outriggers are not fixed parallel to the main hull, and the deviation is deliberate. A report from Uvea in Polynesia gives a separation of six feet, nine inches at the front of the outrigger and seven feet at the back. In Futuna (Melanesia) an account is given by Edwin Burrows of the outrigger being lined up visually before lashing 'an inch or two nearer the hull at the bow [front] end ... but this was not so in all of them.' Te Rangi Hiroa even reports an instance in Kapingamarangi (Polynesia) where the outrigger pointed out at the bow, but this practice was unusual. A more complete account by Kennedy of canoes in Vaitupu (Tuvalu, Polynesia) sheds some light on the deviations. Having the outrigger on the left, for example, makes it hard to turn to the right. This difficulty in steering is alleviated somewhat by pointing the outrigger in at the bow. However, if the outrigger is pointed in too sharply there is noticeable resistance in sailing straight ahead. Thus, in Vaitupu, outriggers are about one hand-breadth closer at the bow end, and ease in sailing explains the general tendency to point outriggers in at the front.

Thus Mokil is something of an exception with its care in measuring to make the outrigger parallel to the main hull. Yet Micronesians generally are excellent sailors and canoe builders and surely would know that canoes are more maneuverable when the outrigger is pointed in at the bow. The explanation for the unusual Mokil design lies in the way sailors from Mokil trim and sail their boats. They always sail with the outrigger to the downwind side of the boat, getting better stability. However, with a west wind, this would mean that the front end of the canoe when sailing north would become the rear end when the canoe is turned around to go south. With no fixed front to the vessel the outrigger cannot be permanently mounted to point in at the front, and thus a parallel mounting – the kind used – is optimal.

Constructing right angles

Many Pacific groups traditionally built rectangular houses. One would think, in laying out the plan for such a building, that visual alignment of the wall lines would suffice, along with measuring the lengths of sides to get opposite sides of equal length. This is indeed the situation in most cases. Detailed accounts describing home building omit any description of constructing right angles in the traditions of the southeast Solomon Islands (Melanesia), Hawaii, and Borneo. It is interesting, then, that in accounts by Best and Te Rangi Hiroa from New Zealand, by Te Rangi Hiroa from Kapingamarangi, and by Lebar from Truk (Micronesia), special constructions are described for getting right angles at the corners of houses. The same basic construction is used in all three cases.

According to Lebar's account, a construction chief would be in charge of the project and would supervise the measuring, positioning, and marking of the materials. Positions for the walls of the structure would be set by placing foundation beams ('plates') in location. The side plates could be measured and cut to the same length, as would the end plates. There were standard proportions to be observed for the ratio of length to width. Then, as Lebar describes the process,

> In order that the building may assume the correct shape and proportions, the *souiim* [construction chief] makes a series of measurements and readjustments both before the building is started and during construction. First the wall and end plates are placed on the ground in the relative positions they will assume in the completed building. To ensure that all corners of this frame will be right angle corners the *souiim* stretches a line of coconut fiber cord diagonally the length of the frame, from one corner to the other, and the timbers are readjusted until the same distance obtains between the remaining two corners.

Basic geometry verifies that this construction indeed gives a rectangle.

Accurately measuring an angle

Often the need for accurate calendars led to developments in mathematics. The fact that in temperate climates many activities are seasonal – planting, harvesting, hunting, migration – led people to want to determine the seasons accurately. Seasons are also important in parts of the tropics, and some methods of determining seasons are quite interesting geometrically.

One important season change on Borneo is the start of the short dry season, a time when many people clear land for planting. The change from wet to dry there is a relative, not absolute, difference: both seasons have rains, sometimes heavy, and both seasons have times of clear skies. But overall the dry season has less rain and more clear sky. The change, however, cannot be reliably determined by daily moisture. Rather, tribes learned to anticipate that change, as well as other important season changes, by astronomical observations.

The Kenyah people had a special astronomical observer to determine the seasons, and he did this by observing the angle of elevation of the midday sun (Figure 3). Preparations were elaborate. He positioned a specially prepared staff vertically in the ground. The staff had a collar which showed how far it was to be pushed into the ground, and from that point to the top it measured a full span from outstretched fingers to outstretched fingers plus a span from thumb to first finger. Vertical alignment was obtained using a plumb line. The ground around the base of the pole was smoothed and leveled. The other part of the device was a flat stick with notches cut at different positions. The flat piece was laid horizontally extending from the vertical shaft, and whenever possible the position of the midday shadow was noted. Once the shadow reached its greatest length and started to shorten, when it again reached the notch corresponding to the distance from armpit to the middle of the upper arm, the sowing season was announced. Other seasons corresponded to the other notches on the horizontal piece.

The Kayan tribe also determined seasons using the sun's midday elevation, but they determined it differently. A hole was made in the roof of the observer's house. Distances on a special plank on the floor from the point below the hole to the spot of sunlight on the plank were then used to determine seasons. The

Figure 3 Kenyah determination of the angle of elevation of the midday sun. Photograph by Charles Hose, from Hose, Charles, and William McDougall. *The Pagan Tribes of Borneo*. London: Frank Cass & Co. Ltd., 1912, reprinted New York: Barnes & Noble, Inc., 1966.

plank was adjusted to horizontal by using smooth round stones and noting if they rolled.

Sea Dayaks also determined seasons astronomically, but used the angle of elevation of the Pleiades at dusk for their observations. A long bamboo tube, closed at one end, was used. First it was filled with water, then tipped to point at the constellation. When tipped back the level of water was noted; when it agreed with a predetermined mark from previous observations the correct time of year was recognized.

All three of these methods for telling seasons involve accurately recognizing particular angles which had been determined in the past, and the different means of doing so show the tribes' geometric ingenuity.

Curves – string figures

A common pastime in Pacific cultures, as in many other parts of the world, is the forming of string figures or cat's cradles. Typically such a design is created by catching a loop of string over the fingers of both hands, then twisting it in various ways, catching it at other points, repeating the process until very elaborate patterns are formed. Frequently two or more people will work on a pattern, one person holding a configuration while the other does the looping and reconfiguring. Also, the teeth and occasionally the feet are used to hold parts of the loop while it is being reshaped. There are usually some standard forms (which have descriptive names) that one tries to reproduce; the loops and techniques allow for experimentation and variation as well of course. Occasionally patterns are made using more than one loop. Sometimes the loops are large; in one account the participants started with a six-foot loop. Dickey reports that over four hundred different standardized string figures were recorded from Pacific island cultures, including one hundred fifteen from the Hawaiian Islands alone. Handy reported thirty figures from the Marquesas and twenty-five from the Society Islands, and gives the origin of string figures on the Marquesas as lashing patterns used on canoes, tools, house posts, coffins, and other commonly needed items (Dickey, 1928; Handy, 1925) (Figure 4).

Curves – knots

Western societies have an oversupply of fastening devices: nails, screws, staples, tape, wire, pins, Velcro, and of course string. Before they were so generously

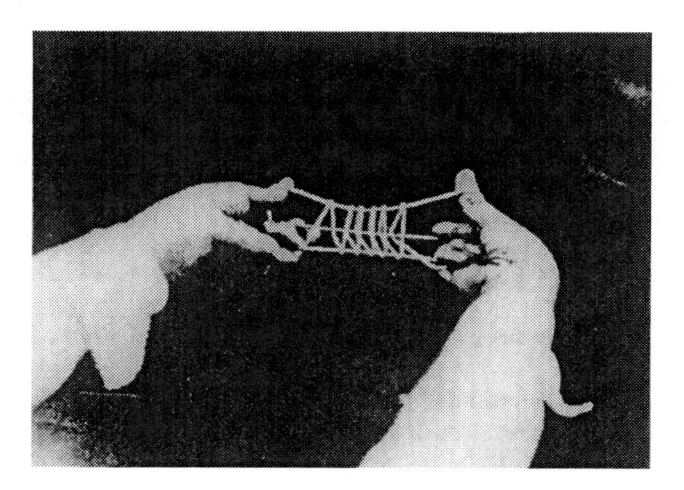

Figure 4 String figure from the Marquesas Islands. From Handy, Willowdean Chatterson. *String Figures from the Marquesas and Society Islands.* Bernice P. Bishop Museum Bulletin 18, 1925, reprinted New York: Krause Reprint, 1971. Published with the kind permission of the Bernice P. Bishop Museum.

provided with alternatives, people made extensive use of rope and string, and were familiar with and could tie many different forms of knots and lashings. Today it is a rare person in a Western society who can tie five different kinds of knots.

Traditional Pacific island cultures use knots and lashings extensively to fasten things: lashings to hold supports together in house construction and canoe building, lashings to hold stone heads on tools, knots to attach hooks on fish lines, to tie sails to masts or booms on boats, to tie thatch onto roofs, and to make fishing nets. Different uses generally required different knots or lashings, so that in building a house or canoe, for example, many different knot and lashing patterns would be required. Each design shows geometric awareness in its configuration. Some notions of geometry are also contained in the block and cord puzzle described below in the section on games.

Curves – sand drawing or ground writing

From what is now central Vanuatu northwest to the southeast Solomon Islands the Pacific Islanders have a traditional practice of ground writing as they call it or sand drawing as others have dubbed it. Individuals will trace a pattern on the ground with a finger, in general completing a complex design without lifting the finger off the ground between start and finish. The patterns vary in complexity; they can be used to impart simple messages, tell stories, convey sacred meanings, recount a historical event, or just for play. Usually they are drawn and then quickly wiped out. Early in the twentieth century Bernard Deacon collected more than a hundred of these patterns, as did John Lanyard (Deacon, 1934; Huffman, 1996). Ground writing designs varied considerably in style. One pattern, representing the nest of the cardinal honeyeater, was composed of straight lines and angles (Figure 5). More typical patterns, like the mourning of Uripiv, may have had some straight sections but were composed mainly of loops and curves (Figure 6).

The origin of ground writing and the importance of learning to trace the figures correctly in one instance is traced to a legend involving two brothers and the wife of one of them. One brother leaves to go on a trip, first closing his wife in his house with a web-like design made of vine across the door. When he returns he notes a difference in the vine's pattern. He has the local men draw the pattern he left in the ashes, notes that his brother instead draws the figure he found on returning, and thus deduces that his brother had secretly visited his wife. To take revenge he then kills the brother.

Of greater importance as a motivation for correctly learning the sand patterns are two other cultural linkages. One has to do with social position and rank of the men and women in society. Different positions are associated with different designs, in addition to other visible signs of rank, so these patterns are important as indicators of social position.

The other importance of certain sand tracings has to do with admission to the afterworld after death. According to tradition the route to the afterworld is guarded by a being who controls the passage. This being sits behind a sand

Figure 5 Malakulan ground writing – 'cardinal honeyeater' design. From Huffman, Kirk W. '"Su tuh netan 'monbwei: we write on the ground" sand drawings and their associations in northern Vanuatu.' In *Arts of Vanuatu*, Joel Bonnemaison *et al.*, eds. Bathurst: Crawford House Publishing, 1996. Reproduced with the kind permission of Crawford House Publishing.

tracing, and when the dead one approaches, the guardian erases part of the figure and requires the person to complete the figure. If he completes the figure properly, passage is granted; otherwise the guardian devours him. The way to redraw the figure correctly should have been learned during the individual's life (Ascher, 1991).

Symmetry

Both mathematically and aesthetically people tend to look for symmetries in designs. In this context, a symmetry of a design is any motion which leaves the pattern unchanged. Many medallion patterns, by way of illustration, exhibit rotational symmetries: a rotation about the center of a quarter turn, for example, gives exactly the same pattern.

Geometers have made an analysis of symmetries for designs in a long narrow strip and for designs in a large flat surface (in the plane). Recently the artwork of various groups has been studied to identify which sorts of symmetry the designs exhibited. One thorough description of such a study, of Maori rafter patterns, is given in Ascher's book *Ethnomathematics*.

Patterns which are thought of as repeating along a strip (extending indefinitely in both directions) can be classified mathematically into one of eight groupings, according to what symmetries they possess. These are illustrated by

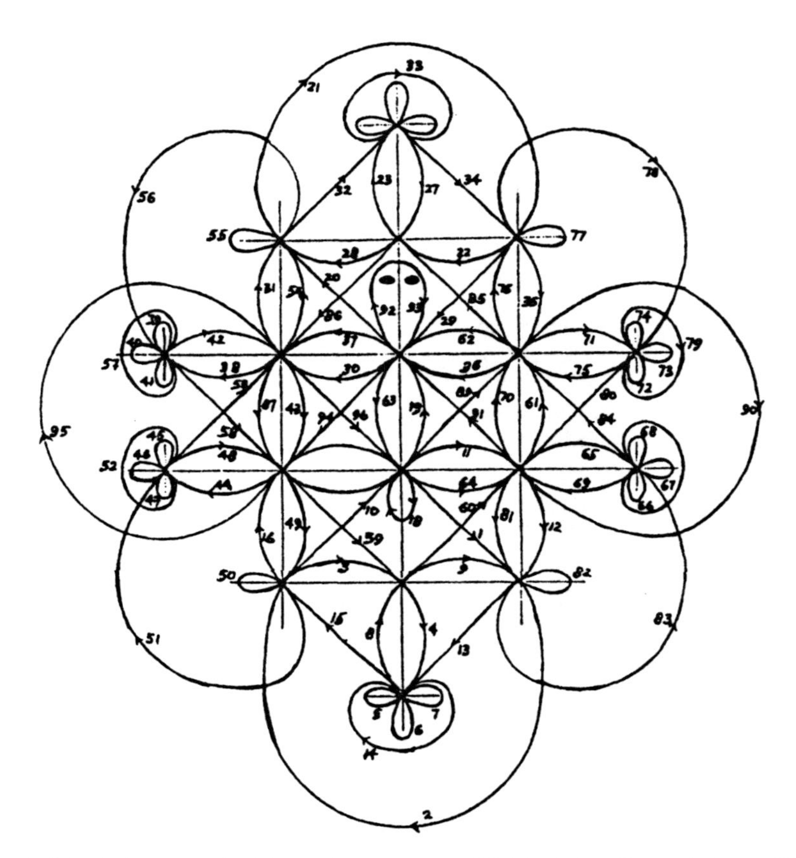

Figure 6 Malakulan ground writing – the mourning of Uripiv design. From Huffman, Kirk W. "'Su tuh netan 'monbwei: we write on the ground" sand drawings and their associations in northern Vanuatu.' In *Arts of Vanuatu*, Joel Bonnemaison *et al.*, eds. Bathurst: Crawford House Publishing, 1996. Reproduced with the kind permission of Crawford House Publishing.

examples in Figure 7. Some strip designs, like (o), have no symmetries; the only way to have the pattern seem unchanged is to leave it unchanged. This possibility is usually omitted in discussions of symmetry, and we shall consider only the other seven cases. Design (i) exhibits only translational symmetry: moving the design to the left (or right) allows it to match its original outline. Design (ii) allows translational symmetry and a reflection in a horizontal axis through the center of the strip. The basic symmetry in design (iii) is a translation combined with a reflection in the horizontal axis. In design (iv) there are two basic symmetries, a translation and a reflection in a vertical axis. Design (v) also has two basic symmetries, a translation and a rotation of 180 degrees about a center point in the design. Design (vi) shows all these kinds of symmetries: translation, horizontal reflection, vertical reflection, and rotation. The last design (vii) has as basic symmetries the translation-horizontal reflection combination, the vertical reflection, and the 180 degree rotation.

A study of Maori rafter carvings from the point of view of mathematical

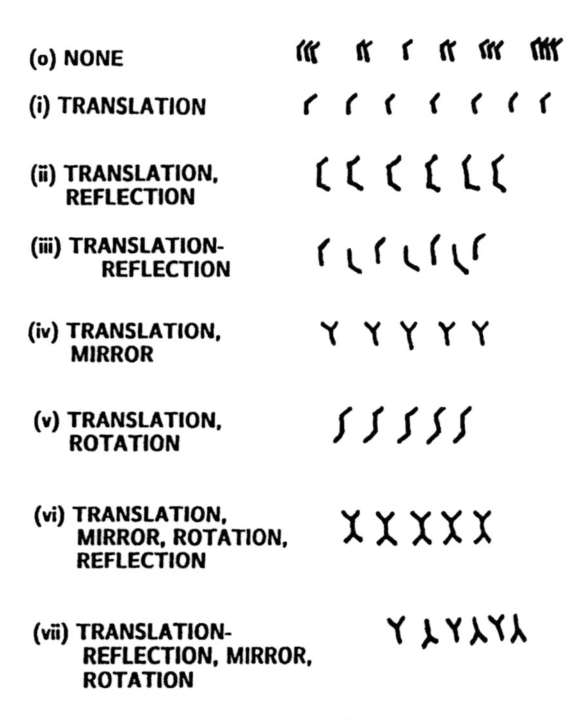

Figure 7 Examples of symmetry classes (author's sketches).

symmetry reveals that carvings in one collection show examples of each of these symmetry groups. But Maori rafters, in addition to showing carved strip patterns, are also painted in three colors: black, white and red. The introduction of color into the designs allows identification of other patterns, referred to as antisymmetries. In grouping (i), for example, if red and black symbols alternate on a white background, then a single translation and color reversal gives the original pattern. Including symmetry and antisymmetry groupings, the number of classifications of pattern types goes to twenty-four. Seventeen of these twenty-four pattern types have been identified in the sample of Maori rafter patterns studied.

An analysis of symmetry in Hawaiian stamps for imprinting strip patterns on *tapa* (bark cloth) shows that the designs used contain examples in all seven of the basic symmetry groupings listed above. The presence of these different sorts of symmetry in design suggests some understanding and appreciation of symmetry among the artists.

For designs which are repeated both horizontally and vertically on a flat surface more patterns of symmetry can occur. Knight identifies twelve patterns which might be possible in woven designs, having ruled some plane symmetry patterns out on the grounds that they could not occur in standard rectilinear weaving. He then analyzes some Maori woven work and notes that ten of these possible twelve patterns occur in his samples.

Gladys Reichard in work early in the 1900s studied selective crafts in Melanesia in detail. She noted that some cultures, like the Solomon Islanders, sought to have symmetry in their carvings. She observes, 'The Solomon Island artist lays out his fields in such a way as to show his liking for symmetry.' (p. 113). In other cases, however, Reichard detected a strong tendency to avoid symmetry. In describing a Tami carved decoration on a wooden bowl, she notes, 'The banding on both sides further substantiates the theory that the Tami artist, not only does not demand, but actually avoids, absolute symmetry. It is incredible that an artist, master of his technique as evidenced by this bowl, could not have divided his space evenly had he so desired. He preferred rather to preserve balance while at the same time avoiding exact repetition.' (p. 53). The practice of this Tami carver was general to others in his culture, and some artists in other places also at times demonstrate a desire to avoid symmetry.

Patterns in art: curvilinear or rectilinear

There are many places for decorative arts in the lives of Pacific peoples – in carved decorations on houses or canoes, in designs on pottery or gourds, in patterns stamped on cloth, in combs, bangles and other items of adornment, in ceremonial or religious pieces, and in tattoos, for example. Even where the design depicts a person, animal, or plant it is interesting to see the combinations of straight and curved features used, but this is even more the case in purely geometrical designs, for example, around the borders of other figures. With the representational forms, it is possible to depict human figures using straight lines and angles for the most part, or it is possible to use mostly curves.

In studying Oceanic art one is tempted to identify certain cultures as preferring rectilinear forms: straight lines, angles, polygons like diamonds and triangles, star shapes, zig-zags, and meanders. This seems to be the preference for most of Polynesia except for New Zealand and for much of Melanesia outside of Papua/New Guinea. Some tribes of Borneo also show aspects of this style.

In contrast are styles one could call curvilinear. Here curved is better than straight, and one sees circles, spirals, crescents, scrollwork, and other curved shapes which defy naming in English. Reichard says of Maori design, 'Even where straight lines would be technically more convenient and artistically more appropriate, the Maori carver uses curves.' (p. 149). These styles are predominant among the Maori in New Zealand and in many parts of Papua/New Guinea.

A closer look at craftwork from the Pacific, however, shows considerable overlap. Many works show both rectilinear and curvilinear features in the same piece, and artists' descriptions further indicate this mix. Reichard, in describing Tami carved bowls, writes, 'The Tami carver uses curved as frequently as straight lines, and ... he has no preference for either.' (p. 37). Her list of customary Tami elements in decorative carving includes 'ellipses, circles, semicircles, rectangles or trapezoids, straight lines, triangles, ribbons, the cross

Figure 8 Designs from Pacific cultures (author's sketches). (a) Samoa (detail, based on Barrow: 31); (b) Iban, Borneo (detail, based on Hose, 1912: 243); (c) Tami, Melanesia (detail, based on Reichard, 1933: 80); (d) Hawaii (detail, based on Te Rangi Hiroa, 1957: 199); (e) Cook Islands (based on Te Rangi Hiroa, 1944: 130); (f) Samoa (detail, based on Stanley, 1993: 176); (g) Papua (from Bernice P. Bishop Museum exhibit); (h and i) Maori, New Zealand (detail, based on Donnay and Donnay, 1985: 26, 27); (j) Bouganville (detail, based on Stanley, 1993: 417).

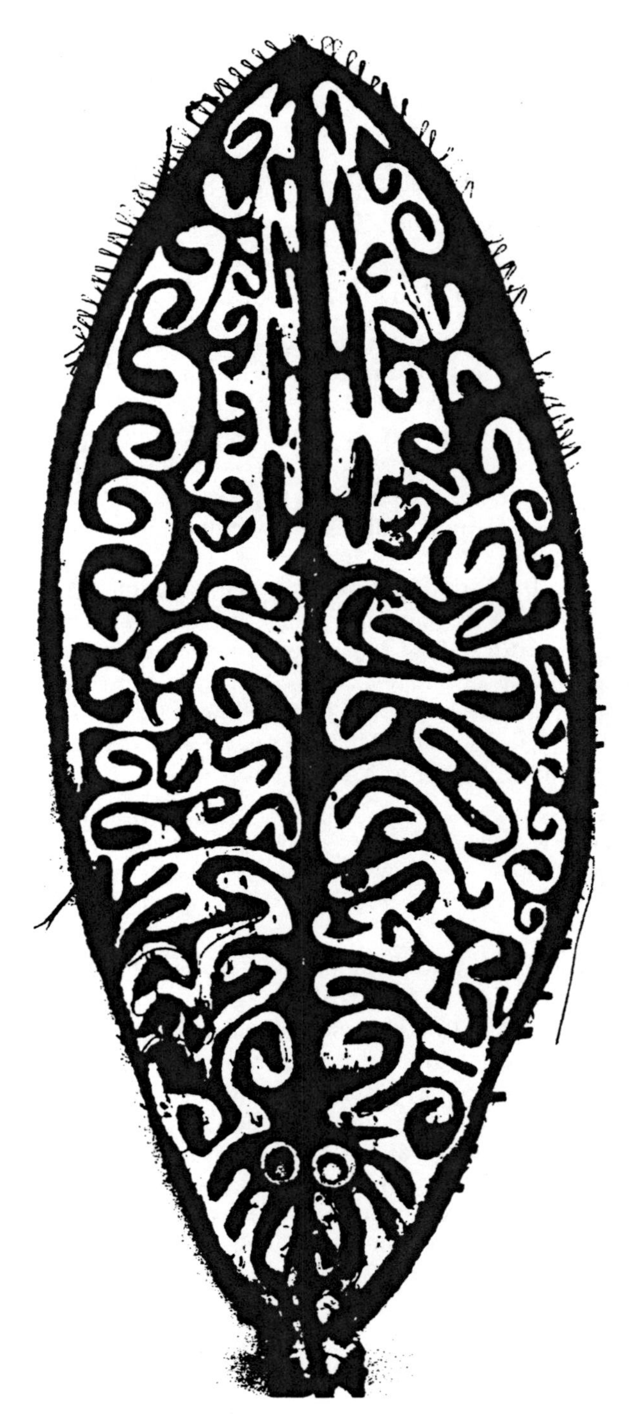

Figure 9 Mask, Papua. National Museum of Ireland, Dublin.

element with triangles.' (p. 41). Borders of decorations on Palauan clubhouses also show both rectilinear and curvilinear features, with the rectilinear predominant. Shapes here include zig-zags, diamonds, squares and checks, as well as circles and arcs of circles and ovals. May and Tuckson, in describing patterns on pottery among the Mailu of central New Guinea, comment, 'Decorative patterns, generally geometric [rectilinear] but some curvilinear, are incised into the top quarter of the vessel.' (p. 62).

Noting position: charts and maps

Almost every culture that relies on travel develops some way of noting positions of geographic features to give directions. Whether travel is for trade, hunting or fishing, social or religious purposes, or for exploration and adventure, people want to be able to find their way from one place to another reliably and want to describe to others how to make journeys and find particular locations. In Pacific cultures people made charts and maps as aids in travel. Some of these showed geographic features, while equally important were others showing positions of stars in the sky.

Among the Madang tribe of Borneo it was common to make temporary maps of a region using twigs, stones, and other objects to represent locations of landmarks, and to lay these objects out on the ground or another surface for illustration. When a Madang chief was telling Western visitors of tribal beliefs concerning the afterworld, he illustrated his statements about the journey after death with a rough map of the afterworld and the land which must be traversed to get there. Hose and McDougall write

> This was done in the way maps of their own country are always made by the Borneans, namely he laid on the floor bits of stick and other small objects to represent the principal topographical features and relations. We tested the trustworthiness of his account by asking him to repeat it on a subsequent occasion; when he did so without any noteworthy departure from the former description.

It is reasonable to surmise that similar map sketching practices exist in most hunting cultures.

Maps or charts of various kinds are also an essential feature for navigators in numerous cultures. Haddon reports accounts of Torres Straights natives (Melanesia) drawing maps of reefs as navigational instructions and also maps showing the positions of nearby islands. The natives of the Caroline Islands (Micronesia) also make stone diagrams to teach swell and wave patterns.

More durable charts or diagrams were made by the residents of the Marshall Islands to help with navigation (Ascher, 1995) (Figure 11). These charts were usually made on a thin palm rib frame, with the strips lashed together with threads of coconut fiber. To the frame were attached other wood strips and small pieces of coral or small shells. There were three kinds of navigation charts. The first, called *mattang*, showed the patterns of waves around an island, including typical patterns of reflected and refracted waves. Next in scope were charts of several islands showing their relative positions and other local features of the seas around them. These were called *meddo*. The most inclusive charts,

Figure 10 Asmat paddle blade. Used with the permission of the Koninklijk Instituut voor de Tropen, Amsterdam.

rebbelith, showed the whole island group, giving relative positions of the islands but little information about the surrounding seas.

Another instance of charting or graphing positions shows up in the traditional astronomy of the natives in Kiribati, where astronomy was studied as an aid to navigation. A few members of the population would go through a thorough training program to learn star positions. Such training started in a community shelter called the *maneaba* with a large rectangular roof supported on a ridgepole, bounded on the east and west by roof plates and divided by rafters and crossbeams. This grid represented the night sky, with the ridgepole representing the meridian and the roof plates the eastern and western horizons. The night sky was thus divided into strips, and would-be navigators would learn the relative positions and movements of stars by picturing them against this grid. Only after learning under the roof the positions of over 150 heavenly bodies and the changes of position which occurred with different times of the year would the novice navigator study the stars in the actual night sky. In this way navigators learned the astronomy necessary to navigate canoes on voyages to destinations over a thousand miles away. (See Grimble, 1972: 215–218.) This account is of particular mathematical interest because in going from the heavens to the *maneaba* ceiling the natives projected the hemisphere of the heavens onto a rectangular region.

An account from Hawaii describes star charts made on the surface of a gourd which had been marked off with burned lines, and similar globes were used in the Caroline Islands to indicate star positions (Lewis, 1973). On the northwest coast of Easter Island at Matariki a grouping of fifty-six marks carved in the rock is thought by some to be a chart of the heavens (Bahn, 1992).

Sailors from Kiribati also learned and passed on from generation to generation the relative positions of other features (*betia*) which would serve as aids to navigation. Francis Grimble termed such features sea-marks. These were particular features of the sea apt to occur in certain locations, which thus served as aids in determining one's position on what might otherwise seem an endless sea. Sea-marks included such features as areas where fish swarmed or birds congregated (further distinguished by the particular species involved), areas of mist or of unusual wave patterns, places where the prevailing winds changed, and regions of unusual driftwood concentrations. Such sea-marks also played a role in some Polynesian navigation.

GAMES

Games relate to mathematics in two ways. Games of chance lead to notions of probability, expectation, and fairness; one wants to know what the chances are of winning, what one can expect to win on the average in the long run, and whether each player has the same chance of winning. Games of strategy lead to logical thinking about the best course of action in competitions and how different choices of action would lead to different responses by the opponent. We will examine briefly two games of chance, two games of strategy, and a puzzle from the Pacific region.

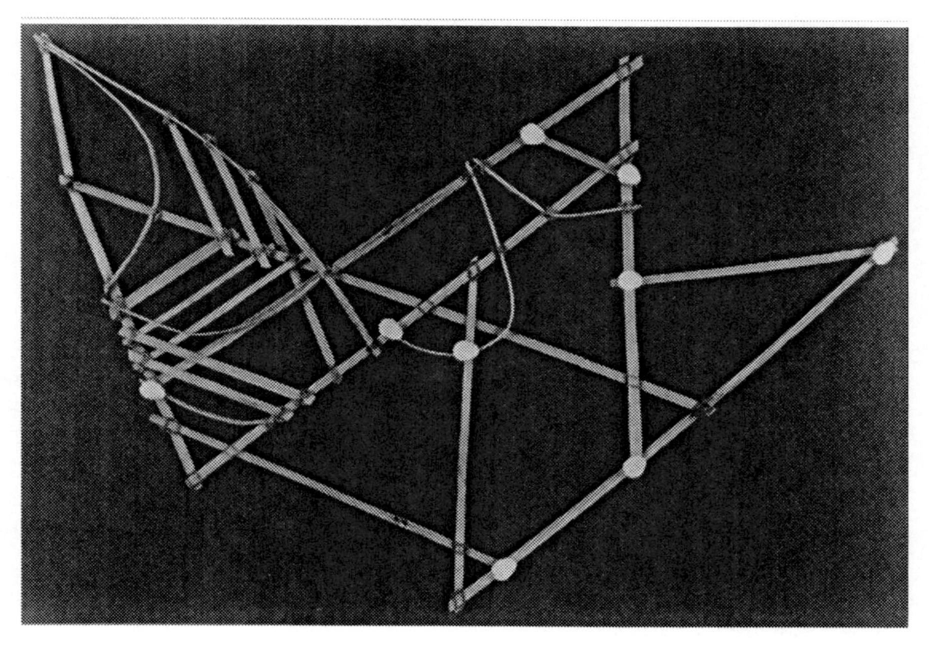

Figure 11 Marshall Islands chart. Bernice P. Bishop Museum photograph. Used with the kind permission of the Bernice P. Bishop Museum.

Lu-lu

The Hawaiian game of *lu-lu* is played with four disks which traditionally were made of stone. The disks are blank on one side and have one, two three, and four dots respectively on the other side. A player tosses the four disks to start a turn, then can toss a second time any which landed blank side up on the first toss. After the second toss, the player scores a point for each dot showing. If all four disks show dots, the player gets another turn right away; otherwise play passes to the next player. The winner can be the one with the highest point total in a round, or the game can continue until a predetermined number is reached. Bets are frequently placed on the outcome of the game (Culin, 1899).

A Zambales dice game

The Negritos of the Zambales region of the Philippines play a dice game with six-sided dice. Traditional dice are marked with the six symbols I, II, III, X, +, #. A player gets up to five tosses of a pair of such dice, and wins by having both dice come up the same on one of the tosses. Again, bets are placed on the game (Reed, 1904).

Konane

The Hawaiian game of *konane* is played on stone slabs or gameboards marked with indentations in a rectangular pattern (Figure 12). The indentations are intended to hold a single pebble or marker each. Boards come in varying dimensions: one has 14 rows of 17 indentations each, another shows a 9 by 9 pattern, and 9 by 13 and 13 by 20 arrangements are also used. Konane is a two player game, with one player using black pebbles or markers and the other using white. To start, pieces are arranged alternately to fill all the board spaces. The player using black pieces removes one from a corner or from one of the center spaces, and the other player takes a white piece from an adjacent spot. Play then alternates. On a turn, a player uses one of his pieces to jump over an adjacent opponent's piece to a vacant space on the other side, removing the jumped piece. More than one jump can be made in a turn, provided all jumps are made by the same piece and follow one after the other in a straight line. Play continues until a player cannot make a move, at which time that player loses. A few unusual konane boards are laid out with alternate rows one of which is longer or shorter than the next. One such example has 10 rows, with 6 and 7 indentations alternately. On such boards moves are made along diagonals, but otherwise play is as described above (Emory, 1924).

Main Machan

The Iban tribe of Borneo plays a board game called *main machan*. The board has thirty-seven positions, twenty-five of which form a square to which two triangles are attached at the mid-points of opposite sides (Figure 13). One player has twenty-eight pieces (*anak* or children) and the other has two (*endo*

Figure 12 Konane board. Used with the kind permission of the University of Hawai'i Committee for the Preservation and Study of Hawaiian Language, Art and Culture.

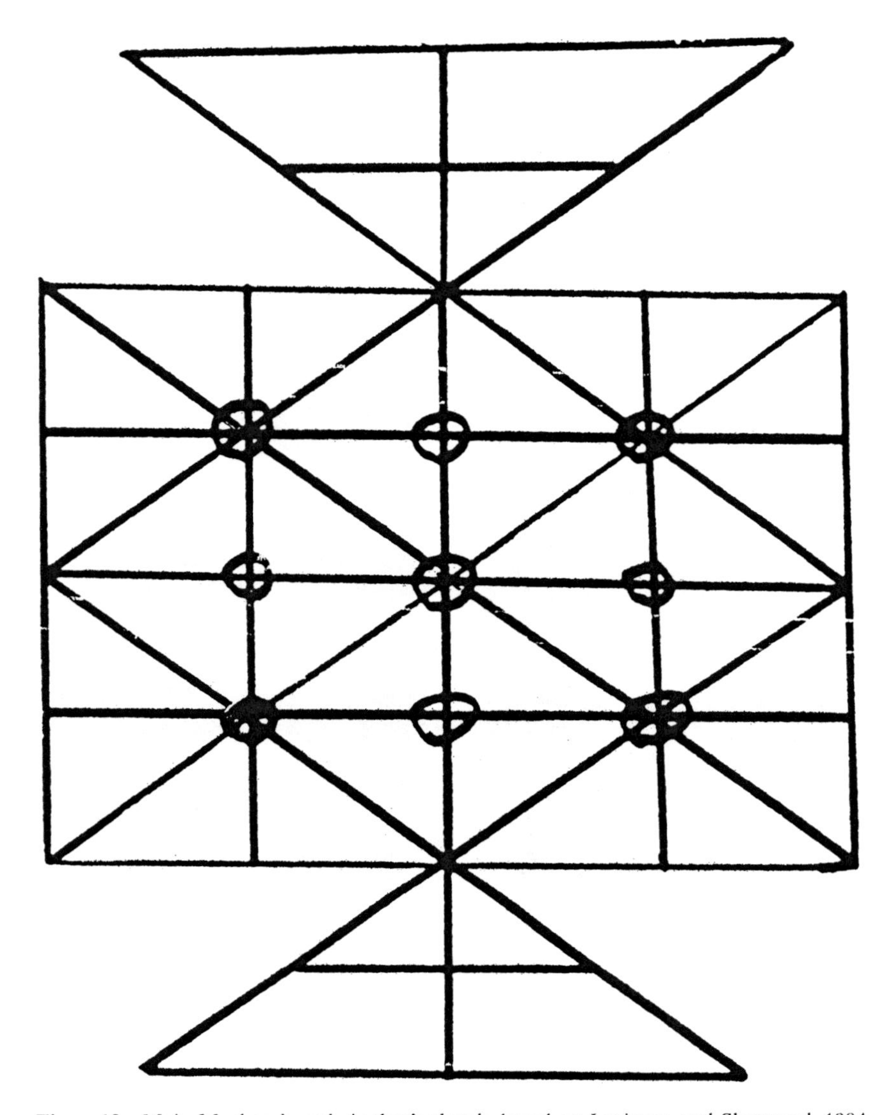

Figure 13 Main Machan board. Author's sketch, based on Levinson and Sherwood, 1984.

or women). The player with twenty-eight pieces starts by putting nine pieces on the nine center points. The other player removes three of these and puts the two *endo* pieces on any vacant spaces. Play then alternates. The player with the twenty-eight pieces can place a piece on any empty space on the board, until all twenty-eight have been played. Once all pieces have been put on the board, a play consists of moving a piece to a vacant adjacent space or jumping a piece over one of the opponent's pieces (but not removing the jumped piece). The player with two pieces, on a turn, can move to a vacant adjacent space or can jump an odd number of the opponent's pieces and remove the jumped pieces. The object for the player with the twenty-eight pieces is to hem in his opponent's two pieces so he cannot move; the goal of the other player is to remove enough of the opponent's pieces to avoid being trapped. This game is similar to the European game of fox and geese and to the Hawaiian game of *ma-nu*. Board configurations and number of pieces for each player vary with these other games, as might details of the playing rules (Levinson, 1984).

Pu-waa-pa, a cord and block puzzle

Many cultures developed puzzles consisting of a wooden block or blocks with holes through them, through which a cord passes while intertwining with itself. The object of these puzzles is to remove the cord from the blocks without untying the cords. One such puzzle from Hawaii is called *pu-waa-pa* or the canoe puzzle (Figure 14). A legend which accompanies this puzzle tells of a

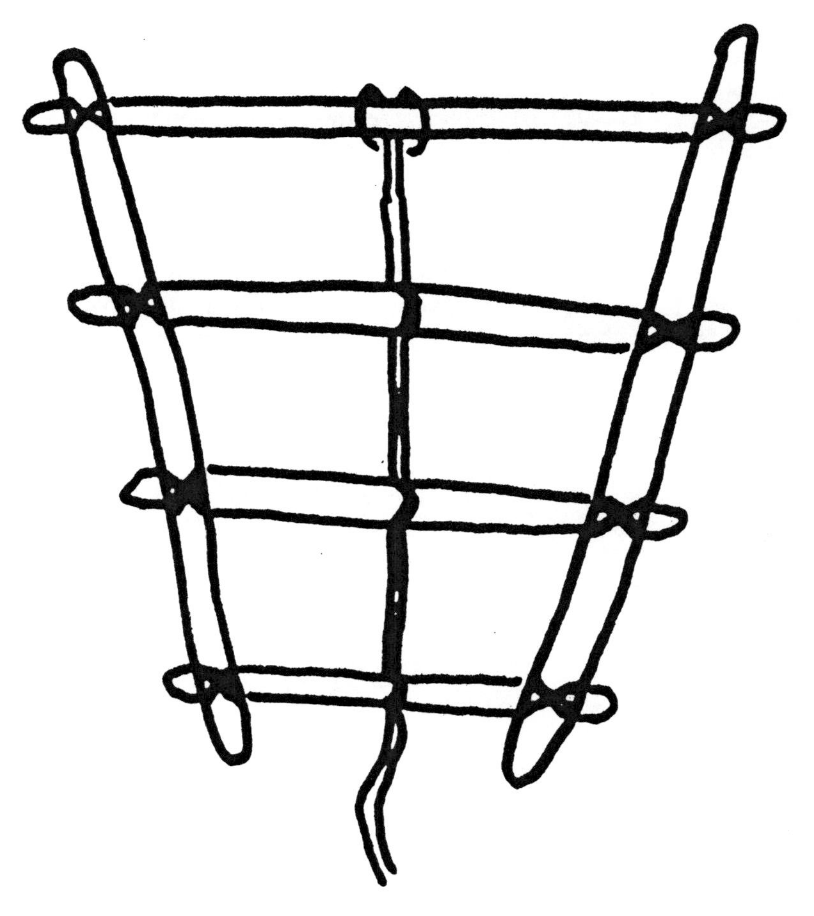

Figure 14 Pu-waa-pa. Author's sketch, based on Culin, 1899.

king whose daughter was in love with a man of lower rank. The king, disapproving, had his daughter put in a canoe which was secured to the shore by a long rope, expecting her to die there. The king was thwarted when the lover swam to the canoe and freed it from the rope, allowing the couple to escape to a nearby island (Culin, 1899).

RELATIONS

The mathematical connection

The concept of relation is important in mathematics. In general one considers a particular way that two items might be related and one asks whether they are related in that way or not. The primary example is perhaps the relation of equality for numbers: we say, for example, that 2/4 is equal to 3/6 – they satisfy the relation of equality – even though conceptually 2/4 is what one gets when one takes two of four equal parts, and this is not exactly the same as taking three of six equal parts (3/6). Other common mathematical relations include the relation of one number's being less than another and the relation of one whole number's evenly dividing another.

Relations are studied in more abstraction in mathematics and arise in numerous other settings as well. In introducing relations in mathematics, to emphasize the generality of the principles involved and yet to give familiar examples, sometimes kinship relations among people are cited as examples. For example, we use the relation among people of 'A is the mother of B' or 'X is the brother of Y'. In our terminology it is interesting to note what grouping, if any, takes place in the use of common kinship terms. 'My mother' is a designation which fits only one person; 'sister' may apply to more than one person, but the relationship is unambiguous; 'aunt' applies equally to four groups of people – my father's sisters, my mother's sisters, my mother's brothers' wives, and my father's brothers' wives. In other societies more or less grouping takes place in using kinship terms, and examples of such terminology for Pacific cultures will be given.

A second kind of relation occurs in some cultures where people are assigned to different sub-groupings (totems, clans, moieties) and marriages have to be arranged between subgroupings according to certain rules – for example, a man from clan 1 must marry a woman from clan 2, but a woman from clan 1 must marry a man from clan 3. This sort of structure is present in Melanesia, and we will examine it there.

Kinship terms

Frequently kinship terms in Oceania group people together more than Western terms do. In Samoa, for example one gets the following groupings:

Samoan term	Meanings
tina or *matua*	mother, father's sister, mother's brother's wife
uso	brother, father's sister's child, mother's brother's child
tama or *matua*	father
matua or *tama*	mother's brother
tama	grandchild
matua, tama, or *tupunga*	grandfather

In Hawaii there is less grouping than in Samoa, and one obtains the following partial list.

Hawaiian term	Meanings
makua	mother's brother, father's sister
kaiki	child, sister's son, brother's son
wahine	wife, wife's sister
kane	husband, husband's brother

Looking at Melanesia, again one gets some interesting groupings. In Guadalcanal one gets the following examples.

Guadalcanal term	Meanings
nia	sister's son, mother's brother, father's sister's husband, father-in-law
iva	wife's brother, wife's sister, husband's brother, husband's sister, mother's brother's child, father's sister's child
tarunga	mother's brother's wife, mother-in-law

Among the Mota of the Banks Islands one gets some grouping also.

Mota term	Meanings
tamai	father, father's sister's son
natui	mother, father's sister's daughter
tupui	grandparent, grandchild

On the other hand, with the Mota, some relatives we group together under one term are called by different terms depending on the precise relationship (adapted from Rivers, 1968).

Mota term	Meanings
veve vus rawe	aunt (father's sister)
mateima	(mother's brother's wife)
tamai or *veve*	cousin (father's sister's child)
natui	(mother's brother's child)

Idealized marriage patterns

Some duplication of kinship terms, using the same term for people one is related to in different ways, can be explained by looking at marriage customs and patterns. Rivers has done much work along these lines, and two examples relating to Pentecost Island (Vanuatu) will illustrate how marriage customs affect language. In this situation, marriage classes or moieties existed, and sometimes a kinship term refers not only to a particular person but by extension to anyone in that class.

One common marriage was between a man and his mother's brother's widow (wife). If such a marriage was arranged, the term for a man's child should be the same as the term used for his mother's brother's child, and this is in fact the case in the Pentecost language, *nitu* being used in both instances. Other kinship term duplication shows this sort of marriage scheme – for example, the word *sibi* both for grandfather and for father-in-law, perhaps indicating that a woman's father-in-law would also be her grandfather (-in-law).

Another marriage practice on Pentecost Island was for a man to marry his daughter's daughter (or his brother's daughter's daughter, a person of the same marriage class). This sort of alliance is reflected linguistically in the use of the word *tuaga* for mother's mother; it is the common term used for an older sibling of the same sex as the speaker. Also, a woman calls the brother of her maternal grandmother *hagosi*, the same term she uses for her own brother. Likewise, the terms for a man's wife's father and his daughter's husband are identical, both *bwaliga*, and indeed with the indicated marriage pattern these individuals would be the same person.

Another idealized marriage system is described in Malakula (Vanuatu) by Layard and in coastal Viti Levu and Guadalcanal, Tanna, and Anaiteum by Rivers. In this system brother and sister marry sister and brother; most ideally, a man marries his mother's brother's daughter, who is also his father's sister's daughter. Such a pattern is called cross-cousin marriage, and of course leads to combined terms for relatives. One factor encouraging a brother and sister to marry a sister and brother would be a strong matriarchal (or patriarchal) family orientation: if the convention is for a man to go to live with the wife's family, then when he marries his family loses a male contributor to his wife's family. Balance is maintained if another marriage is arranged which will transfer a male the other way – for example, between the bride's brother and the groom's sister.

Marriage classes

Throughout Oceania there were marriage restrictions based on kinship. Frequently these took the form of a prohibition of marriage between siblings and between parents and children, and in many cases went further to prohibit marriages in cases of other close blood ties. In most cases these prohibitions included severe penalties for incestuous relations outside of marriage as well. In many parts of Melanesia further structure was imposed, dividing a society into several hereditary classes and requiring people to marry outside their class, frequently specifying the class a person had to marry into. An example showing the marriage class structure of a group on Pentecost Island will be given in detail. This example illustrates a system with six sub-groupings. Another example with four sub-groupings will also be examined. In other situations the number of classes and the rules governing marriages between classes differ, so that, for example, in Vanikolo (Banks Islands) there were ten groups, and in other cases twelve.

Among the Tahau on the northern part of Pentecost Island the first division was into two groups or moieties, the *tagaro* and the *malau*. Each moiety was further divided into three subdivisions, which Rivers labels merely A, B, C in one case and D, E, F in the other. Two of the groupings for the *tagaro* may have been the *matan dura* (sow) and the *matan talai* (giant clam), and for the *malau* the subdivisions may have been the *matan avua* (turtle), the *matan tabwatabwa* (a flower), and the *matan bweta* (taro), but we will use the labels A, B, C, D, E, F. The group a person belonged to was the same as the group of that person's mother. Marriages could be arranged in accordance with the following table.

man	A	B	C	D	E	F
woman	D	E	F	B	C	A

Thus a man from group A would marry a woman from group D and all his children could belong to group D. His son, a D, would marry a woman from group B. If we start with a man in group A and trace his male descendants for the next six generations, we see that they cycle through all six of the groups in the order D B E C F A, and it would take six generations before a male descendant would repeat a class of a male ancestor. A similar pattern is observed starting with a man in any of the other groups and looking at male descendants. Of course, according to the rule, any time a woman is included in a lineage, her class is repeated for her children.

Another example comes from Fiji (Layard, 1942). In this case, a man was in the same kinship class as his paternal grandfather and a woman in the same class as her maternal grandmother. All children of the same set of parents were in the same group, which was different from the group of either parent. A man could only marry into his mother's brother's daughter's group, which was the same as his father's sister's son's group. Such rules led to a four-fold division of society into kinship groups. Group one marries into group two and vice versa; children whose father is in group one and whose mother is in group two belong to group three; children whose mother is in group one and whose father is in group two belong to group four. Groups three and four intermarry; children whose father is in group three and whose mother is in group four belong to group one; children whose mother is in group three and whose father is in group four belong to group two. A quick check, using the rules just given, shows that indeed a man's son's son is in his same group and a woman's daughter's daughter is in her same group, as claimed.

Marriage classes impose another sort of relationship on society. One can think of members of the same class as being related, if not indeed by blood then by virtue of their class membership. Sometimes language reflects this relationship, and for example all male members of the same class as one's father might be referred to as father.

A quite different analysis of the mathematics of kinship relations is given in Ascher's book, *Ethnomathematics*. Much of her focus is on an Australian aboriginal system, but she looks at the system on Malakula as well. See also Helen Verran's article on Yolngu mathematics in this volume.

* * *

In these topics of mathematical interest we see in Pacific cultures what one expects in general. People develop the concepts they need for their lives (some numeration, some symmetry, parallelism, describing location). Sometimes the development goes beyond the necessary, as when number systems extend beyond the necessity of ever counting objects. Further concepts are developed in areas which enrich life, beyond the level of necessity (games, design features). The merging of mathematics and recreation or art, in fact the merging of mathematics and the rest of life, is evident in the many examples that arise.

BIBLIOGRAPHY

Ascher, Marcia. *Ethnomathematics: A Multicultural View of Mathematical Ideas.* Pacific Grove, California: Brooks/Cole Publishing Company, 1991.

Ascher, Marcia. 'Models and maps from the Marshall Islands: a case in ethnomathematics.' *Historia Mathematica* 22: 347–70, 1995.

Bahn, Paul and John Fenley. *Easter Island Earth Island.* London: Thames & Hudson, 1992.

Barrow, Terence. *The Art of Tahiti and the Neighboring Society, Astral, and Cook Islands.* New York: Thames & Hudson, 1979.

Bentzen, Conrad. *Land and Livelihood on Mokil, an Atoll in the Eastern Carolines.* Washington, DC: Pacific Science Board, National Research Council, 1949.

Best, Elsdon. *The Maori as He Was.* Wellington: Dominion Museum, 1924.

Bibina, Raurenti. *Numbers in the Gilbert and Ellice Islands.* Tarawa: Tarawa Teacher's College, 1974.

Bonnemaison, Joel, *et al.*, eds. *Arts of Vanuatu.* Honolulu: University of Hawaii Press, 1996.

Burrows, Edwin. *Ethnology of Futuna.* Bernice P. Bishop Museum Bulletin 138. Honolulu: Bernice P. Bishop Museum, 1936.

Burrows, Edwin. *Ethnology of Uvea.* Bernice P. Bishop Museum Bulletin 145. Honolulu: Bernice P. Bishop Museum, 1937.

Clark, E. W. 'Hawaiian method of computation.' *Hawaiian Spectator* 2: 91–94, 1839.

Codrington, R. H. *The Melanesian Languages.* Oxford: The Clarendon Press, 1885.

Culin, Stewart. 'Hawaiian games.' *American Anthropologist*, New Series 1: 201–247, 1899.

Deacon, Bernard A. 'Geometrical drawings from Malekula and other islands of the New Hebrides.' *Journal of the Royal Anthropological Institute* 64: 19–143, 148–75, 1934.

Dickey, Lyle A. *String Figures from Hawaii.* Bernice P. Bishop Museum Bulletin 59, 1928, reprinted New York: Krause Reprint, 1971.

Donnay, J. D. H. and Gabrielle Donnay. 'Symmetry and antisymmetry in Maori rafter designs.' *Empirical Studies of the Arts* 3: 23–45, 1985.

Emory, Kenneth P. *The Island of Lanai.* Bernice P. Bishop Museum Bulletin 12, 1924, reprinted Honolulu: Bishop Museum Press, 1969.

Franklin, Karl and Joice. 'The Kewa counting systems.' *Journal of the Polynesian Society* 71: 188–91, 1962.

Gardner, Robert and Karl G. Heider. *Gardens of War: Life and Death in the New Guinea Stone Age.* New York: Random House, 1968.

Grimble, Arthur. *Migrations, Myth and Magic from the Gilbert Islands: Early Writings of Sir Arthur Grimble.* Boston: Routledge and Kegan Paul, 1972.

H. M. Nagata *English-Maori Dictionary* (on-line) http://www.learningmedia.co.nz/nz/nd/ndindex.htm

Haddon, A. C. and James Hornell. *Canoes of Oceania* (vols. 1–3). Honolulu: Bernice P. Bishop Museum Special Publications, 1936, pp. 27–29.
Handy, Willowdean Chatterson. *String Figures from the Marquesas and Society Islands.* Bernice P. Bishop Museum Bulletin 18, 1925, reprinted New York: Krause Reprint, 1971.
Heider, Karl G. *The Dugum Dani.* Chicago: Aldine Publishing Co., 1970.
Hose, Charles and William McDougall. *The Pagan Tribes of Borneo.* London: Frank Cass & Co. Ltd., 1912, reprinted New York: Barnes & Noble, Inc., 1966.
Huffman, Kirk W. '"Su tuh netan 'monbwei: we write on the ground" sand drawings and their associations in northern Vanuatu.' In *Arts of Vanuatu*, Joel Bonnemaison *et al.*, eds. Honolulu: University of Hawaii Press, 1996, pp. 247–253.
Kennedy, Donald. *Field Notes on the Culture of Vaitupu.* Memoirs of the Polynesian Society, vol. 19. New Plymouth, New Zealand: Thomas Avery & Sons Ltd., 1931.
Knight, Gordon. 'The geometry of Maori art–rafter patterns.' *The New Zealand Mathematics Magazine* 21: 36–40, 1984.
Knight, Gordon. 'The geometry of Maori art–weaving patterns.' *The New Zealand Mathematics Magazine* 21: 80–86, 1984.
Layard, John. *Stone Men of Malekula Vao.* London: Chatto & Windus, 1942.
Lean, Glendon A. *Counting Systems of Papua New Guinea.* Lae, Papua New Guinea: Department of Mathematics and Statistics, Papua New Guinea University of Technology, 1991.
Lebar, Frank. *The Material Culture of Truk.* Yale University Publications in Anthropology 68. New Haven: Department of Anthropology, Yale University, 1964.
Levinson, David and David Sherwood. *The Tribal Living Book.* Boulder: Johnson Books, 1984.
Lewis, David. *We, the Navigators.* Honolulu: University Press of Hawaii, 1973.
Mariner, William. *An Account of the Natives of the Tonga Islands* (1816). New York: AMS Press, 1979.
May, Patricia and Margaret Tuckson. *The Traditional Pottery of New Guinea.* Kensington, New South Wales: Bay Books Pty. Ltd., 1982.
Montgomery, James, ed. *Journal of Voyages and Travels by the Rev. Daniel Tyermanand George Bennet.* Boston: Crocker and Brewster, 1832.
Newton, Douglas. *Art Styles of the Papuan Gulf.* New York: University Publishers, 1961.
Panoff, Michael. 'Father arithmetic: numeration and counting in New Britain.' *Ethnology* 9: 358–365, 1970.
Powdermaker, Hortense. *Life in Lesu.* London: Williams & Norgate, Ltd., 1933.
Reed, William Allan. *Negritos of Zambales.* Manila: Bureau of Public Printing, 1904.
Reichard, Gladys. *Melanesian Design.* New York: Columbia University Press, 1933.
Rivers, W. H. R. *The History of Melanesian Society.* Oosterhout, The Netherlands: Anthropological Publications, 1968.
Stanley, David. *South Pacific Handbook.* Chico, California: Moon Publications, 1993.
Te Rangi Hiroa. *Arts and Crafts of the Cook Islands.* Bernice P. Bishop Museum Bulletin 179. Honolulu: Bishop Museum Press, 1944.
Te Rangi Hiroa. *Arts and Crafts of Hawaii.* Bernice P. Bishop Special Publication. Honolulu: Bishop Museum Press, 1957.
Te Rangi Hiroa. *The Coming of the Maori.* Wellington, New Zealand: Whitcombe & Tombs Ltd., 1949, reprinted 1962.
Te Rangi Hiroa. *Material Culture of Kapingamarangi.* Bernice P. Bishop Museum Bulletin 200, 1950, reprinted New York: Kraus Reprint, 1971.
Trowell, Margaret and Hans Nevermann. *African and Oceanic Art.* New York: Harry N. Abrams, Inc., 1968.
Turner, George. *Samoa, A Hundred Years Ago and Long Before.* London: Macmillan, 1884, reprinted New York: AMS Press, 1979.

HELEN VERRAN

ABORIGINAL AUSTRALIAN MATHEMATICS: DISPARATE MATHEMATICS OF LAND OWNERSHIP

MATHEMATICS AND POLITICS

A front page report in *The Australian* newspaper of 10th August 1994 began

> Cape York pastoralists and Aborigines have jointly called for state and federal governments to legislate ... a form of statutory co-existence of title on pastoral leases ... A marathon seven-hour meeting in the Queensland town of Coen last week ... sought to address the uncertainty and financial difficulties flowing to Cape York pastoralists from the Wik people's claim to a large area of Cape York, including 12 pastoral leases.

These negotiations, initiated by the Cape York Aborigines and cattle station owners, and later joined by representatives of local tourist operators, did in due course lead to a preliminary form of agreement between these disparate interests. The agreement, and the negotiations out of which it grew, attempted a loose working together of disparate mathematics of land owning. As such they are puzzling, and even threatening, to many Australians, Aboriginal and non-Aboriginal alike. In 1994 the Wik people's claim to a large part of Cape York had every chance of being recognised in law; in consequence new sorts of negotiating and possible agreements between Aborigines and pastoralists emerged. These were negotiations over logic of knowing and mathematics of owning land claimed by both these politically opposed groups. Politics, logic and mathematics are inseparable.

The full bench of the High Court of Australia delivered a significant ruling relating to land ownership in Australia in June 1992: the land of Australia and the surrounding islands had been owned by indigenous peoples before 1770 when British officials claimed the land for the British Crown. With subsequent legislation, this meant that most land not specifically claimed in the past under freehold title became eligible for claims by indigenous communities under the *Native Title Act of 1993*. A central provision of this Act, which was formulated in the public gaze under intense pressure from indigenous Australians, is multiple possibilities and opportunities for negotiation between the parties involved.

Subsequently the High Court of Australia, in their decision on the Wik

H. Selin (ed.), Mathematics Across Cultures: The History of Non-Western Mathematics, 289–311.

Native Title case on 23 December, 1996, ruled that native title could coexist with pastoral rights on leases. This was significant because it followed that native title had not been extinguished on the mainland of Australia in pastoral areas where many Aboriginal people live, maintaining their traditional association with their lands. The scene was set for the new and unfamiliar negotiations such as those begun in Coen in 1994, to spread quickly and widely, generating fear and confusion on all sides. This in turn had the (by now conservative) Federal Government introducing what has become known as its 'Wik Bill' to amend the 1993 act. The whole thrust, and the explicit central provision of this highly controversial amendment, was the abolition of the rights of parties to negotiate.

As things currently stand the amendment remains contentious, and is currently under challenge in several international forums. Despite it, Aborigines and others continue to negotiate, attempting the working together of disparate forms of logic. For, as evidenced by the negotiations in Coen, some non-Aboriginal Australians still have strong reasons to find ways to work the mathematics of modern[1] land ownership, largely originating in colonising European traditions, together with the very different logic and mathematics in the knowledge traditions of indigenous Australians. If they do not, they are likely to suffer the 'uncertainty and financial difficulties' mentioned in the quotation on the first page.

In earlier times, for colonising Europeans such negotiations would have been unthinkable. In the eyes of the grandfathers and grandmothers of those cattle-run owners who in 1994 sat around the table negotiating over ways of knowing and owning the land they both claimed, the Wik people were, in public contexts (as distinct from private involvements of many sorts), part of the undifferentiated black 'other'. Being part of 'the natural', Aborigines could be studied by anthropologists and shot and poisoned, led away in chains from their land and imprisoned elsewhere – to be civilised, to become part of 'the social' (Latour, 1993). Only when they became modern individuals, only when they stopped being Wik, could they know and act as knowing agents, although even that would not necessarily prevent their being discriminated against and denied social justice.

The *Native Title Act of 1993* seeks to deal with an institutionalised injustice at the heart of the Australian polity, yet a colonising frontier mentality continues to thrive in Australia. Indeed the gradual watering down of the *Native Title Act* attests the continuance of this attitude. The denial and denigration which characterises the official face of covert colonising everywhere has had Australian courts up until very recently denying that Aboriginal Australians owned the land at all before the British arrived (see Verran, 1997). It has also had scholars denying that Aboriginal Australian communities had a mathematics (see Watson-Verran, 1991).

Perhaps the most recent overt denial that Aborigines own their lands came some twenty five years ago when a group of Aboriginal clan leaders made a High Court challenge to the Federal Government's plan to develop a huge bauxite mine and build a large mining town on land they and other members of their clans believed they owned. In 1970 Justice Blackburn handed down the judgement, which went against the Yolngu people. In the judge's view, although the claimants established the existence of a subtle and powerful knowledge system, recognisable as a system of law, by which particular clans were linked through identifiable sites both to sections of land and to each other, '... [it] did not provide for any proprietary interest in the clans in any part of the areas claimed' (*Millpirrum and others Vs Nabalco Pty Ltd and the Commonwealth of Australia*, 1971: 143). Rather, '... it seems easier on the evidence, to say that the clan belongs to the land than that the land belongs to the clan' (*Millpirrum and others Vs Nabalco Pty Ltd and the Commonwealth of Australia*, 1971: 271).

The judge was correct in noting that quite different notions of agency are involved in Aboriginal knowledge systems. But whether he was correct in thus concluding that the clans had no proprietary interest is quite another matter. What did the judge consider the evidence for the land not being property of the clans? The crux of it was ownership based in observance of ritual and the limited rights of exclusion that clans had.

> The clan's right to exclude others is not apparent ... [T]he greatest extent to which this right can be said to exist is in the realm of ritual. But it was never suggested that ritual rules ever excluded members of other clans completely from clan territory; the exclusion was only from sites (*Millpirrum and others Vs Nabalco Pty Ltd and the Commonwealth of Australia*, 1971: 272).

Justice Blackburn's ruling is now all but irrelevant; his conclusions were overturned in the June, 1992 High Court ruling. However it makes a useful place to start in elaborating a mathematics of Aboriginal land ownership. Thirty years later we see that his notion of ritual was inappropriately limited, but perhaps more significantly his failure to recognise that there is more than one basis for a valid mathematics of location is now obviously a mistake.

Justice Blackburn's inability to recognise an alternative mathematics in Aboriginal Australian knowledge traditions was far from unique. It was an inability he shared with many others, including linguistic and anthropological scholars. We can see this for example in the ways linguists and anthropologists discuss the formalised cognitive elements identifiable in various language communities. That elaborated and formalised kinship systems, which are a feature of the life of many non-Western communities, are logically analogous to number systems, has only recently been generally recognised. Perhaps this attests a low mathematical sensitivity on the part of anthropologists and comparative linguists, or perhaps anthropologists were blinded by their use of 'kinship systems' as evidence in their search for the origins of society.

This failure of recognition is despite the fact that comparative linguistics has had a long-standing interest in the relation between language and the formal cognitive area of number (see Verran this volume). This interest dates back well over a century when the endeavour was understood as developing a hierarchy of the world's languages – and incidentally races – by analysis of the cognitive resources they exhibited. Back in the 1890s the 'tribes of Australia' did not even make it into this hierarchy, having 'failed to develop a numeral system' (Conant, 1896). And still, as lately as 1982, this prejudice formed part of Chomsky's famous theory about the existence of a biologically 'wired in' language acquisition device, in which arguments over characteristics of numeral systems formed part (Chomsky, 1980). Completely ignoring the remarkable mathematics of their elaborated and formalised kinship systems, Chomsky is quite unequivocal in identifying a specific biological lack in the intelligence of Aboriginal Australians. He implies that if Aboriginal Australians could count they would.

> ... [I]t is just extraordinarily unlikely that a biological capacity that is highly useful and very valuable for the perpetuation of the species and so on, a capacity that has obvious selectional value, should be latent and not used. That would be amazing if it were true (Chomsky, 1982: 18–19).

In Chomsky's view Australian Aborigines, having no elaborated number system, must be considered biologically less than truly human. Others have somewhat softened Chomsky's position but many would still see the following as racial vilification.

> The lack of numeral systems in Australian aboriginal [sic] languages is attributable ... to the fact that no aborigine [sic] ancestor ever invented a numeral system. ... Though all humans appear to have the capacity to acquire a numeral system, only some humans have the attributes or the opportunities which give rise to the development of a numeral system *de novo*. ... Some may be richly endowed with the relevant inventive capacity; others possibly not at all (Hurfurd, 1987: 74).

In Australia the *Native Title Act of 1993* has begun a process of institutional recognition of Aboriginal Australian knowledge traditions. Their logical system, and specifically the mathematics of their land ownership, have been legally recognised. This has led to negotiations which aim at developing land titles working disparate mathematics of land ownership together. Yet such negotiations are likely to lead to much more than new sorts of land title. Along with novel forms of land ownership, these new sorts of negotiations will have the effect of generating new understandings of what logic and mathematics are. In these negotiations we can confidently expect that understandings will change on both sides. Questions arising in working logic and mathematics across cultures are not only of concern for a few marginal 'others'; they are central concerns in many contemporary times and places.

MAPPING THE ELEMENTS OF DISPARATE MATHEMATICS OF LAND OWNERSHIP

I will return to the issue of negotiations between pastoralists and Aborigines in the concluding section of this paper, where I present a suggestion on how negotiations over land title might proceed more easily and openly. In leading up to that, to help identify what the content of such negotiations might be, I lay out a negotiated mapping of mathematics in another Aboriginal Australian community across the years 1987–1996. This story emerges from the work of a group of Yolngu Aboriginal and non-Aboriginal teachers and researchers located in Yirrkala in northeast Arnhem Land in Australia's Northern Territory. Their work was developing a school mathematics curriculum, and, working in the community as a teacher educator, I was a part of the group. This work (Yirrkala Community School, 1996a–e) helps in recognising similarities and differences in the mathematics of land ownership across the cultural boundary between Aboriginal and non-Aboriginal Australia.

The motivation for our endeavours arose in efforts to re-make schooling in ways that would serve the interests of the Yolngu community. In mapping logics, we were not legislating or generating another official version of Aboriginality; the solutions are local. Indeed in a robust Yolngu sense they are owned by particular people at particular places. Telling stories of this work, as I do in this paper, re-presents Yolngu community life in particular ways. To get the lines of authority correct it is important to conceptualise this text as a performance. From my work in Yolngu schools, I tell of the translations which were effected in making that new curriculum. Recognising this text as a performance grounded in my long participation in this work, we can then ask: What is the relation between this text and others in northeast Arnhem Land? What are the power relations between the community of scholars for whom this paper is of interest and the community of Yolngu scholars? Our notions of logic need to contain possibilities for addressing such questions (see Verran, forthcoming).

In the mid-1980s this group of teachers set out to reconsider the understanding of the disparate kinds of logic which inform a mathematics curriculum in contemporary Yolngu Aboriginal schools. We were thoroughly immodest despite the gloomy prognoses of orthodox accounts of reasoning and logic in cross-cultural settings. From the beginning we understood our work as dealing with different kinds of knowledge systems. This immediately involved us in the difficult work of considering what useful accounts of logic and mathematics in this context might be. That question is pursued in another chapter of this book (Verran, this volume); the conclusion is that we need to understand logic as constituted both as a system of rule following and as a system of metaphor. Understanding things this way we are able to recognise logic as multiple logics. In turn this has us accepting multiple mathematics.

In the mid-1980s the current Western philosophical views on our endeavour were less than helpful. They either condemned us to paralysis or reasoned our puzzlements and pain out of existence. They gave us nothing to go on in attempting to develop reasoned courses of action in our cross-cultural situation. In contrast Yolngu traditions helped us imagine new possibilities.

In Western philosophical discussions of the issues we met in our work, two doctrines were much in evidence: incommensurability and indeterminacy of translation. Both use the metaphor of conceptual scheme mapping onto reality, a notion that goes back at least to Kant. Quine's indeterminacy thesis (Quine, 1960) says there are too many possible translations between conceptual schemes to make the reasoned use of both a viable proposition. Pulling in the opposite direction Feyerabend (1975), and to a lesser extent Kuhn (1970), say there are no reasonable translations at all. For them disparate systems of thought are not mutually expressible. In a different way, Davidson (1984), in inveighing (correctly it seems to me) against 'the dogma of dualism between scheme and reality', and Rorty (1979) with his liberal ethnocentrism, seem not even to see our dilemmas.

Yolngu Aboriginal traditions were more helpful. In them we found conceptual resources we could use to support our work in direct ways. We began with one particular metaphor, *ganma*, which calls up a deep pool of brackish water whose surface is sometimes marked by lines of foam, at Biranybirany in northeast Arnhem Land. (Biranybirany has recently become a Gumatj Clan Yolngu settlement, in part because of the intellectual significance of this waterhole.) The Gumatj clan uses ganma in negotiations with other clans over sacred and other matters. The concept is owned, like the place in which it is located, by particular groups of people. The mathematics curriculum is thus grounded in the practical political life of the community. This metaphor now has a robust life both in Yolngu and in English discourse (Watson with the Yolngu Community and Chambers, 1989). But its life as English language discourse was, and occasionally still is, hotly contested. There are all sorts of occasions in which it is proper for Yolngu to dispute the legitimacy of non-Yolngu people using ganma.

In transferring this metaphor to our cross-cultural endeavour, great care is needed, for Westerners are wont to misunderstand it. While for many non-Aboriginal people the notion of a deep pool of brackish water is an object of distaste, for Yolngu it expresses a profound account of knowledge. European traditions identify essences and purity as worthwhile; pollution is to be avoided. Brackish water seems to express pollution to many Westerners. In contrast, for Yolngu people brackish water is a source of inspiration expressing an admirable balance of natural processes – a truth – which people should seek to emulate in their life of reason.

Ganma calls up a deep waterhole where the paperbarks of the land give way to the mangroves of the saltwater. Here amongst other things crocodile life – understood as Gumatj life – is remade. Crocodile life is a rhythmic hatching and feeding in the context of the balanced interactions of fresh water and salt water. Fresh water streams from the escarpment flow into this pool, and saltwater flows in along salty rivers with the tide. Their meeting has produced a vortical flow of currents of salty and fresh water. Fingers of salt water mix with fingers of fresh water, creating turbulence and foam where the surfaces of the two meet.

The streams retain their integrity but the force of their meeting winds deeply into the land which itself shapes the meeting. The pool is a balance between two quite different cycles: the tidal flow and ebb of salt water, and the cycle of variation in the flow of fresh water streams across the wet and dry seasons. The deep pool of brackish water is a complex dynamic system constituted as a balanced entity, embedded in a particular context and embedding other forms of life. Ganma is a view of how traditions can work together. It expresses the notion of truth as balance over time. Ganma was given to us by elders of the Gumatj clan to guide us in our mapping. In taking up this working image we should not forget that 'mapping kinds of logic' is also a metaphor. Negotiating new ways to identify similarities and differences in the mathematics of land ownership across the cultural boundary between Aboriginal and non-Aboriginal Australia involves developing new conventions of what it is that is being mutually mapped here and what are the relevant features.

In my next two sections I tell of two exercises which formed part of the negotiations out of which arose a new mathematics curriculum for Yolngu schools. First I tell of mapping the modern number system with the formal kinship system which orders contemporary Yolngu life, known by Yolngu as *gurrutu*. Then I elaborate a mapping of two logical orders located in the land itself: the order of *djalkiri* in Yolngu life and the abstract system of categories assumed by quantification in the modern world. Taken together number and quantification constitute a mathematics of modern land ownership, and similarly *gurrutu* and *djalkiri* together make a mathematics of Yolngu Aboriginal land ownership.

The mapping implicit in the curriculum takes both logic and mathematics as objects. This assumption is not innocent (Verran, this volume). For Yolngu, gurrutu is generally considered given in the Yolngu world, yet the idea of gurrutu as an *object* would be rather an uncomfortable notion for most Yolngu most of the time. In contrast having number as an object (natural number) is natural to modern understanding; it is a notion many children pick up in primary school, despite the fact that in most uses of number it is not an object but a process. Similarly, having the land as an object with its given order, djalkiri, separated from the life of Yolngu people, does not sit at all comfortably with Yolngu orthodoxy. Yet to have the land, the environment as something 'out there', which is quantified on the basis of attributes or qualities, is a most unremarkable fact of modernity.

GURRUTU AND NUMBER

I move on now to consider the first element of the mapping work which eventually gave rise to a new mathematics curriculum developed in the classrooms of Yolngu schools. Since it challenges accepted understanding on both sides, it should not surprise us that the validity of the similarities and differences argued here is still contested both by Yolngu and non-Aboriginal Australians.

In Yolngu life the concerted use of a standardised knowledge form makes it possible for all people and places to be joined in a formally related whole. Yolngu, like many contemporary Aboriginal peoples, use a formalised recursive representation of the material pattern of kinship as an integrative form of knowledge. Yolngu know their system as gurrutu. In much the same way the formalised representation of the material pattern of tallying on fingers – number – constitutes an integrative form of knowledge in modern life. The use of number makes it possible to know where we stand in a myriad of situations. This section of the paper makes the argument that the domains of gurrutu and number are analogous in a strict sense.

I became involved with the Yolngu community at Yirrkala in Australia's Northern Territory on returning to Australia in the late 1980s, after many years as a lecturer in a Nigerian University, where I had worked with Yoruba teachers of mathematics (see Verran, this volume; Watson, 1990; Watson, 1987). In Yirrkala, as the teacher of several Yolngu people engaged in tertiary education at Deakin and Melbourne Universities, I was adopted into the Rirratjinggu Clan, as a member of the Marika family. It was made clear to me that this adoption was to be understood on my part as a commitment to learn and on the Clan's part as a commitment to teach.

Adoption means that I have a place in the complex web of genealogical relations known as gurruṯu. On adoption I found that I have a specifiable relation to every Yolngu person and place and to every significant element in the Yolngu world. The terms of the specification are kin terms. The most important boundary in this exhaustive set of relations is that between what Yolngu (taking up an anthropological term) have come to name in English moieties: Yirritja and Dhuwa. Every Dhuwa person has a Dhuwa father and a Yirritja mother. Every Yirritja person has a Yirritja father and a Dhuwa mother. I am Dhuwa, along with one half of everything that is the Yolngu world. The rest of the people, lands, words, songs, designs, concepts, eternal beings are Yirritja. My children, my mothers and my potential husbands are Yirritja. My fathers, brothers and sisters are Dhuwa, like me.

I soon learned to recognise a number of sisters and brothers, fathers and mothers, all of varying ages, some much older and some much younger. While this was important, much more important was memorising the names of 16 categories of relationship, and the characteristics of each member of this set of relational categories. With respect to me and my gurruṯu placement, every Yolngu person, and more surprising every part of the Yolngu lands and their flora and fauna, and the eternal beings, concepts and designs associated with the constitution of these places, now fitted into one of these categories. It was most important to know how to engage with a person or a place when I was instructed on the name of the category they were placed in with respect to me.

In contemporary Yirrkala, as in other Aboriginal communities, the gurruṯu system as a scheme of ordering social exchanges exists alongside and works with numbers. In contemporary Yolngu life the use of number predominates when the community or individuals are dealing with the official social order. But in exchanges involving only Yolngu, number, as expressed say in the use of money, is subordinate to the working of gurruṯu.

It is perhaps difficult for us to understand gurruṯu as a domain of logic. Many researchers are used to thinking of kinship as precisely the opposite of logical. It is supposed to be a romantic pre-modern form of social arrangement; it appears irrational, because it seems to take subjective feelings as the basis for arranging things. Some might ask how a community where everyday life is primarily mediated by position in a system of kin relations could be understood as logical. It seems to some to be the epitome of a community life *not* based in logic. These are precisely the prejudices that this section of my paper aims to unravel.

All over the world parents and teachers train their children to say the number names and show them how to use them. In English speaking communities babies learn to chant 'one, two, three ...' and to do the activities that go along with saying the words – pointing to or placing pegs, or people, or pieces of toast just so. They learn the number names in songs, 'One, two, three, four, five, Once I caught a fish alive,...'. Learning to use the number names in sequence and how to carry out the gestures of number use is considered one of the most important things that children learn at primary school.

The familiar story about origins of numbers sees them as originating in the practices of material tallying. Patterns in most number systems derive from the patterns of human digits (Menninger, 1969; Ifrah, 1985). We can imagine this as using a finger as the passing of a sheep through a gate, or the placing of a pebble as the pointing at a soldier, or the engraving of a line on a piece of bone or wood, to record the filling of a vessel with grain.

The number system which has developed in association with Indo-European languages like English has ten as its base; in other words ten is the point in the series which marks the end of the set of numbers about which iteration occurs. As each ten is reached, the series is started again, each time indicating in the number how many tens have been passed. The rule by which new elements are devised is addition of single units and single base ten derived units.

We easily imagine the involvement of fingers and toes in tally keeping: one separated digit as one separated item. But if we then extend the operation and have the saying of a word as finger or toe, we end up with a system which is much more flexible and robust than the material one of fingers and toes or stones or shells. In saying a number as a finger is held up as an item is involved in some event, we understand number is neither the item nor the finger. It is a position in a progression. The system of number names is a way of using the members of a small set over and over again to keep the series going infinitely.

Numbers constitute a linear hierarchical system. Each succeeding number has a higher value than all antecedent numbers. Notions of ratio, equivalence and hierarchy are enabled through the system of numbers. The value of something, its total size or quantity, is revealed when a number is used. Numbers can be just about anything. The working of the number system carries a powerful ideology – a particular image of orderliness of a linear hierarchy is conjured up.

A story of origins similar to the story of origins of number can be told also for gurruṯu. We can imagine Yolngu plotting, on a smooth sandy patch, the pattern of related groups of people. Stones and shells might represent positions of the mesh of family relations, placed to show a network of linkages. On biological grounds we might expect a manageable set of positions to recognise the presence of three generations, differentiated along matrilineal and patrilineal lines.

If the operations were extended and the relations between the positions plotted as saying words, we would have eight pairs of reciprocal names pointing across generations or clans, along either matrilineal or patrilineal lines (two parental lines raised to the power of three generations give eight reciprocal pairs). These reciprocal word pairs name relative positions in the kinship mesh. They do not name clans or individual people.

Right from the beginning of their lives, Yolngu babies are instructed on the relation that this or that person has to them. The set of relations conjured up in these activities involves notions of hierarchy but not a single linear hierarchy. Each position plays different roles which are woven together to form a decentralised orderly mesh. Gurrutu, like number carries a powerful ideology.

In many communities where systematising kin relation is primary in community ordering we see the ideal of what anthropologists call 'cross cousin marriage'. We might understand this ideal as generating a dynamic image, just as the image of going along the fingers of the hands constitutes the image of tallying. In this ideal arrangement women marry their father's sister's sons, or to put it the other way, men marry their mother's brother's daughters.[2] This has the effect of marrying out of one's descent group while still marrying kin. There are three distinct positions in the cycle (Figure 1).

In this cycle the woman and the man she marries constitute one position, her father and his sister another, and the husband of her father's sister (and his sister) a third position. In the Yolngu case, the pattern of this genealogical ideal implies two different clans (the marrying woman's clan and the marrying man's clan) – a Yirritja clan and a Dhuwa clan.

Exhaustively specifying the primary positions generated in this genealogical ideal constitutes eight reciprocal pairs, sixteen positions: two groups (clans A and B) and three reciprocal positions in each: two sets of two to the power of three. Naming this set of sixteen positions constitutes a primary template, just like naming the set of ten fingers constitutes a primary template in tallying. The eight pairs of reciprocals name across generations or across matrilineal or patrilineal lines. The brother/sister reciprocal pair completes the set of named relations and enables continuing iteration.

Gurrutu names are names of positions in a formal series but not a series with a linear form as in numbers. This is a series with the form of a matrix.

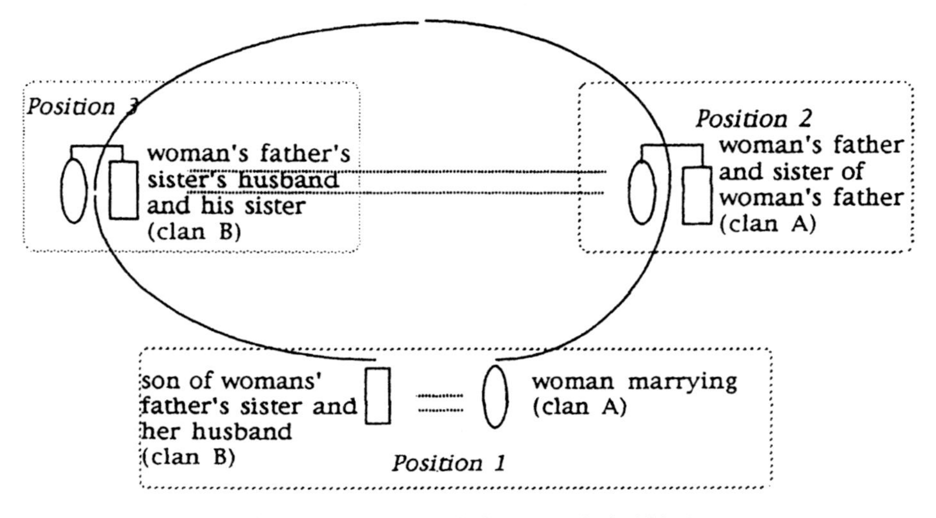

Figure 1 Pattern on a Yolngu genealogical ideal.

Just as the number series is used to reveal value in the material world, the gurrutu matrix is used to reveal relative location in the material world. For example it formally locates places in the landscape.

We can list the eight reciprocal pairs which constitute the base of the gurrutu system. The easiest way to begin to understand these categories is to personalise them in terms of my kin relations.

Or we could do it the other way around: the person I call *gäthu* (child) calls me *mukul bäpa* (aunty – father's sister) and so on through to the person I call *dhuway* (husband) calls me *galay* (wife). The dual names in some rows on the left-hand side of the pairs represent male and female holders of a position, necessarily brother and sister. Each of those pairs has the reciprocal relation of *yapa* (sister) and *wäwa* (brother). The reciprocal pair *yapa/wäwa* enable the pattern of names to work as a true recursion; they are not themselves members of the base set of reciprocal name pairs. Elaborating this list is analogous to listing the set of numbers from one to ten. In that set, zero holds the null position. Zero is not itself a member of the base set of names and can be understood as having a similar function to *yapa/wäwa*.

In just the same way as number, the name series of gurrutu constitutes a system where elements, i.e. reciprocating positions, constituted in the series provide the basis for further constitution of elements. Together the names and rules of generation form an infinite series.

One way to indicate the structure of the system of names is to map them in the conventional form that anthropologists map genealogies. The basic unit in this map is the hypothetical ideal family tree as below:

Table 1 Kin relations

The person I call:	calls me:
•*bäpa* (man) – father •*mukul bäpa* (woman) – father's sister	•*gäthu* – child
•*ngändi* (woman) – mother *ngapipi* (man) – mother's brother	•*waku* – child
•*märi'mu* – father's father and father's father's sister	•*marratja* – grandchild
•*ngathi* (man) – mother's father •*momu* (woman) – mother's father's sister and simultaneously father's mother	•*gaminyarr* – grandchild
•*märi* – mother's mother and mother's mother's brother	•*gutharra* – grandchild
•*mumalkur* (woman) – mother's mother's brother's wife •*ngathiwalkur* (man) – mother's mother's brother's wife's brother	•*dhumungurr* – mother's brother's mother-in-law
•*mukul rumaru* (woman) – mother's brother's wife •*maralkur* (man) – mother's brother's wife's brother	•*gurrung* – mother-in-law
•*galay* – mothers' brothers' daughters and sons simultaneously brother's wife and her brothers	•*dhuway* – sister-in-law

Table 2 A conventional genealogical map of the basic elements in Yolngu gurrutu

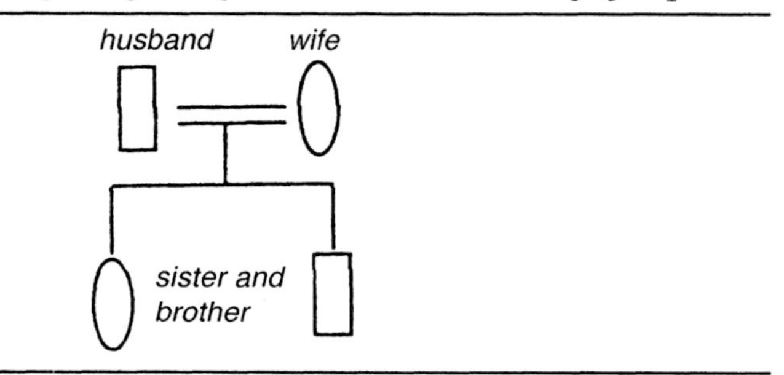

This ideal unit illustrates the fundamental contrast between the two different sorts of relations which constitute the gurrutu system of relations – that between wife and husband and brother and sister. When we map out the series of eight reciprocal kin relation pairs using this unit we get the diagram shown in Table 3.

Gurrutu and number can be understood as analogous. Like number, gurrutu has a set of basic names and a set of rules for devising further names from any one name. Both are prescriptive. Although at first they look very different, from a mathematical point of view these differences are trivial. In numbers the set of rules mimic the patterning of counting on fingers. In gurrutu it is the patterning in mapping kin relations across three generations in two moieties that is mobilised.

I am presenting number and gurrutu as analogous, but they are very different in their content. Should that worry us? These two great patterns – tallying codified in counting, and genealogy codified in ordering descent and ancestry – are of a kind, albeit different in form. They differ in structure: each number has one direct antecedent and one direct successor, and each gurrutu position has two direct antecedent positions and one successor position. However the two systems are characterised by recursive definition, a form of mathematical induction. Number can be defined recursively:

One is a number and successors of numbers are natural numbers.

We can set strict limits with:

There are no numbers but what this requires.

Similarly gurrutu can be defined recursively:

Bäpa is a gurrutu position and successors of gurrutu positions are gurrutu positions.

This too can be limited:

There are no gurrutu positions but what this requires.

Using only English words this definition would come out something like:

The parents of x are ancestors of x, and the parents of ancestors of x are ancestors of x.

The limitation is that:

There are no ancestors of x but what this requires.

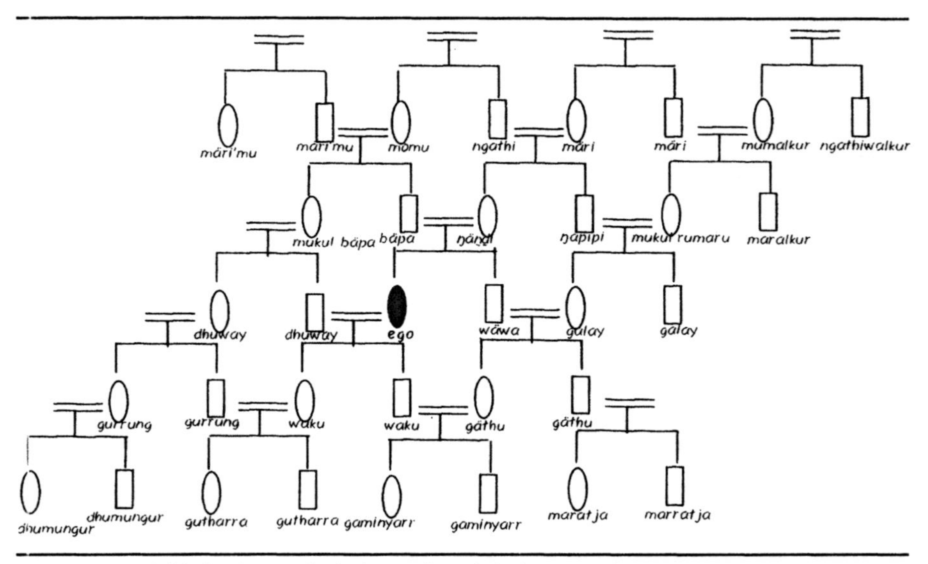

Table 3 A genealogical mapping of the base set of gurruṯu elements

The recursive form of this argument was used in mathematics as far back as Euclid; Fermat recognised it as a form of mathematical induction three centuries ago. A recursive definition formalises the idea of 'and so on'. It is a mathematical definition in the weaker sense, in being indirect. However it can be given a more direct form as was elegantly demonstrated a little over a hundred years ago by Frege (1879). We can define gurrutu positions as the members shared by all classes that contain a gurrutu position and the successors of all their own members. Similarly we can define the numbers as the members shared by all classes that contain one and the successors of all their own members.

I am not saying here that tallying is absent from Yolngu life; it is not. Nor, in noting the predominance of tallying in modernity, am I saying that genealogy (arranging matters on the basis of close and distant kinship) is absent. What I am saying is that in Yolngu life tallying does not carry the organising burden that genealogy does. Juxtaposing number and gurrutu the primary principle embedded in these contrasting forms becomes evident. In numbers we can see that the primary rule is a version of the process of holding up another finger as another sheep leaps through the gateway. The important thing about this operation is that each finger is exactly equivalent to each other finger. And any collected set of fingers has a precise and identifiable relation to any other collections of fingers; they relate to each other as a specific ratio. The primary principle of the number system is the principle of ratio: the ratio of any term to another is determined by the number of times one contains the other.

In gurrutu the principle is quite different. Going back to the ideal marriage arrangement which underlies gurrutu we can see that an inherent reciprocity marks the form. Husband-taker and wife-taker are reciprocally related. Clan A and Clan B together constitute a unity – the offspring of the marriage. This is the principle of reciprocity – two opposing elements constituting unity. A reciprocal is an expression of elements so related that their product is unity. It is the inherent opposing duality constituting unity which is the fundamental principle of gurrutu. Every position in the gurrutu matrix is necessarily constituted by a two way naming. The elements must be specified from both ends before they come to life as an entity.

Reciprocity can be pictured as two sides of a coin. On the old Australian penny (which had the head of the current king or queen of England on one side and a kangaroo on the other), head and tail constitute reciprocals, and together constitute the unity of the coin. The notion of reciprocal is often difficult for learners of mathematics centred around numbers to understand; they need to understand fractions before it is possible to understand reciprocals in a rigorous manner.

In gurrutu reciprocity is primary. If you take any base family relation as a unit, for example *märi-gutharra* – the matrilineal relation across two generations (mother's mother or mother's mother's brother) – you can understand it as constituted by the mutual engagement by both sides. The descendant and the grandmother/great uncle are equally important in constituting the unity described by *märi-gutharra*. Each side is necessary to constitute the unity held between them.

And just as reciprocity can be secondarily derived in the system of numbers pivoting around ratio, similarly the notion of ratio can be secondarily derived in gurruṯu. *Milmarra* is a name given to this way of rendering gurruṯu. A simple ratio involves juxtaposing a *yothu-yindi* (mother-child) relation with a *märi-gutharra* (child-mother's mother) relation: 'If the two clans of the *yothu-yindi* are A and B respectively, what is the *mari* clan?' There is always only one correct answer which can be understood as expressing the ratio of the two reciprocals. Such puzzles are often set for learners of the gurruṯu system.

ANALOGOUS LOGICAL ORDERS IN THE LAND

I now turn my focus to the analogy between the logical orders implied by quantification on the one hand and *djalkiri* on the other. Practices of number and quantification in modern life are tightly linked, and similarly gurruṯu and djalkiri work in concert in Yolngu life. Gurruṯu-djalkiri together constitute a mathematics of land ownership in the Yolngu community, just as number-quantification constitute the mathematics of modern land ownership. We could say that the order implied by quantification enables the world to be worked as number, and that the order implied by djalkiri allows the world to be worked as gurruṯu. In the case of quantification land is rendered so as to manifest in linear arrayed units evoking the form of an infinite linear extension of integers. In the case of using djalkiri land is rendered as a set of interconnected places evoking the form of an idealised network of kin relations.

The practices involved in working the world as number pivot around the understanding that various sorts of objects, like tracts of land, exhibit qualities. Numerosity, length, and area, for example, are taken to be inherent in something like a bounded plot of land. These qualities are imaged as linear extensions, so that plots of land manifest these various qualities to varying extents. A collection of plots of land has numerosity to a certain extent, and boundary lengths and area of varying extent. Qualities and their units are evoked in routine gestures of counting and measuring. Sets of units like plots, metres, hectares, for example, emerge from rigorously set routines, and are taken as corresponding to an integer of the number series. This is a scalar mathematics of land ownership. Qualities, imaged as linear extensions, emerge numbered as scalar quantities. Each objective value stands in a specific ratio to other objective values.

It is difficult to see quantifying something like land as ritual. So familiar are the routine gestures of quantifying that we feel that they *must* have inherent meaning. But learning to see the rituals by which modern quantification proceeds is necessary to appreciate fully how multiple mathematics are generated and sustained. To help in this I ask you to imagine some scenes acted out near Yirrkala during the 1950s.

For most of Australia's short history, to most Australians, northeast Arnhem Land and its Aboriginal inhabitants seemed impossibly remote. All this changed however with World War II when the area was bombed by Japanese planes and was readied to block the expected Japanese invasion. One effect of all the focus on this area was the official realisation that while the soils of the area were too poor to sustain agriculture, nevertheless they held a bounty in the form of bauxite. Thus it was that Yolngu men and women came to witness some curious rituals.

White men would turn up in trucks and jeeps from which they would take out various objects. There was a tripod with a small barrel on top. One man would put his eye to the barrel and wave his arms about meanwhile uttering loud calls. The partner, standing some distance away carrying a vertical pole, danced backwards and forwards and from side to side in apparent response to the calls from his mate. In another performance they would pull out a long rolled-up tape with black or red markings. With one person holding the end this would be laid carefully on the ground, with care taken to flatten any intruding shrubs and moving stones to get the tape flat and straight. But perhaps the most comical of all was the wheel on a stick which clicked. This would be pushed along through the bush in a straight line. Then the person would abruptly stop, turn either to the left or the right and head off again, rolling and clicking through the scrub. Then a little later a post would appear with an inscribed paper displayed. It was only many years later that Yolngu men and women came to understand that these rituals were the preliminaries of their dispossession by the Australian Government.

We who live our lives in modern worlds can read my descriptions as three examples of the routines and gestures of quantification. As rituals they are entirely transparent. As we follow the story picturing the scene we can see the lengths being laid out. They were almost totally opaque to most Yolngu men and women at that time.

For moderns, attributes like the length or area of an individual plot are assumed to have existed prior to any human practices. The modern assumption is that qualities are merely revealed and numbered in quantifying. Such qualities are taken as inherent in matter and generally justified through appeal to sensory evidence. Forgetting years of training in mathematics classes, it is assumed that length can be seen and area experienced in walking around. 'Qualities are characteristics that land really has; that is the way it is', is the type of explanation favoured by many if they are challenged to explain the certainty they feel in the qualities which underlie numbers. In numbering the land we understand ourselves as revealing objective value.

Without bothering about any of this complicated cognitive justification, those who use qualities every day take it for granted that the landscape is somehow naturally meaningful in that it contains qualities. Yet anyone who has tried to describe and explain qualities to small children will realise that they are rather mysterious things. Moderns tend not to admit this, preferring to hide this mysterious making of the land and the material world in general as inherently meaningful. Through practices of quantification, denied as ritual, modern knowing and owning land maintains the myth that knowable space is at once ordered and empty. An outcome of this is that all space is equivalent. Yet it is the very practices of quantification which constitute, and in turn are constituted by, that uniform frame of empty, ordered space.

These sets of routines associated with numbering look and feel very different to the sets of practices associated with working the world through gurrutu. They are not easily understood as analogous, and the ways that they are justified as reasonable by those who use them are also quite different. There is a very different way of dealing with the paradox inherent in land having inherent order in Aboriginal Australia. There land exists primarily as sets of sites; there are inherent foci in the land connected in particular ways. Through these interconnected sites, often called 'sacred sites' in English, land is meaningful. Djalkiri, the practices and justifications involved in working the world as gurrutu, hinge on the understanding that the world manifests itself as particular places linked in particular ways. Particular groups of people are linked in differential ways with specific sites.

Part of the routines of djalkiri are the dance and song cycles, the large ceremonies regularly performed in many places by Yolngu people working in their gurrutu groups. Ceremonies, called 'business' in English by Yolngu, are often associated with the occasion of a funeral or a circumcision, but not always so. Groups of people located in particular relations to the places which are the focus of the ceremony organise or perform particular parts of the ceremony.

Places have their origins in the activities of idealised Ancestors, and the ceremonies which (re-)constitute place re-enact those activities of the Ancestors. Actual locations are related to other actual locations in a series of gurrutu names. Places are linked differentially and reciprocally just as people are linked through gurrutu. Places emerge as such, as effects of these links, just as significant groupings of people emerge as effects of links. The primary focal sites through which land is meaningful are connected in recursive logic modelled on kin relations. Gurrutu translates inherent meaning of the land in just the way the logic of number translates qualities. But for Yolngu the mystery of the origins of these sites is celebrated in mythical stories.

According to these stories eternal beings made the people in clan groups and particular places in the land as they went about their living: hunting, eating, defaecating, urinating, having coitus, menstruating, crying, and having babies. This is understood to have occurred in what is known in English as 'the dreamtime', *Wangarr* in Yolngu language. This is often considered, incorrectly, as the far distant past, but a contrast between time as secular and as eternal is probably a better way to explain it. Wangarr is time of a different sort (something like eternal time) to that in which we live our everyday lives (secular time); it is not time only of the far distant past. It is a time which we can find here and now and will be able to continue to find in the future. Boundaries between these different sorts of time are continually maintained and celebrated in Yolngu life.

The knowledge of sites and their connections is contained in a large corpus of stories and the songs, dances and graphic designs which go along with the ceremonial elaboration of these stories (hence the popular English language notion of Aboriginal songlines). These are performed in ceremonies where both the complex logic of gurrutu and particular land sites are re-presented. The words of the songs which celebrate this imaginary world are not memorised. It is the general picture of the network of places and their interconnections that is memorised.

This is a complex set of spatial images, a cognitive map which can be understood as quite analogous to the modern imaging of qualities in material objects as held in varying extents pictured as infinite linear extensions, which can be made analogous to the infinitely extending line of integers. In contrast to the scalar mathematics of modern land ownership, the Yolngu Aboriginal system is a vector mathematics. Every place emerges as a vector, understood only in relation to an orderly matrix of places.

It is knowing the map, which we can understand as a matrix of vectors with each place defined through relations of varying intensity and direction, and coming up with metaphoric insights to express this map in performing songs and stories, that is valued as Yolngu intellectual work. There is a correct map which everyone knows in greater or lesser detail, and the map may be expressed in more and less elegant ways. This matrix of sacred sites is not all there is to land ownership in Aboriginal Australia. The Ancestral Beings designated particular places as special for engaging in particular acts. And in a secondary way these actions constitute the space between these places: the blood, or the tears, or the urine, or the honey flowed in a particular way, so that ownership radiates out from sacred sites encompassing the broadlands between. These broadlands can be used by other clans for hunting and gathering, and for passage, by negotiation; exclusion is only from sacred sites. Apart from this eternally justified ownership, clans may also own, in a secondary way, particular sites within the broadlands of other clans – these are known as *ringgitj* by Yolngu and likened to clan embassies. Ownership of these subsidiary sites has been constituted historically within secular time; they are important in the on-going collective life of the group of Yolngu clans. Parties of clan members camp at these sites during ceremonial times.

For Aboriginal Australians space is primarily constituted as live or enlivened, and not all places are equivalent. In ceremony this eternal constitution of the land through the living eternal Ancestral Beings is re-presented. Particular sacred sites and associated broadlands are ritually evoked in representing the acts of the Ancestors: making fire, collecting honey, having sex, etc. This differentially focussed space is also maintained in everyday activities: hunting and gathering of particular foods at particular places at particular times by particular people is also seen as important in the on-going constitution of space.

Moderns own land through the mediation of number. Number is a logic which translates qualities *in* the land. Area and length are the qualities conventionally used as the basis for quantifying land, that is, publicly knowing land, and the basis for owning particular sections of land. Qualities come to life as images, and are conceived of in a way which echoes the structure of separated integers of the number system: an infinite extension of units (metres, hectares, etc.). The notion of an ordered landscape existing prior to any human practices and as characteristic of the land itself which is implicit in ordering through djalkiri is justified by Yolngu through appeals to quite different sorts of explanations to those favoured by those who explain the order mobilised by quantifying as just the way the (natural) world is. For Yolngu the order in land was made in another order of time by another order of beings. It is the responsibility of contemporary Yolngu to remake the order by venerating the Ancestor Beings.

In understanding these disparate explanations I suggest that we can understand both sorts of justification as valid ways to attempt explanation of the inexplicable. There is a sense in which we can understand the practices of quantification and the practices of djalkiri as metaphors. The genius of the notion of inherent qualities in the land, and of the land as *a priori* places connected in particular ways, is that they both capture the ineffable with possibilities for specificity and precision.

WORKING DISPARATE MATHEMATICS OF LAND OWNERSHIP TOGETHER

I have mapped the elements of a particular Aboriginal Australian mathematics of land ownership – that of the Yolngu Aboriginal people who already have freehold ownership of their lands in northeast Arnhem Land in Australia's Northern Territory. The details of gurrut̲u and djalkiri as a working logic (which necessarily includes working images) are specific to that Yolngu world. We can, however, cautiously generalise to other Australian Aboriginal peoples. It seems that amongst such peoples two facets are common: land is owned by collective groups of kin, and it is owned through what are known in English as sacred sites. The systems look and feel very different to modern land ownership. At first these differences are likely to overwhelm us. But let us attend to the ways in which they are similar.

Perhaps the first difficulty for modern people attempting to come to terms with Aboriginal land ownership is its collective nature. Land is owned by clans not by individuals. This is not as much of a problem as it might first appear: the precision and specificity in defining a kin group amongst a people who organise themselves through a formalised kin system can be quite analogous to the precision and specificity of defining any given individual person. The problem is already quite familiar through the notion that corporations – for example a pastoral company – owns land. The compelling nature of the relation between land and person invested in the notion of ownership implies only that behaviour is regulated. Regulation of behaviour can as easily be an internal matter for a collective.

The next difficulty that might arise is the notion that Aboriginal ownership proceeds through sites and not bounded areas. How can exploitation of resources of land be regulated if ownership is solely vested in particular sites? The answer here too is fairly readily available. The *meaning* in the broad acres around sites derives from the particular meanings of the sites. This particular meaning tells which clans can take which resources from this area and whose acknowledgment is required. Ownership through sites in fact accomplishes very specific possibilities for controlled access to resources.

Perhaps most difficult for those of us used to a scalar mathematics of land ownership is the vectorial mobilizing of land which systems like that of gurrut̲u-djalkiri effect. Land *is* a matrix of vectors. Places and areas exist *only* in relation to each other. This might appear to make the system inherently unstable. Yet the vectors through which sites are accomplished are precisely specifiable. Sets of people (precisely specifiable) have differentiated interests (again precisely specifiable) in sites set in relation to each other in precisely specifiable ways.

This vector mathematics of land ownership is more complex than the scalar mathematics of modern land ownership, but maybe that is not necessarily a disadvantage. Given the complexities of today's world, having a scalar mathematics of land ownership as the only possible mathematics of regulation access to and control over land is perhaps beginning to look far too simplistic. The vector mathematics of Aboriginal land ownership is a manageable complexity because of its precision and specifiability. In that the system is equivalent to scalar mathematics of modern land ownership.

I return now to those Australians meeting at Coen in 1994, the Wik Aborigines and the non-Aboriginal pastoralists who sought to negotiate their differences over the land that both groups loved and claimed as theirs. How could they, struggling to imagine what a new sort of land title might look like, be inspired by the contingent mapping which emerged from the work of those developing a mathematics curriculum at Yirrkala in the Northern Territory? First they would need to accept that they are dealing with disparate mathematics, appreciating that this will enable them to develop titles that perform in different ways, that is, connect particular people with particular places in ways other than the simplistic scalar mathematics of modern land ownership.

In extending my comparisons to explore the notion of a joint land title which mediates the cultural boundary between Aboriginal and non-Aboriginal Australia, I briefly sketch out a consideration of how negotiations might be framed. Specifically I am looking to develop a vision of land title which is compatible with the notion of a joint Native Title–Pastoralist Lease title. It is important to recognise that at present this is no more than an idea. I am not aware that any of the many negotiations over land title currently occurring around Australia have gone this far in the re-imagining of land title.

A joint title could try to represent Aboriginal land ownership within the conceptual framework of modern land ownership. It could retain the notion of empty space, trying to impose some Aboriginal boundaries on this space. These boundaries could represent the limits of influence of sacred sites present within these boundaries. This is in fact is the route taken in the 1976 Northern Territory Land Rights Legislation under which my Yolngu friends hold freehold title of their lands. Using this means of translation, areas of crown land can be declared as the freehold property of various Aboriginal clans.

From the point of view of joint ownership between pastoralists and Aborigines however, this leaves us with the problem that two opposing groups claim one area. Adjudication in favour of one or the other is the only possible way of settling the dispute, although caveats giving limited access and use for specific purposes to the other group might be appended to the title. This solution is however unlikely to please either group. One group sees itself as unfairly disadvantaged by the arrangement, and the other – the owning party

– sees itself as being set up to bear the bitterness and grievance of their opponents. This is a lose-lose arrangement in which the uncertainties and financial difficulties which inspired the impulse to negotiate in the first place continue and increase.

There is also a further problem in going this way. Under the *Native Title Act of 1993*, Native Title must be the basis of joint title negotiations. This is understood to have existed prior to any British style land titles in Australia. Native Title of course can only be legitimately understood in the terms of the natives' mathematics, and (on mainland Australia) these terms do not include notions of bounded area as a primary category in land ownership. An alternative way to proceed would be to follow the Aboriginal conceptual scheme of understanding foci in the land as the basis of the meaning of the land, and hence as a basis for ownership. Can we imagine a pastoral lease in these terms?

We first need to see that in a pastoral lease not all places and lands are equivalent, just as for the Aboriginal clans who own the land, not all places and lands are equally and similarly important. We can understand a pastoral run as constituted by a hub site – the homestead, the machinery sheds, the yards and perhaps some home paddocks where the milking cows are maintained. This is the site of primary significance in a pastoral run and might be understood by analogy to Aboriginal sacred sites – those places where the ancestors lived, doing things which made the site and its surroundings meaningful.

Meaning flows from Aboriginal sacred sites into the surrounds. Similarly we can understand that the significance of the pastoral lease hub site can be seen as flowing into the surrounding lands. The tending and care of cattle and the lands on which they thrive flow from the hub. With respect to these broadlands, what pastoralists need is the possibility of negotiating rights for harvesting these broadlands (in the form of beef cattle) and this possibility is constituted by the care that exudes from the station. And then there are the out stations, the huts, the bores with their pumps, the watering holes along river courses – the subsidiary sites. The use and maintenance of these sites too can be negotiated over. Claims for use of these subsidiary sites can be seen as growing out of the caring which gives meaning to the land.

Working along with this we can imagine an Aboriginal ownership which similarly proceeds through specific sites from which meanings flow. These sacred sites have a different ontological status to the functionally contingent status of the pastoral run hub, but from the point of view of practical regulation, this difference in ontological status is irrelevant. What is important is that all sorts of further negotiations over land management, carried out between joint owners of equal status, would become possible once a title which engages these sorts of understandings is in place.

What I have laid out here is a hybrid land title which we might imagine as a stable oscillation effected between a vector mathematics and a scalar mathematics. It helps us understand that mathematics across cultures is not only a matter of dominations and/or incommensurability. We can escape from the colonising stories that mathematics has told of itself in the past. Imaginatively working mathematics across cultural boundaries contains possibilities for making different futures.

ACKNOWLEDGEMENTS

I owe a large debt to the teachers (both Yolngu and non-Aboriginal), the children, and the Yolngu elders who make up the extraordinary Yolngu educational community in northeast Arnhem Land. I have been changed in my learning with them. Since 1987 this intellectual community has struggled with profound conceptual and political issues in generating the Garma Maths Curriculum, currently in the process of achieving official status under the Northern Territory Board of Studies. Tragically, in 1999, that same Government announced that it plans to abandon the policy of bilingual education for Aboriginal children, after some 25 years of successful operation. This retrogressive policy jeopardises the learning of children in many Aboriginal communities.

NOTES

[1] I use the term 'modern' in a precise way in this paper, implying the set of conventional understandings that inform the workings of modern life and are generally associated with scientific ways of understanding. In the past I have used 'Western' where I now use 'modern', but since many non-Western polities work with this set of modern understandings, the term Western just does not do the job. By contrasting Aboriginal and modern in this way, I certainly do not mean to imply that somehow Aborigines are not modern in the sense of contemporary. Aboriginal life is as contemporary as modern life. 'Traditional' is often contrasted to modern in accounts of Aboriginal-Australian life. But this causes problems too, since modernity has traditions just as much as Aboriginal Australia has them.

[2] The study of kinship is in many ways the central endeavour in anthropology, and the meaning of the term kinship continues to be hotly contested. All sides agree on the existence of this genealogical ideal in 'descent group societies', but have radically different interpretations about its meaning in terms of a general theory of society. *Kinship*, by C. C. Harris, Open University Press, 1990 is a useful introductory text here.

BIBLIOGRAPHY

Conant, Levi Leonard. *The Number Concept: Its Origins and Development*. New York and London: Macmillan and Co., 1896.

Davidson, Donald. *Inquiries into Truth and Interpretation*. Oxford: Clarendon Press, 1984.

Chomsky, N. *Rules and Representations*. Oxford: Basil Blackwell, 1980.

Chomsky, N. *The Generative Enterprise: A Discussion with Riny Huybregtsand Henk van Riemsdijk*. Dordrecht, The Netherlands: Foris Publications, 1982.

Feyerabend, Paul. *Against Method*. London: Verso, 1975.

Frege, Gottlob. *Begriffsschrift, eine der arithmatischen nachgebildete Formelsprache des reinen Denkens*. Halle: Nebert, 1879. Reprinted as *The Basic Laws of Arithmetic: Exposition of the System*, M. Furth, trans. and ed. Berkeley: University of California Press, 1964.

Harris, C. C. *Kinship*. Milton Keynes: Open University Press, 1990.

Hurford, James. *Language and Number: The Emergence of a Cognitive System*. Oxford: Basil Blackwell, 1987.

Ifrah, Georges. *From One to Zero: A Universal History of Numbers*, Lowell Blair, trans. New York: Viking Penguin, 1985.

Kuhn, Thomas. *The Structure of Scientific Revolutions*. Chicago: University of Chicago Press, 1970.

Latour, Bruno. *We Have Never Been Modern*. Cambridge, Massachusetts: Harvard University Press, 1993.

Menninger, Karl. *Number Words and Number Symbols.* Paul Broneer, trans. Cambridge, Massachusetts: MIT Press, 1969.

Millpirrum and others Vs Nabalco Pty Ltd and the Commonwealth of Australia. Melbourne: The Law Book Co., 1971.

Morphy, Howard. *Ancestral Connections: Art and An Aboriginal System of Knowledge.* Chicago: University of Chicago Press, 1991.

Quine, W. V. O. *Word and Object.* New York: Wiley, 1960.

Rorty, Richard. *Philosophy and the Mirror of Nature.* Princeton: Princeton University Press, 1979.

Verran, Helen. 'Imagining ownership working disparate knowledge traditions together.' *Republica* 3: 99–108, 1995.

Verran, Helen. 'Re-imagining land ownership in Australia.' *Postcolonial Studies* 1(2): 237–254, 1998.

Warner, W. Lloyd. *Black Civilization: A Social Study of an Australian Tribe.* New York: Harper and Row, 1935.

Watson, Helen. 'Learning to apply numbers to nature. A comparison of English speaking and Yoruba speaking children learning to quantify.' *Educational Studies in Mathematics* 18: 339–357, 1987.

Watson, Helen. 'Language and mathematics education for Aboriginal Australian children.' *Language and Education* 2(4): 255–273, 1988.

Watson, Helen with the Yolngu Community at Yirrkala and D. W. Chambers. *Singing the Land, Signing the Land.* Geelong: Deakin University Press, 1989.

Watson, Helen. 'The Ganma project: research in mathematics education by the Yolngu Community in the schools of Laynhapuy (N.E. Arnhem Land).' In *Language Issues in Learning and Teaching Mathematics*, Gary Davis and Robert Hunting, eds. Melbourne: LaTrobe University, 1990, pp. 33–50.

Watson, Helen. 'Investigating the social foundations of mathematics: natural number in culturally diverse forms of life.' *Social Studies of Science* 20(2): 256–283, 1990.

Watson-Verran, Helen. 'Review of *Language and Number: The Emergence of a Cognitive System*, by J. Hurford.' *Australian Journal of Linguistics* 11: 113–120, 1991.

Watson-Verran, Helen and Leon White. 'Issues of knowledge in the policy of self-determination for Aboriginal Australian communities.' *Knowledge and Policy* 6(1): 67–78, 1993.

Wertheim, Margaret. 'The way of logic.' *New Scientist* 148(2006): 38–41, 1995.

Williams, Nancy. *The Yolngu and their Land. A System of Land Tenure and the Fight for its Recognition.* Canberra: Australian Institute of Aboriginal Studies, 1986.

Yirrkala Community School. *Living Maths 1 Djalkiri – Space Through Analogs.* Boulder Valley Films. Distributed by Australian Film Institute, Melbourne, 1996a.

Yirrkala Community School. *Living Maths 2 Space – The Grid Digitised* Boulder Valley Films. Distributed by Australian Film Institute, Melbourne, 1996b.

Yirrkala Community School. *Living Maths 3 Gurruṯu – Recursion Through Kinship.* Boulder Valley Films. Distributed by Australian Film Institute, Melbourne, 1996c.

Yirrkala Community School. *Living Maths 4 – Tallying Number.* Boulder Valley Films. Distributed by Australian Film Institute, Melbourne, 1996d.

Yirrkala Community School. *Living Maths – The Book of the Videos.* Boulder Valley Films. Distributed by Australian Film Institute, Melbourne, 1996e.

PAULUS GERDES

ON MATHEMATICAL IDEAS IN CULTURAL TRADITIONS OF CENTRAL AND SOUTHERN AFRICA

In this paper I will first present some evidence for early mathematical activity in Central and Southern Africa. Then follows a short overview of fields of mathematical activity which have been recently studied.[1] Special attention will be given to geometrical explorations in Central and Southern Africa. In the third part I will present, as an illustration, geometrical ideas involved in the traditions of the Yombe mat weavers, the Sotho house wall decorators, and the Chokwe sand drawers. The paper concludes with a reflection about ethnomathematics and mathematical thinking.

EARLY EVIDENCE

Historians generally assume that Africa, in particular Eastern or Southern Africa, constitutes the cradle of mankind. In a lecture at the University of the United Nations, the first president of the African Mathematical Union (1975–1986), Henri Hogbe-Nlend of Cameroon, asked whether Africa could also be considered the cradle of world mathematics. His response was an unequivocal yes. He gave particular attention to ancient Egyptian mathematics (Hogbe-Nlend, 1985). More recently, in 1995, the Congolese Egyptologist and historian of science, Théophile Obenga, stressed the philosophical and demonstrable character of ancient Egyptian geometry and related it to geometrical ideas in Africa south of the Sahara (cf. Gerdes, 1985, 1995a: vol. 3, 1997b). What can be said about the emergence of mathematical activity in Central and Southern Africa?

As evidence for early mathematical activity in Central Africa, Zaslavsky (1973) presented a bone initially dated at 9000–6500 BC, unearthed at Ishango (Congo/Zaire). The bone has what appear to be tallying marks on it, notches

---

[1] For a more extended overview, I refer the reader to a paper published in the journal *Historia Mathematica* (Gerdes, 1994), where I presented an overview of research findings and of sources on or related to the history of mathematics in Africa south of the Sahara, particularly studies that have appeared since the publication of Claudia Zaslavsky's classic study *Africa Counts* (1973).

H. Selin (ed.), Mathematics Across Cultures: The History of Non-Western Mathematics, 313–343.

carved in groups. Its discoverer, De Heinzelin, interpreted the patterns of notches as an 'arithmetical game of some sort, devised by a people who had a number system based on 10 as well as a knowledge of duplication and of prime numbers' (Heinzelin, 1962). Marshack (1972), however, explained the bone as an early lunar phase count. Later, he reevaluated the dating of the Ishango bone, setting it back to 20,000 BC (Marshack, 1991). Zaslavsky asked, 'who but a woman keeping track of her cycles would need a lunar calendar?' and concluded, 'women were undoubtedly the first mathematicians!' (Zaslavsky, 1991). A still much older mathematical artifact from southern Africa was reported in 1987: 'A small piece of the fibula of a baboon, marked with 29 clearly defined notches, may rank as the oldest mathematical artifact known. Discovered in the early seventies during an excavation of Border Cave in the Lebombo Mountains between South Africa and Swaziland, the bone has been dated to approximately 35,000 BC.' (Bogoshi, Naidoo and Webb, 1987: 294). The bone resembles calendar sticks still in use today by San hunters in Namibia and Botswana. Counting, measurement, time reckoning, classification, tracking and some mathematical ideas in technology and craft among the surviving San hunters have been studied (Lea, 1990). The San, one of the earliest peoples in Southern Africa, developed very good visual discrimination and visual memory as needed for survival in the harsh environment of the Kalahari desert (Lea, 1990; Liebenberg, 1990).

In general, early humans learned geometry in the context of their daily activities (Gerdes, 1990, 1999b). Rock paintings and rock engravings have been reported from several parts of Central and Southern Africa (Willcox, 1984). Some date from several hundred years ago, others from several thousand. Often they have a geometric structure. Figure 1 presents two petroglyphs from Calunda (extreme east Angola) with an interesting geometrical structure (cf. below '*Sona*'). Other archaeological finds that may give an indication of geometrical exploration by African hunters, farmers and artisans are stone and metal tools and ceramics. Archaeological finds of perishable materials, like baskets, textiles, and wooden objects, are exceptional. Ethnomathematical research may constitute an alternative way to reveal mathematical ideas in the history of Central and Southern Africa, as the following sections will show.

Figure 1 Petroglyphs from Calunda. Reproduced from Gerdes, 1999a, Fig. 1.1c.

COUNTING AND NUMERATION SYSTEMS

Obenga (1973) analyzes the numeration system and arithmetic (including the use of fractions) and what he calls the 'cosmic numbers' of the Mbosi (Congo-Brazzaville). Mfika (1988) analyses oral and possible graphic numeration systems from Congo/Zaire. In particular, he deals with the symbolic expression of numbers in Luba cosmogony, e.g., the significance of even and odd, the use of 'numbers of peace': 4 and 12, 24, 48, 96 .... The author stresses that 'the explanation of the origin of life [among the Luba] by numbers [is] practically equal to that of Pythagoras' (p. 153). Mantuba-Ngoma (1989a: 61) describes the binary system used among the Yombe (see below) in the lower Congo to measure *lubongo* (plural: *zimbongo*, traditional money made out of cloth. One *lubongo* corresponds to a piece of cloth, made out of raffia or pineapple strands, of a length of about 2 yards. The next units are: 1 *kindela* = 2 *zimbongo*; 1 *nlabu* = 2 *bindela*; 1 *babu* = 2 *minlabu*.

Numeration and counting systems from Mozambique are discussed in Gerdes (1993), where, along with the presentation of historical written sources, Ismael and Soares analyze popular counting systems, Mapapá and Uaila present a comparative overview of numeration systems, and Draisma reflects on mental arithmetic. For an early overview of numeration systems in Africa, see Schmidl (1915).

CALCULATION GAMES

Several studies are dedicated to 'board' games played in Central and Southern Africa which display certain mathematical aspects. Mve Ondo (1990) published a study on two *mancala* (cf. Russ, 1984) or calculation games, i.e. *owani* (Congo) and *songa* (Cameroon, Gabon, Equatorial Guinea) (cf. Mizone, 1971). Silva (1995) analyses *mancala* games from Angola. Townshend (1977) presents *mancala* games from Congo/Zaire, Rwanda and Burundi (cf. Huylebrouck, 1996). *Omweso*, an analogue four-row board game from Uganda, is presented by Nsimbi (1968). An analysis of mathematical aspects of a Mozambican version, *ntchuva*, and a Tanzanian version, *bao*, have been initiated by Ismael (1999) and De Voogt (1998).

ALGEBRAIC ALGORITHMS IN DIVINATION

Ascher (1997) explores ethnographic and written historical sources in her study of mathematical aspects embodied in *sikidy*, a system of divination in Madagascar. For instance, formal algebraic algorithms are applied to initial random data, and knowledge of the internal logic of the resulting array enables the diviner to check for and detect errors.

GEOMETRICAL EXPLORATIONS

It seems that geometrical exploration constitutes the prime area of mathematical activity in the history of Central and Southern Africa. It is often closely connected to artistic activity, as the first example of the Mangbetu in northeastern Zaire/Congo may underscore. Eglash (1998, 1999) analyzes an ivory hatpin from the Mangbetu and the geometric algorithm involved in its production. The top of the pin is composed of four scaled, similar heads. Other activities in which Mangbetu interest in artistic-geometric structures is manifest (cf. Gerdes, 1999a, ch. 1), are body painting (Figure 2), barkcloth painting, wood carving, mural painting (Figure 3), and basket weaving. Figure 4 displays a detail of a Mangbetu basket, with a two-color rotational symmetry.

The weaving of decorated mats and baskets is indeed one of the spheres of

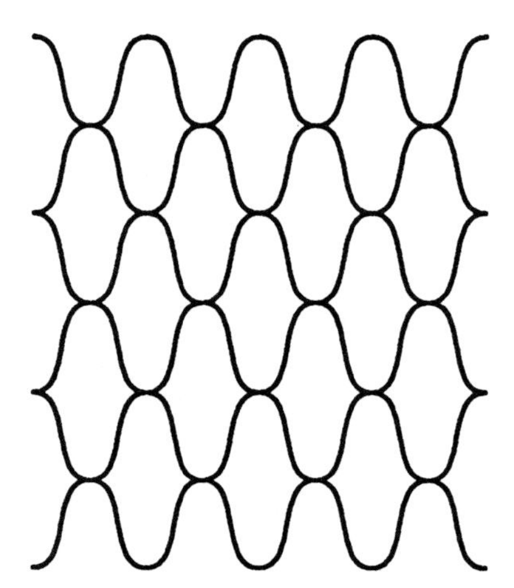

Figure 2 Mangbetu body painting. Reproduced from Gerdes, 1999a, Fig. 1.42.

Figure 3 Mangbetu mural painting. Reproduced from Gerdes, 1999a, Fig. 1.47.

Figure 4 Detail of a Mangbetu basket. Reproduced from Gerdes, 1999a, Fig. 1.62.

life in Central and Southern Africa that most stimulated geometrical considerations in association with arithmetical calculations. Figure 5 presents a detail of a beautifully plaited strip pattern that decorates part of the wall above the door of the house of a Bamileke chief in Cameroon. The Tonga women in Southern Mozambique, who weave *sipatsi* handbags diagonally (see Figure 6), have to perform several mental calculations before they can start their weaving. The total number of strips has to be an even multiple of each of the periods of the decorative motifs in order to guarantee that each decorative motif appears a whole number of times around the cylindrical wall of the handbag. The creativity of the Tonga weavers manifests itself also in the fact that they invented strip patterns of all the seven theoretically possible classes, as Figure 7 illustrates. Mat weavers in Tanzania and Uganda make mats by sewing together diagonally woven strips. Each strand used in this weaving process zigzags along the strip (see the example in Figure 8). By alternating groups of naturally

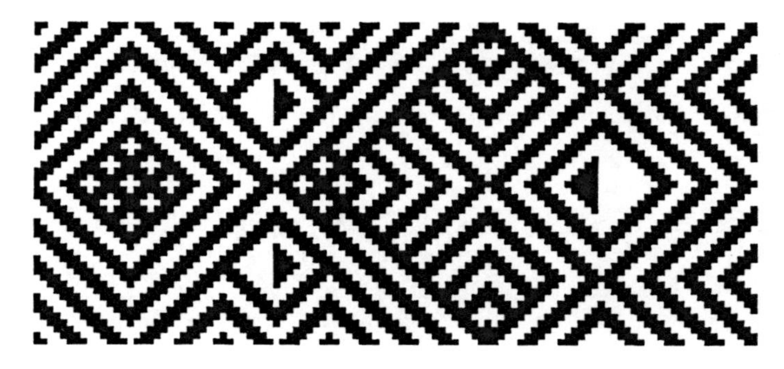

Figure 5 Bamileke mat design. Reproduced from Gerdes, 1999a, Fig. 1.53.

Figure 6 Example of a *sipatsi* handbag. Author's collection.

Figure 7 Tonga strip patterns. Reproduced from Gerdes & Bulafo, 1994, ch. 2.

Figure 8 Zigzagging strand. Reproduced from Gerdes, 1999a, Figure 3.150a.

colored strands (white) with colored strands (black) decorative motifs are introduced. By alternating two black groups of four strands each with two white groups of four and five strands respectively (see Figure 9a), an Ugandese strip pattern is obtained that is invariant under a glide reflection (Figure 9b). When black and white strands alternate, attractive strip patterns like the ones from the island of Zanzibar (Tanzania) represented in Figure 10 may be constructed (cf. Gerdes, 1999a, ch. 15). Decoration of coiled baskets, for instance

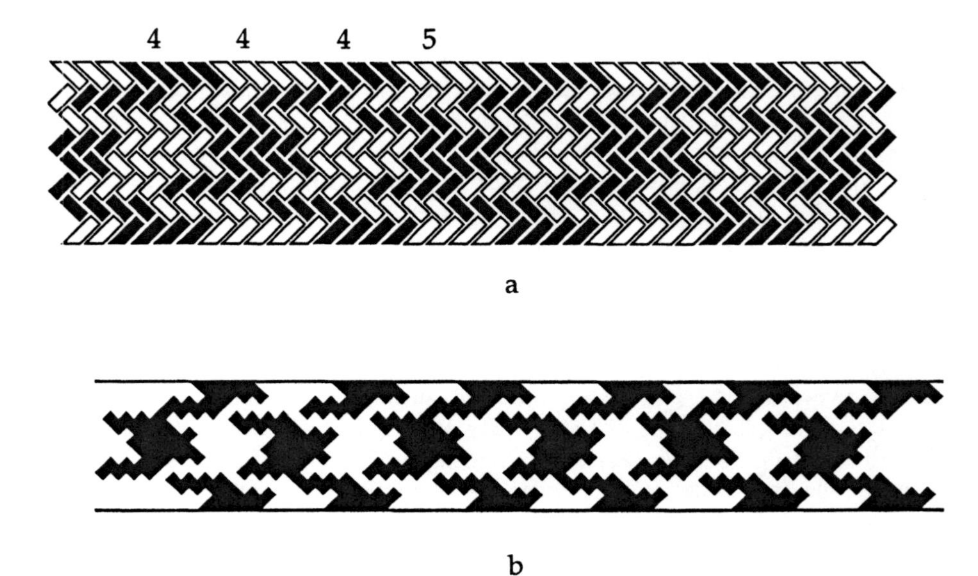

Figure 9 Strip pattern from Uganda. Reproduced from Gerdes, 1999a, Figure 3.152.

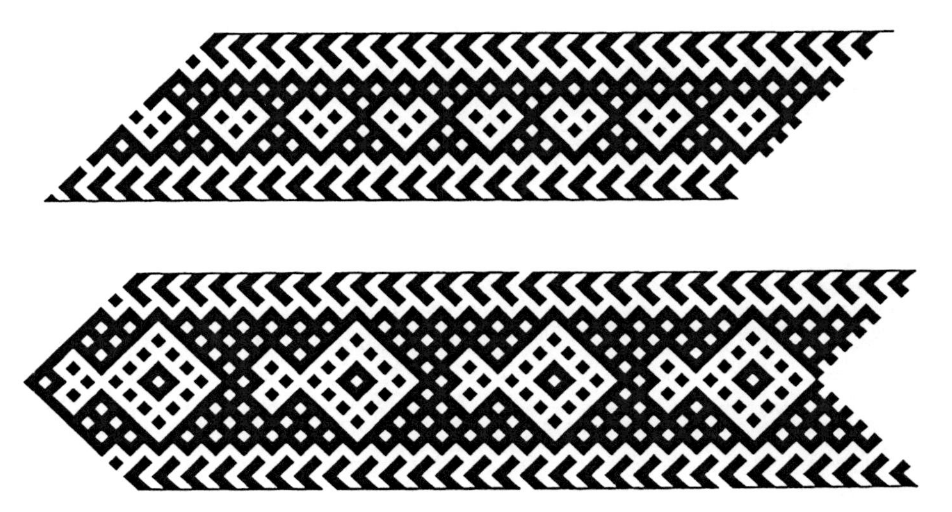

Figure 10 Strip patterns from Zanzibar. Reproduced from Gerdes, 1999a, Figure 3.154c, 3.155.

in Swaziland and Botswana, poses geometrical problems such as how (approximate) rotational symmetry may be introduced. Figure 11 presents a basket bowl (seen from above) with ten-fold symmetry and a small pendant with five-fold and axial symmetry produced by Swazi women.

Artisans of several peoples in Central and Southern Africa explore the basket weaving technique whereby the strands are woven one-over-one-under in three directions leading to a very stable fabric with hexagonal holes (Figure 12). Pygmies in the forests of Congo/Zaire and Cameroon use the technique to produce light transport baskets, as is done in northern Mozambique and Madagascar. In northeastern Congo/Zaire the technique is used among the Meje for covering pots and among the Mangbetu for weaving hats. The Makhuwa in northeastern Mozambique make fish traps with the hexagonal hole pattern (Gerdes, 1988, 1999a, 1999b). Today, the same hexagonal weaving technique may be explored in the study of geometrical models for fullerenes, large cage-like carbon molecules (Gerdes, 1998c, 1999c).

Examples of other cultural activities which bear a strong artistic and geometrical character are ceramics, beading, painting, mural decoration, hair braiding, tattooing, wood carving and architecture (cf. Gerdes, 1996, 1998b, 1998a; Denyer, 1978). In particular, various dihedral, strip and plane patterns are explored by the artisans (Gerdes, 1996a, 1998b). The books by Carey (1986) on beadwork, by Meurant (1986) and Washburn (1990) on Kuba textile design (Congo/Zaire) (cf. Crowe, 1975), and by Meurant and Thompson (1995) on Mbuti paintings (Congo/Zaire), present beautiful examples of geometrical decorations. In the next section, illustrative examples of geometrical exploration will be presented.

THE YOMBE *MABUINU* TRADITION

Female Yombe mat and basket weavers from the lower Congo area call their decorative motifs *mabuinu* (singular: *dibuinu*) (Mantuba-Ngoma, 1989a: 179). They have invented designs and patterns and experimented with combining and transforming them. A beautiful collection of Yombe mats (see the example in Figure 13), many from the end of the 19th century and the beginning of the 20th (cf. Coart, 1927) is kept at the Royal Museum for Central Africa (Africa-Museum, Tervuren, Belgium) in Tervuren (Belgium). Mantuba-Ngoma (1989a) studied the art and craft of the Yombe women in its cultural context, saying that it has still to be studied concerning its inherent mathematical ideas. He regrets that lately the quality of woven mats is degrading. In order to survive, weavers have to produce more mats more quickly, using strands of a width of about 10 mm instead of 3 mm as before, making it impossible to reproduce various intricate *mabuinu*. Many younger weavers have not learnt the meaning of the motifs (Mantuba-Ngoma, 1989a: 312). Let me explain some mathematical aspects of the weaving of *mabuinu*.

The mats are plaited with white strands in one direction (let us say horizontal) and black strands in the orthogonal direction (vertical). Yombe mat weavers know that the best way to give an extended patch of a mat a gray

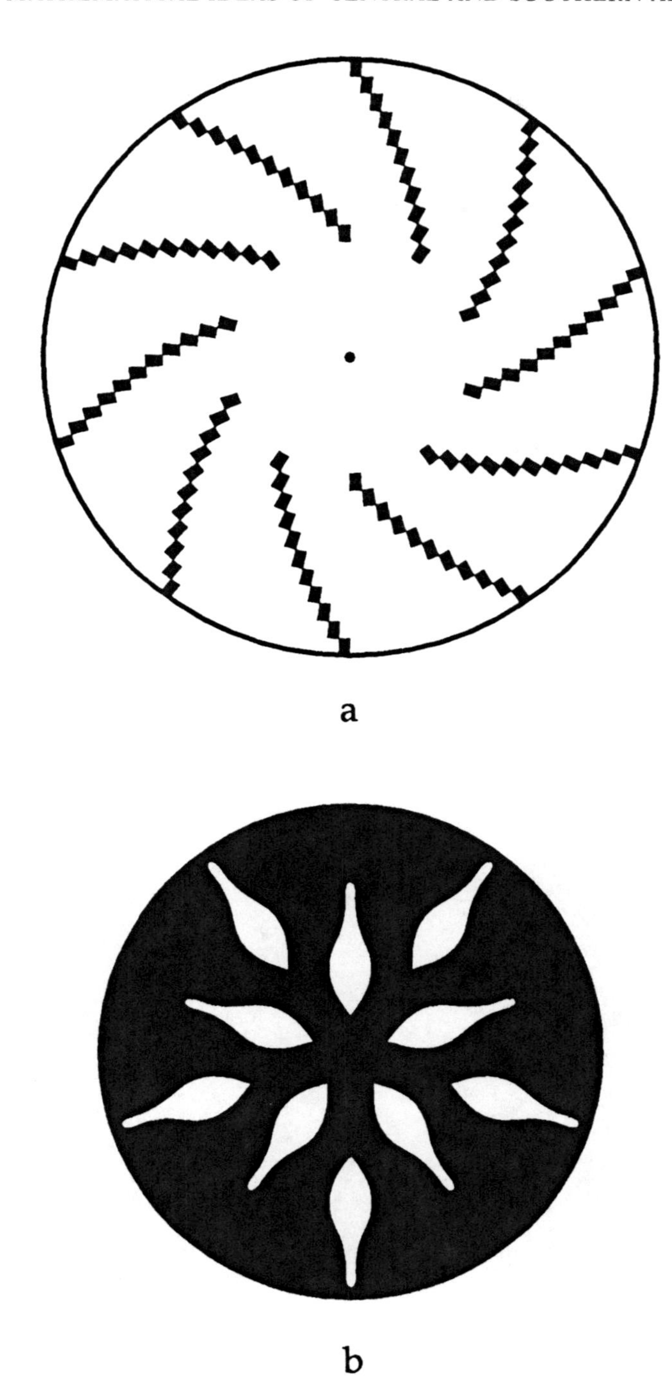

Figure 11 (a) Swazi basket bowl, seen from above. Reproduced from Gerdes, 1996, 1998b, Figure 2. (b) Swazi pendant. Author's collection.

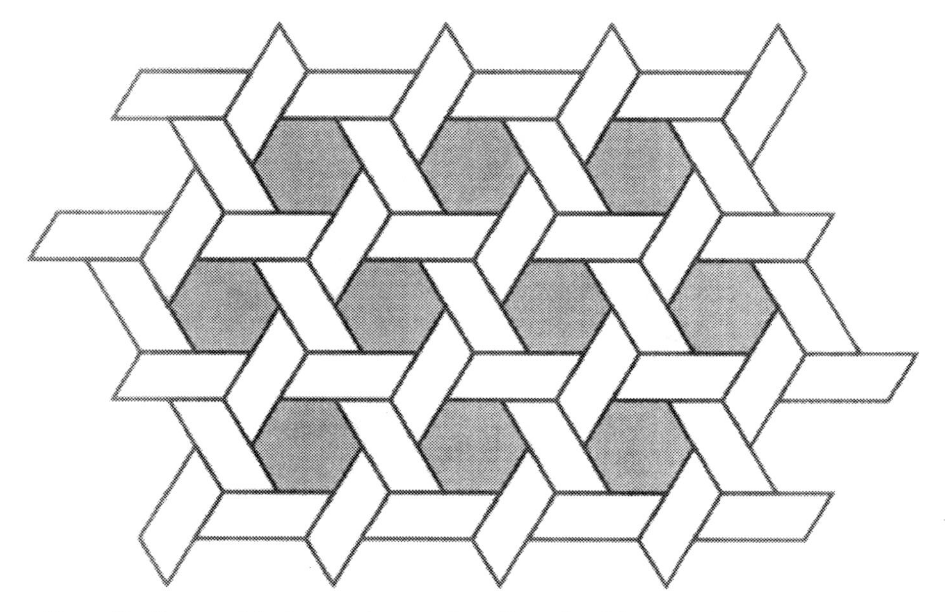

Figure 12 Hexagonal basket weaving technique. Reproduced from Gerdes, 1999a, Fig. 3.74.

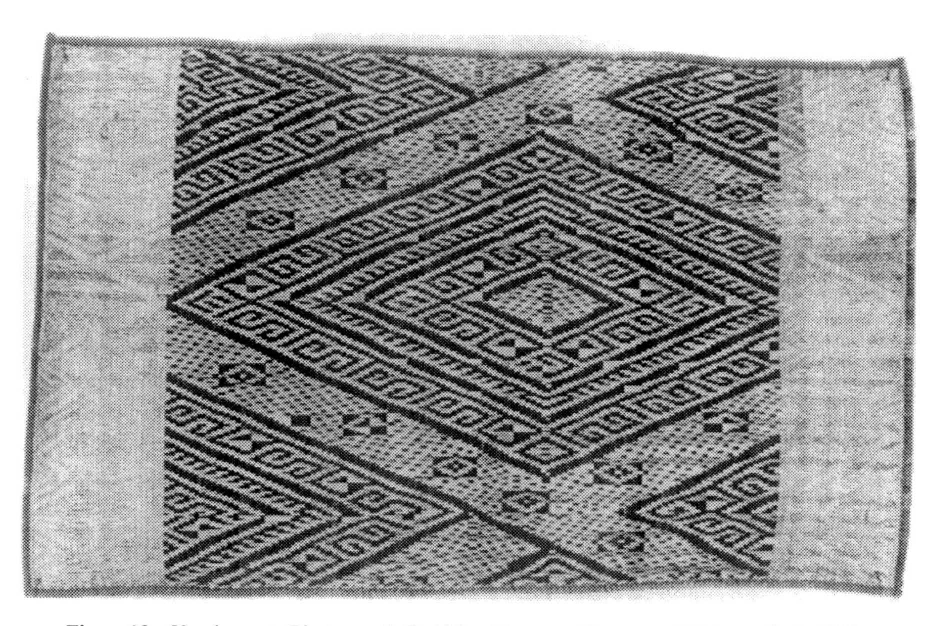

Figure 13 Yombe mat. Photograph © Africa-Museum, Tervuren, Belgium, cl. G. 4569.

image without a black strand crossing over too many white strands is to introduce white dots on the black patch. Each vertical black strand goes repeatedly over four white strands and then under one white strand, in such a way that from one dot to the nearest on the first column to its right, one goes two rows lower (see Figure 14). Associating this idea with that of the plaiting pattern in Figure 15, the Yombe women invented the pattern in Figure 16, whereby the toothed squares (see Figure 17) are transformed into toothed rhombi (see Figure 18). Figure 19 displays a variant that is smaller on the one side (8 dots instead of 12 dots along the toothed hypotenuse), but, on the other side, belongs also to a different symmetry class. In this, a half turn about the center of the fundamental rectangle of the first pattern (Figure 20) inverses the colors, whereas a half turn about the center of the fundamental rectangle of the second pattern (Figure 21) preserves color. In both cases, reflections about the diagonals of the rhombi inverse the colors. Figure 22 presents a Yombe pattern similar to the one in Figure 16, but this time with 5 dots along the hypotenuse. Its meaning is *mabaka*, a two-sided knife (Mantuba-Ngoma, 1989a: 210).

When the Yombe women wanted to produce patterns similar in a certain way to the previous ones, but where the diagonals of the rhombi do not invert the colors, they invented patterns like the one presented in Figure 23. Similar

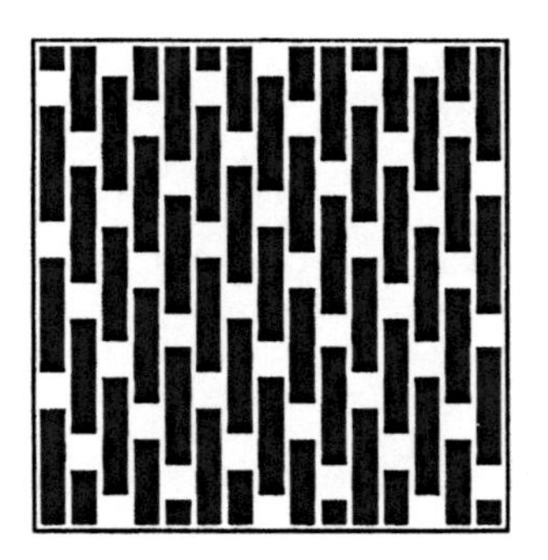

Figure 14 Yombe design of square dots. Drawn after Africa-Museum, Tervuren, Belgium, RG 80.12.1.16.

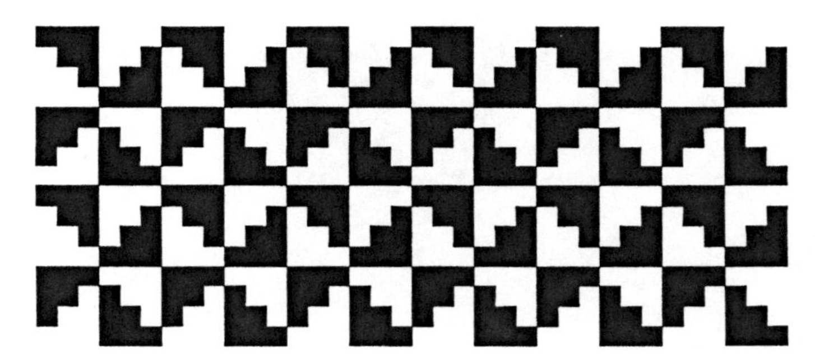

Figure 15 Yombe design. Drawn after Africa-Museum, Tervuren, Belgium, RG 80.12.1.46.

Figure 16 Yombe design. Drawn after Africa-Museum, Tervuren, Belgium, RG 80.12.1.19.

Figure 17 Toothed square. New drawing by the author.

Figure 18 Toothed rhombus. New drawing by the author.

Figure 19 Yombe design. Drawn after Africa-Museum, Tervuren, Belgium, RG 80.12.1.20.

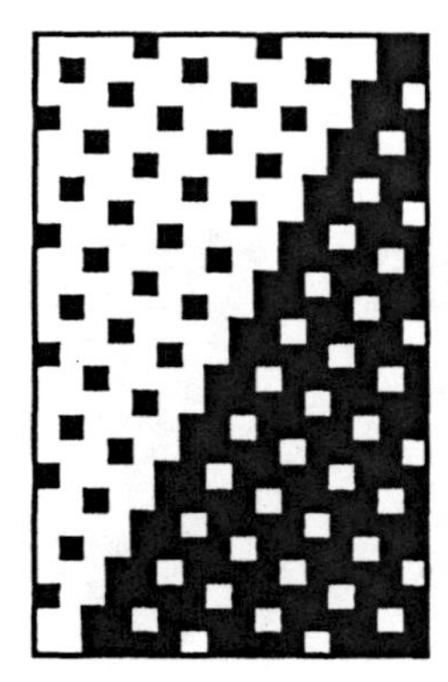

Figure 20 Fundamental rectangle of the design in Figure 16. New drawing by the author.

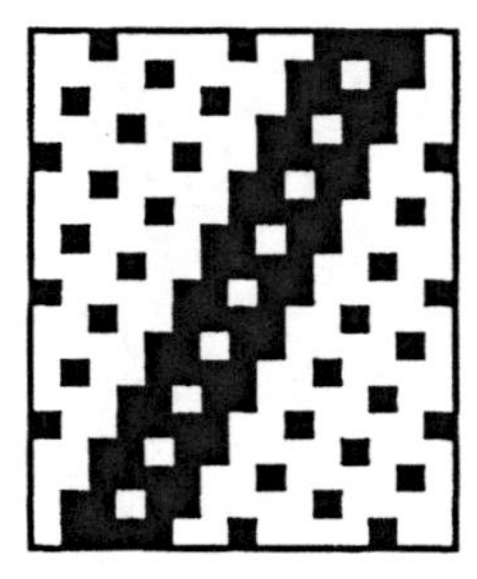

Figure 21 Fundamental rectangle of the design in Figure 19. New drawing by the author.

Figure 22 Yombe design. Drawn after Africa-Museum, Tervuren, Belgium, RG 80.25.1.37.

Figure 23 Yombe design. Drawn after Africa-Museum, Tervuren, Belgium, RG 80.12.1.22.

Figure 24 Yombe design. Drawn after Africa-Museum, Tervuren, Belgium, RG 80.12.1.5.

Figure 25 Yombe design. Drawn after Africa-Museum, Tervuren, Belgium, RG 80.12.1.71.

Figure 26 Yombe design. Dimensions 30 × 29. Reproduced from Gerdes, 1999a, Fig. 3.108b.

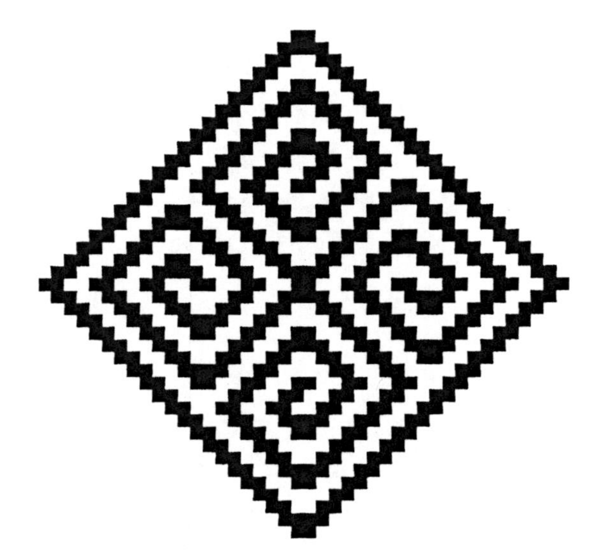

Figure 27 Yombe design. Dimensions 42 × 41. Reproduced from Gerdes, 1999a, Fig. 3.127.

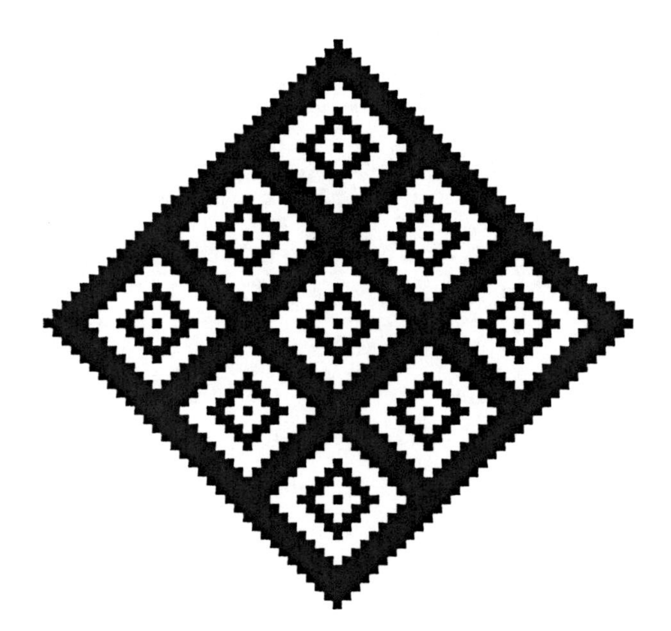

Figure 28 Yombe design. Dimension 65 × 65. Reproduced from Gerdes, 1999a, Figure 3.109e.

to the transformation of the pattern in Figure 19 into the one in Figure 18 is the transformation of the rectangular zigzag pattern in Figure 24 into the acutangular zigzag pattern in Figure 25. One of the composition rules the Yombe mat weavers discovered is that in order to cover their mats with a composition of designs using the basic over-two-under-two plaiting technique, the dimensions of their toothed square designs had to be consecutive odd and even numbers (see Figure 26). Although both dimensions of these woven squares have to be different, the weavers were capable of producing *mabuinu* which give the impression of having fourfold rotational symmetry. Figure 27 presents an example. Another solution the Yombe weavers found consists in having both two-over-two-under weaving (or more general, even-over-even-under) combined with three-over-three-under (or more general, odd-over-odd-under). Figure 28 presents an example with (exact) fourfold symmetry (cf. Gerdes, 1999a, ch. 3.9).

THE SOTHO *LITEMA* TRADITION

In Lesotho and neighboring zones of South Africa, Sotho women developed a tradition of decorating the walls of their houses with designs. The walls are first neatly plastered with a mixture of mud and dung, and often colored with natural dyes. While the last coat of plastered mud is still wet, the women engrave the walls, using their forefinger. Their art is seasonal: the sun dries it and cracks it, and the rain washes it away. An entire village is redecorated before special occasions such as engagement parties, weddings, and important religious celebrations.

The Sotho women call their geometric patterns *litema* (singular: *tema*). The books *The African Mural* by Changuion, Matthews and Changuion (1989) and *African Painted Houses: Basotho Dwellings of Southern Africa* by Van Wyk (1998) contain beautiful collections of photographs of *litema*. The National Teacher Training College of Lesotho published a collection of *litema* patterns collected by its mathematics students (Mothibe, 1976). In his presentation, the coordinator states that, 'Like other national traditions this one is in danger of dying out as more and more houses are built of concrete walls which are usually painted or white-washed. Also a growing number of women no longer like or know the art anymore' (Mothibe, 1976: 2).

Symmetry is a basic feature of the *litema* patterns. Figure 29 presents part of a *tema* pattern. As is often the case, this pattern is built up from a basic square that constitutes the (unit) cell of the pattern. Figure 30 displays the cell for the *tema* in Figure 29. The Sotho women lay out a network of squares and then they reproduce the basic design in each square. The number of reproductions or repetitions of the unit cells depends, in practice, on the available space on the wall to be decorated. As in Figure 29, a whole pattern is built up out of repetitions of a 2×2 square, in which the unit cell appears in four positions, obtained by horizontal and vertical reflection about the axes of the 2×2 square. The symmetries of a whole pattern depend on the symmetries of the unit cell. The unit cell in Figure 30 has two diagonal axes of symmetry. The unit cell of the *litema* pattern in Figure 31 has no axial symmetry; however it is invariant under a half turn. The unit cell of the *litema* pattern in Figure 32 has one axis of symmetry.

Several painted *litema*, and others in which changes in the relief of the dung surface of the plaster suggest two distinct colors, may be represented on paper

Figure 29 *Tema* pattern. Reproduced from Gerdes, 1996, 1998b, Fig. 9.13.

Figure 30 Unit cell of the *tema* pattern in Figure 29. Reproduced from Gerdes, 1996, 1998b, Fig. 9.13.

Figure 31 *Tema* pattern. Reproduced from Gerdes, 1996, 1998b, Fig. 9.8.

as black-and-white patterns. Some, as the one in Figure 33, are built up in the same way as the earlier patterns considered. Others are two-color patterns, where in each horizontal or vertical reflection of the unit cell the colors are reversed. The image of a unit cell is the negative (in photographic terms) of the reflected cell (see the example in Figure 34, leading to the *tema* pattern in

Figure 32 *Tema* pattern. Reproduced from Gerdes, 1996, 1998b, Fig. 9.16.

Figure 33 *Tema* pattern. Reproduced from Gerdes, 1996, 1998b, Fig. 9.36.

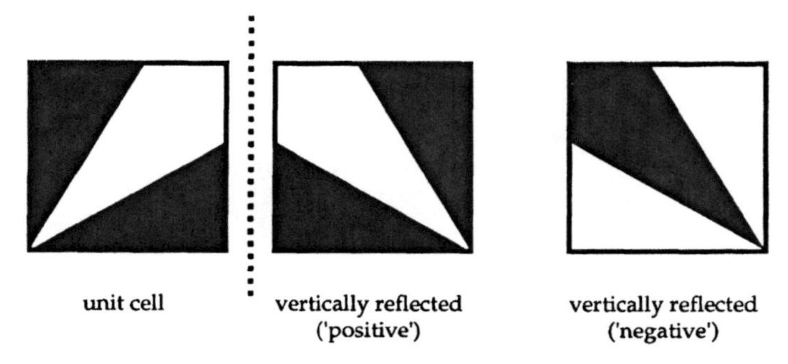

Figure 34 Building up of the *tema* pattern in Figure 35. Reproduced from Gerdes, 1996, 1998b, Fig. 9.38.

Figure 35 *Tema* pattern. Reproduced from Gerdes, 1996, 1998b, Fig. 9.38.

Figure 35). Figure 36 presents further examples. A different and colorful style of geometric wall decoration has been developed by Ndebele women in South Africa (cf. Courtney-Clarke, 1986; Powell and Lewis, 1995; Gerdes, 1996, 1998b, ch. 10).

THE CHOKWE *SONA* TRADITION

The *sona* tradition was developed among the Chokwe of Northeastern Angola and related neighboring peoples in Angola, Zambia and Congo/Zaire. The Chokwe culture is well known for its decorative art that ranges from ornamentation on woven mats and baskets, iron works, ceramics, sculpture and engravings on calabash to tattooing, paintings on the walls of houses, and *sona* sand drawings (singular: *lusona*). Boys learnt the meaning and execution of the easier *sona* during initiation rites. The more complicated *sona* were transmitted by drawing experts (*akwa kuta sona*) to one or some of their male descendants. These drawing experts were at the same time the story tellers who used the sona as illustrations made in the sand, referring to proverbs, fables, games, riddles, and animals. The drawings were executed in the following way: After cleaning and smoothing the ground, the drawing experts first set out with their fingertips a net of equidistant points and then they drew a line figure that embraced the points of the network. The experts executed the drawings swiftly. Once drawn, the designs were generally immediately wiped out. Figure 37 presents some examples of sona.

Colonial occupation provoked a cultural decline and the loss of knowledge about sona. On the basis of an analysis reported by missionaries, colonial administrators and ethnographers (e.g. Hamelberger, 1952; Dos Santos, 1961; Pearson, 1977; Fontinha, 1983; Kubik, 1988), I tried to contribute to the reconstruction of mathematical elements in the sona tradition. As the examples in Figure 37 suggest, symmetry and monolinearity played an important role as

Figure 36 *Litema* patterns. Reproduced from Gerdes, 1996, 1998b, Fig. 9.46, 9.49, 9.41.

cultural values: most Chokwe sona are symmetrical and/or monolinear. Monolinear means composed of only one (smooth) line; a part of the line may cross another part of the line, but a part of the line may never touch another part. Figure 38 illustrates the execution of a monolinear lusona.

The drawing experts developed a whole series of geometric algorithms for the construction of monolinear, symmetrical designs. Figure 39 displays two monolinear sona belonging to the same class in the sense that, although the dimensions of the underlying grids are different, both sona are drawn applying the same geometric algorithm.

The drawing experts also invented various rules for the building up of

Figure 37 Examples of symmetrical, monolinear sona. Reproduced from Gerdes, 1997b, pp. 39, 101, 52, 88.

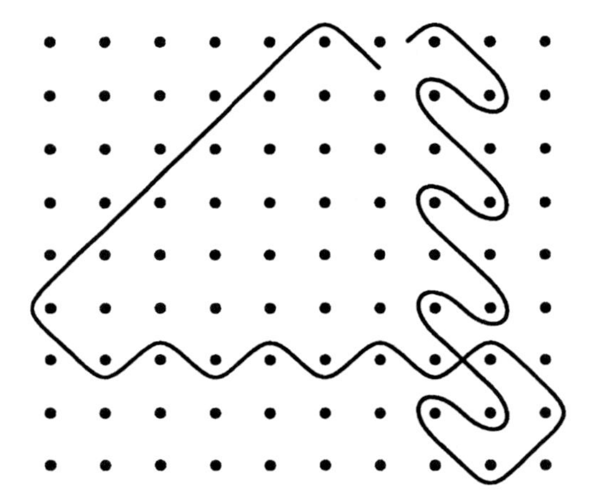

Figure 38 Execution of a monolinear lusona. Reproduced from Gerdes, 1997b, p. 77.

Figure 39 Two sona. Drawn with the same geometric algorithm. Reproduced from Gerdes, 1997b, pp. 77, 76.

monolinear sona. Let me present an example (Figure 40). The three sona are similar to each other: each is characterized by a reference point design of triangular form. I suppose that the drawing experts who invented these probably began with triangular line designs and transformed them into monolinear designs with the help of one or more loops (see the example in Figure 41). The monolinear drawings so obtained were adapted topologically so that they could express the ideas the drawers wanted to transmit by means of them.

The sona experts also discovered various rules for chaining monolinear sona to form bigger monolinear ones. Figure 42 illustrates an example of a chain

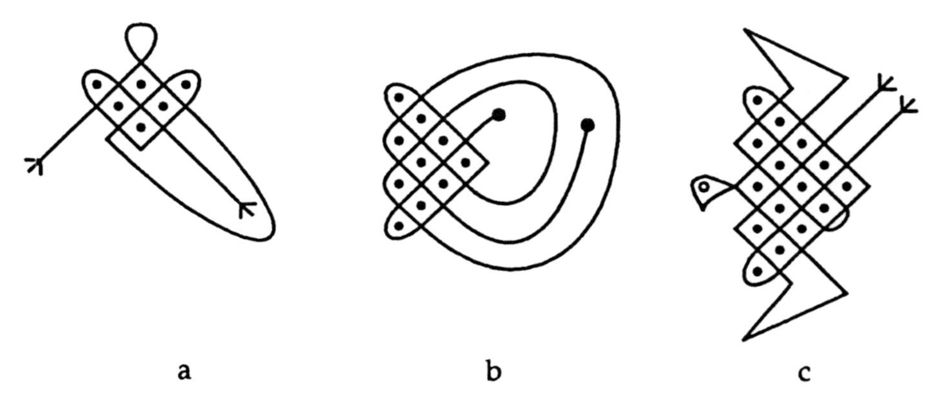

Figure 40 (a) An eagle carrying a chicken; (b) a person dead from fatigue; (c) a *thimunga* bird in flight. Reproduced from Gerdes, 1997b, pp. 114, 115, 115.

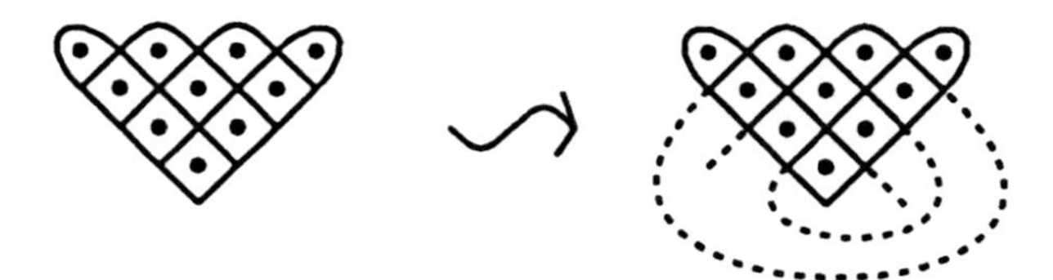

Figure 41 Transformation of a triangular design into a monolinear design. Reproduced from Gerdes, 1997b, p. 117.

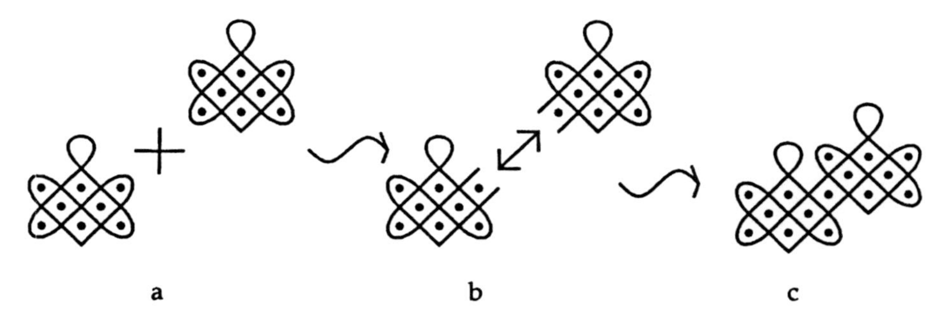

Figure 42 (a, b, c) Example of the application of a chain rule. Reproduced from Gerdes, 1997b, p. 138.

rule. It indicates how the appearance of the monolinear drawing (Figure 42c) may be explained on the basis of the monolinearity of the two drawings (Figure 42a) and the way they have been chained together (see Figure 42b). The monolinearity of the lusona in Figure 43a may be explained on the basis of another chain rule: When one joins square grids (in the example: two grids of dimensions 2 × 2) to a rectangle with dimensions which are relative prime (in the example: 3 × 5), then the resulting grid leads to a monolinear drawing if one applies the same 'plaited mat' algorithm.

When analyzing and reconstructing elements of the sona tradition, I found that there are many reported sona which clearly do not conform to the cultural values of symmetry and monolinearity. Sometimes the symmetry or monolinearity was broken in order to give the drawing a specific meaning. Figure 44 gives an example of a monolinear lusona, where the symmetry of the grid is broken by the drawing. More often we seem to be dealing with mistakes or errors (Figure 45 gives an example). The drawing experts may have committed some of these mistakes, because they were contacted when they were already old men. They said that as young men they had been much better *akwa kuta sona*. We may be dealing with errors in the transmission of knowledge from one generation to the next or with mistakes on the part of the reporter, who had little time to make his copies, as traditionally the drawing experts wipe their drawings out immediately after concluding their story. The wiping out was one way to maintain the monopoly of sona knowledge. Here we have another probable reason for mistakes in the reporting; the experts may have made conscious mistakes to deceive the reporter and so protect their knowledge. This secret character of the sona tradition provides one reason for its gradual extinction. For instance, as soon as a drawing expert was enslaved, the knowledge might disappear from his community.

The analysis of mathematical ideas incorporated in the sona tradition may stimulate the reflection on traditions – both from other parts of Africa (Ghana, Nigeria, or Ethiopia) and other regions (see the examples in Figure 46) – that from a mathematical-technical point of view display certain similarities with the sona tradition. Some examples are geometrical algorithms in ancient Egypt

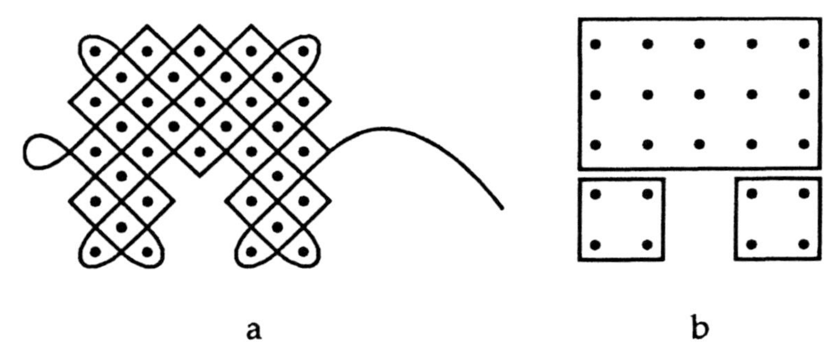

Figure 43 (a) Representation of a monkey-dog; (b) grid structure. Reproduced from Gerdes, 1997b, p. 147.

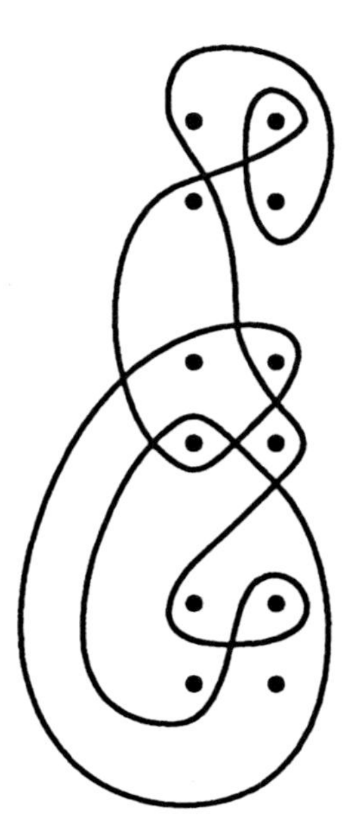

Figure 44 Lusona representing *kuku* or *kalamba*, the thinker. Reproduced from Gerdes, 1997b, p. 45.

used for the decoration of scarabs (Gerdes, 1995a: vol. 3, 1997b), designs from ancient Mesopotamia, threshold designs drawn by Tamil women in India (Gerdes, 1989, 1995b), Celtic knot designs (Cromwell, 1993; Gerdes, 1999d), sand drawings from the Vanuatu islands (Oceania) (Struik, 1948; Ascher, 1988; Gerdes, 1997b, 1995a: vol. 3), and Hausa embroidery designs (Nigeria). Knowledge of sona or of sona-like drawings may have survived in one way or another in the New World (see the example in Figure 46g, cf. Fauvel and Gerdes, 1990).

MATHEMATICAL THINKING AND ETHNOMATHEMATICS

Are the examples given an expression of mathematical thinking? M. Locke (1918–1994), writing on the weaving of basketry patterns among the Matabele women in Zimbabwe, would answer no: 'The tribal craftswomen generally had no knowledge of mathematics. They were able to count, but could not divide. They judged the layout of patterns by eye and, as far as can be ascertained, used no form of measuring. Their lack of mathematical knowledge, however, was no hindrance to a skillful distribution of design elements around the bodies of the baskets' (Locke, 1994: 15). For Locke mathematics seems to be only arithmetic and measurement. If one reads Dos Santos (1960) on the mathematical knowledge of the Chokwe, his answer once more would be that the Chokwe knew some arithmetic, some time reckoning and some geometrical vocabulary (line, curve, point ...), but they did not know mathematics. The same author published an interesting study on the sona sand drawings, but he did not see

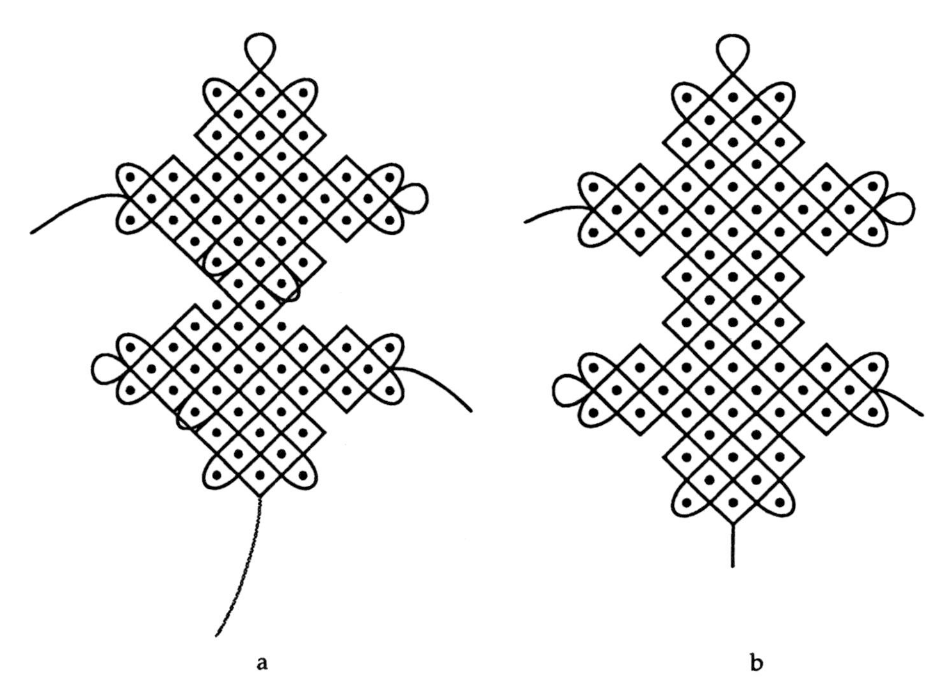

Figure 45 (a) Reported *lusona*, representing a lioness with her two cubs. The drawing is neither symmetrical nor monolinear. (b) Reconstructed symmetrical and monolinear line drawing (the tails are added at the end). Reproduced from Gerdes, 1997b, pp. 162, 163.

that they had any relationship with mathematics. In both cases, the 'no' answers reflect the mathematics education of the authors. But does this give a real picture of what mathematics is all about? What is mathematical thinking?

Ethnomathematics (or ethnomathematicology) is the field of inquiry that studies mathematical ideas in their historical-cultural contexts. For instance, an ethnomathematician is interested in understanding the role and embodiment of mathematical thinking both in the invention and (re)production of decorative patterns. If no direct dialogue, physically or historically, with the inventors and/or (re)producers is possible, the researcher tries to reconstruct elements of mathematical thinking probably involved in the invention and reproduction processes. These reconstructed elements may be called 'frozen' or 'hidden' mathematics (cf. Gerdes, 1990, 1999b). The reconstruction is not completely impossible as the researcher may have developed some feeling for mathematical ideas, just like any musician (or even any human being) may develop a certain understanding of and feeling for musical expressions, and any linguist (or even any human being) may develop a certain understanding and recognition of language phenomena. In this sense, mathematical thinking is as panhuman as using a language or involvement in music (playing, listening, etc.). Mathematics becomes the product of all cultures, the school mathematics experience of a researcher being only one form of mathematical experience. Mathematics is not the product of a particular culture sphere but a common human experience.

Figure 46 Examples of symmetrical and monolinear designs in various cultures. Reproduced from Gerdes, 1997b, pp. 331, 343, 369, 387, 389; Fig. 46g is drawn after information given to the author by Susan Enger (1997). (a) Ancient Egyptian scarab design; (b) Ancient Mesopotamian cylinder design; (c) Tamil *kolam* design; (d) Celtic knot design; (e) Sand drawing design from Vanuatu; (f) Hausa decorative motif; (g) a lusona type drawing among people of African descent in the Mississippi Delta (USA).

In the process of studying mathematical ideas in diverse cultural contexts, the understanding of what mathematics is, or better of what constitutes mathematical activity, may be deepened. The concept of mathematics becomes broadened and deepened (cf. Ubiratan D'Ambrosio, 'A Historiographical Proposal for Non-Western Mathematics' and Ron Eglash, 'Anthropological Perspectives on Ethnomathematics' in this volume), and other ethnocentric views may be gradually surpassed (cf. Bishop, 1988, 1990; Powell and Frankenstein, 1997; Ascher, 1991). Mathematical thinking is only interculturally intelligible.

* * *

Although mathematical ideas in the history of Central and Southern Africa were not written down in articles and books, they were registered in other graphical systems (cf. Kubik, 1986; Mantuba-Ngoma, 1989a), such as sona sand drawings, intricate plaited patterns, and wall painting and beadwork designs. As is so often the case, the mathematical ideas embedded in them are expressed through formulas. Their meaning can only be understood in their context. Frequently, women played a leading role in Central and Southern Africa in inventing and developing mathematical ideas and expressions.

I hope this paper gave the reader some feeling for mathematical ideas involved in cultural activities in the history of Central and Southern Africa, where the connection between geometrical ideas, optimization, symmetry and beauty is very strong.

ACKNOWLEDGMENTS

I would like to thank Arthur B. Powell (Rutgers University, Newark, New Jersey, USA) for his comments on the first draft of this chapter, and the Research Department (SAREC) of the Swedish International Development Agency for supporting financially the Mozambican Ethnomathematics Research Project and History of Mathematics in Africa Project.

BIBLIOGRAPHY

Ascher, Marcia. 'Graphs in cultures: a study in ethnomathematics.' *Historia Mathematica* 15: 201–227, 1988a.

Ascher, Marcia. 'Graphs in cultures (II): a study in ethnomathematics.' *Archive for History of Exact Sciences* 39(1): 75–95, 1988b.

Ascher, Marcia. 'A river-crossing problem in cross-cultural perspective.' *Mathematics Magazine* 63(1): 26–29, 1990.

Ascher, Marcia. *Ethnomathematics, a Multicultural View of Mathematical Ideas*. Pacific Grove, California: Brooks/Cole, 1991.

Ascher, Marcia. 'Malagasy *sikidy*: a case in ethnomathematics.' *Historia Mathematica* 24: 376–395, 1997.

Bishop, Alan. *Mathematical Enculturation*. Dordrecht: Kluwer, 1988.

Bishop, Alan. 'Western mathematics: the secret weapon of cultural imperialism.' *Race and Class* 32(2): 51–65, 1990.

Bogoshi, Jonas, Kevin Naidoo and John Webb. 'The oldest mathematical artifact.' *The Mathematical Gazette* 71: 294, 1987.

Careccio, John. 'Mathematical heritage of Zambia.' *The Arithmetic Teacher*, 1970, pp. 391–395.

Carey, Margret. *Beads and Beadwork of East and Southern Africa*. Aylesbury, Bucks.: Shire, 1986.

Changuion, P., T. Matthews and A. Changuion. *The African Mural*. Cape Town: Struik, 1989.

Courtney-Clarke, Margaret. *Ndebele: The Art of an African Tribe*. New York: Rizzoli, 1986.

Cromwell, Peter. 'Celtic knotwork: mathematical art.' *The Mathematical Intelligencer* 15(1): 36–47, 1993.

Crowe, Donald. 'The geometry of African art I. Bakuba art.' *Journal of Geometry* 1: 169–182, 1971.

Cunnington, W. 'String figures and tricks from Central Africa.' *Journal of the Royal Anthropological Institute* 36: 121–131, 1906.

D'Ambrosio, Ubiratan. *Ethnomathematics*. Albuquerque, New Mexico: ISGEm, 1997.

Denyer, Susan *African Traditional Architecture*. London: Heinemann, 1978.

Dowson, T. *Rock Engravings of Southern Africa*. Johannesburg: University of Witwatersrand Press, 1992.

Eglash, Ron. 'Geometric algorithms in Mangbetu design.' *Mathematics Teacher* 91(5): 376–381, 1998.

Eglash, Ron. *African Fractals: Modern Computing and Indigenous Design*. New York: Rutgers University Press, 1999.

Fauvel, John and Paulus Gerdes. 'African slave and calculating prodigy: bicentenary of the death of Thomas Fuller.' *Historia Mathematica* 17(2): 141–151, 1990.

Fontinha, Mário. *Desenhos na areia dos Quiocos do Nordeste de Angola*. Lisbon: Instituto de Investigação Científica Tropical, 1983.

Gerdes, Paulus. 'Three alternate methods of obtaining the ancient Egyptian formula for the area of a circle.' *Historia Mathematica* 12: 261–268, 1985.

Gerdes, Paulus. 'On culture, geometrical thinking and mathematics education.' *Educational Studies in Mathematics* 19(3): 137–162, 1988.

Gerdes, Paulus. 'Reconstruction and extension of lost symmetries: examples from the Tamil of South India.' *Computers and Mathematics with Applications* 17(4–6): 791–813, 1989.

Gerdes, Paulus. *Ethnogeometrie. Kulturanthropologische Beiträge zur Genese und Didaktik der Geometrie*. Bad Salzdetfurth: Verlag Franzbecker, 1990.

Gerdes, Paulus. 'On mathematical elements in the Tchokwe drawing tradition.' *Discovery and Innovation: Journal of the African Academy of Sciences* 3(1): 29–36, 1991a.

Gerdes, Paulus. 'On mathematical elements in the Tchokwe *sona* tradition.' *Afrika Mathematika, Journal of the African Mathematical Union* Series 2, 3: 119–130, 1991b.

Gerdes, Paulus, ed. *A numeração em Moçambique*. Maputo: Ethnomathematics Research Project, 1993.

Gerdes, Paulus, ed. *Explorations in Ethnomathematics and Ethnoscience in Mozambique*. Maputo: Ethnomathematics Research Project, 1994a.

Gerdes, Paulus. 'On mathematics in the history of sub-Saharan Africa.' *Historia Mathematica* 21: 345–376, 1994b.

Gerdes, Paulus. *Sona Geometry: Reflections on the Tradition of Sand Drawings in Africa South of the Equator*. Maputo: Ethnomathematics Research Project, 1994c.

Gerdes, Paulus. *Une tradition géométrique en Afrique – Les dessins sur le sable*. Paris: L'Harmattan, 1995a (3 volumes).

Gerdes, Paulus. 'Extension(s) of a reconstructed Tamil ring-pattern.' In *The Pattern Book: Fractals, Art, and Nature*, C. Pickover, ed. Singapore: World Scientific, 1995b, pp. 377–379.

Gerdes, Paulus. *Femmes et Géométrie en Afrique Australe*. Paris: L'Harmattan, 1996.

Gerdes, Paulus. *Recréations Géométriques d'Afrique – Lusona – Geometrical Recreations of Africa*. Paris: L'Harmattan, Paris, [1991] 1997a.

Gerdes, Paulus. *Ethnomathematik dargestellt am Beispiel der Sona Geometrie*. Berlin, Heidelberg, Oxford: Spektrum Verlag, 1997b.

Gerdes, Paulus. 'On some geometrical and architectural ideas from African art and craft.' In *Nexus II: Architecture and Mathematics*, Kim Williams, ed. Fucecchio: Edizioni dell'Erba, 1998, pp. 75–86.

Gerdes, Paulus. *Women, Art, and Geometry in Southern Africa*. Lawrenceville, Asmara: Africa World Press, [1995] 1998b.

Gerdes, Paulus. 'Molecular modeling of fullerenes with hexastrips.' *The Chemical Intelligencer* 4(1): 40–45, 1998c.

Gerdes, Paulus. *Geometry from Africa: Mathematical and Educational Explorations.* Washington, DC: Mathematical Association of America, 1999a [Preface by Arthur B. Powell].

Gerdes, Paulus. *Culture and the Awakening of Geometrical Thinking. Anthropological, Historical, and Philosophical Considerations. An Ethnomathematical Study.* Minneapolis: MEP Press, 1999b [Preface by Dirk Struik].

Gerdes, Paulus. 'Molecular modeling of fullerenes with hexastrips.' *The Mathematical Intelligencer* 21(1): 6–12, 27, 1999c.

Gerdes, Paulus. 'On the geometry of Celtic knots and their Lunda-designs.' *Mathematics in School* 28(3): 29–33, 1999d.

Gerdes, Paulus, and Gildo Bulafo. *Sipatsi: Technology, Art and Geometry in Inhambane.* Maputo: Ethnomathematics Research Project, 1994.

Haddon, A. 'String figures from South Africa.' *Journal of the Royal Anthropological Institute* 36: 142–149, 1906.

Hamelberger, E. 'A escrita na areia.' *Portugal em Africa* 53: 323–330, 1952.

Hauenstein, Alfred. *Examen de motifs décoratifs chez les Ovimbundu et Tchokwe d'Angola.* Coimbra: Instituto de Antropologia da Universidade de Coimbra, 1988.

Heinzelin, J. de. 'Ishango.' *Scientific American* 206: 109–111, 1962.

Hogbe-Nlend, Henri. *L'Afrique, berceau de la mathématique mondiale?* Nairobi: United Nations University, 1985 (mimeo).

Horton, Robin. *Patterns of Thought in Africa and the West. Essays on Magic, Religion and Science.* Cambridge: Cambridge University Press, 1993.

Huylebrouck, Dirk. 'Puzzles, patterns, drums: the dawn of mathematics in Rwanda and Burundi.' *Humanistic Mathematics Network Journal* 14: 9–22, 1996.

Ismael, Abdulcarimo. 'On the origin of the concepts of even and odd in Makhuwa culture.' In *Explorations in Ethnomathematics and Ethnoscience in Mozambique*, Paulus Gerdes, ed. Maputo: Ethnomathematics Research Project, 1994, pp. 9–15.

Isoun, T. 'Mathematics and Africa.' *Discovery and Innovation: Journal of the African Academy of Sciences* 4(1): 4–6, 1992.

Kubik, Gerhard. 'African graphic systems.' *Muntu, Revue Scientifique et Culturelle du Centre International des Civilisations Bantu* 4–5: 71–135, 1986.

Kubik, Gerhard. *Tusona-Luchazi Ideographs, a Graphic Tradition as Practiced by a People of West-Central Africa.* Fohrenau: Verlag Stiglmayr, 1988.

Kubik, Gerhard. 'Visimu vya mukatikati – dilemma tales and 'arithmetical puzzles' collected among the Valuchazi.' *South African Journal of African Languages* 10(2): 59–68, 1990.

Lea, Hilda. 'Spatial concepts in the Kalahari.' *Proceedings of the 14th International Conference on Psychology of Mathematics Education* 2, 1990, pp. 259–266.

Leakey, M. and L. Leakey. *Some String Figures from North East Angola.* Lisbon: Museu do Dondo, 1949.

Levinsohn, Rhoda. *Art and Craft of Southern Africa.* Craighall: Delta Books, 1984.

Liebenberg, Louis. *The Art of Tracking: The Origin of Science.* Claremont: David Philip, 1990.

Lindblom, K. *String Figures in Africa.* Stockholm: Riksmuseets Etnografiska Avdelning, 1930.

Locke, Marjorie. *The Dove's Footprints. Basketry Patterns in Matabeleland.* Harare: Baobab, 1994.

Mantuba-Ngoma, Mabiala. *Frauen, Kunsthandwerk und Kultur bei den Yombe in Zaire.* Göttingen: Edition Re, 1989.

Mantuba-Ngoma, Mabiala. *Losa: Flechtwerke der Mbole.* München: Verlag Fred Jahn, 1989b.

Marshack, A. *The Roots of Civilization.* Mount Kisco, New York: Moyer Bell, 1972, reprinted 1991.

Meurant, Georges. *Shoowa Design: African Textiles from the Kingdom of Kuba.* London: Thames and Hudson, 1986.

Meurant, Georges and Robert Thompson. *Mbuti Design: Paintings by Pygmy Women of the Ituri Forest.* London: Thames and Hudson, 1995.

Mfika, Mubumbila. *Sur le sentier mystérieux des nombres noirs.* Paris: L'Harmattan, 1988.

Mizone, M. 'Les jeux stratégiques camerounais et leurs aspects mathématiques.' *Annales de la Faculté des Sciences du Cameroun* 6: 19–38, 1971.

Mothibe, B., ed. *Litema, Designs Collected by Students of the National Teacher Training College of Lesotho.* Maseru: NTTC, 1976.

Mve-Ondo, Bonaventure. *L'Owani et le Songa: Deux jeux de calculs africains. Découverts du Gabon.* Libbreville: Centre Culturel Français Saint-Exupéry, 1990.

Njock, Georges E. 'Mathématiques et environnement socio-culturel en Afrique Noire.' *Presence Africaine* 135(3): 3–21, 1985.

Nsimbi, M. B. *Omweso: a Game People Play in Uganda.* Los Angeles: African Studies Center, 1968.

Obenga, Théophile. 'Système opératoire négro-africain.' In *L'Afrique dans l'Antiquité. Egypte pharaonique-Afrique noire.* Paris: Présence Africaine, 1973.

Obenga, Théophile. *La Géométrie Égyptienne – Contribution de l'Afrique antique à la Mathématique mondiale.* Paris: L'Harmattan, 1995.

Pearson, Emil. *People of the Aurora.* San Diego: Beta Books, 1977.

Powell, Arthur B. and Marilyn Frankenstein, eds. *Ethnomathematics: Challenging Eurocentrism in Mathematics Education.* Albany: State University of New York Press, 1997.

Powell, Ivor and Mark Lewis. *Ndebele: A People and their Art.* Cape Town: Struik, 1995.

Raum, Otto. *Arithmetic in Africa.* London: Evans Brothers Ltd., 1938.

Russ, Lawrence. *Mancala Games.* Algonac, Michigan: Reference Publications, 1984.

Santos, Eduardo dos. 'Sobre a matemática dos Ouiocos de Angola.' *Garcia da Orta* 3(2): 257–271, 1960.

Santos, Eduardo dos. 'Contribuição para o estudo das pictografias e ideogramas dos Quiocos.' *Estudos sobre a etnologia do ultramar português* 2: 17–131, 1961.

Schmidl, Marianne. 'Zahl und Zählen in Afrika.' *Mitteilungen der Anthropologischen Gesellschaft in Wien* 45: 165–209, 1915.

Seidenberg, A. 'On the eastern Bantu root for six.' *African Studies* 18(1): 28–34, 1959.

Seidenberg, A. 'km, a widespread root for ten.' *Archive for History of Exact Sciences* 16(1): 1–16, 1976.

Silva, Elísio Santos. *Jogos de quadrícula do tipo mancala com especial incidência nos praticados em Angola.* Lisbon: Instituto de Investigação Científica Tropical, 1995.

Ssembatya, Vincent and Andrew Vince. 'Mathematics in Uganda.' *The Mathematical Intelligencer* 19(3): 27–32, 1997.

Stott, Leda and Hilda Lea. *Common Threads in Botswana.* Gaborone: British Council, 1993.

Struik, Dirk. 'Stone age mathematics.' *Scientific American* 179: 44–49, 1948.

Townshend, Philip. *Les jeux de mankala au Zaire, au Rwanda et au Burundi.* Brussels: Cahiers du CEDAF, 1977.

Townshend, Philip. 'African mankala in anthropological perspective.' *Current Anthropology* 20: 794–796, 1979a.

Townshend, Philip. 'Mankala in Eastern and Southern Africa: a distributional analysis.' *Azania* 14: 108–138, 1979b.

Voogt, Alex de. 'Seeded players.' *Natural History* 107(1): 18–22, 1998.

Washburn, Dorothy. *Style, Classification and Ethnicity: Design Categories on Bakuba Raffia Cloth.* Philadelphia: American Philosophical Society, 1990.

Willcox, A. R. *The Rock Art of Africa.* New York: Holmes & Meier, 1984.

Wyk, Gary van. *African Painted Houses. Basotho Dwellings of Southern Africa.* New York: Harry Abrams, 1998.

Zaslavsky, Claudia. *Africa Counts: Number and Pattern in African Culture.* Brooklyn: Lawrence Hill Books, 1973, reprinted 1979, 1999.

Zaslavsky, Claudia. 'Women as the first mathematicians.' *Women in Mathematics Education Newsletter* 14(1): 4, 1991.

Zaslavsky, Claudia. 'Mathematics in Africa: explicit and implicit.' In *Companion Encyclopedia of the History and Philosophy of the Mathematical Sciences*, I. Grattan-Guinness, ed. London: Routledge, 1994a, pp. 85–92.

Zaslavsky, Claudia. '*Africa Counts* and ethnomathematics.' *For the Learning of Mathematics* 14(2): 3–8, 1994b.

HELEN VERRAN

ACCOUNTING MATHEMATICS IN WEST AFRICA: SOME STORIES OF YORUBA NUMBER

The meeting of the Royal Anthropological Institute of Great Britain and Ireland, held in London on 9th March 1886, passed a motion that the paper, 'Notes on the Numeral System of the Yoruba Nation', prepared by Adolphus Mann Esq., be taken as read. The published report of that meeting, which includes the text of Mann's paper, does not enable us to be entirely sure of how Adolphus Mann Esq. came to make his notes, but there are indications that he had made personal observations of people actually using Yoruba numerals in the course of their trading. Adolphus Mann is an enthusiastic presenter, giving every sign of taking great delight in Yoruba numbers. Advising his readers that

> A superficial knowledge, with a slight attempt at praxis, suffices to understand the peculiarities in the arrangement of these numerals ... [and with this] we light, as it were on a building, which, when viewed from base to summit is not behind our European systems in regularity and symmetry, while the system surpasses them in the aptitude of interlinking the separate members; it stands to them in the same relation as the profusely ornamented Moorish style stands to the more sober Byzantine (Mann, 1887: 60).

Mann speculates that this wonderful system has its origins in the ways cowrie shell currency was counted in Yoruba trading.

> When a bagful [of cowries] is cast on the floor, the counting person sits or kneels down beside it, takes 5 and 5 cowries and counts silently, 1, 2, up to 20, thus 100 are counted off, this is repeated to get a second 100, these little heaps each of 100 cowries are united, and a next 200 is, when counted swept together with the first. Such sums as originate from counting are a sort of standard money, 20, 100 and then especially 200, and 400 is 4 little heaps of 100 cowries, or 2 each of 200 cowries, representing to the Yorubas the denominations of the monetary values of their country as to us 1/2d., 1d., 3d., 6., 1s., &c. (Mann, 1887: 62).

A little over one hundred years later I too developed an account of Yoruba numbers. My interest was sparked by puzzles arising in my work as a teacher educator in mathematics and science at a Nigerian university. In studying the system of Yoruba numbers I was looking for understandings that would enable me to do things differently in my work with Yoruba teachers. I wanted to find

H. Selin (ed.), Mathematics Across Cultures: The History of Non-Western Mathematics, 345–372.

ways to help my students deal with classroom tensions which I intuited as arising in differences between Yoruba and English numbers.

In developing my account of Yoruba numbers I used details available from Mann and those who came after him (Johnson, 1921; Abraham, 1962; Armstrong, 1962), and I developed further material by consulting Yoruba language grammars and informants. The text of my account put together some 15 years ago (see Watson, 1986) forms the first part of this chapter. In the second section I consider this text, asking what it actually tells us about Yoruba and English numbers. I conclude that although pointing to interesting contrasts in symbolising, the text actually assumes that Yoruba and English numbers are the same, and provides no basis for grasping difference. In my third section I adopt a new understanding of what numbers are which enables me to elaborate how Yoruba and English numbers are different in ways that are useful when it comes to considering the workings of contemporary Nigerian classrooms.

INTRODUCING THE YORUBA NUMERATION SYSTEM

In expressing the fundamental practice of tallying recursion, numerals constitute an infinite series by having a set of base elements about which repetition occurs and rules by which any element may be derived from the element which precedes it. Any particular numeration system can be characterised by considering the elements in which progression is expressed, the base around which this progression is organised, the rules for generating successive elements of the progression from their predecessors, the ways in which the elements of the system are used in reckoning operations and the grammar of using numerals in talk.

The Yoruba series of number names uses a base of twenty. With a secondary base of ten and then a further subsidiary base of five, integers emerge. There are fifteen basic numerals from which the infinite series is devised: *ọ̀kan (ení)*, one; *èjì*, two; *ẹ̀ta*, three; *ẹ̀rin*, four; *àrùún*, five; *ẹ̀fà*, six; *èje*, seven; *ẹ̀jo*, eight; *ẹ̀sán*, nine; *èwá*, ten; *ogún*, twenty; *ọgbọ̀n*, thirty; *igba*, two hundred; *irínwó*, four hundred; *ọ̀kẹ́*, twenty thousand. The core of working the system is progression from vigesimal (twenty) to vigesimal. Multiplication generates multiples of *ogún* (twenty), or to be more exact: 'multiple placing out' as the Yoruba verb involved has it. About the vigesimals, intermediate numerals are then woven through tens and fives.

Assuming vigesimals at 20, 40, 60, etc. in Yoruba numeration, we have:

- the first four numerals of each vigesimal as generated through addition of ones, say 40 plus 1 (41), 40 plus 2 (42), etc., a process which is fairly familiar to base ten users.
- after 44 we 'leap' to 60 take away 10 take away 5 (60 – 10 – 5) to generate 45; 46 is (60 – 10 – 4), and so on to 49, progressively taking away one less at each step.
- fifty is 60 take away 10 (60 – 10), 51 is 60 take away 10 add 1 (60 – 10 + 1); this continues up to 54.
- fifty-five is 60 take away 5 (60 – 5), 56 is (60 – 4) and so progressively taking away one less at each step to 59.

I have expanded this way of presenting the Yoruba numeration system in Table 1 using the base ten graphic symbols to illustrate the pattern implicit in each Yoruba number name[1].

Table 1 Yoruba language numerals, showing their mode of formation (From Abraham, 1962)

1	*kan*	1
2	*méjì*	2
3	*mẹ́ta*	3
4	*mẹ́rin*	4
5	*márùún*	5
6	*méfà*	6
7	*méje*	7
8	*mẹ́jọ*	8
9	*mẹ́ṣan*	9
10	*mẹ́wa*	10
11	*mọ́kọ̀nlaa*	$(+1+10)$
12	*méjìlàá*	$(+2+10)$
13	*mẹ́tàlàá*	$(+3+10)$
14	*mẹ́rìnlàá*	$(+4+10)$
15	*mẹ́ẹ̀ẹ́dogún*	$(-5+20)$
16	*mẹ́rìndínlógún*	$(-4+20)$
17	*mẹ́tàdínlógún*	$(-3+20)$
18	*méjìdínlógún*	$(-2+20)$
19	*mọ́kọ̀ndínlógún*	$(-1+20)$
20	*ogún*	(20)
21	*mọ́kọ̀nlélógún*	$(+1+20)$
22	*méjílélógún*	$(+2+20)$
23	*mẹ́tàlélógún*	$(+3+20)$
24	*mẹ́rìnlélógún*	$(+4+20)$
25	*mẹ́ẹ̀ẹ́dọ́gbọ̀n*	$(-5+30)$
26	*mẹ́rìndínlọ́gbọ̀n*	$(-4+30)$
27	*mẹ́tàdínlọ́gbọ̀n*	$(-3+30)$
28	*méjìdínlọ́gbọ̀n*	$(-2+30)$
29	*mọ́kọ̀ndínlọ́gbọ̀n*	$(-1+30)$
30	*ọgbon*	(30)
31	*mọ́kọ̀nlélọ́gbọ́n*	$(+1+30)$
32	*méjìlélógbón*	$(+2+30)$
33	*mẹ́tàlélọ́gbọ́n*	$(+3+30)$
34	*mẹ́rìnlélọ́gbọ́n*	$(+4+30)$
35	*marùúndínlógójì*	$(-5+(20\times 2))$
36	*mérìndínlógójì*	$(-4+(20\times 2))$
37	*mẹ́tàdínlógójì*	$(-3+(20\times 2))$
38	*méjìdínlógójì*	$(-2+(20\times 2))$
39	*mọ́kòndínlógójì*	$(-1+(20\times 2))$
40	*ogójì*	(20×2)
41	*mọ́kọ̀nlógójì*	$(+1+(20\times 2))$
42	*méjìlógójì*	$(+2+(20\times 2))$
43	*mẹ́tàlógójì*	$(+3+(20\times 2))$
44	*mẹ́rìnlógójì*	$(+4+(20\times 2))$
45	*márùúndínláàádọ́ta*	$(-5-10+(20\times 3))$
46	*mẹ́rìndínláàádọ́ta*	$(-4-10+(20\times 3))$
47	*mẹ́tàdínláàádọ́ta*	$(-3-10+(20\times 3))$
48	*méjìdínláàádọ́ta*	$(-2-10+(20\times 3))$
49	*mọ́kọ̀ndínláàádọ́ta*	$(-1-10+(20\times 3))$
50	*àádọ́ta*	$(-10+(20\times 3))$
51	*mọ́kọ̀nléláàádọ́ta*	$(+1-10+(20\times 3))$
52	*méjìléláàádọ́ta*	$(+2-10+(20\times 3))$
53	*mẹ́tàléláàádọ́ta*	$(+3-10+(20\times 3))$
54	*mẹ́rìnléláàádọ́ta*	$(+4-10+(20\times 3))$
55	*márùúndínlọ́gọ́ta*	$(-5+(20\times 3))$

Table 1 Continued.

56	*mérìndínlọ́gọ́ta*	(−4 + (20 × 3))
57	*mẹ́tàdínlọ́gọ́ta*	(−3 + (20 × 3))
58	*méjìdínlọ́gọ́ta*	(−2 + (20 × 3))
59	*mọ́kọ̀ndínlọ́gọ́ta*	(−1 + (20 × 3))
60	*ọgọ́ta*	(20 × 3)
61	*mọ́kọ̀nélọ́gọ́ta*	(1 + (20 × 3))
62	*méjìlélọ́gọ́ta*	(2 + (20 × 3))
63	*mẹ́tàlélọ́gọ́ta*	(3 + (20 × 3))
64	*mẹ́rìnélọ́gọ́ta*	(4 + 20 × 3))
65	*márùúndínláàádọ́rin*	(−5 − 10 + (20 × 4))
66	*mẹ́rìndínláàádọ́rin*	(−4 − 10 + (20 × 4))
67	*mẹ́tàdínláàádọ́rin*	(−3 − 10 + (20 × 4))
68	*méjìdínláàádọ́rin*	(−2 − 10 + (20 × 4))
69	*mọ́kọ̀ndínláàádọ́rin*	(−1 − 10 + (20 × 4))
70	*àádórin*	(−10 + (20 × 4))
71	*mọ́kọ̀nléláàádọ́rin*	(+1 − 10 + (20 × 4))
72	*méjìléláàádọ́rin*	(+2 − 10 + (20 × 4))
73	*mẹ́tàléláàádọ́rin*	(+3 − 10 + (20 × 4))
74	*mẹ́rìnléláàádọ́rin*	(+4 − 10 + (20 × 4))
75	*márùúndínlọ́gọ́rin*	(−5 + (20 × 4))
76	*mẹ́rìndínlọ́gọ́rin*	(−4 + (20 × 4))
77	*mẹ́tàdínlọ́gọ́rin*	(−3 + (20 × 4))
78	*méjìdínlọ́gọ́rin*	(−2 + (20 × 4))
79	*mọ́kọ̀ndínlọ́gọ́rin*	(−1 + (20 × 4))
80	*ọgọ́rin*	(20 × 4)
90	*àádọ́ran*	(−10 + (20 × 5))
100	*ogọ́rùún*	(20 × 5)
110	*àádófà*	(−10 + (20 × 6))
120	*ọgófà*	(20 × 6)
130	*àádóje*	(−10 + (20 × 7))
140	*ogóje*	(20 × 7)
150	*àádọ́jọ*	(−10 + (20 × 8))
160	*ọgọ́jọ*	(20 × 8)
170	*àádọ́san*	(−10 + (20 × 9))
180	*ọgọ́san*	(20 × 9)
190	*igba ódin mẹ́wa*	(−10 + 200)
200	*igba*	(200)
300	*ọ̀ọ́dúnrún(ọ̀ọ́dún)*	(300)
400	*irinwó*	(400)
500	*ẹ̀ẹ́dẹ́gbẹ̀ta*	(−100 + (200 × 3)
600	*ẹgbẹ̀ta*	(200 × 3)
700	*ẹ̀ẹ́dẹ̀gbẹ̀rin*	(−100 + (200 × 40))
800	*ẹgbẹ̀rin*	(200 × 4)
900	*ẹ̀ẹ́dẹ̀gbẹ̀rùún*	(−100 + (200 × 5))
1000	*ẹgbẹ̀rùún*	(200 × 5)
1100	*ẹ̀ẹ́dẹ́gbẹ̀fà*	(−100 + (200 × 6))
1200	*ẹgbẹ̀fà*	(200 × 6)
1300	*ẹ̀ẹ́dẹ́gbẹ̀je*	(−100 + (200 × 7))
1400	*egbẹ̀je*	(200 × 7)
1500	*ẹ̀ẹ́dẹ́gbèjọ*	(−100 + (200 × 8))
1600	*ẹgbèjọ*	(200 × 8)
1700	*ẹ̀ẹ́dẹ́gbẹ̀sán*	(−100 + (200 × 9))
1800	*ẹgbẹ̀sán*	(−200 × 9)
1900	*ẹ̀ẹ́dẹgbàá*	(−100 + (200 × 10)
2000	*ẹgbàá*	(200 × 10)

GENERATING NUMBER NAMES

We can appreciate the architecture of a numeral system better if we recognise how the number names are generated. The etymology of numerals is helpful. This is equivalent for example, to recognising that in English eleven derives from the Old Teutonic *ainlifun* (one remaining after ten) using the Aryan verb *leip, lieq* (to leave or remain). Similarly, twelve comes from (*twen* + *liban*: two remaining after ten). The pattern continues through thirteen (three + ten) to nineteen (nine + ten) and to twenty *twen* (two) + *tig* (decade from Old Norse) (Dantzig, 1954: 12).

Understanding that verbs are essential parts of deriving Yoruba numerals is important, and to do the etymology one needs to know the conventions of vowel harmony and elision in Yoruba, which combine the basic numeral names and verbs in particular ways.

The verb *nọ̀n*, as in *ónọ̀n*, 'the state of being placed out or arranged', is involved in formulating numerals. When used with *ogún* (twenty), *ónọ̀n* shortens to *ọ̀*. Thus in sixty, (the mode of twenty placed three times), the numeral is *ogún-ọ̀-ẹta*, which becomes *ọgọ́ta* (sixty) by elision and vowel harmony. In creating multiples of *ogún* the form of the numerals often changes in line with obligatory elision and vowel harmony conventions so that the etymological origins of numerals are not always obvious to those unfamiliar with those rules. Vowel harmony conventions acknowledge an affinity between *e* and *o* and *ẹ* and *ọ*. For example, in the two cases where the element multiplying *ogún* begins with *e*, like *èjì* (two) and *èje* (seven), the names of multiples of *ogún* begin with *o* as in *ogóji* (twenty placed two times – forty) and *ogóje* (twenty placed seven times – one hundred and forty). In the case where the initial vowel of the multiplicand is *ẹ*, the vowel at the beginning of the vigesimal is *ọ* as in *ọgọ́ta* (twenty placed three times – sixty) and *ọgọ́rin* (twenty placed four times – eighty).

After multiplication of twenties, subtraction is the next most important process in Yoruba numeral formation. The involvement of subtraction is indicated by the inclusion of the verb *ó dín* (it reduces) in various forms. Subtraction is used for deriving numerals from *àrùún* (five) to *ọkan* (one) below a vigesimal or decimal point. For example there is *mẹ́rìndínlógún* (twenty it reduces four – sixteen), and *mọ́kàndínlógún* (twenty it reduces one, nineteen). *Ó dín* (subtraction) is used also for deriving odd tens from the next highest vigesimal as in *igba ódin mẹ́wà* (two hundred it reduces ten – one hundred and ninety). For deriving odd hundreds from even hundreds, odd thousands from their even counterparts, and odd tens of thousands in the same way, see Table 2.

Addition is of relatively minor importance in the system. It is confined to involvement with the first four numerals of each decade. The verb *ó lé* (it adds) is used in various forms. For example, numerals eleven through fourteen incorporate the element *láà*, a contraction of *ó la ẹ́wá*, as in *méjìlàá* (twelve). The verb is also incorporated (in a different form) in *mọ́kọ̀nlélógún* (twenty-one).

Putting the step by step operations with twenty, ten and five, together with the use of the three verbs *ó nọ̀n*, *ó dín* and *ó lé*, with the conventions of vowel harmony and elision, it can be seen that there are four basic patterns in numeral formation. I illustrate these four patterns in Table 2.

While frequently used lower numerals have accepted forms of derivation, multiple versions of larger numbers are possible. Table 3 shows seven different numerals for the English language numeral nineteen thousand, six hundred and sixty nine (19,669) (Ekundayo, 1975). And there are still more ways of arriving at satisfactory numerals which would be the equivalent of 19,669. To generate large numerals one must first be able to render the number as factors, using twenty, ten and five. Familiarity with factors is what matters most when generating a number name; multiple forms of any one

Table 2 Summary of the four patterns of number name formation in the Yoruba language

Numeral formation pattern 1

Forty one: *mọ́kọ̀nlógójí* (1 + (20 × 2))

- *m̲´*: mode grouped
- *okòn*: mode one
- *l̲´*: elision of *ó lé* (it adds to)
- *ogóji*: elision of *ogún ọnọ̀n eji* (twenty placed two ways)

Numeral formation pattern 2

Forty five *márùúndínláàádọ́tà* (− 5 − 10 + (20 × 3))

- *m̲´*: mode grouped
- *àrùún*: mode five
- *dín*: elision of *ó dín* (it reduces)
- *láàád̲´*: elision of *ó l'ẹ̀wá ó dín* (add ten it diminishes)
- *ọtà*: elision of *onọ̀n ogún ẹ̀ta* (twenty placed three ways, sixty)

Numeral formation pattern 3

Fifty one *mókọ̀nléláàádọ́tà* (+ 1 − 10 + (20 × 3))

- *m̲´*: mode grouped
- *ọ́kòn*: mode one
- *lé*: elision of *ó lé* (it adds to)
- *láàád̲´*: elision of *ó l'ẹ̀wá ó dín* (add ten it diminishes)
- *ọtà*: elision of *onọ̀n ogún ẹ̀ta* (twenty placed three ways, sixty)

Numeral formation pattern 4

Fifty six: *mérìndínlógótà* (− 4 + (20 × 3))

- *m̲´*: mode grouped
- *ẹrìn*: mode four
- *dínl̲´*: elision of *ó dín* (it reduces)
- *ọ́tà*: elision of *onọ̀n ogún ẹ̀ta* (twenty placed three ways, sixty)

Table 3 Eight ways of deriving the number 19,669 (From Ekundayo, 1975)

(a) *ọ̀kẹ́ kan ó dín erinwó ó lé okaàn dínláàádọ́rin*
$(((20{,}000 \times 1) - 400) + (-1 - 10 + (20 \times 4))) = 19{,}669$

(b) *ọ̀kẹ́ kan ó dín òódúnrúnó dín ọ́kànlélọ́bọ́n*
$(((20{,}000 \times 1) - 300) - (1 + 30)) = 19{,}669$

(c) *ẹ̀ẹ́dẹ́gbàáwàá ó lé ẹgbẹ̀ta ó lé ọ́kàndínláàádọ́rin*
$((20{,}000 - 1{,}000) + (200 \times 3) + (-1 - 10 + (20 \times 4))) = 19{,}669$

(d) *ẹ̀ẹ́dẹ́gbàáwàá ó lé ẹ̀ẹ́dẹ̀gbẹrin o´dín ọ́kànlélọ́gbọ́n*
$((20{,}000 - 1{,}000) + (-100 + (200 \times 4)) - (1 + 30))$

(e) *ẹ̀ẹ́dẹ́gbàáwàá ó lé ótalélegbèta ó lé mésán*
$((20{,}000 - 1{,}000) + ((20 \times 3) + (200 \times 3)) + 9) = 19{,}669$

(f) *ẹ̀ẹ́dẹ́gbàáwàá ó lé ọ̀rinlélegbẹ̀taò dín ókànlàá*
$((20{,}000 - 1{,}000) + (-(20 \times 4) + (200 \times 3)) - (1 + 10) = 19{,}669$

(g) *ọ̀kẹ́ kan ó dín ọ́tadínírinwó ó lé mẹ́sán*
$((20{,}000 \times 1) - (-(20 \times 3) + 400) + 9) = 19{,}669$

position in the series are possible since there are multiple ways of combining the bases to achieve the same result. The particular numeral used in any context will depend primarily upon the arithmetic predilections of the numerator. Some numerals are regarded as more elegant than others, so that some numerators could be regarded as superior to others. In large scale reckoning, operations on each of the bases is dealt with in turn, starting with the twenties and carrying as necessary into the tens and fives. Reckoning with large numerals requires profound familiarity with the system, and in former times this familiarity was fostered during the long apprenticeship which was a necessary part of becoming a numerator in Yorubaland (Fadipe, 1970).

ORIGINS OF NUMERATION

We might say that the Yoruba language numeral generation rules can be taken as modelling the following sequence of specifications: specifying the finger/toe complement of a person gives the major base *ogún* (twenty). Shifting from one vigesimal to the next codes for starting a new set of fingers and toes – literally 'placing out a new set of twenty' in Yoruba, for example sixty (20×3) – *ọ́ta* is twenty placed three times. Specifying *l'ẹ́wa ó dín* (reduce by ten) signifies 'hands only', and the additional specification, *l'árùún í dín* (reduce by five), signifies 'one hand only'. At this point we have $(-5 - 10 + (20 \times 3))$, *márùúndínláàdọ́ta*. Continuing in this vein the next specification is 'reduce the reduction by ones' $(-4 - 10 + (20 \times 3))$ *mẹ́rìndínláàádọ́ta*; and along the fingers and so on to the next hand and along its fingers $(+1 - 10 + (20 \times 3))$ *mọ́kònléláàádọ́ta*. Next specify 'toes (one foot)' $(-5 + (20 \times 3))$ *márùúndínlọ́gọ́ta*; then 'reduce the reduction by ones, working along the toes' to 'add the toes of the other foot by ones' $(+4 + (20 \times 3))$ *mẹ́rìnlélọ́gọ́ta*. (Notice that the first *ogún* is not specified before starting; one works towards the first *ogún*, not back from it. Presumably the numerator starts with his/her own fingers/toes). Yoruba numeral generation seems to imply a picture of numerals nested within each other, as toes nested in feet and fingers in hands are nested in the human body.

This is very different to the picture we get with the base ten numerals used in modern quantifying and mathematics, say in English. Here, with twenty implying the fingers of two hands gone through twice, twenty one begins again: thumb of left hand held up; twenty two: forefinger of left hand; twenty three: tall man of left hand and so on to twenty six: thumb of right hand; and on up to thirty which is three sets of two hands. What we see in English is a linear passing along the fingers until the end and then keeping tally and doing it again. Each integer is related to the one before and the one after it in a linear array. Demonstrably both modern base ten numeration in English and Yoruba numeration are tallying recursions.

Interestingly we find that these contrasting patterns that we see generated in English and Yoruba language numerations echo a contrast we find in the accounts of number given by Western mathematicians. Von Neuman and Zermelo seek to explicate number by identifying the referents of number words, and they have come up with quite different explications. Theirs is an enterprise very different to the one that engages me. They are trying to establish the true nature of the referent of number names – the true structure of the abstract entity number. In contrast I am content to limit myself to studying the representations of numbers. It seems to me that attempts to identify the referents of number words are as misguided as attempts to identify the referent of a part of a ruler. What is important about number names is that they form a sequence. Within any numbering system it is the place marked and how this is achieved that matters. Despite this fundamental contrast in our endeavours, the differing accounts of number presented by these mathematicians are useful in helping us see the contrast between English language and Yoruba language numeral recursions more clearly.

For me, a native English language speaker, von Neuman's account seems to be correct. Cardinality in this account seems to encode a one-by-one collection of predecessors. Von Neuman has the members of an n membered set paired with the first n

numbers of the series of numbers. For von Neuman 0 is $\varnothing$, the empty set. The set which contains the empty set as its sole member is $\{\varnothing\}$, one; the successor of this number is $\{\varnothing\{\varnothing\}\}$, or two; and three is the set of all sets smaller than three $\{\varnothing, \{\varnothing\}, \{\varnothing, \{\varnothing\}\}\}$. The successor of any number in von Neuman's version is generated by adding the successor of 0, that is 1. A number in this version is the last number of the series reached through one-by-one progression. A numeral names the point at which the progression ceases.

In contrast I suggest that a Yoruba speaker would choose Zermelo's account of number as correct. Here the number n is a single membered set, the single member of the set is $n-1$. Zermelo has 0 as $\varnothing$, a set with no members. Then 1 has the empty set as its sole member, $\{\varnothing\}$. The set which contains the unit set, 1, as its sole member is 2, $\{\{\varnothing\}\}$. Three is the set which has two as its sole member, $\{\{\{\varnothing\}\}\}$. In this version each number is totally subsumed by its successor, and any one number has a unified nature. I contend that for a Yoruba speaker, the model of number which would jump out as the intuitively correct account would be this one, for Yoruba language numbers carry the flavour of a multiply divided whole; there is no sense of a linear stretching towards the infinite here.

The two models agree in overall structure in identifying number; each model is demonstrably a recursive progression. But, importantly, for von Neuman's model and I suggest English language number, the set 14 has 14 members, while for Zermelo's model, and I suggest Yoruba language number, 14 has one member only.

THE WORKINGS OF NUMERATION

The Yoruba numeral scale is oral in expression and associated with the actions of counting. Graphic forms did not develop in association with the use of this numeration. And, related to this, there is no parallel to zero in the system, although more generally the notion in 'nought' or 'nothing' in Yoruba is *àìwà* or *àìkanwà*. In these words the *àì* indicates negation and *wà* is a verb (to manifest); *kan* indicates 'the mode of being one'. Literally *àìkanwà* means 'not in the mode one in a state of manifesting', and *àìwà* means 'not in a state of manifesting'. It is not considered a numeral. Why would this be so? I suggest that the oral nature of the Yoruba numerals explains the absence of zero[2].

At first glance the absence of zero seems crippling in a numeral system. Many of us are so used to the change between ninety-nine and one hundred being the change between 99 and 100, that we think we need zero to make the change. We do not. The words alone mark the fact that the series has leapt to a new base ten level. We are forgetting about the ways words perform and thinking only that written symbols do the work of a numeral system.

The invention of zero in the Indo-European tradition is considered by many historians of number to have been an event of the utmost importance.

> The discovery of zero marks a decisive stage in a process of development without which we cannot imagine the progress of modern mathematics, science, and technology. The zero freed human intelligence from the counting board that had held it prisoner for thousands of years, eliminated all ambiguity in the written expression of numbers, revolutionised the art of reckoning, and made it accessible to everyone (Ifrah, 1987: 433).

Yet most agree that zero was generated in practice, constituted by the ingenuity of those who needed to keep track in their reckoning. In recording a system of numerals as graphs, a problem is encountered which does not occur when numerals are words. The problem centres on the graph which marks the decimal place. In graphic numeration the reaching of the decimal place is recorded by placing a 1 in a new column created to the left of the column under consideration. To be workable in reckoning the space in the original right hand column must also be occupied – otherwise the reckoner can easily make mistakes.

It is useful to indicate that the numeral is a double column, triple column numeral, etc. In the Hindu world, where the modern graphic numeral system developed, numerators came to represent this emptiness in right hand columns of figures with a commonplace cipher '0' – probably just an enlarged dot – which helped prevent muddles. The Arabic word *cifr* became 'zero' in Europe when Europeans adopted the system of graphic numerals in the eleventh century. Zero, the name of a graphic symbol, gradually came to mean nothing or nought. It came to be a numeral in its own right. The naming of a cipher thus eventually resulted in the modern numeral sequence being given a new point of beginning. And even more, a new number was added to the domain of abstract number. Seeing the modest origins of zero-the-numeral, we see the origins of number through the inventions of numerals. This origin story for zero-the-number is in fact suggestive of the processes by which the social practices of arranging and patterning, formalised in recursion, constitute the domain of natural number.

Associated with its multibase nature, the Yoruba numeration scale has a capacity to handle calculation with large numbers which is far superior to the single base ten numeral scale. While for those of us who grew up in worlds ordered by the modern base ten numeration it is hard to imagine using the Yoruba numeration scale for reckoning, a Yoruba numerator, with a well honed memory of factors, would scorn the simplistic and cumbersome system of adding by ones and tens of the decimal system. Writing things down for him would constitute a significant interference in working the system. Practitioners speak a word as they do something with their hands; the two actions go together. Oral numerals are much more explicit. Only in very odd circumstances would there be a point in speaking a word (like *àìkanwà*) to announce my not doing something. The absence of zero is exactly what we would expect of an oral numeration; it is redundant. Zero may be an essential element in the ways natural number works in modernity, but it would be counter-productive in the working of Yoruba numeration.

There is a second group of numerals in English which name not the sum which has been reached at any point in the numeral sequence, but specifically the position in that sequence. These are the numerals first, second, third, fourth, fifth, and so on, often called 'the ordinal numerals'. These are represented graphically by combining the cardinal graphic numeral with the ending of the linguistic ordinal numeral, as in 1st, 2nd, 3rd, 4th, or 5th. The names cardinal and ordinal are themselves interesting. As names of groups of numerals they arose relatively recently, after Europeans started to write commentaries on number systems. It was thought that the group named by one, two, three, etc. were more important than the group named by first, second, third, etc. and the name 'cardinal' was coined for this set. It has its origin in the Latin word *cardinis* (a major pivot or hinge). The name 'ordinal' derives from the Latin *ordo, ordinis* (sorting into ranks, rows or sequences). The sense of linear spatial ordering implied by the name ordinal is already present in the numerals of this sequence. 'Second' derives from *secundum*, gerund of the verb *sequi* (to follow in the sense of along a spatial extension), and the '-th' of fourth, fifth etc. derives from the Greek suffix, *-tos*, used to denote attributes that have degrees of extension as in the English words width and length.

NUMERALS IN LANGUAGE USE

Yoruba numerals are elided verb phrases, and as such they occur in Yoruba sentences in ways that are very different to modern English numerals and are in no way adjectival.

Earlier I elaborated the sixteen basic numerals in Yoruba and we saw how the verbs *ó nọ̀n, ó dín* and *ó lé* are an integral part of their working (Table 2). This form of numerals is used only for the actual performance of counting; using Yoruba numerals in sentences, to evoke a notion of specific and precise value, always involves elision with another verb depending on what the numeral is being used for.

In considering how this is achieved I begin by looking again at the primary numeral forms: *ọ̀kan (ení)*, one; *èjì*, two; *ẹ̀ta*, three; *ẹ̀rin*, four; *àrùún*, five; *ẹ̀fà*, six; *eje*, seven; *èje*, eight; *ẹ̀sán*, nine; *èwá*, ten; *ogún*, twenty; *ọgbọ̀n*, thirty; *igba*, two hundred; *irínwó*, four hundred; *ọ̀kẹ́*, twenty thousand. These numerals are ancient (apart from *ọ̀kẹ́* which is a fairly recent addition) and etymologically their origins are hazy, though understanding the rules for elision and vowel harmony in Yoruba means that significant insights can still be gained. We can say that they all have the form of nominalised verb phrases. All these number names appear to be elisions of introducers and verbs, although the verb forms from which they derive are no longer to be found in the language. Saying that the numerals are elided phrases means that grammatically the numerals function as mode nouns. Thus *ọ̀kan* is better translated as 'mode one', and *èjì* as 'mode two' – they imply a particular form or arrangement. These mode nouns are not used in conjunction with other nouns; they constitute the counting series. A sequence of modes is named by the sequence of Yoruba numerals in counting.

The primary numeral form (already a mode noun – an elided phrase with verb and introducer) is further modified when used in quantification statements in Yoruba talk. The primary numeral name is prefixed with *m* and a high tone. The form of the prefix implies that there has been an elision of the primary numeral noun with a verb of the form '*m* + (high tone vowel)'. The most likely verb is *mú* (Bamgbose, 1974; Awobuluyi, 1978) an obsolete verb related in meaning to the present day *mún* (to take or pick up several things in a group or as one) (Abraham, 1962). This elision results in a secondary form of numerals (one to ten) as follows: *kan, méjì, mẹ́ta, mẹ́rin, márùún, mẹ́fà, méje, mẹ́jọ̀, mẹ́sàán, mẹ́wá.*

The '*m* + (high tone)' form of numerals implies the mode of being grouped. This is the nominalisation of a verb phrase which contains a noun previously derived in the same way. We could say that this form of the numerals names a mode of a mode. Thus one may say in Yoruba, '*ó fún mi ni òkúta mẹ́rin*', which is conventionally translated as 'he gave me four stones'. A more literal translation is 'He gave me stonematter in the mode of a group in the mode of four'. Similarly we may have '*ó rí afá mẹ́ta'* (he saw three dogs) which is precisely translated as 'He saw dogmatter in the mode of a group in the mode of three'. The numeral *ọ̀kan* does not accept the prefix '*m* + (high tone)'; instead the first vowel of the primary numeral *ọ̀kan* is dropped to form *kan* which may be used as follows: *Eja kan kò tó* (one fish is not enough) literally 'fishmatter in the mode one does not reach'. Perhaps the failure of *ọ̀kan* to accept combination with the verb *mú* relates to *ọ̀kan* already being in the mode of a group. This set of numerals has been called 'the multiplicity set' (Fadipe, 1970).

Another set of secondary Yoruba language numerals is often referred to in English as the 'ordinal numerals' (Abraham, 1962). These are used to indicate position in a sequence. In this set of secondary numerals the verb *kó* (to collect items severally) seems to be involved (Bamgbose, 1974; Awobuluyi, 1978). This set seems to be prefixed with '*i* + *k* + (high tone vowel)' (introducer + verb). However it is only in the northern dialects of Yoruba that the elision is completely regular. Except in these northern dialects, the high tone is not assigned to the initial vowel of the original numeral. The most common version of the position numerals (first to tenth) is as follows: *ìkìínní (àkọ́kọ́), ìkeji, ìkẹta, ìkẹrin, ìkarùún, ìkẹf'a, ìkeje, ìkẹjọ, ìkẹs'an, ìkẹwàá.* A literal translation of *Ó gb'a ìwé ìkẹta ni* (he took the third book) illustrates the double modal nature of this form of numeral: 'he took bookmatter in the mode of collected individual items, in the mode three'. *Ìkẹta* is a double mode noun formed by nominalising a verb phrase which already contains a mode noun, the primary numeral. But in contrast to the implication in the multiplicity form that the individual elements lose their individuated existence, the implication here is that the constituent elements retain their oneness.

The third modified form in which the Yoruba language numerals exist is the form used specifically in connection with currency. In this case the word for money, *owó*, is incorporated into the numeral. With elision this results in the first vowel of the numeral being elongated or duplicated, and a high tone imposed. The first five numerals in this set are *ọkan, ééji, ẹ́ẹ́ta, ẹ́ẹ́rin*, and *árùún*. With the primary numerals *ogún, ogbọ̀n, igba*, the word *owó* is suffixed rather than prefixed, resulting in the numerals *okòó* (twenty units of currency), *ọgbọ̀n òó* (thirty units of currency) and *igbiyó* (thirty units of currency)[3].

What is remarkable about Yoruba numerals when we see how they are inserted into ordinary language use is that they imply modes of presenting. In use they are modes of modes, further material arrangements of arranged presentations. Grammatically there is no adjectival element in Yoruba numeral use, implying that those mysterious abstract entities, qualities or properties, so central in modern quantifying, do not appear to be part of Yoruba quantifying at all. This represents a profound difference between modern base ten numerals as used say in English and Yoruba numerals. Whereas Yoruba numerals are a second order form of modifying, English language numerals are a form of second order qualifying.

In modern English usage 'ten' in the sentence 'I cooked ten potatoes' is taken as telling about the extent of the quality of numerosity held by that group of potatoes boiling in the pot. Numerosity is a quality or property of the collection of potatoes held to a particular extent – the ten single brownish lumps. Similarly in 'She is five feet tall' the 'five' tells about the extent of the quality or property of length in the body of a particular woman. In English, number names work as adjectives; they qualify in a second order way. Despite this, most English grammars class number names as abstract nouns, following the official line that numbers are naturally occurring entities.

It is this characteristic of working through modes that has an analyst like Hallpike (1979) classifying Yoruba numeration as primitive. In his account of quantifying, processes which proceed in ways which fail to evoke the abstract element of qualities are primitive. Similarly Carnap's positivist account has the evoking of qualities as absolutely central in the development of quantitative concepts (Carnap, 1966). And this is where we see the difference between an account of quantifying which like Bloor's (1991) has social practices of ordering constituting the foundation for the domain of number. The latter allows for the possibility of numerals naming modes as easily as it accepts the possibility for them to evoke qualities.

WHAT DOES THIS TEXT TELL US ABOUT YORUBA AND ENGLISH NUMBERS?

This comparative account of English and Yoruba numerations lays out their significant characteristics in impressive detail. Yet fifteen years later I find myself quite dissatisfied with the presentation. By this I do not mean to imply that it is worthless – understanding the details of comparison is important. It is the framing of this text with which I now take issue. Examining the assumptions out of which it grows can help explain why this account is not useful when we want to consider in a practical way the everyday negotiation of using English and Yoruba numbers together.

My study of Yoruba numbers was motivated by the confusions and discomforts I experienced watching my Yoruba students struggling to teach in sparsely resourced, overcrowded classrooms (Verran, 1999; Verran, this volume; Verran, forthcoming). I sought to describe and explain these tensions and to point to them in ways that would enhance my students' understanding of their work as Yoruba teachers. I was seeking, too, to change my own teaching practices, to develop routines in my classrooms as exemplars for my students. I intuited that as teachers in the Yoruba classrooms of contemporary Nigeria, we needed to train pupils in two sets of routines *and* be able to do, and explain, translations between the two sets of routines.

With this in mind I set out on a study of quantification in contemporary Yoruba life (Watson, 1986; Watson, 1987; Watson, 1990). This study culminated in my preparing a book manuscript, from which the text above is drawn. In attempting to conclude this book manuscript I went back to the puzzles and discomforts I had experienced in Nigerian classrooms and the perceived need to devise new ways to teach mathematics. I wanted to elaborate new classroom practices which would be justified and supported by my study of Yoruba quantification. But what became obvious when I attempted to do this was that my painstaking study had proved that, mathematically, Yoruba quantification and modern Indo-European derived quantification were the same. I had actually explained *away* the difference. I had left myself absolutely nothing to work with in devising ways to do mathematics lessons in contemporary Yoruba classrooms which recognised and used the tensions between Yoruba and English quantifying that I and my students felt so strongly.

Realising this I recognised that I needed a new framework for my study. I needed an interpretive frame, which, accepting that quantification in Yoruba and English were mathematically the same, could also show how they were mathematically different. This turned out to be no simple matter. It led me to a search for a new description of mathematics and logic, one that was more useful for times and places struggling for a postcolonial life (see Verran, this volume).

Emblematic of my discomfort with my old text is the way it obsessively distinguishes between numbers and numerals. Most people use 'numbers' and 'numerals' quite interchangeably, but in my account I insist on precise usage. The significance of the distinction is not made clear in my presentation. The precision mobilises the orthodox view that numerals are symbols while numbers are far more than this. Naively adopting this orthodox picture of what numbers and numerals are, and are not, in the end I show that Yoruba and English numbers are the same thing, albeit that the symbolising differs. The differences that my students and I felt so intensely in our practices could not be grasped with this conventional understanding of what numbers are.

In an attempt to reveal what lies inside this conventional understanding, assumptions which I must examine if I am to find a way of grasping difference, I now develop a historical perspective on that old study. This is helpful in recognising why conventional foundationist understanding of numbers has me explaining difference away.

My old text is a display, and in making that display I made many contrivances which are almost invisible in the text. I made translations. Quite obviously, since I am writing in English, I have translated Yoruba words into English, struggling to find adequate ways of conveying meanings that are not too clumsy when rendered as English words. In showing the structure of Yoruba numeration I also translated between the patterns of generating new positions in a base 20 – 10 – 5 – 1 recursion and the pattern of a base 1 – 10 recursion with which most of my readers are familiar. My translating, like that of Adolphus Mann, can be understood as a form of standardising, as a way of finding a form to bring the numeral system home and display its beauty and its unusual features to my colleagues.

This double translating, between languages and between recursions, is an integral part of showing the Yoruba numeration system, helping it achieve a standardised form for the necessary display. As such it can be compared to the drawings of objects collected in earlier colonising endeavours. For example as Nicholas Thomas points out (1991: 134), in Captain James Cook's *A Voyage Around the South Pole* we see standardised graphic depictions of artefacts collected from various Pacific islands as the expedition wended its way across the Pacific Ocean. Irrespective of the islands from they were collected, Maori, Malekula and Tannese axes and spears, flutes and fans, are rendered seemingly transparent as material objects in the arrangement of drawings on the pages of Cook's journal.

In that display much attention has been paid to scale of presentation and the placement of the diagrammatic images on the page, to create a pleasing, quite standard aesthetic. The collected objects have been translated into the form of a journal, and in this the islands from which the objects were removed, and the people who fashioned the objects, figure hardly at all; they are reduced to annotations in small print placed just so, as a title. The implements emerge with a pleasing aesthetic, giving an aura of beauty to the tools, although as Nicholas Thomas points out, this standardised rendering as beautiful did not extend to the makers of the tools (1991: 133). Thomas' point is that in the exchanges of colonialism, the taking, theft or purchase of objects leads to creative recontextualization and even rewriting history. Museums were and are prominent in that rewriting. Extending this to something like a numeration system allows us to see a side of this recontextualisation which can easily be missed when the focus is on actual objects picked up in the arms of sailors, carried on board a ship, sketched, crated, and later off-loaded in England to be transported to the British Museum.

When we are focussing on a numeration system, the disembedding/re-embedding, disentangling/re-entangling, and disembodying/re-embodying work of the translations to generate standard forms becomes more obvious. We might understand this as a transfer between different ways of knowing. Spears, axes and carrying baskets, flutes, fans and hats, have meanings and are participants in everyday life amongst the Maori, Tannese etc. men and women who made them. They have quite different meanings and participate in quite different ways when encased in glass and wood cabinets, watched over and guarded by uniformed attendants. There is a de-objectifying/re-objectifying occurring here as well, which is easily missed because in an experiential way fingers can be enclosed over them as objects in both contexts.

Adolphus Mann and I have translated between languages and recursions and in another way too. As well as translating Yoruba to English, and base 20 – 10 – 5 to base 10, we have disentangled the Yoruba numeral system from its embedded and embodied way of contributing to the on-going life of Yorba trading. By getting it onto pages and into books in re-entangling and re-embodying it, we have literally objectified it; we have rendered it as an object – the Yoruba numeration system – whereas before our interventions there was not an object at all. Before it was made into charts printed on pages, and paragraphs of descriptive and explanatory English words in books, Yoruba numerators and ordinary Yoruba people knew their counting in particular ways. The patterns made in spoken words were part of the life of the markets of Yorubaland, so that it was scarcely possible to separate human body and numeral. Each person had his own way of working the pattern, according to his own aesthetic, even though the numeral words when used had a specific and precise meaning.

In getting onto paper and into books, actions of hands and feelings over patterns expressed in spoken words become a thing – the Yoruba numeration system. We understand it now as a separated object in space and time – standardised and objectified. As Adolphus Mann tells us, it is 'a building' which may be 'viewed from base to summit'. As a standard object the Yoruba numeration system has become a particular example of the class of all numeration systems. All examples of this class have a history – they came to life and were born as objectified participants – in discourse. In discourse, numerals no longer participate through human embodiment.

As an example of the group of entities which constitute the class of numeration systems, the Yoruba system is a particular which has certain qualities. As Adolphus Mann points out, it has 'regularity' and 'symmetry'. And having qualities the Yoruba numeration system can be compared to other numeration systems, other instances of the class of numeration systems. It 'surpasses our European systems in the aptitude of interlinking' and has 'a profusely ornamented Moorish style' compared to the sober Byzantine style of European examples of the class of numeration systems (Mann, 1887).

What is the significance of this objectification, so pervasive that we fail to notice that *this* is what our studies achieve? In answering this I want to put another question. Is it not a little puzzling that a century ago Adolphus Mann, perhaps in Lagos, squatted down uncomfortably with his notebook, sweating in a crowded room in a market? Possibly causing much hilarity and not a little embarrassment to himself and others, he made careful notes on the actions of the numerator who had been called in to count publicly the cowries to be handed over. What motivated Mann to weather this discomfort? And, back in London, what moved him to transcribe and expand his notes and put together a written presentation to be given to his friends of the Royal Anthropological Institute of Great Britain and Ireland on 9th March 1886? What led the Royal Anthropological Institute of Great Britain and Ireland officially to 'take the presentation as read'?

In 1886 Adolphus Mann's work was part of a vast cataloguing endeavour. This was the arena in which the relatively newborn object, the Yoruba numeration system, came to life. As Mann helpfully noted in beginning his presentation:

> ... of late the nations and languages of West Africa have largely occupied the attention of the learned linguists of Europe, and grammars and vocabularies are being published in considerable number .... the classification of four or five hundred languages has been advanced to such an extent as could not some years ago have been expected. Perhaps the following notes of the numeral system of the Yoruba nation may ... be of some use in investigating the nature of the mind that can form such an unusual, yet regular structure (Mann, 1887: 60).

It is an endeavour which is by no means confined to Europe. A few years earlier, C. H. Toy, Professor of Jewish Studies at the Southern Baptist Theological Seminary, Louisville, Kentucky, U.S.A., had addressed 25 other philologists at the Tenth Annual Session of the American Philological Association, in the Opera Hall of the Grand Union Hotel in Saratoga Springs, New York on Tuesday evening July 9, 1878. His talk, entitled 'The Yoruban Language' was, we are told in the record of the transactions of that meeting, remarked upon by Professor S. S. Haldeman, University of Pennsylvania, Philadelphia (Toy, 1878: 3). And later a longer paper with a section focussing specifically on Yoruba numerals was published in the record of the transactions of the Association as 'The Yoruban Language', along with such titles as 'Contributions to the History of the Articular Infinitive', 'Influence of Accent in Latin Dactylic Hexameters' and 'Elision, Especially in Greek' by his fellow philologists. Toy explains his motivations in this way

> The main body of African languages ... fall into three groups: The Hottentot in the south, the Bantu, occupying the whole center ... and the Negro lying in Senegambia and Soudan, the last of which has as yet received little attention, while the structure of the others has been carefully studied and satisfactorily exhibited. On the Guinea coast, however there is found a group of dialects wholly different in vocabulary and structure from all these and offering interesting linguistic features ... [T]he most important member [of this group] is the Yoruban which is spoken by a partially civilized population of about two million people (Toy, 1878: 19).

Mann and Toy are interested in the characteristics of language systems, considering each of them as an instance of the class of systems representing the real. Each language has particular characteristic ways of representing the real, an array of categories realised as material and abstract objects, which lies behind and below representations in language. And they are interested in the evolution of languages. Their displaying, classifying and cataloguing of languages is a way of exhibiting a hierarchy which displays the various stages in the progressive evolution of languages. The Classical languages of Greek and Latin serve as reference points in this enterprise, and necessarily languages are constructed as standardised objects.

Numeral systems are of particular interest in this endeavour because numbers are a specific and crucial element of the real. The type of numeral system a language system incorporates is an important sign of its degree of development. It is a convenient basis, a significant indicator, we would say these days, for classifying languages on a scale of adequacy in giving access to the real. However in neither Mann nor Toy is this privileged status for number and the associated special interest in numerals explained. The universal of natural number occupies the unmarked position in their texts, a universal presence marked by absence.

In scrutinising and displaying the qualities, properties or attributes of various numeration systems, many particular instances of the class of numeration systems are characterised in a descriptive, exhibiting enterprise, explicitly or implicitly compared to some ideal system of representation – a system so perfect, so transparent, as to allow the domain of natural number to present itself. Constituting numerations as particulars, standardised and objectified in the various translations which render them analogous to spatiotemporal particulars, also has the effect of constituting the universal, natural number. Each particular remakes the universal of natural number projecting into the domain of the abstract where universals reside. The whole point of rendering numeration in discourse is to show it as a particular 'particular' and hence an instantiation of a 'universal', a given category of the physical foundation, the real world upon which knowledge rests.

The writing of Mann and Toy presents the particular, 'Yoruba numeration', as a found entity pointing to, and constituting, a universal, the ideal and eternal category, 'natural number'. In contrast the counting person ordering cowries on the floor and the market women who, 100 years later, laughed merrily at my attempts to use Yoruba numbers, would have been very unlikely to think of them as a numeral scale, an object with particular characteristics, which in turn points to the existence of a universal, natural number.

Valued among other things for their capacity to contribute to the vast endeavour of producing a catalogue of world languages, explorers and adventurers in the 19th century keenly sought numeration systems and collected them for display back home. Brought back to the metropole, demonstrated at anthropological society meetings and such, these collected items became available to be taken up in further enterprises. One such was, and is, an endeavour largely carried out by mathematicians: to detail the history and origins of the objects of mathematics – an enterprise not far removed from that of Mann and Toy. Sure in the knowledge that the inferior systems of 'others' are various stages in the development of 'our' advanced modern representations of mathematical universals, these books sought, and still seek (Menninger, 1969; Ifrah, 1987; Crump, 1990), empirical, as distinct from philosophical, answers to questions over the origins and development of mathematical objects.

Most regard the base ten numeration system of modern mathematics as a perfected tool, whose perfection has been achieved through the dual determinants of a real world of number and a superior rational mind particularly attuned to that real, though abstract domain. Less advanced are those systems developed by other cultures using inferior languages and minds less well attuned to that shadowy domain of number. In this genre of storytelling there is continuing disagreement over whether or not all peoples have at least some degree of natural affinity to the abstract domain of natural number and over whether the domain is accessible to non-human animals (Davis and Hersh, 1983; Crump, 1990).

Levi Conant's 1896 book, *The Number Concept: Its Origins and Development*, is a rather extreme example of the genre. The author was an American mathematician working in a worthy New England academy with a good library at his disposal. He is sure of the existence of the abstract domain of natural number and other mathematical universals. Despite his title, on page 2 he notes that there is nothing to be gained in investigating the origins of number in any absolute sense

> Philosophers have endeavoured to establish propositions concerning this subject, but, as might have been expected, have failed to reach any common ground of agreement ... Why this question should provoke controversy, it is difficult for the mathematician to understand ... The origin of number would in itself, then, appear to lie beyond the proper limits of inquiry (Conant, 1896: 3).

Having thus dealt with the universal number, unproblematically establishing its otherness, Conant is free to devote the next 200 or so pages to considering numerations, the various particulars representing this universal, almost without mentioning it again; its pervasive presence in the text is marked through its absence.

Sure as he is of the existence of number in that 'other' domain, Conant is equally sure of the extraordinary capacity of modern man to apprehend that domain and of the perfection of the modern numeration system (Conant, 1896: 33). Animals he thinks might have a hazy awareness of that number as might some 'others'. He approvingly quotes

> ... an amusing and suggestive remark in Mr Galton's interesting *Narrative of an Explorer in Tropical South Africa*. After describing the Demara's weakness in calculation he says: 'once while I watched a Demara floundering hopelessly in a calculation on one side of me I observed 'Dinah' my spaniel, equally embarrassed on the other; she was overlooking half a dozen of her new-born puppies, which had been removed two or three times from her, and her anxiety was excessive as she tried to find out if they were all present, or if any were still missing. She kept puzzling and running her eyes over them backwards and forwards, but could not satisfy herself. She evidently had a vague notion of counting but the figure was too large for her brain. Taking the two as they stood, dog and Demara, the comparison reflected no great honour on the man (Conant, 1896: 4).

Conant presents a hierarchy of numeration systems and capacities to comprehend number, which is worked out in impressive detail. Moderns at the top only have need of a numeration system as an aid to memory (Conant, 1896: 8), and the Japanese numeration system 'is the most remarkable [he has] ever examined, in the extent and variety of the higher numerals with well-defined descriptive names' (Conant, 1896: 94).

Below this the Russians, Welsh and Catalan are somewhat ahead of Eskimos, North American Indians and Pacific Islanders. While the Africans lie only just up from South American Indians at the bottom of the scale, the Australians, with their mere two or three number words, only just make it into the hierarchy. Such a fixed and detailed framework of interpretation would render many of

Conant's assessments amusing a century later were it not that many contemporary analysts share his frame of analysis, presenting their rather similar hierarchies in more sanitised terms.

Conant comments extensively on Yoruba numeration, noting that a

> ... species of numeral form, quite different from any that have already been noticed, is found in the Yoruba scale which in many respects is the most peculiar in existence ... the words for 11 and 12 are formed by adding the suffix *-la*, great[4], to the words for 1 and 2 etc., ... The word for forty was adopted because cowrie shells which are used for counting were strung by forties; and *igba* for 200, because a heap of 200 shells was five strings, a convenient higher unit for reckoning. Proceeding in this curious manner [it was reported by a Yoruba that] – the king of the Dahomans, having made war on the Yorubans, and attacked their army, was repulsed and defeated with a loss of 'two heads, twenty strings, and twenty cowries' of men or 4820 (Conant, 1896: 70).[5]

It is clear that for Conant the Yoruba system is 'most peculiar and curious' because it does not fit with his hierarchy. He seems in fact to be quite put out by it.

> ... the development of a numeral system is [not] an in fallible index of mental power, or of any real approach to civilization [since] a continued use of the trading and bargaining faculties must and does lead to some familiarity with numbers sufficient to enable savages to perform unexpected feats in reckoning. Among some of the West African tribes this has actually been found to be the case; and among the Yorubas of Abeokuta the extraordinary saying 'You may seem very clever but you can't tell nine times nine', shows how surprisingly this faculty has been developed, considering the general condition of savagery in which the tribe lived. There can be no doubt that in general, the growth of the number sense keeps pace with the growth of the intelligence in other respects (Conant, 1896: 32).

Like many curators Conant wants his exhibition to be tidy. And he has other problems as well, troubles growing from the diverse origins of the objects he is displaying. Not only were there glaring inaccuracies in what was collected, as we saw in one of the quotations above, but provenances too were not properly attended to. Many collectors of numeral scales, whom Conant dismissively refers to as 'vocabulary hunters' (Conant, 1896: 85), too often were content to collect merely words for the numerals, making no systematic enquiries as to original meanings of the number names (Conant, 1896: 48).

I take it that Conant's complaint here points to a tension in the colonial enterprise noted by Thomas, 'a tension between a scientific controlled interest in further knowledge and an unstable 'curiosity' which is not authorized by any methodological or theoretical discourse, and is grounded in passion rather than reason' (Thomas, 1991: 127).

As Thomas goes on to point out, the 'giddy passion' of collecting and the scientific ordering tended to coincide with a difference in placement in the colonial enterprise. The sailors on ships in the Pacific (Thomas, 1991: 128) and the vocabulary hunters of Africa were perhaps strongly moved by an ambiguous curiosity. A desire to contribute to the scientific imperialist ordering of those at home collating the objects was only part of this. The need to assess the degree of development of the object of their desire, with respect to its capacity to represent a universal category, did not mean much to many of those sufficiently moved by the passion of curiosity to withstand the discomforts of collecting. Those impressed by the need for methodological classification were not much inclined to move from their comfortable studies.

Conant is using members of the class of numeration systems, particulars generated in the standardising and objectifying translating work of collectors like Adolphus Mann, to assemble an exhibit which we might understand as a zenith (or nadir) in the huge cataloguing task that Mann and Toy see themselves as part of. The Yoruba numeration system began its career as a modern object as an irritating anomaly in a hierarchy of natural languages-and not incidentally, races. The conventional understanding that numerations are found objects is an outcome of a project which most people nowadays would find utterly abhorrent.

Reacting against these sorts of endeavours, the account of Yoruba numbers with which I began this chapter adopts a relativist framing, although it does not overtly announce itself as such. In refusing the universalist understandings of Mann, Toy and Conant, I hoped to distance myself from their by now discredited enterprise. What I failed to notice is that my relativist frame retains many of the assumptions inherent in their work. We all took numeration systems to be found objects. We all failed to notice the translations which achieve both standardisation and objectification. Our Yoruba numeration system, like all other members of the class 'numeration systems' is rendered a particular object in our work. These particulars, coming to life, as they do in discourse, are taken to be particular products of particular societies, and as such we all understand that they exhibit various qualities in variable degrees, and we are all convinced that they have been found and not made.

Despite the obviousness of the ways numeration systems come into being as particulars, through translations which both standardise and objectify, it is crucial for a foundationist vision of knowledge (universalist and relativist alike) to fail to notice that this is worked up in translation. The credibility of foundationist visions can only be maintained by an absolute separation of the domains of particulars and the domain of the universal (or general) category. It is crucial to abide by the convention of finding particular instances which attest the existence of those universal (general) categories. The particulars must be seen to have been *found* as standard objects, and not made in translation. And in this my presentation of the Yoruba numeration system is quite exemplary.

In my presentation I showed a found pattern, yet in finding that pattern I did not notice that I could only see and show the pattern (as distinct from feeling it when I uttered words) by working up a systematic translation of one set of patterns into another (the base 20 – 10 – 5 recursive patterns into the base 10 recursive patterns). I also found that Yoruba numerals of the Yoruba counting sequence were mode nouns, but I could only show this through elaborate translations of single Yoruba words into clumsy English phrases. Next I found that in numerals in Yoruba sentences are double mode nouns, and I had to show that by even more elaborate and clumsy English phrases.

Yoruba numerals exist as words spoken in Yoruba markets for example, and as spoken words they weave patterns which people inhabit and which inhabit people. But the object Yoruba numeration system is born in the work of translation. Perhaps it was my (and others') enthusiasm and delight in the patterns that blinded us. The scam seems so obvious: conflating and eliding a set of felt patterns made in spoken words in the markets of Yorubaland with something literally objectified as a chart printed on pages and paragraphs of descriptive and explanatory English words in books, contributing to the life of the academy and official discursive life, is quite an accomplishment.

For years as I puzzled about the differences between Yoruba and English numbers, I remained convinced that I had accidentally stumbled across all the signs of the existence of the object – Yoruba numeration. I failed to notice that if I had just used numbers, say in buying provisions for my family in the market, concentrating only on the price of the bowl of rice (insisting loudly that the vendor heap it up) or pile of tomatoes (noting in my negotiations their over-ripeness), bargaining with the vendors to get a better price, there would have been no resources to make an object with. It is precisely that I am bringing the object of the Yoruba numeration system home to the academy, rather than just the rice and tomatoes home to my family, which provides the materials to generate the object 'Yoruba numeration system'.

Achieving these translations, I remake the necessary boundary between particulars and general without hindrance. Numerations easily fall on the side of particular symbolic representations; recursive tallying as a general process generating cardinal number sits in that 'other' domain. The foundationist vision of a set of given categories (given either by the structure of the real world, or by the common patterning of a general social process) attested by the characteristics of found objects, remains intact. All four of us whose work I have shown here are making knowledge claims within this vision. As authorities we propose these stories as true accounts of the real world. And what authorises us in making these claims is the evidence we present of having glimpsed the real nature of things, of having seen the universal or general categories in their true light. Our claims to truth are based on evidence. What is the nature of this evidence? It is presented in our descriptions of the particulars we have found.

Making the claim that the common social practice of recursive tallying is variously instantiated in numeration involves me in displaying the found numeration systems of contemporary English and Yoruba as my evidence. Developing their theory of the evolution of languages as symbolic apparati by which the structure of the real is revealed and apprehended, Conant, Mann and Toy present the characteristics of a series of found particulars – languages and their numeration systems. This is all thoroughly orthodox negotiation over knowledge claims in the modern academy.

All these knowledge claims can be evaluated as examples of inductive reasoning – the process of drawing conclusions about the nature of a general category on the basis of observation of particular instances. In an assessment, a would-be evaluator can question whether the evidence is sufficient and relevant in establishing the proposition concerning the general category. All four of us assemble our evidence in the light of this evaluative regime. It is in our interest to fail to notice the duplicity upon which our knowledge claims rest. If it is generally noticed, or even if we notice, that the particular objects are generated in the endeavour of research, in our translations, there would be no point to any of it. Making knowledge claims within the foundationist style of reasoning would be revealed as self-vindicating activity.

As we go on, systematically ignoring evidence of our own participation in generating the scenario, we are remaking universals and the notion of a domain of the real over there, and a domain of representative particulars over here. We are re-generating ourselves as authorities who have privileged access both to the domain of particulars and that 'other' domain of the general. We are re-generating the universal spacetime of the real, in contrast to the 'here and now' or 'then and there' of the particular domain. And in pointing to a foundation, we are regenerating a view of separate knowledge worlds as differing ways of symbolising. When the foundation is taken as naturally given categories, it carries the implication that only one of these knowledge worlds of representative particulars is a true representation of these universal categories. Others' knowledge worlds are, at best, partial representations of the true state of things. In the alternative relativist version of foundationism, where the foundation is taken as originating in general social practices, the knowledge worlds are understood as competing accounts.

These traditions of making knowledge claims are different forms of the generic 'othering' of colonialism. The universalist prescribes that those other, partial forms of representing and symbolising should be eschewed in favour of the truer system of modern science and mathematics. Difference is abolished through the application of standards of legitimacy deriving from modern science. A set of proper categories brings with it a set of proper knowledge authorities – a particular group of actual people who attribute praise and blame in interpreting found problems and prescribing solutions, their authority resting on their privileged view of both the universal and the particular.

The relativist dispenses with difference through making it ineffable. 'Our' knowledge and its authorities are in competition with 'theirs'. We evaluate our knowledge claims and attribute praise and blame in our way, and they do in theirs. Difference here is pushed beyond our ken and just as effectively removed from consideration. With a relativist framework I must assume the sameness of Yoruba and English number, and I become unable to credit difference.

DOING THINGS ANOTHER WAY

How can I account for things like numbers in a way that would allow me to render difference generatively? How could I grasp both sameness and difference in Yoruba and English number in ways that are useful? My suggestion is that we give up notions of foundations and their imaging in knowledge. I want to refuse the vision that has my old text insisting, and assuming as self evident, that numerals are fundamentally distinct from numbers as representer and represented. This will have us also refusing the notion that numbers and numerations are found objects. It will enable us to recognise and celebrate the ingenuity and creativity of our translations, and instead of denying them significance we can see that the difference we seek to grasp lies in the translation.

Giving up notions of foundations brings with it a focus on the doing, the enacting. It also involves the recognition that sometimes numbers and numerations can be enacted as, accomplished as, objects, as they are for example in some ways of doing mathematics. This last recognition brings with it the corollary that as objects, numbers and numerations are not (not ever) complete objects; their making as objects is a project that can not be completed.

This focus on the doing of numbers is both a simplifying and a complicating move. It is simplifying because we now have only one domain where everything is only its (incomplete) self, not a symbol or a thing, but always and already both a symbol and a thing. It complicates because now we have a myriad thing–symbols, a glorious profusion of participants, all with ways of their own, to deal with in our interpretations. No longer can we restrict ourselves to the few neat generic categories of the foundation, categories which are easily manipulated in their purity.

In beginning to understand something of the significance of this move I turn now to considering a letter received by a district officer representing the British Colonial Office in Ibadan in the early 1920s.[6]

> ... the recent 1921 census was not correct on the whole for two reasons: 1. It is a customary instinct in Yorubaland to be disinclined to tell or count the number of people in a town, District or Household for fear they will die – this instinct finds expression in the proverbial saying 'A ki ikà ọmọ f'olọmọ, A ki ikà ọmọ f'obi'. 'It is not customary to count children for, or in the presence of their parents.' 2. The recent census unfortunately coincided with the tribute collection. This makes our ignorant people strongly suspect, though wrongly, that the intention was to extend tribute-payment to women & children. The last census was correct only where the minds of the Heads of Houses had been prepared by men intelligent & well known to them. So, I would humbly suggest:
>
> (i) Either that your respected self or one acquainted with, or well-known to our people like yourself again undertake to recount all quarters whose Bales were not previously instructed before the last counting; or (to save time and money) call upon the supervisors who did not proceed upon that line suggested to make re-count, with the advice to instruct the Bales before recounting.
>
> The method we adopted in our quarter was to call the Headman with another man & a woman; to ask them to give the *names* of men, each represented by a stone; after all the names were exhausted, the stones were counted; so for women & children. If some such method were adopted with the preparation spoken of, a recount in 2 or 3 adugboo or quarters would, I presume, show that an increase of nearly 25% would not be an exaggerated estimate (Akinyele, 1921).

The letter was penned on 18 August 1921, and the writer, Reverend A. B. Akinyele, later Bishop Akinyele, was one of the Yoruba Christian clergymen committed to the cultural work of Yoruba ethnogenesis as part of their Christianity (Peel, 1989). No doubt it was helpful to the British colonial officer, showing as it does a way to mediate between the Yoruba understanding of numbers and their significance and British visions of numbers and their roles in running an empire (see Apparadurai, 1993; Porter, 1995).

We can understand the work that the stones do as translating. They are translating between numbers which have life in the *adugboo* or quarters of 1920s Ibadan and the numbers which have life in the British Empire. What are the stages in this translating? First there is the repeated making of *singularities*: a person steps forward, or a name is uttered and a stone is placed. The routine is repeated until the set of people is exhausted. Second a *plurality* is made as the stones are taken as a collection. Third, this number is rendered a *singularity*, the population of a compound. Fourth, a further *plurality* is made as the numbers from many compounds are collected, to become a further *singularity*, the population of Ibadan. And so on. In a series of recursive switches between singularities and pluralities, a child asleep on his mother's back in a compound in Ibadan on a particular day in 1921 enters the ledger books of the British Empire.

Each of these makings of singularities and pluralities is achieved in repetition of a series of routine gestures and utterances. In the first making of singularity, a name is uttered or a person pointed to, and a stone is placed; the routine is repeated until all persons have been so rendered. This singularity is translated as a plurality by the picking up of a stone accompanied by the utterance of a word, repeating until all stones are replaced, and a further singularity constituted. With repeated inscriptions in a listing activity, the populations of the many compounds become a plurality. And in a mathematical process of addition which we can understand as analogous to a series of routine gestures and utterances, a further singularity emerges. When we do numbers, when we relate one to many, we do it in routine small bodily actions.

My point in elaborating this with a detail that is apparently redundant is to point out that what we might call the holographic effect of numbers (Strathern, 1995) – their capacity seamlessly to connect a child sleeping on his mother's back in Ibadan with the ledger books of the British Empire, and at the same time negotiate a scale shift from the minutiae of family life to the macro-social category of British Empire, depends on figuring and a recursive re-figuring. The figures and their re-figuring emerge in repeated activities with hands and words. The precision and specificity of some routine gestures, and the precision and specificity of figuring – the image of singularity continually re-accomplished from a background of plurality – are mutually enabling. This is a quite different view of what numbers and numerations are. The bodily doings of number, and the pictures that are integral to those doings, re-emerge as significant in any account of numbers.

The series of routine actions which generate a sleeping child in Ibadan as enumerated-boy in the census are generating a relation: a relation, not in the sense that the child, as son of Kehinde and 'Diran, is a kin relation – the embodiment of his parents' relationship – but rather as relation of the sort one-to-many. The enumerated boy of the census is a relation of the sort which recursively juxtaposes singularity and plurality.

Most people are used to thinking of themselves as kin, as 'having relations' as we say in English. But they are probably less familiar with thinking of themselves *as* relation, as embodying the juxtaposition of their parents who relate reciprocally as husband and wife. Akinyele probably did not think his actions with pebbles rendered this boy asleep on his mother's back as the relation one-to-many. But as a relation of singularity and plurality, the embodied number/enumerated boy has the remarkable properties that Marilyn Strathern elaborated for the relation, using kin as her material (Strathern, 1995). The relation, according to Strathern, has two properties: it contains itself within itself, and it requires some criterion to complete it (in the case of numbers the criterion defining singularity which in turn enables plurality). Strathern glosses these two properties as holography and complexity respectively.

> ... being an example of the field it occupies, every part containing information about the whole, and information about the whole being enfolded in each part, [the relation] produce[s] instances of itself ... [T]hrough relational practices ... relations can be demonstrated.
>
> [The relation] has the power to bring dissimilar orders or levels of knowledge together while conserving their difference.
>
> [And] what happens when we bring these two properties (holography and complexity) together, when we consider the facility of The Relation to both slip across scales and keep their distinctiveness? In late twentieth-century parlance, our little construct starts looking like a *self-organising device* (Strathern, 1995: 18–20).

Strathern's mention of scale here does not refer to the scale we (English speakers) see in numbers (the ladder-like ascent through integers embodied in the school child's ruler). Rather, scale here implies the notion of scale as in mapping: larger scale/smaller scale, macro level/micro level. The relation, and numbers as relations are exemplary in this, has the capacity to elide the macro and the micro, all the while leaving the distinctions between the levels intact. Large puzzles and problems concerning the organisation and ordering of worlds are solved in ordering at the most micro-level when relations get into the act. As we have already seen, Rev. Akinyele solves the problem of how the British Empire is to order its subjects with the banal repetitious act of saying a name and dropping a pebble.

When we recognise that doing numbers is doing the relation one-to-many, what becomes clear is the significance of figuring and how figuring is translated through series of routine gestures in doing number. I now find myself able to recognise and give play to a contrast I identified in my old text on Yoruba and English numbers. Within the framework of that old text the contrast could not take on any significance. I am referring here to the contrasting figures that emerged when I took the two systems back to fingers and toes. In my old text I understood that process as the first step in symbolising. 'Yoruba numeral generation seems to imply a picture of numerals nested within each other'. I suggested that

> ... what we see in English is a linear passing along the fingers until the end and then keeping tally, and doing it again. Each integer is related to the one before and the one after it in a linear array.

I also suggested that this difference could be glimpsed when the alternative forms were represented with the symbolic conventions of mathematicians.

> For me as a native English language speaker, von Neuman's account seems to be correct. Cardinality in his account seems to encode a one-by-one collection of predecessors. Von Neuman has the members of an *n* membered set paired with the first *n* numbers of the series of numbers.

For von Neuman 0 is $\varnothing$, the empty set. The set which contains the empty set as its sole member is $\{\varnothing\}$, one; the successor of this number is $\{\varnothing\{\varnothing\}\}$, or two; and three is the set of all sets smaller than three $\{\varnothing, \{\varnothing\}, \{\varnothing, \{\varnothing\}\}\}$. The successor of any number in von Neuman's version is generated by adding the successor of 0, that is 1. A number in this version is the last number of the series reached through one-by-one progression. A numeral names the point at which the progression ceases.

In contrast I suggest that a Yoruba speaker would choose Zermelo's account of number as correct. Here the number n is a single membered set, the single member of the set is $n-1$. Zermelo has 0 as $\varnothing$, a set with no members. Then 1 has the empty set as its sole member, $\{\varnothing\}$. The set which contains the unit set, 1, as its sole member is 2, $\{\{\varnothing\}\}$. Three is the set which has two as its sole member, $\{\{\{\varnothing\}\}\}$. In this version each number is totally subsumed by its successor, and any one number has a unified nature. I contend that for a Yoruba speaker, the model of number which would jump out as the intuitively correct account would be this one, for Yoruba language numbers carry the flavour of a multiply divided whole; there is no sense of a linear stretching towards the infinite here.

The two models agree in overall structure in identifying number; each model is demonstrably a recursive progression. Importantly for logicists, for von Neuman (and I suggest English language number) the set 14 has 14 members, while for Zermelo (and I suggest Yoruba language number) 14 has one member only.

In an informal and pictorial sense we might imagine the first pattern, von Neuman's linear picture of number, and what I intuit as English language number, as a journey. The relation of one and many can be shown as a journey where each single step adds to those already taken. A journey is at once *both* the singularity of each step *and* the plurality of many steps. Zermelo's number and the figure I intuit in Yoruba number can also be a journey. This image begins as the singularity of the journey: an arrival here after a leaving from there. Yet this singular journey is opposed by the plurality of its steps which might be infinitely added to as the journey, a singular entity, continues. Putting things this way we can see immediately that both these figures contain within them a recursion, the incipient paradox of juxtaposing plurality and singularity. And moreover, each figure can be retrieved from the other by extending the recursion of singularity and plurality in one direction or the other. Yoruba and English numbers differ in expressing the contrasting figures, the necessary opposed moments of doing numbers.

This contrast of figures variously contained in English and Yoruba numbers seems to me to be the most enlightening juxtaposition of my old comparison between English and Yoruba numbers. Yet in my old foundationist frame it can have no significance in number *use*, being merely a difference in the form of cardinality, merely a difference in ways of symbolising that can carry no weight at all in the real world. Though I felt it significant at the time I wrote that text, I had no way to recognise it other than perhaps as a characteristic of individual psychology linking it up with some weak notion of how individuals might see the world. With a different way of understanding numbers we can now take these contrasting figures as significant, recognising how they are connected to what is done with hands and words. At last I can connect my study of Yoruba numbers to what happens in Yoruba classrooms.

The figures in Yoruba and English mathematics can be retrieved from each other in the flick of a relation. Ways of being in English and ways of being in

Yoruba are connected. From one moment to the next, depending on what is done, number is either a singularity or a plurality. Yoruba and English numbers exemplify the differing moments of doing mathematics and can be translated one to the other in ways that those who use number already are already familiar with.

This is quite a different way to credit recursion than in my relativist writing. There the recursion is objectified. As a foundation established in a historical act, recursion as tallying generates numbers. In turn these are represented in various numerations, which presumably are available for bilinguals to flick between in their heads. That old way of telling things is a politics of difference which, paradoxically, by insisting that underneath all logic is the same, remakes difference as separation.

In my new way of telling, numbers as singularity and as plurality are both located in particular times and places. They coexist as multiple ways of doing numbers. The difference is kept and tolerated, in routine acting here and now. Having the recursion, the relation one-to-many which is number, in this set of actions or that, makes a particular politics of difference. We can do number this way; we can do it that way. We do it both ways and just manage, as Yoruba teachers have managed now for over a hundred years.

NOTES

[1] In commenting on this manuscript in 1986, Karin Barber noted that, 'The education authorities have already tried to simplify the Yoruba numeral system by removing one stage in the process, i.e. the fives up-and-down between the decades. Now they are copying the Indo-European system and adding from one to nine from the previous decade: *mejelogun* instead of *metadinlogbon*. At the same time they've simplified the construction of the words: *metalogun* instead of *metalelogun*. As they've left the construction of decades intact I don't know whether the new system will make it easier or just confuse things further .... The bridge between the two conceptual systems they were beginning to teach [in Okuku] before I left is only a bridge ... at a very superficial level'. The scheme of reform that Karin Barber reports here seems to derive from a proposal made in 1962 by Robert Armstrong when he was Director of the Ibadan Institute of Education in which, '... all subtractive numerals are abolished, and a zero is added at the beginning ... [so that] any of the operations of arithmetic can be easily expressed verbally in this [proposed] decimal system' (p. 21).

[2] Karin Barber (personal communication, 1986) noted 'Zero – the word used for this in schools (e.g. for 'nought out of ten') is *odo*, and there is a slang word for dunce, *olodo*. I guess this word (i.e. *odo)* was made up to fill the gap, literally, but have no idea of its etymology. *Aiwa* also exists, I think more commonly, as *aisi*, because the negative of *o wa* is *ko si*. For example, in *oriki* I have heard '*Aisi Ajayi nile, ilu toro'* (Ajayi's not being at home, the town was quiet). Robert Armstrong used *odo* as his 'added zero'.

[3] Karin Barber (personal communication, 1986) noted that 'The numbers associated with counting currency seem actually to be the *first* ones that many Yoruba children learn. I suspect that the connection with cowries has more or less been forgotten, and that *ookan, eeji, eeta* ... are now interchangeable with *eni, eji, eta*...What I heard them chanting all day long in Okuku was this: 'ONE – *ookan* ! *TWO-eeji* ! *THREE* – *eeta* ! They were learning English language, English counting, and Yoruba counting all at once'.

[4] This is an incorrect translation of Yoruba to English.

[5] The quote is from Burton, R. F. *Memoires of the Royal Anthropological Society of London*, vol. 1, p. 314, but Conant notes Burton made a miscalculation and had the Yorubans or the Dahomans – it is not clear which – losing 6820 men.

[6] I am grateful to Ruth Watson for thinking of me when she came across this letter during her own research on the history of colonial Ibadan.

BIBLIOGRAPHY

Abraham, R. C. *Dictionary of Modern Yoruba*. London: Hodder and Stoughton, 1962.

Armstrong, R. G. *Yoruba Numerals*. Ibadan: Oxford University Press, 1962.

Akinyele, A. B. *Papers*. Box 10 – Duplicates of Letters, Letter Book April 1921 – January 1922. Maps and Manuscripts Collection, Kenneth Dike Memorial Library, University of Ibadan.

Apparadurai, Arjun. 'Number in the colonial imagination.' In *Orientalism and the Postcolonial Predicament: Perspectives on South Asia*, Carol Breckenridge and Peter van der Veer, eds. Philadelphia: University of Pennsylvania Press, 1993, pp. 314–340.

Awobuluyi, Oladele. *Essentials of Yoruba Grammar*. Ibadan: Oxford University Press, 1978.

Bamgbose, Ayo. *A Short Yoruba Grammar*. Ibadan: Heinemann, 1974.

Bloor, David. *Knowledge and Social Imagery*. Chicago: University of Chicago Press, 1976, 2nd edition, 1991.

Carnap, Rudolf. *Philosophical Foundations of Physics: An Introduction to the Philosophy of Science*, Martin Gardner, ed. Chicago: University of Chicago Press, 1966.

Conant, Levi Leonard. *The Number Concept: Its Origins and Development*. New York and London: Macmillan and Co., 1896.

Crump, T. *The Anthropology of Numbers*. Cambridge: Cambridge University Press, 1990.

Dantzig, Tobias. *Number: The Language of Science*. New York: The Free Press, 1954.

Davis, P. J. and R. Hersh. *The Mathematical Experience*. London: Pelican, 1983.

Ekundayo, S. A. 'Vigesimal numeral derivational morphology: Yoruba grammatical competence epitomized.' Paper presented in the Linguistics Department, University of Ife, 1975.

Fadipe, N. A. *The Sociology of the Yoruba*. Ibadan: University of Ibadan Press, 1939, reprinted 1970.

Hallpike, C. R. *The Foundations of Primitive Thought*. Oxford: Clarendon Press, 1979.

Ifrah, Georges. *From One to Zero*. New York: Viking, 1987.

Johnson, Samuel. *The History of the Yorubas, From the Earliest Times to the Beginning of the British Protectorate*. Lagos: CMS (Nigeria) Bookshops, 1921.

Mann, Adolphus. 'Notes on the numeral system of the Yoruba nation.' *Journal of the Royal Anthropological Institute* 16: 60, 1887.

Menninger, Karl. *Number Words and Number Symbols: A Cultural History of Numbers*. Cambridge, Massachusetts: MIT Press, 1969.

Peel, J. D. Y. 'The cultural work of Yoruba ethnogenesis.' In *History and Ethnicity*, Elizabeth Tonkin, Maryon McDonald, and Malcolm Chapman, eds. London and New York: Routledge, 1989, pp. 198–215.

Porter, Theodore. *Trust in Numbers, The Pursuit of Objectivity in Science and Public Life*. New York: Princeton University Press, 1995.

Strathern, Marilyn. *The Relation Issues in Complexity and Scale*. Cambridge: Prickly Pear Press, 1995.

Thomas, Nicholas. *Entangled Objects Exchange: Material Culture and Colonialism in the Pacific*. Cambridge, Massachusetts: Harvard University Press, 1991.

Toy, C. H. 'The Yoruban language.' *Transactions of the American Philological Association*, 1878.

Watson, Helen. 'Applying numbers to nature: a comparative view in English and Yoruba.' *The Journal of Cultures and Ideas* 2(3): 1–26, 1986a.

Watson, Helen. 'The types of objects that English speakers and Yoruba speakers talk about.' In *Science and Society*. Geelong: Deakin University Press, 1986b, pp. 420–423.

Watson, Helen. 'Investigating the social foundations of mathematics: natural number in culturally diverse forms of life.' *Social Studies of Science* 20: 283–312, 1990.

Watson, Helen. 'Learning to apply numbers to nature: a comparison of English speaking and Yoruba speaking children learning to quantify.' *Educational Studies in Mathematics* 18: 339–357, 1987.

JEAN-CLAUDE MARTZLOFF

CHINESE MATHEMATICAL ASTRONOMY

THE HISTORIOGRAPHICAL CONTEXT

The literature in Western languages on the history of Chinese mathematical astronomy is generally more sophisticated than that on Chinese mathematics; it is also more ancient and can be traced as far back as the end of the 16th century if we consider the study of Chinese chronology.

In the beginning, extravagant theories on Chinese science and chronology were held. For example, in his *De emendatione tempore* (1583), the well-known Protestant scholar Joseph Scaliger (1540–1609) echoed the allegations of an imaginary but very influential traveler, the Augustinian priest Juan Gonzalez de Mendoza (Mackerras, 1991: 24–26). Later the French Calvinist Isaac de La Peyrère (1594–1676) claimed on such a basis that the Chinese empire had begun 880,073 years before AD 1594 (Grafton, 1991: 210 and 1993: 406), long before the Genesis, a scandalous proposition in the eyes of Catholics. But for many centuries, mythology was not distinguished from history and Chinese chronology remained shrouded in mystery. As for Chinese science, it was believed to be deeply buried in the mysterious diagrams of the *Yijing* and Chinese characters.

In 1729, Antoine Gaubil (1689–1759), from the China mission, wrote a pioneering history of Chinese astronomy which is still of great interest, the *Histoire de l'astronomie chinoise* (Gaubil, 1729–1732, t. 2). Exclusively using primary sources and extensive word-for-word translations of his own, the learned Jesuit translated into French many sophisticated Chinese mathematical techniques. Using the *Shoushi li* method, he gave for the first time numerical examples of predictions of solar and lunar eclipses computed the Chinese way. He based his translation on the numerical examples he had detected in the *Gujin Lüli kao* (Researches into Calendars, Computistics[1] and Musical Tubes, Ancient and Modern) (*WYK*, vol. 787, ch. 48–50), a work compiled by a mathematician from the late Ming dynasty, Xing Yunlu (? – *ca.* 1620). The *Shoushi li* is a set of mathematical techniques from the Yuan (Mongol) dynasty, which were used in China for calendrical, astronomical and astrological purposes between 1281 and 1367 and later during the whole of the Ming dynasty up to

H. Selin (ed.), Mathematics Across Cultures: The History of Non-Western Mathematics, 373–407.

1644, with some minor modifications. It is believed to represent the summit of Chinese mathematical astronomy and is still not fully understood.

But the *Histoire de l'astronomie chinoise* probably came too early and Gaubil had no successor of the same caliber. At the beginning of the 19th century, the famous historian of astronomy Joseph Delambre (1749–1822) could not understand Gaubil's work and found Chinese astronomy obscure (Delambre, 1817, vol. 1, p. 398).

During the 19th century and the first half of the 20th, with the remarkable development of sinology, many historians became obsessed by the intractable problem of the origins of Chinese astronomy. Mostly basing their studies on pre-Han literary sources, and often baffled by chronological and textual pitfalls, they reached contradictory conclusions. Some thought Chinese astronomy had developed in isolation from other cultures, while others considered its origin Babylonian (Needham, 1959: 182–186). Nowadays, the question tends to be viewed as pertaining more to archaeoastronomy than to the history of astronomy as such. As Nathan Sivin puts it, 'It seems that a number of strands – the observational, the computational, the metaphysical, the mythological, the ritual – began to evolve very gradually, each at its own pace, and only came together to form the complex of what we think of as Chinese astronomy about 100 BC' (Sivin, 1989: 58). To be sure, prior to that date, Chinese mathematical astronomy can only be described at best as limited to rudimentary calendrical computations. In such a setting, the antique Chinese computational science of the heavens appears utterly insignificant. It is thus little wonder that, with the unsolvable and ill-defined problem of its origins, Chinese astronomy was then almost entirely studied from the point of view of astronomical instrumentation, uranography and observation.

Judging from the voluminous number of publications which have accumulated since then, the history of Chinese astronomy has evidently progressed. On the whole, it still appears to be a pristine science more than three thousand years old which was obsessed with the exhaustive recording of all possible celestial and meteorological phenomena. Thus it would seem that ancient Chinese astronomers attained a legendary level of accuracy and reliability utterly unavailable in other cultures practically without mathematics worthy of the name. The acme of such a conception was attained when Pierre-Simon Laplace (1749–1827), the famous author of the *Système du Monde*, tried to find a historical confirmation of his theoretical computations establishing the secular variations of the obliquity of the ecliptic, by means of an analysis of ancient Chinese and Islamic records. But all this was done without any Chinese mathematical astronomy and without ever suspecting that Chinese results could have been calculated rather than observed (Laplace, 1904).

During the last sixty years, a handful of Chinese and Japanese historians began studying less accessible and thinly documented periods of Chinese history, especially the Han and later periods. In 1960, two research centers, one in Kyoto and the other in Tokyo, were created. Confronted with real technical texts, Japanese researchers opened entirely new perspectives and became involved with Chinese mathematical astronomy rather than mythology, literature or non-mathematical chronology. Among them, the Japanese historian Yabuuchi Kiyoshi (born in 1906) from Kyoto should be mentioned first. Unfortunately, his seminal work *Chūgoku no tenmon rekihō* (Chinese Computistics), first published in 1969, has never been translated into any Western language and has hardly reached the scholarly community beyond Japan. Even Chinese historians of astronomy quote him parsimoniously if ever. However, the same year, Nakayama Shigeru (Tokyo University) published *A History of Japanese Astronomy: Chinese Background and Western Impact*, a work which gives a succinct but technical idea of the fundamentals of the subject. More recently, one might also mention Hashimoto Keizo and his research into the late history of Chinese mathematical astronomy, mainly during the 17th century (Hashimoto, 1988), a work published from a Ph.D. dissertation under the direction of Joseph Needham. Sometimes, results presented as new discoveries by the Japanese disciple of the British historian and 'inventor' of Chinese science already appear in Gaubil.[2]

In China, results more or less similar to those of Japanese scholars of the generation preceding Yabuuchi and Nakayama were obtained independently by scholars such as Gao Pingzi (1888–1969) – an astronomer trained by the French Jesuit F. S. Chevalier from Zi-ka-wei [Xujiahui] near Shanghai (Gao Pingzi, 1987) – or the astronomers Zhu Wenxin (1883–1939) and Chen Zungui (1901–1991). Simultaneously, Qian Baocong (1892–1974), Li Yan (1892–1963) and above all Yan Dunjie (1917–1988) have published articles of the highest interest, notably on interpolation, astronomical tables, reconstruction of ancient Chinese computi and original Chinese calendars recently discovered in tombs by archeologists. But curiously, all three are better known for their numerous publications on the history of Chinese mathematics. More recently, in a series of brilliant articles, Chen Meidong (Academia Sinica, Beijing) has shown for the first time the crucial importance of approximation formulas in Chinese mathematical astronomy (Chen Meidong, 1995). The astronomer Zhang Peiyu (Zijinshan Observatory, Nanking) has attacked the difficult question of the dating of recently discovered ancient Chinese calendars from the Han, Tang and Yuan dynasties. Taking into account Chinese traditional computistics he contributed to the technical history of the subject from the point of view of Chinese techniques, minutely followed step by step. In another direction, Huang Yilong from Taiwan has dated many manuscript calendars from the Dunhuang caves, using all available tools, from philology to ancient Chinese mathematical astronomy. On the whole, these scholars give detailed descriptions of the contents of the Chinese texts rather than interpretations and translations based on modern astronomical and mathematical practice.

In Europe, quite surprisingly, Joseph Needham (1900–1997), the editor of the monumental collection *Science and Civilisation in China*, had nothing to say about Chinese mathematical astronomy, save a few casual remarks on simultaneous cycles (Needham, 1959, 406–408). In the better informed, but older and pioneering works of Wolfam Eberhard or Henri Maspero, the subject is at best confined in footnotes or appendixes, and the analysis is not always exact (Maspero, 1939; Eberhard, 1970).

Despite the very limited number of specialists on the subject and whatever the vicissitudes of future scholarship, recent studies show beyond doubt that Chinese mathematical astronomy is massively mathematical and at the same time considerably less ancient than usually believed. Given the present state of the documentation, it could hardly be much older than the Han dynasty (206 BC–220 AD). On the whole, if Greek and Chinese mathematical astronomy are compared from a merely technical and non-historical point of view, as an abstract catalogue of isolated results, it clearly appears that Chinese developments, such as those connected with the discovery of the precession, or the solar and lunar inequality, occur much later than the former, usually five or more centuries later.

THE SOURCES

The most ancient Chinese treatise of mathematical astronomy is probably the anonymous *Zhoubi suanjing* (Canon of the Gnomon of the Zhou Dynasty), a work which can be best understood as a product of the Han age (Cullen, 1996: 1). Partly literary and partly technical, it develops a cosmography based on the idea of a flat earth and on an universe measurable from the recording of noon shadows cast by a gnomon. Its mathematical apparatus intensively uses Pythagoras' theorem and its variations as well as the similarity of right-angled triangles. It also contains a theoretical list of gnomon shadows in the form of an arithmetical progression obtained by linear interpolation for each of the 24 solar nodes (*jieqi*) dividing the Chinese solar year into 24 equal intervals of time. Because of the limited interest of the Chinese for cosmological speculation, however, Chinese computistics developed arithmetically, independently from cosmology, so that space, evaluations of distances and more generally deductive geometry are utterly out of the scope of the question. Chinese calendrical astronomy was concerned not with space, but with time and angular positions of celestial bodies, the heavens being 'a source of series of events ordered in time rather than a spatially integrated whole' (Cullen, 1996: 39). Thus, the *Zhoubi suanjing* and other such treatises appear more like an interesting curiosity rather isolated from the main stream of Chinese science than a work of any importance for Chinese mathematical astronomy. Xu Guangqi (1562–1633), an influential Christian scholar and high official of the end of the Ming dynasty well acquainted with the Western science of his time, even wrote of the book that it was 'the greatest stupidity of all times' (Cullen, *ibid.*, xii).

Apart from the *Zhoubi suanjing*, there also exists a manuscript on divination by the five planets which has been entitled *Wuxing zhan* by modern historians after its discovery in a tomb at Mawangdui, near Changsha, Honan province, in 1973. Judging from the text which has been recently published, the *Wuxing zhan* is more quantitative than really mathematical (Ren Jiyu, 1998, 1: 79–94). It contains numerical data for the periods of visibility, invisibility, station or retrogradation of the five classical planets, studied in non-classical order – Jupiter, Venus, Mars, Saturn and Mercury.

In fact, the bulk of Chinese literature on mathematical astronomy has been preserved in almost one kind of source: the *lifa* treatises incorporated into most Chinese official histories. As can be surmised from their title – *li* means 'calendar' and *fa* 'methods' – the lifa treatises are composed, among other things, of ready-made mathematical techniques for the determination of the quantitative elements of the Chinese calendar without the least logical indication about their rationale. And the scope of these mathematical methods covers a broader spectrum than that usually associated with the idea of calendar. Besides purely calendrical elements questions of computational astronomy, such as coordinate conversions, algorithms for the position of the sun, the moon and the planets and eclipse prediction are also included.

In addition, contrary to an extremely widespread belief, the lifa treatises do not present any specific almanac with its corresponding rituals, festivals or usual hemerological prescriptions concerning activities to be avoided on certain days. Rather, they are heavily computational and correspond to some extent to what the Western tradition calls the computus or computistics (Butzer et al., 1993) but with a wider scope. A comparison with medieval astronomical tables or the Islamic *zīj* is also helpful, but there exists a fundamental difference because the latter are based on tight geometrical and logical deductions of Greek origin, whereas nothing of the sort exists in the Chinese context.

In view of the importance of mathematics in the lifa treatises many historians also use 'mathematical astronomy' as an equivalent to lifa. This translation has its merits because it emphasizes the fact that the scope of the Chinese computus was not limited to the calendar but was more generally concerned with a broad spectrum of astronomical questions dealt with mathematically. But it should be emphasized that if 'mathematical astronomy' ever existed in traditional China, it was akin to computistics.

Whatever the scope of mathematics, the lifa texts have in fact a still more complex structure apparently never seen outside China: they also cover more or less systematically all sorts of subjects beyond mathematics, notably historical questions concerning the reform of the Chinese calendar, philosophical digressions, evaluations of the merits and defects of specific astronomical techniques, metrology, music, and even what we would be tempted to call 'mathematics' but which in fact corresponds more to 'numerical computation' in the Chinese historical context. In particular, the few known details on the approximation $3.1415926 \leq \pi \leq 3.1415927$, discovered by Zu Chongzhi (429–500), are consigned in the lifa treatise of the *Sui History* (*Suishu*, ch.16: 409) and not in any mathematical treatise, even though such a treatise, only known under the obscure title of *Zhuishu*, has probably really existed.

In view of the variety of their content, the lifa treatises are frequently divided into several subtreatises. Some of them, especially the later ones, are extremely prolix. All in all, the full text of the lifa treatises, from the Han to the Ming dynasty, amounts to approximately 2500 pages. Unfortunately, their process of compilation is badly understood since all original archives have been lost or destroyed. Since the lifa treatises were composed up to several centuries after the fall of a dynasty by the following dynasty, their real subject is the history of mathematical astronomy rather than mathematical astronomy proper. Even in their time, the mathematical techniques they describe were rather irrelevant for contemporary Chinese astronomical practice, but certainly not as much as Ptolemy's *Almagest* or Copernicus's *De revolutionibus* would presently appear in the eyes of a contemporary professional astronomer not interested in historical matters. In fact, the Chinese considered ancient techniques as a repository of historical precedents intended for the technical education of the Confucian elite at the highest level. Consequently, certain aspects of ancient mathematical techniques were presented as positive or negative examples to be followed or avoided in astronomical practice. But nothing was ever rejected completely and no rigid paradigm was ever constituted: the Kuhnian notion of scientific revolutions does not make much sense in such a context. In a non-specialized world where scholars were trained in literature to the exclusion of other subjects, they were also supposed to be able to master all sorts of technical subjects when necessary. In a world where history was so to speak 'the queen of sciences', the angle of attack was historical above all.

Other treatises relevant to the study of Chinese mathematical astronomy are Chinese translations or better, adaptations, of foreign astronomical texts. These are extremely important for historical research since they offer a more direct insight into real history rather than reinterpreted history, as is systematically the case with official histories. With all the difficulties inherent in the acquisition of the specialized linguistic, technical and interdisciplinary skills needed for their study, our knowledge of the field is still limited and progresses very slowly. The conclusion is generally true not only in the case of contacts between India and the Islamic world but also of those between Europe and China from the beginning of the 17th century. Contrary to the former, however, these are much better documented and have left a deep imprint on Chinese mathematical astronomy and mathematics. In particular, Qing scholars left hundreds of publications representing not only slavish translations of European works but also original Chinese works, official or non-official, triggered by the official adoption of European mathematical astronomy (cf. the huge bibliography of Chinese astronomy in Ding Fubao and Zhou Yunqing, 1957.) Owing to this, the documentation relating to the late Ming and to the whole Qing period (1644–1911) is rather rewarding, historically speaking. Simultaneously, a renewal of interest of Chinese scholars in the history of their own computistic tradition became more and more prominent. Among the most interesting Chinese works of the early 19th century, one must mention the *Chouren zhuan* (Intellectual Biographies of Computists), a work finished in 1799 but first published in 1810 by Ruan Yuan (1764–1849), the patron of Chinese scholars of the lower Yangzi region. As in the astronomical works of the Jesuits translated into Chinese, the work was highly technical as well as mathematical and characterized by a rejection of astrology; it relied on extensive but biased patchworks of quotations culled from official histories. Biographies of European astronomers were generously incorporated into a separate chapter of the text. The purpose of the author was to 'prove' the Chinese origin of astronomy by means of all sorts of philological and technical comparisons not wholly unlike those Joseph Needham and his disciples extensively developed one century and a half later (Martzloff, 1997: 166–173).

Other sources are also found outside China, mainly in Korea and Japan. The most important of these is probably the extensive set of manuscripts on the *Shoushi li* known as the *Jujireki gikai* (A commentary on the *Shoushi li*) still preserved at Tokyo University[3] and composed by the Japanese Takebe Katahiro (1664–1739), a private servant of the shogun Ienobu. Written in classical Japanese and never published subsequently, Takebe's study is particularly useful because of its countless numerical examples of calendrical computations from all periods of Chinese history, fully mastered and worked out without neglecting details; by contrast, Chinese sources are particularly wanting in this respect. Moreover, Takebe is particularly reliable and his manuscript is composed with an admirable meticulousness; it contains practically no error of any kind. Rather surprisingly, historians of Chinese mathematical astronomy, even the Japanese, have apparently never used it. Similar remarks also apply in the case of Korean sources.

In a different order of ideas, a few Chinese historical calendars have been discovered as a result of archeological research from the beginning of the 20th century on. The most ancient of these are carved on wooden or bamboo tablets from the Han dynasty. Those from the Dunhuang caves, Western Gansu, are paper manuscripts. They have been tentatively dated (Huang Yilong, 1992) and critically edited (Deng Wenkuan, 1996). All of these belong to the period 450 AD–993 AD, and most are authentic.

More than two million copies of the calendar were officially printed each year during the Qing dynasty (1644–1911) (Smith, 1991: 75), but contrary to what might be expected, the number of extant calendars from this recent period is extremely limited, perhaps on the order of a few hundred copies scattered all over the world.

Whatever the period, the calendars which have been preserved are all the more precious as they can be used to check the very numerous calendrical and mathematical algorithms of the official histories. As a result, it can be shown that calendars can be divided into two categories: official calendars, which conform to some version of the Chinese computus, and non-official calendars (*xiao li*, 'little calendars'), generally slightly at variance with official chronology and mathematical techniques. As a rule, the difference between both sorts of calendars appears most obvious in the location of leap months and the repartition of short and long lunar months, respectively composed of 29 and 30 days.

Last but not least, various other sources liable to contain dates are potentially important for researches into Chinese computistics. Among these one might list encyclopedias and, more generally, all sorts of printed documents, epitaphs and other lapidary inscriptions.

THE HISTORICAL AND EPISTEMOLOGICAL CONTEXT

The importance of mathematical astronomy in the Chinese world is a direct consequence of the tight association of politics with astrology and astronomy, an association which was maintained intact until the end of imperial China, in 1911 (Jiang Xiaoyuan, 1991).

Political theories from antiquity, developed by influential Han thinkers such as Dong Zhongshu (*ca.* 179–104 BC), and later incorporated into neo-Confucianism by Zhu Xi (1130–1200), assumed the existence of a resonance (*ganying*) between the celestial and terrestrial realms through the intermediary of the emperor, depository of the heavenly mandate. With the interpretation of portents, the orientation of buildings according to cosmological rules, ritual symbolism, and the monopolistic diffusion of the calendar, the monarchy lost no occasion to render the astrological connection manifest.

Under such circumstances, the desire to predict future events in order to avoid them, or inversely, to justify them in terms of political failures or successes, led to the establishment of a permanent Bureau of Astronomy and to a division of labor according to what was needed: astronomical observation, astrology, divination, time keeping or computistics. Beyond the observation of all sorts of phenomena, regular or not, mathematical computations imposed their necessity in the light of a belief in an all-pervasiveness of number, manifesting itself in all sorts of numerological correlations between phenomena. But above all, the ancient and all-encompassing practice of divination became instrumental in the formation of the belief in the possibility to predict future events.

As with astrology, mathematical predictions were not always successful and the Chinese discovered that the calendrical systems elaborated by their astrologers and astronomers were not perfect: the astronomical and the calendrical new moons frequently differed from one another, the calendrical dates of solstices and equinoxes were off by several days, eclipses failed to be predicted or were predicted but not seen, planets did not show up where they should have, if the mathematical deductions had been correct. As constantly noted by ancient Chinese historians, Chinese calendrical techniques were generally inaccurate and continuously in need of revision from the very beginning. The idea of the intrinsic limitation of mathematics was repeatedly asserted in the treatises on the computus of the official histories. On the one hand computational techniques were considered unreliable in the long run, because systems of predictive astronomy were based on the belief that a projection into the future of mathematical regularities based on the fulfillment of initial conditions was possible, a belief contradicting Chinese metaphysics. In addition, the necessarily limited precision of observations onto which mathematics was attached was felt as an aggravating circumstance. Thus Chinese traditional astronomy was in some sense more comparable with actual meteorology than ancient Western astronomy whose fundamental concern was the explanation rather than the prediction of the phenomena.

Contrary to the Platonic and Ptolemaic assumption of an eternal, unchanging and perfect heaven, the Chinese believed that the universe was in a state of constant evolution while mathematics was merely a temporary human construction, fixed once and for all and inherently approximative when couched in terms of computistics. The contradiction between an ever changing universe and mathematical models independent in some sense of physical time entailed various degrees of unpredictability of celestial positions from the knowledge of their initial state at a particular instant, or even during spans of time accessible to human experience. In other words, the future state of the heavens was felt no more exactly deductible from their present than their past one. Moreover, this conception was accompanied by the idea that temporarily undetectable discrepancies were bound to become observable only after a very long lapse of time. These were seen as the result of the slow and continuous accumulation of infinitesimal inaccuracies, unobservable and undetectable at the start, but more and more apparent with the passage of time (Martzloff, 1993–94: 66–92). Thus, if the universe were compared with a clock, it had to be slow or fast but never wholly exact.

The introduction of new mathematical techniques based on the idea of secular variations of astronomical constants under the Song and later dynasties springs from this conception of the relation between mathematics and physical reality. But the new technique did not convince the Chinese of the possibility of perfectly reproducing astronomical reality through mathematics. The calendar reformer Zhu Zaiyu (1536–1611) argued that the technique of secular variations could not faithfully represent reality because if astronomical constants were to decrease continuously, even extremely slowly, they would vanish after an immensely long but nonetheless finite time (*Siku quanshu*, vol. 786, 455–456). Consequently, experimental and mathematical knowledge of the motion of the heavens was surreptitiously but continuously bound to deviate from astronomical reality with the passage of time. Predictive systems were thus doomed to remain temporary and irremediably linked to a particular historical period.

A quantitative evaluation of the alleged rate of obsolescence of calendrical systems based on the influential prophetic literature (*chanwei*) was eventually given during the later Han dynasty; it asserted that the period of validity of a given computus was limited to three centuries (Dull, 1966: 276). Simultaneously, politics and astrology required astronomical predictions deviating as little as possible from reality, with an increasing degree of precision. This of course does not amount to saying that faked records or choices of astronomical constants determined by motives quite foreign to 'objective' astronomy did not exist. But on the whole, a tendency towards the construction of mathematical models in accord with astronomical reality increased more and more.

Under these circumstances, reforms of the calendar, or more exactly reforms of the computus, were so often promulgated that their frequency is perhaps one of the most striking characteristics of Chinese civilisation. From the initial reform promulgated in 104 BC until the end of imperial China in 1911, approximately 50 reforms of the computus were officially confirmed[4]. This amounts to approximately one reform per 40 years. But this is only an average and while many computi were short-lived and abandoned after a few years, some others lasted several centuries. Of course, not all reforms were justified by astrological or astronomical reasons. The symbolic meaning of the calendar was such that quite often the promulgation of a new computus was a major step in dynastic legitimization.

The preparation of a reform was accompanied by a competition between rival systems of mathematical astronomy, and Chinese original sources list approximately 200 such systems. Very often, more than ten years elapsed before disputes were settled by the supreme political authority of the emperor. Technical points, relating for example to eclipse cycles or the adequate choice of an epoch for the calendar, were much debated for several years, and even after a reform, controversies did not come to an end. In order to decide between rival systems, the predictive power of mathematical techniques was systematically tested using comparative statistics. Predictions (*tuibu*, lit. 'deductions concerning the pace [of celestial bodies]) were called 'close' (*qin*) when the error between reality and calculations was less than one *ke* (14 mn., approximately), 'less close' (*ci qin*) when the fall in an error interval was 2 *ke*, 'rough' (*shu*) in the case of 3 *ke* and 'off' when the prediction was still worse (*Yuanshi*, ch. 52, 1161–1170). Sometimes, the evaluation was restricted to a binary scale and predicted events were said to be conforming (*he*) or non-conforming (*bu he*) to astronomical reality.

Some techniques of evaluation were of a purely observational character. As early as the Han dynasty, lunar eclipses were taken as objective indicators of new moons and the observation of the apparent new moon (defined by the exact instant of the luni-solar conjunction and not of the appearance of the first crescent as in the Islamic world) was systematically compared with calendrical new moons. Lunar and solar eclipses were also used in the same way and planetary positions were similarly checked.

But other comparative techniques were based on records of past astronomical events, preserved in the Confucian historical literature, especially the *Zuo zhuan* (Zuo Qiuming's Commentary on the Spring and Autumn Annals). Thus, besides the future, the past itself, or rather the canonical past of China, served as a yardstick for testing predictive systems of mathematical astronomy. Consequently, computi were used both in their 'normal' and 'proleptic' form; i.e. they were not only used to predict future events but also to retrodict past events. The situation was somewhat comparable to that which would have existed if the Gregorian calendar had been considered superior to the Julian calendar on the ground of some alleged better ability to retrodict the dates of past events mentioned in the Bible. Yet, the validity of ancient astronomical records was taken for granted and never questioned before the development of textual criticism in the 18th century.

As commentaries on eclipse prediction developed in the *Xin Tang Shu* (New Tang History) or the *Yuanshi* (History of the Yuan Dynasty) show, the question of what was to count as a success or failure was not settled by merely classifying eclipses into two categories: seen or not seen. For example, the *Xin Tang Shu* (New Official History of the Tang Dynasty) notes the case of a solar eclipse predicted but not seen. The event was interpreted as a success by explaining that the moon had changed its usual behaviour: the eclipse should have been seen but had actually been avoided as the result of the emperor's adequate and 'virtuous' response to previous heavenly warnings. Similar remarks about the unpredictable behaviour of the planets were also explicitly made (*Xin Tang*

Shu, ch. 17b: 625). Likewise, the *Yuanshi* (Official History of the Yuan Dynasty), mentions the case of a wrong prediction of a solar eclipse and interprets the event by saying that the sun had changed its rate of motion whereas the mathematical computations of the *Shoushi* computus were nonetheless wholly correct (*Yuanshi*, ch. 52: 1139). But the wrong predictions were not necessarily interpreted in such a way, and conversely, verified predictions were not evaluated as proving the correctness of a given computus but were sometimes attributed to hazard (Chen Meidong, 1995: 30). The underlying notion of prediction was thus not deterministic in any sense of the word but rather astrological, political and mathematical at the same time. Precisely for these reasons, the idea of mathematical laws of nature does not belong to the intellectual framework of Chinese mathematical astronomers. But with the successive reforms of the calendar, the precision of mathematical techniques was believed to be ever increasing and the importance of astrology diminished accordingly.

CONTACTS BETWEEN CHINA AND OTHER CULTURES

Chinese mathematical astronomy presents structural features analogous to those of Babylonian astronomy. In particular, the treatment of the 'solar equation of the center' in the two cultures appears particularly striking: both rely on what the historian of astronomy Otto Neugebauer has called step functions or zig zag functions[5] or generalizations of these (Jiang Xiaoyuan, 1988) (Figure 1). However, the time lag of approximately 800 years between the appearance in Mesopotamia and China of these similar techniques excludes direct contacts between both civilizations. Indirect influences remain of course possible but have not been established hitherto.

The consequence of the diffusion of Buddhism in the Chinese world during the first millennium of our era is better documented. Especially during the Tang dynasty (618–907), a considerable number of astronomical and astrological texts were brought into China by religious groups of various creeds: Buddhists, Zoroastrians, Nestorians, Manicheans and Muslims.

Among the few texts which have survived, one first notes the *Jiuzhi li* (Computus of the 'Nine Demons', i.e. the sun, the moon, the five classical planets and the two nodes of the orbit of the moon, *rāhu* and *ketu*), an

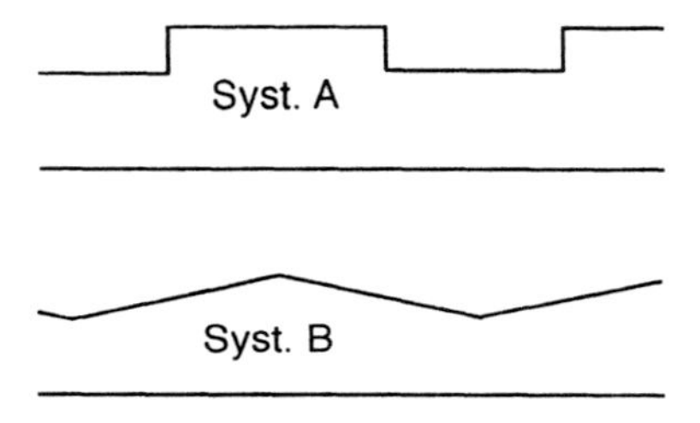

Figure 1a Graphs for the annual variations of the solar velocity in Babylonian astronomy. From: Otto Neugebauer, *Ancient Mathematical Astronomy*, 3 vols., Berlin, Heidelberg, New York: Springer-Verlag, 1975, vol. 3, fig. 7, p. 1317.

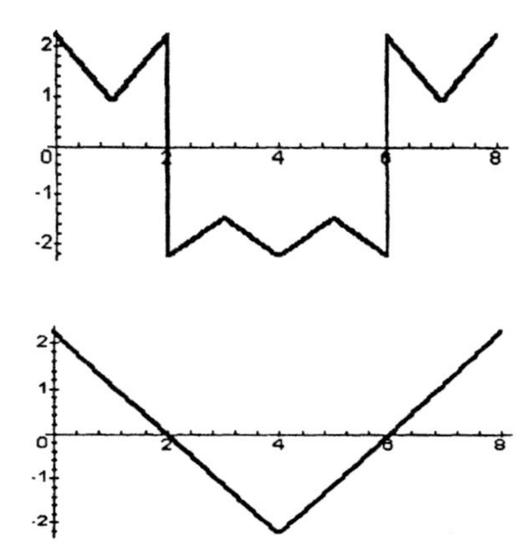

Figure 1b Graphs for the variation of the solar velocity in two Chinese computi from 600 and 724 AD. From Qu Anjing *et al.*, 1994: 232. Used with the kind permission of the author.

astrological text representative of the Indian *Ārdharātrika* school of astronomy. The question of sources of the *Jiuzhi li* is open and it is not even known whether it is a translation or a compilation of Indian texts. Clearly, however, the Chinese text preserves the Indian methods of astronomical calculations (*karaṇa*), though the motions of the five planets are omitted. Is has been translated into English by Yabuuchi with an introduction by Michio Yano, one of the best scholars in the field, able to master Chinese astronomy as well as Sanskrit and Arabic sources (Yabuuchi, 1988: 1–42). Furthermore, from a mathematical point of view, it has often been noted that the text of the *Jiuzhi li* contains a table of sines apparently based on an ancient table of chords by Hipparchus and an unambiguous mention of the Indian zero (see, for example, Martzloff, 1997: 100–101 and 207).

Another astrological manual, the *Qiyao ranzai jue* ('Tricks' to Prevent Disasters Provoked by the Seven Luminaries) was introduced into Japan in 865; it contains crude planetary ephemerides for casting horoscopes and is notable for its adoption of the solar *saura-māsa* rather than the synodic month as a fundamental time of time. Thus, the proposal to replace the Chinese lunar calendar with a purely solar calendar advocated by the high official and astronomer Shen Gua (1031–1095) in notice no. 545 of ch. 2 of his *Bu bitan* (Supplementary Jottings) is perhaps not so original as it seems to many historians of Chinese science. The *Qiyao ranzai jue* is also related to the *Futian li* (The Heavenly Tally), an unofficial computus compiled in China between 780 and 783 which expresses the solar equation of the center by means of a parabolic function perhaps of Central Asian origin but differing both from its Indian and Chinese analogs (Nakayama, 1964).

The consequences for the history of mathematical astronomy of the Mongol conquest of China and of the concomittant massive emigration of tens of thousands of Central Asians and Muslims into China are not well understood (see however Yabuuchi, 1997). After the fall of the Mongol dynasty, however, a set of astronomical tables of Islamic origin was introduced into China. Less than a century later, the remaining fragments of theses tables were completed and a full translation was eventually achieved in 1477. Known as the *Qizheng tuibu* (Computational Techniques of Predictive Astronomy for the 'Seven Governors', i.e. the sun, the moon and the five classical planets), the version of the text incorporated into the *Mingshi* (Official Ming History) is entitled *Huihui li* (Muslim Computistics). The text begins with an explanation of the Muslim calendar and contains, in particular, a Chinese transliteration of the Persian names of the days of the week. The other chapters are purely prescriptive and contain no proofs or logical deductions of any kind, not even a single geometrical figure, in total contrast with the *Almagest* and later treatises of mathematical astronomy from the Islamic world. The *Huihui li* also contains a large number of astronomical tables, among which are tables for lunar, solar and planetary positions, in latitude and longitude. Tables for the prediction of solar and lunar eclipses take into account the parallax. A table of stars, not identical with that of Ptolemy, is also inserted at the end of the volume.

The text has been handed down to us in several editions; Chinese, Korean and Islamic astronomy remained in official usage together with Chinese autochthonous astronomical techniques until 1659. It is clearly related to a *zīj* composed by a certain al-Sanjufīnī, from the region of Samarkand, who had been in the service of the Mongol viceroy of Tibet, a direct descendant in the seventh generation of Genghis Khan. The manuscript contains anonymous notes in Chinese, marginal notes in Mongolian as well as an explanatory text and no less than 50 astronomical tables. A unique copy is still extant (Manuscrit arabe 6040, Bibliothèque Nationale, Paris) (van Dalen, 1996). Yano Michio and Benno van Dalen are both conducting research on the technical aspects of the subject.

From the end of the sixteenth century, mathematical astronomy was totally renewed in China as a result of the intense scientific activity of the European Jesuit mission. The scientific impact of the Fathers was so great that traditional Chinese astronomy almost sank into oblivion. At first, the Jesuits had no intention of introducing European mathematical astronomy into China, but realizing that foreign religions were generally rejected and even considered subversive by the Chinese, they constructed a better evangelizing policy. Rather than trying to convert the Chinese at all costs, they made themselves indispensable in matters of importance for Chinese leaders. One of these was the reform of the calendar, a reform which had been on the agenda of the monarchy for many years but which had never been taken seriously for want of sufficiently plausible projects. Since the educational program of the Jesuits insisted not only on theology but also on mathematics and astronomy, the Fathers did their utmost to fulfill the Chinese demand. The first proponent of such a policy was Matteo Ricci (1552–1610), a former student of Christopher Clavius (1537–1612), the celebrated reformer of the calendar who was successful in arguing for the pedagogical indispensability of mathematics to philosophy and the other disciplines in the Jesuit educational program (Westman, 1986: 93).

The predictive power of European mathematical astronomy proved so supe-

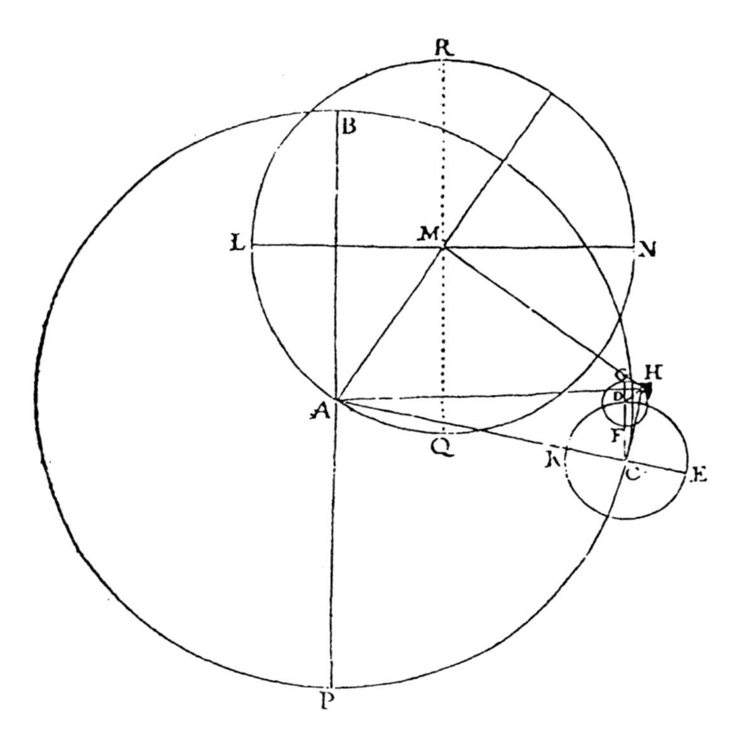

Anomalia Ec-
centri ultra ſemi-
circulum, nempe
arcus P C, cui
æqualis K D eſt
79 gr. 42 min. 20
ſec. Proinde hu-
ius duplum 159
gr. 24 mi. 40 ſec.
angulum H D
C ex lege revo-
lutionum menſu-
rat. At quoniam
ſimul latera eun-
dem hunc angu-
lum H D C
comprehendẽtia
ſuperiori capite
dantur, utputa E-
picyclorum orbis
ſemidiametri, &
quidem non ſo-
lummodo dantur
ſed in certa quo-
que ad invicem
hoc eſt quadru-

Figure 2a The theory of Mars according to Longomontanus's *Astronomia Danica*, ch. 9, p. 224. From: Longomontanus, *Astronomia Danica*, Amsterdam: G.I. Caesii, 1622, p.224.

rior to autochthonous and sinicized Muslim techniques, especially for eclipses, that despite long standing conflicts between Chinese traditionalists and modernists, Chinese computistics were completely replaced by European mathematical astronomy. But the fundamental structure of the Chinese calendar remained unmodified: the reform aimed at a better agreement between the Chinese calendar and astronomical reality, not at the replacement of the Chinese calendar by another kind of calendar. Consequently, from 1644 on, the calculation of the Chinese calendar was performed according to European methods, and ancient Chinese techniques began to sink into oblivion. Simultaneously, Jesuits were promoted to the head of the Bureau of Astronomy (list in Dehergne, 1973, 307–308); Chinese or Manchus were sometimes vice-directors but never replaced them until the suppression the Company of Jesus by Pope Clement XIV, in 1773.

Even though the number of publications on the Jesuit mission has attained a respectable size, the history of mathematical astronomy in China during the 17th and 18th centuries is not really well known. Despite considerable scholarly efforts, the historiography of the subject has not yet succeeded in wholly extricating itself from endless debates on the negative or positive role of Christianity in the process of the diffusion of science in China. Most promi-

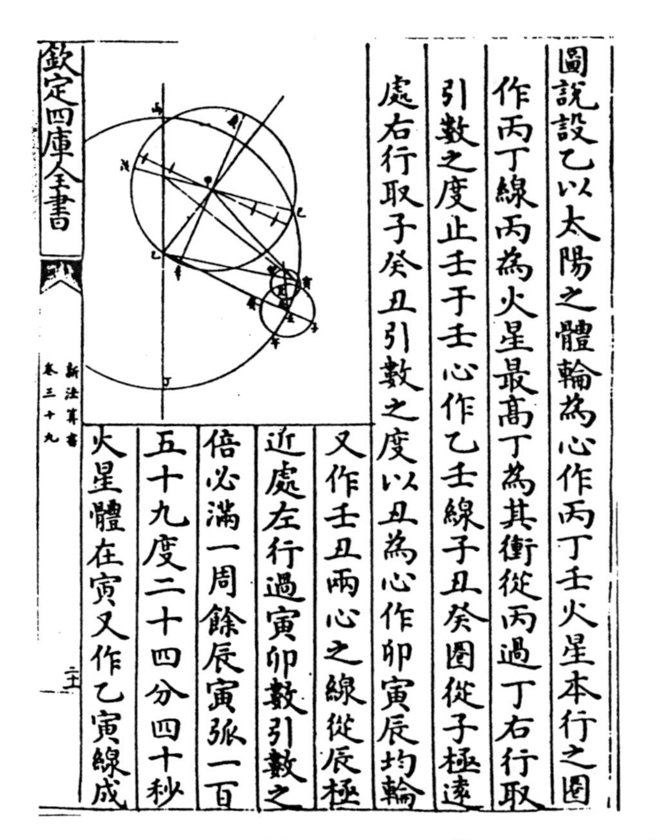
欽定四庫全書

新法算書 卷三十九

圖說設乙以太陽之體輪為心作丙丁壬火星本行之圈作丙丁線丙為火星最高丁為其衝從丙過丁右行取引數之度止壬于壬心作乙壬線子丑癸圈從子極遠處右行取子癸丑引數之度以丑為心作卯寅辰均輪又作壬丑兩心之線從辰極近處左行過寅卯數引數之倍必滿一周餘辰寅弧一百五十九度二十四分四十秒火星體在寅又作乙寅線成

Figure 2b Giacomo Rho (1592–1638)'s Chinese translation of Longomontanus in *Xinfa suanshu*, ch. 39. From: *Xinfa suanshu*, ch. 39, p. 21a, reproduced from the copy of the text preserved in the *Wenyuange* collection of the *Siku quanshu*, vol. 788, p. 695, Taipei: Shangwu Yinshuguan, 1986.

nently, the cosmological question of the diffusion or non-diffusion of heliocentrism into China has been in the foreground, where only a question of reform of the calendar, utterly foreign to cosmology, was at stake. Taking notice of the condemnation of Galileo and of Copernicanism by the Church, many secular historians have rightly or wrongly interpreted Sino-Western scientific exchanges in light of a belief in an inherent incompatibility between science and religion, without always realizing the complexity of the European situation and the non-monolithic character of the Society of Jesus. Likewise, some have also developed the idea of the antiquarian character of Jesuit science and of 'the limited ability of the transmitters [i.e. the Jesuits] themselves to communicate the best contemporary understanding of science' (Sivin, 1995). On the contrary, religious historians have sometimes cultivated paradoxes by replying that certain missionaries were overtly Copernican, especially those from Poland or, somewhat more convincingly, by arguing that the Copernican hypothesis had been left experimentally unproved, even in Europe, for a long time after

Table XXV. Parallaxe de la Lune de 5 en 5 degrés de la hauteur de ſon centre ſur l'horizon.

Haut. ☾																		
D.	M.	S.	M.	S.	M.	S.	M.	S.	M.	S.	M.	S.	M.	S.	M.	S.	M.	S.
0	54	6	54	32	55	25	56	20	57	18	58	16	59	16	60	19	61	23
5	53	53	54	20	55	11	56	9	57	4	58	2	59	3	60	5	61	10
10	53	16	53	45	54	56	55	30	56	26	57	22	58	22	59	24	60	27
15	52	14	52	43	53	32	54	22	55	21	56	22	57	13	58	16	59	17
20	50	50	51	18	52	8	52	57	53	50	54	46	55	42	56	41	57	41
25	49	1	49	30	50	16	51	3	51	54	52	48	53	43	54	40	55	38
30	46	56	47	18	48	0	48	48	49	37	50	27	51	20	52	14	53	10
35	44	18	44	40	45	26	46	10	46	58	47	44	48	33	49	24	50	17
40	41	26	41	48	42	29	43	10	43	53	44	42	45	26	46	12	47	2
45	38	15	38	30	39	15	39	52	40	32	41	12	41	57	42	41	43	24
50	34	46	35	5	35	40	36	13	36	50	37	28	38	6	38	47	39	28
55	31	2	31	18	31	48	32	20	32	52	33	25	34	0	34	35	35	13
60	27	2	27	19	27	43	28	10	28	37	29	8	29	38	30	10	30	42
65	22	52	23	2	23	27	23	49	24	13	24	37	25	2	25	29	25	53
70	18	30	18	40	18	58	19	18	19	38	19	55	20	17	20	37	20	0
75	14	12	14	20	14	36	14	53	15	9	15	20	15	36	15	52	16	8
80	9	32	9	37	9	47	9	57	10	7	10	17	10	28	10	39	10	50
85	4	48	4	50	4	55	5	0	5	5	5	10	5	15	5	20	5	25
90	0	0	0	0	0	0	0	0	0	0	0	0	0	0	0	0	0	0

Figure 3a La Hire's table for the lunar parallax. [The title reads: Parallax of the moon from 5 to 5 degrees of the height of its center on the horizon.] From: Philippe de la Hire, *Tables astronomiques*, 3rd ed., Paris: Chez Montalant, 1735 (the table is identical with that of the Latin first edition, Paris, 1702).

the arrival of the Jesuits in China. Even now, polemical debates along these lines are not extinct and are always subject to ideological taboos on both sides.

A more solid knowledge of the historical context of Chinese mathematical astronomy and of the Jesuit mission during the 17th and 18th centuries has been triggered by long-term research projects involving international cooperation, especially in Belgium (Golvers, 1993; Witek, 1994). Together with the patient accumulation of scholarly publications from the dual points of view of European and Chinese sources (Zürcher *et al.*, 1991), a more objective picture of the scientific intercourse between Chinese and Europeans begins to emerge slowly.

It appears that at the beginning, Jesuit adaptations of European works borrowed much from Clavius. In particular, the first six books of Clavius's extensive commentary on Euclid's *Elements* were translated into Chinese as early as 1607. More significantly perhaps, the astronomical encyclopedia translated into Chinese under the aegis of the German Jesuit Adam Schall von

自地平向天頂於每五度太陰地半徑差之表

月心高度	分	秒	分	秒	分	秒	分	秒	分	秒	分	秒	分	秒	分	秒	分	秒
〇	五四	六	五四	三二	五五	二五	五六	二〇	五七	一八	五八	一六	五九	一六	六〇	一九	六一	二三
五	五三	五三	五四	二〇	五五	一一	五六	九	五七	四	五八	二	五九	三	六〇	五	六一	一〇
一〇	五三	一六	五三	四五	五四	五六	五五	三〇	五六	二六	五七	二二	五八	二二	五九	二四	六〇	二七
一五	五二	一四	五二	四三	五三	三二	五四	二二	五五	二一	五六	二二	五七	一三	五八	一六	五九	一七
二〇	五〇	五〇	五一	一八	五二	八	五二	五七	五三	五〇	五四	四六	五五	四二	五六	四一	五七	四一
二五	四九	一	四九	三〇	五〇	一六	五一	三	五一	五四	五二	四八	五三	四三	五四	四〇	五五	三八
三〇	四六	五六	四七	一八	四八	〇	四八	四八	四九	三七	五〇	二七	五一	二〇	五二	一四	五三	一〇
三五	四四	一八	四四	四〇	四五	二六	四六	一〇	四六	五八	四七	四四	四八	三三	四九	二四	五〇	一七
四〇	四一	二六	四一	四八	四二	二九	四三	一〇	四三	五三	四四	四二	四五	二六	四六	一二	四七	二
四五	三八	一五	三八	三〇	三九	一五	三九	五二	四〇	三二	四一	一二	四一	五七	四二	四一	四三	二四
五〇	三四	四六	三五	五	三五	四〇	三六	一三	三六	五〇	三七	二八	三八	六	三八	四七	三九	二八
五五	三一	二	三一	一八	三一	四八	三二	二〇	三二	五二	三三	二五	三四	〇	三四	三五	三五	一三
六〇	二七	二	二七	一八	二七	四三	二八	一〇	二八	三七	二九	八	二九	三八	三〇	一〇	三〇	四二
六五	二二	五二	二三	二	二三	二七	二三	四九	二四	一三	二四	三七	二五	二	二五	二九	二五	五三
七〇	一八	三〇	一八	四〇	一八	五八	一九	一八	一九	三八	一九	五五	二〇	一七	二〇	三七	二〇	〇
七五	一四	一二	一四	二〇	一四	三六	一四	五三	一五	九	一五	二〇	一五	三六	一五	五二	一六	八
八〇	九	三二	九	三七	九	四七	九	五七	一〇	七	一〇	一七	一〇	二八	一〇	三九	一〇	五〇
八五	四	四八	四	五〇	四	五五	五	〇	五	五	五	一〇	五	一五	五	二〇	五	二五
九〇	〇	〇	〇	〇	〇	〇	〇	〇	〇	〇	〇	〇	〇	〇	〇	〇	〇	〇

Figure 3b Jean-François Foucquet (1665–1741)'s manuscript Chinese translation of this table. From: Foucquet's *Lifa wenda*, Vatican Library, Borgia Cinese 319(4), p. 15.

Bell (1592–1666) and of the Italian Giacomo Rho (1592–1638) from 1628 on, and later known as the *Xinfa suanshu* (Computational Techniques [of Astronomy] According to the New [European] Methods) relies on the works of many famous astronomers from various periods of European history, such as Ptolemy, Regiomontanus, Copernicus (from the mathematical point of view[6]), Galileo, Tycho Brahe, Longomontanus (Figure 2), Viète and even Kepler (Martzloff, 1998b). The encyclopedia also contains a full exposition of trigonometry, plane and spherical, sets of astronomical tables and more generally all that was needed at the time for the practice of mathematical astronomy in tune with the contemporaneous European applied mathematical astronomy. Even the edition of Kepler's *Rudolphine Tables* published in 1627 reached China in 1646 at the latest (Mission Catholique, 1949: item no. 1902).

During the 18th century, the rivalry between French and other European Jesuits increased. On the one hand some of the former advocated a reform astronomy based on the most recent advances. In particular, the French Jesuit Jean-François Foucquet (1665–1741) wrote a new astronomical treatise in Chinese for the Kangxi emperor himself. Taking note of the most recent discoveries from the French Académie des Sciences, he expounded the most recent advances in the domain (Figure 3) especially the theory of refraction, parallaxes and geodesics (Martzloff, 1998a). On the other hand, proponents of a more conservative approach were more influential and Foucquet's works were never printed. Instead, Kangxi sponsored a treatise on epicyclic astronomy apparently based on the astronomical tables of the Venetian astronomer Nicaise Grammatici (?–1736) (Gaubil, 1970), the *Lixiang kaocheng*. Later, new

1	*jia-zi*	甲子	21	*jia-shen*	甲申	41	*jia-chen*	甲辰
2	*yi-chou*	乙丑	22	*yi-you*	乙酉	42	*yi-si*	乙巳
3	*bing-yin*	丙寅	23	*bing-xu*	丙戌	43	*bing-wu*	丙午
4	*ding-mao*	丁卯	24	*ding-hai*	丁亥	44	*ding-wei*	丁未
5	*wu-chen*	戊辰	25	*wu-zi*	戊子	45	*wu-shen*	戊申
6	*ji-si*	己巳	26	*ji-chou*	己丑	46	*ji-you*	己酉
7	*geng-wu*	庚午	27	*geng-yin*	庚寅	47	*geng-xu*	庚戌
8	*xin-wei*	辛未	28	*xin-mao*	辛卯	48	*xin-hai*	辛亥
9	*ren-shen*	壬申	29	*ren-chen*	壬辰	49	*ren-zi*	壬子
10	*gui-you*	癸酉	30	*gui-si*	癸巳	50	*gui-chou*	癸丑
11	*jia-xu*	甲戌	31	*jia-wu*	甲午	51	*jia-yin*	甲寅
12	*yi-hai*	乙亥	32	*yi-wei*	乙未	52	*yi-mao*	乙卯
13	*bing-zi*	丙子	33	*bing-shen*	丙申	53	*bing-chen*	丙辰
14	*ding-chou*	丁丑	34	*ding-you*	丁酉	54	*ding-si*	丁巳
15	*wu-yin*	戊寅	35	*wu-xu*	戊戌	55	*wu-wu*	戊午
16	*ji-mao*	己卯	36	*ji-hai*	己亥	56	*ji-wei*	己未
17	*geng-chen*	庚辰	37	*geng-zi*	庚子	57	*geng-shen*	庚申
18	*xin-si*	辛巳	38	*xin-chou*	辛丑	58	*xin-you*	辛酉
19	*ren-wu*	壬午	39	*ren-yin*	壬寅	59	*ren-xu*	壬戌
20	*gui-wei*	癸未	40	*gui-mao*	癸卯	60	*gui-hai*	癸亥

Figure 4a The oldest known testimony of the sexagenary cycle (also called sexagesimal cycle), engraved on an oracle bone (*jiaguwen*) from the Shang dynasty (*ca.* 1100 BC). The 60 couples of the cycle are enumerated in 6 successive columns, from right to left. Apart from the graphs of individual characters, the cycle is exactly the same as that still in use in China. From: Li Pu, *Jiaguwen xuanzhu* (Selected oracle bones with annotations). Shanghai: Guji Chubanshe, 1989, pp. 54–55.

European techniques, such as logarithms and developments in infinite series, were added to the list of mathematical tools important for astronomy. These new techniques gave rise to numerous new developments in Chinese mathematics during the 19th century. But the Chinese approach always remained wholly algebraic and independent of infinitesimal considerations.

THE MATHEMATICS OF THE CALENDAR

The Chinese calendar is luni-solar, but its most fundamental component is the 'sexagenary cycle', a cycle composed of 60 elements and obtained by the concatenation of two simultaneous cycles, respectively composed of 10 and 12 elements and called 'the ten celestial trunks' (*tiangan*) and 'the twelve terrestrial branches' (*dizhi*). Although the latter are now usually associated with the cycle of the twelve animals, many other hemerological associations (hours, directions, etc.) no less important also exist.

Figure 4b Transcription of *a.–c.* modern aspect of the cycle. From: Li Pu, *Jiaguwen xuanzhu* (Selected oracle bones with annotations). Shanghai: Guji Chubanshe, 1989, pp. 54–55.

The origin of the individual characters composing the trunks and branches has given birth to numerous imaginative conjectures, but none of them is convincing. It is better to see the trunks and branches as abstract series similar to *i*, *ii*, *iii*, ... or *a*, *b*, *c*, ... The purpose of the system is the enumeration of discrete time units, above all days and then months and years. The system is very ancient and is already present in the oracle bones of the Shang dynasty (1765–1122); its mathematical structure has never changed since (Figure 4). Historians and chronologists generally admit that the sexagenary enumeration of days in the Chinese calendar has never been interrupted since then but there exists no convincing proof of this assertion. Many East Asian countries also use it in their various calendars.

Each unit of the sexagenary cycle is labeled by a couple (a_i, b_j) $1 \leq i \leq 10$ and $1 \leq j \leq 12$, where a_i and b_j respectively represent the ith trunk and the jth branch, and where i and j have the same parity. The cycle begins with the couple (a_1, b_1) where a_1 and b_1 are the first *tiangan* and the first *dizhi*; the nth sexagenary couple is equal to $(a_{n \text{ amod } 10}, b_{n \text{ amod } 12})$ where

$$x \text{ amod } y \stackrel{\text{def}}{=} 1 + [(x-1) \bmod y]$$

represents the adjusted remainder function (as usual, $x \bmod y$ is defined by $x - y^* \lfloor \frac{x}{y} \rfloor$ and $\lfloor x \rfloor =$ the largest integer $\leq x$ (the floor function), for integer values of x and y).[7] In fact, the technique of enumeration of such couples is rather similar to that implicit in the Gregorian calendar, where (Sunday, 31) is followed by (Monday, 1), for example, and where the weekdays and the days of the months are enumerated simultaneously. Apart from the periodicity, the system also has the same structure as that of the Mayan *tzolkin*, a cycle of 260 days resulting from two interlocking cycles of 13 and 20 days. The Chinese have devised special techniques for manipulating these couples from tables and mnemotechnic devices (Chen Zungui, 3, 1984: 1366–1374; Weng Wenbo *et al.*, 1993).

From the point of view of calendrical computations, arithmetical operations are essentially performed on numbers between 1 and 60, representing the rank of some sexagenary couple. For example, (a_4, b_2) was manipulated as if it were identical with its rank, 14 – in order to achieve this, remainders of divisions were numbered as if the remainder n were equal to $n + 1$, so as to give a result always falling in the range 1–60 rather than 0–59. Thus, only arithmetical operations modulo 60 were used, without direct manipulations of couples (a_i, b_j), except at the last stage of calendrical calculations where plain integers were systematically converted back into sexagenary couples because only these were meaningful in the final calendar. In actual practice, the sheer memorization of the bijective correspondence between the 60 elements of the cycle and their respective ranks was sufficient to accomplish the conversion both ways.

Although the day is the most fundamental temporal unit of the Chinese calendar, and although the Chinese calendar seldom records temporal events in terms of time units shorter than a day, it is particularly important to realize that Chinese calendrical calculations do not only manipulate discrete units such as days, but also continuous or better, *fractional* units of time, usually computed with an incredible degree of precision, far surpassing what was attainable with the time-measuring devices available in imperial China.

In order to record precise instants of time, Chinese computists have tried to determine calendrical events by means of couples $\langle x, y\rangle$, where x is an integer reduced modulo 60 denoting the sexagenary rank of the day to which the event belongs, and y a fraction, most often not decimal, indicating the part of a day elapsed between the beginning of the day x, always fixed at midnight, and the event in question. In the Chinese technical terminology, x and y are respectively called 'the big remainder' (*da yu*) and 'the small remainder' (*xiao yu*), the term 'remainder' being justified by the fact that x and y are usually the quotient and the remainder of some division.

The Chinese also introduced new refinements by introducing new numbers for the notation of finer units of time analogous to hours, minutes and seconds, but a little more complex and not standardized. For example, in the *Dayan* computus of the Tantric monk Yixing (683–727) the duration of the anomalistic month *zhuanzhong* (lit. 'a whole revolution') is given as 27 days *ri*, 1685 'left-over' *yu* and 79 seconds *miao* or $27^d\ 1685'\ 79''$ (*Xin Tang Shu*, ch. 28a: 648). In fact, the context shows that in that case 1 day = 3040 *yu* and 1 *yu* = 80 *miao* so that

$$1 \text{ anomalistic month} = 27 + \frac{1685 + \frac{79}{80}}{3040} \text{ days}$$

The actual computations were still more complex, because very often several generalized pseudo-sexagesimal systems were used simultaneously within the same computus; sometimes there was a different one for each specific period of revolution. Moreover, with each reform of the computus, the metrological and numerical basis of the computations was constantly modified in a way which reminds us of ancient metrological practice, in Europe or elsewhere.

Apart from pseudo-sexagesimals, numerous other time units were introduced for special purposes, notably for the measurement of time with the clepsydra, celestial arcs along the ecliptic, the equator or the path of the moon, always considered from the point of view of time and never of space. The Chinese degree was defined in terms of the year and equated to $\frac{365.25}{360}$ degrees, approximately; it had nothing to do with the division of the trigonometrical circumference into 60 degrees or any other unit only dependent on some underlying numeration system. However, the *Shoushi* computus (1281–1367) simplified calculations in a revolutionary way by introducing centesimal fractions. With this new system 1 day = 100 *ke* (lit. 'marks'), 1 *ke* = 100 *fen* (minutes) and 1 *fen* = 100 *miao* (seconds). Since the Chinese numeration system was then fully decimal and positional, numbers between 1 and 100 were written in base ten and not in base 100. The new notation was thus practically equivalent to that which has been universally adopted.

Other fundamental units are the year and the month and their various multiples and submultiples, a bewildering variety of which can be listed. The most well-known example of these is the metonic cycle, a span of 19 years based on the assumption that 19 solar years are equal to 235 synodic months or 19×12 ordinary months + 7 leap months. The synchronisation so realized between the year and the month is usually associated with the name of Meton of Athens (*fl. ca.* 430 BC), although it was already known by the Babylonians. The metonic cycle was used as the framework of many luni-solar calendars, Eastern and Western, Chinese or not. But contrary to widespread opinion, the Chinese have also used generalized 'metonic' relations of their own depending on the equality between $(19n + 11)$ years and $(235n + 136)$ synodic months, $n = 20, 21, 22, 23, 26, 29, 31, 32, 34, 35$ (Chen Zungui, 1984: 1382–1386).

During the Han dynasty and a few centuries later, Chinese calendrical calculations used a period of 76 years, the *bu*, a period identical to that of Callipus of Cyzicos and equal to 4 times the metonic period of 19 years, an eclipse period of 513 years, the *hui*, a cycle of 3 *hui* composed of 1539 years and a great year of 3×1539 years = 4617 years. The purpose of this complicated numerology was to make various astronomical periods mutually commensurable.

These notions, together with astrological issues, gave rise to the notion of the Superior Epoch (*shang yuan*), a fictitious initial instant, corresponding to the epoch, i.e. the start of the calendar and the ultimate origin of all calendrical computations. In the simplest case, the Superior Epoch was defined in such a way that the initial new moon and the winter solstice happened simultaneously at midnight of the beginning of the first day of the calendar, a day numbered no. 1 in the sexagenary cycle. Since the general conjunction of the five planets was believed to be particularly auspicious for the dynasty, the epoch was often conceived, in addition, as the beginning of planetary cycles. Lastly, the beginnings of the anomalistic and nodical months were also set at the instant of the Superior Epoch. In general, the computational structure of a given computus was determined by these various initial conditions, and the calculation of the Superior Epoch became the most fundamental problem of the Chinese computus.

If t designates the number of years between the winter solstice of the Superior Epoch and that of the current year, n the length of the tropical year in days and fractions of day, c_i the various lengths of the cyclical periods involved in the definition of the Superior Epoch and r_i the number of days elapsed between the last beginning of the ith cycle and the current winter solstice, then the following set of equations, where t is unknown, must be satisfied in order to fulfill the various initial conditions inherent in the definition of the Superior Epoch:

$$nt \equiv r_i (\text{mod } c_i) \qquad (1)$$

If we add to these conditions a further one stating that the sexagenary order of the current year with respect to the origin must also agree with some additional numerological condition (stating for example, that the current year should be the first year of the sexagenary cycle in the sexagenary enumeration of years), then it can be shown that, in general, these equations have no solution, since the latter and any two of the former fully determine the unknown t (Qu Anjing, 1990). Consequently, (1) is insoluble in its most general form.

Chinese computists must have been puzzled by such a troublesome problem of simultaneous congruences of the first degree, a problem which was much later discovered and admired by European historians of Chinese mathematics who called it 'the Chinese remainder theorem', at the end of the 19-th century. But there exists no indication that any Chinese mathematician ever solved even particular cases of the question for calendrical purposes, prior to the publication of a correct algorithm for solving the problem, in the *Shushu jiuzhang* (Computational Techniques in Nine Chapters) of Qin Jiushao (1247).

As has been often explained, an easier variant of the so-called Chinese remainder 'theorem' – or, less wrongly, 'algorithm' – involving only small integers and no fractions, already appears in the older *Sunzi suanjing* (Sunzi's Computational Canon) – an elementary arithmetic from the 5th century AD – in the following form: 'An unknown number of objects is such that three by three, 2 remain, five by five, 3 remain, seven by seven, 2 remain. How many objects are there? Answer: $2 \times 70 + 3 \times 21 + 2 \times 30 - 210 = 23$" (Libbrecht, 1973: 269). However, nobody has ever established the slightest historical connection between this recreational trifle and the problem of the Superior Epoch.

The more general question of the history of the remainder problem in various cultures has been explored in great detail by Libbrecht in the same study, where he concludes that the Chinese technique is not connected in the least with the Indian *kuṭṭaka* method, relating to the same problem of remainders. The latter is based on successive substitutions, while the Chinese technique extensively uses calculations of greatest common divisors using the equivalent of Euclid's algorithm in the form of alternate subtractions.

Libbrecht has also analyzed Qin Jiushao's solution, which is more general that the solution usually associated with the Chinese remainder theorem, since it does not impose the usual restriction of pairwise relatively prime modulus but admits all sorts of numbers. More recently, Chinese mathematicians have also provided a good mathematical survey of the subject together with new explanations of some problems from the *Shushu jiuzhang* (Wu Wenjun, 1987; see also Martzloff, 1997). Nonetheless, knowledge of the historical aspect of the question has hardly progressed and it does not seem probable that Libbrecht's seminal work will be superseded for a long time.

Some further remarks about the way the problem raised by the Superior Epoch was settled in the original texts can be inferred from the nature of the astronomical constants chosen in various Chinese computus. For example, it has been shown recently that in his *Huangji* computus, He Chengtian (370–447) introduced a special Superior Epoch for planetary cycles while keeping the usual epoch for other calendrical elements (Qu Anjing, 1990: 33). Hence there was a system with *two* Superior Epochs. Other computists have arbitrarily adjusted the values of the planetary constants independently of observational results so as to fulfill the initial conditions imposed by the Superior Epoch at all costs. Their efforts have resulted in epochs immensely distant from the present, reckoned in millions of years, the superior limit of these being equal to one hundred million years (Qu Anjing *et al.*, 1994: numerical table listing the various epochs adopted throughout Chinese history, 154–156). But the difficult question of the reconstitution of the precise way Chinese computists derived the Superior(s) Epoch(s) in all known Chinese computus still awaits a historian.

The fundamental calculations of the Chinese calendar, which belongs to the luni-solar variety, are easier to master. They begin with the determination of two instants: that of the winter solstice and of the new moon immediately preceding it, both events being taken as the initial points of all subsequent calculations[8].

Once these two elements are computed, the *mean* components of the solar and lunar year (i.e. the 24 mean solar nodes and the successive mean new moons) are obtained by adding multiples of $\frac{a}{24}$ and $\frac{b}{12}$ to these, respectively and lastly by keeping only the integer parts of the results, since divisions of time smaller than the day are irrelevant from the point of view of the final Chinese calendar.

Not less fundamentally, the time lag between the solar year and 12 lunar months was compensated for by the insertion of a 13th leap month using the following definition which was first clarified under the Han dynasty and has never varied since:

> '*The leap month is the first month which does not contain any solar node* (jieqi) *of odd order* ' [the solar nodes count beginning with the winter solstice] (*Han Shu*, 'Lüli zhi'[9], part 1: 984).

In other words, when a leap month occurs, there exist two consecutive odd solar nodes, the first situated just before it and the second immediately after. Equivalently, the leap month, which is approximately 29.53 days long, lies wholly within a solar month – the interval between two consecutive odd solar nodes is precisely equal to $\frac{1}{12}$ year (30.43 days) or one solar month. In order to determine whether a year is liable to contain a leap month, Chinese computists devised simple rules all based on the determination of the age of the moon at the instant of the winter solstice of the year y, in more technical terms the epact $e(y)$. For example, in the *Sifen*[10] computus of the Han dynasty, a given year y is intercalary if $e(y) \geq 12$, because in that case $e(y)$ augments by 7 units per year and $12 + 7 = 19$ units corresponds to one full lunar month (in the *Sifen* computus 19 years have exactly 7 leap months. Hence $e(y)$ increases by $\frac{7}{19}$ months per year and $\frac{7}{19}$ months + $\frac{12}{19}$ months = 1 month). The intercalary nature of a given year being ascertained, $e(y)$was then divided by the monthly increase of the epact, and the resulting quotient was used to find out how many months were left between the month preceding the winter solstice and the sought leap month. The rule, however, was not rigorously exact and calendars makers were in fact obliged to check the position of the odd solar nodes with respect to the leap month so determined.

Some further remarks should be added about the cardinal importance of the winter solstice in the Chinese calendar: in fact, the choice of such an initial point for calendrical calculations was not exactly motivated by astronomy, or a least not only, but also by a metaphysical tenet: as a consequence of *yin*-

yang theories the Chinese believed that the winter solstice coincided with the minimum intensity of the *yang* and correlatively, the maximum of the *yin* (Bodde, 1975: 165). *Yin* and *yang* are the negative and positive principles of the cosmos characterised the one by quiescence, darkness, cold, femininity and earth and the second by light, masculinity, heat, dryness and Heaven. Their attributes are opposites but they are also believed to be complementary and their mutual interplay underlies all cosmic phenomena. For example, the *yin* and the *yang* manifest their influence in the alternating variation of the length of day and night, the variation of shadow lengths over the year, etc.

One method of determining the instant of the winter solstice was based on the measurement of noon shadows of minimum length cast by a standardized gnomon. Since these vary very little around the time of the winter solstice, a more precise method was devised in the fifth century. It consisted of an interpolation between measurements made several days before and after the presumed date of the solstice. The result was an exact instant of time, generally different from noon, of course (Anon., 1981: 89–90). Another method was based on the previous determination of the summer solstice followed by the addition of a constant number of days. Still another method used the weighing of earth and charcoal: the charcoal was believed to become heavier on the eve of the winter solstice (Bodde, *ibid.*: 175). Whatever the method, scientific or not, the idea of the possibility of determining invisible astronomical instants by experiments was reinforced. From a series of determinations of winter solstices, the mean length of the tropical year was deduced. The way the mean length of the synodic month was evaluated is not well documented but the repeated observation of lunar eclipses must have been important.

Once given the epoch, the winter solstice, the length of the solar year and the lunar month, the mean elements of the Chinese traditional calendar might be readily calculated. If $\frac{a}{b}$ = mean length of the tropical year, $\frac{c}{d}$ = mean length of the synodical month, both in days and fractions of day, t = number of mean years elapsed between the winter solstice of the epoch and that of the year under consideration, r_1 = instant of the winter solstice, r_2 = age of the moon at the instant of the winter solstice), then

$$\left(\tfrac{a}{b}\right)t \equiv r_1 (\text{modf } 60) \tag{2}$$

$$\left(\tfrac{a}{b}\right)t \equiv r_2 (\text{modf } \tfrac{c}{d}) \tag{3}$$

where the 'fractional modulo', modf, is defined as

$$x \text{ modf } y \stackrel{\text{def}}{=} x - y \times \left\lfloor \tfrac{x}{y} \right\rfloor \tag{4}$$

and simultaneously allows the calculation of the sexagenary number and the fractional part of the sought day (in practice, slight modifications of (1) and (2) allow calculations with integers only).

Then $(r_1 - r_2)$ mod 60 defines the instant of the *calendrical* new moon immediately preceding the *calendrical* winter solstice. Still, this general formulation is purely theoretical and is not reported as such in the original sources. In fact, for various reasons due to the complexity of the systems of time notation, the actual computations are generally rather more complex than what could be surmised from (2) and (3) but involve integers rather than fractions. For example, in the numerous computus of the Song dynasty (969–1279) the instant of the winter solstice was determined by:[11]

$$\langle \lfloor \frac{(at \bmod 60b)}{b} \rfloor, at \bmod b \rangle \tag{5}$$

and older computus used

$$\langle \lfloor \frac{cm}{d} \rfloor \bmod 60, cm \bmod d \rangle \tag{6}$$

for the instant of the new moon immediately preceding the winter solstice ($m =$ number of *months* between the epoch and the winter solstice of the current year).

Later reforms of the computus, under the Song and Yuan dynasties, introduced an amazing innovation based on the introduction of secular variations of the tropical year. According to Nakayama, the duration of the *mean* tropical year, $\bar{y}$, *in the Tongtian* computus (1199–1207) was taken as equal to $365.2425 - 0.000002t$ per year. In fact, nothing of the sort is explicitly stated in the original sources but Nakayama's conclusion follows convincingly from a rather sophisticated mathematical analysis of the very intricate technique of calculating the winter solstice in the *Tongtian* computus which can be synthesized as follows[12]:

$$\text{Winter solstice } (t) = \frac{\left(4{,}382{,}910t_1 - 237{,}811 - \left[\frac{127t_2}{10{,}000}\right] t_2\right)}{12{,}000} \text{ modf } 72{,}000 \tag{7}$$

where $[x]$ is the nearest integer function and t_1 and t_2 non-independent numbers of years linearly related to t, the number of years elapsed since the Great Origin, reckoned with respect to two different epochs (note also the hidden 60 in the form of $60 = \frac{72{,}000}{12{,}000}$).

In the *Shoushi* computus (1281–1367), however, the same technique was stated quite explicitly and the idea of a remote epoch, fixed at midnight, was abandoned. Moreover, the *mean duration* of the tropical year was supposed to diminish (resp. to augment) at the rate of 10^{-4} days per century towards the future (resp. the past). Correlatively, the secular variations of the sidereal year were considered symmetrical to those of the tropical year (*Yuanshi*, ch. 54: 1191–1192).

The idea of infinitesimal rates of change governing the evolution of the universe during extremely long periods of time was once again advocated by the musicologist Zhu Zaiyu (1536–1611) in his aborted proposal to reform the Chinese calendar at the end of the 16th century (Dai Nianzu, 1986: 170–173). But the existence of the minute quantity involved in the secular variations of the tropical year was not ascertainable from observational data, and in spite of Zhu Zhaiyu's efforts to reintroduce it, it was never used again by Chinese computists after the *Shoushi li*. In particular, the *Datong* (1368–1644) computus adopted a year of constant duration, equal to 365.2425 days, a value which coincides with that of the Gregorian calendar.

Apart from the question of the secular variation of the *mean* year length, many other elements of the Chinese calendar were constantly modified on the occasion of the calendar reforms. First, constants relating to various periods were changed. However, from the Tang dynasty, Chinese computists began using more sophisticated techniques: instead of merely relying on *mean* elements, they introduced *true* elements. This new conception sprang from earlier developments (6th century AD) based on taking into account the 'equation of the center' for the sun and moon.

In order to realize a calendar in better agreement with astronomical reality, special methods of computation of the true instant of the luni-solar conjunction were thus devised. The mathematical processes involved are numerous and particularly involved (see Zhang Peiyu, 1992: 124–125), but it should be noted that the overall principle of the calculation is rather similar to that of Ptolemy's *Almagest*[13] as described, for example, by Petersen (1974: 223) and Yabuuchi (1969: 311–323). A double correction, one due to the motion of the sun and the other to that of the moon, is added to the calculated value of the mean conjunction. Specific calculations rely both on numerical tables and on various interpolation formulas. In the case of the *Dayan* computus[14] (729–761), for example, the position of the sun is interpolated from 3 *true* solar nodes S_i, S_{i+1}, S_{i+2} such that $S_iS_{i+1} = l_1$ and $S_{i+1}S_{i+2} = l_2$ (in days and fraction of a day), $l_1 \neq l_2$. If $f(t)$ represents the longitude of the sun along the ecliptic at time t, then the corrective factor for the calculation of the true new moon is determined by

$$f(t+s) = f(t) + \frac{s\Delta_1}{l_1} + \frac{sl_1}{l_1+l_2}\left(\frac{\Delta_1}{l_1} - \frac{\Delta_2}{l_2}\right) - \frac{s^2}{l_1+l_2}\left(\frac{\Delta_1}{l_1} - \frac{\Delta_2}{l_2}\right) \qquad (8)$$

where s is the number of days elapsed since the initial solar node, S_i, and where Δ_1 are constants obtained from a table (Ang Tian-se, 1976: 146). When adequately rewritten, (8) is identical with the Newton–Gregory formula limited to order 2. Yet, this banal but anachronistic comparison does not lead us very far since such interpolation formulas are also found in the medieval world, especially in India and later in the Islamic countries (Martzloff, 1997: 339).

OTHER ASTRONOMICAL CALCULATIONS

The lifa treatises are particularly notable for the considerable importance they give to various astronomical subjects beyond calendrics. As a rule, the spectrum of subjects dealt with covers the same range as that of Western treatises such as the *Almagest*.

First and foremost, the subject is treated by means of astronomical tables. The oldest of these occurs in the *Hou Han Shu*, 'Lüli', part c, 3077–3079. The 24 solar nodes are taken as arguments of the table and the following data are listed: (*i*) solar positions according to the system of the 28 constellations (*xiu*); (*ii*) polar distances of the sun from the ecliptic, (*iii*) meridian lengths of a standard gnomon, (*iv*) day-night lengths and (*v*) culminating celestial positions at dusk and dawn (*zhongxing*). The data has been masterfully analyzed from an astronomical and statistical point of view in Maeyama, 1975–1976. Regardless of its astronomical meaning, the table is characterized by its linearity.

Later lifa treatises are interspersed with similar tables as well as numerous others specializing in the motion of the sun, moon, the classical and astrological 'planets' (i.e. the nodes of the moon), coordinates, conversions and eclipse prediction. Chen Meidong (1995) has analyzed a significant part of the whole by stressing overall regularities. In this excellent study, the tables exhibit a somewhat simpler structure than those of the Ptolemaic tradition. While the latter are constructed from geometry and trigonometry and involve lengthy iterative calculations which are difficult to subsume under simple algebraic formulations, the former can almost always be readily analyzed using polynomials or rational fractions of degree at most equal to five together with techniques of interpolation, both linear and non-linear (Figure 5).

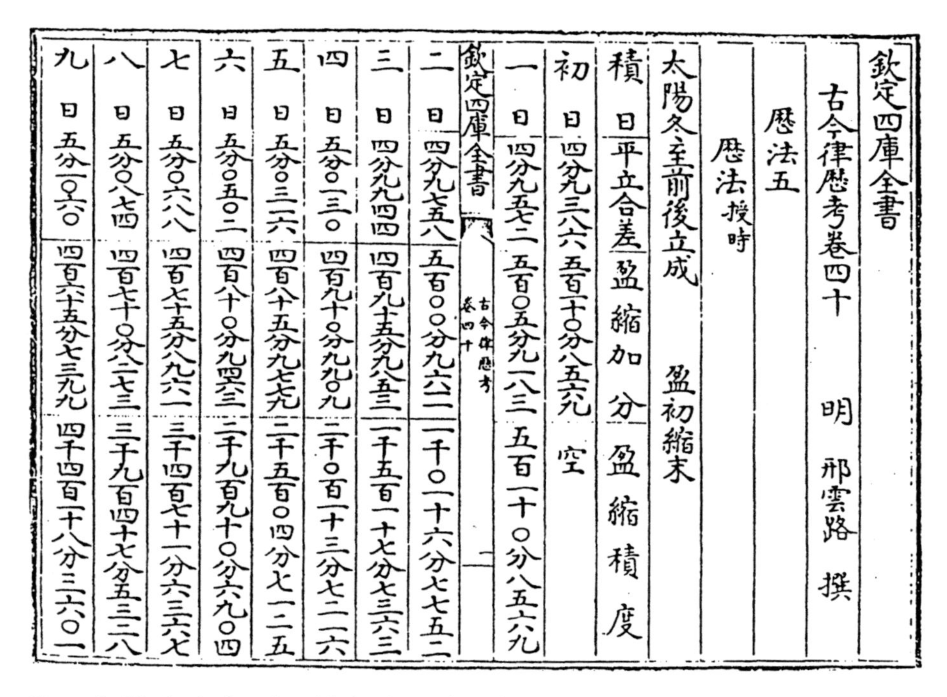

欽定四庫全書

古今律歷考卷四十　明　邢雲路　撰

歷法五

歷法 授時

太陽冬至前後立成　盈初縮末

積日	平立合差	盈縮加分	盈縮積度
初日	四分九三八六	五百一十〇分八五六九	空
一日	四分九五七二	五百〇五分九一八三	五百一十〇分八五六九
二日	四分九七五八	五百〇〇分九六一一	一千〇一十六分七七五二
三日	四分九九四四	四百九十五分九八五三	一千五百一十七分七三六三
四日	五分〇一三〇	四百九十〇分九九〇九	二千〇百一十三分七二一六
五日	五分〇三一六	四百八十五分九七七九	二千五百〇四分七一二五
六日	五分〇五〇二	四百八十〇分九四六三	二千九百九十〇分六九〇四
七日	五分〇六八八	四百七十五分八九六一	三千四百七十一分六三六七
八日	五分〇八七四	四百七十〇分八二七三	三千九百四十七分五三二八
九日	五分一〇六〇	四百六十五分七三九九	四千四百一十八分三六〇一

欽定四庫全書　古今律歷考　卷四十

Figure 5 The beginning of a table for the motion of the sun in the *Shoushi* computus (1281–1367). The table is arranged in successive lines. Let $f(x) = 513.32x - 2.46x^2 - 0.0031x^3$; then the content of the table corresponds to: 1st line – integers from 1 to 89 (x); 2nd line – $\Delta^2 f(x)$; 3rd line – $\Delta f(x)$; 4th line – $f(x)$, where Δ^1 and Δ^2 designate the finite differences of f of the first and second order, respectively. Further details on the astronomical meaning of these computations are given, for example, in Yabuuchi, 1969. From: *Gujin lüli kao*, ch. 40, p. 1a, reproduced from the copy of the text preserved in the *Wenyuange* collection of the *Siku quanshu*, vol. 787, p. 452, Taipei: Shangwu Yinshuguan, 1986.

From the Tang, the mathematical structure of tables was also stated in the form of formulaic prescriptions. In practice, computists were thus given a choice either to use tables or to perform direct calculations according to sequential rules. From these it appears that mathematical rules were most often built using *piecewise functions*. This fundamental characteristic, which has apparently never been noticed by historians of Chinese mathematical astronomy, is in fact closely related to Chinese philosophical conceptions.

According to the most widespread Chinese view of change, developed in the *Yijing* (*I Ching*, Book of Changes) and other works stemming from the same philosophical tradition, all phenomena are liable to analysis in terms of growth, decline, standstill or stagnation. In general, this conception is further subjected to an interpretation in terms of symmetries analogous to those apparent in the figurative structure of the 64 hexagrams of the *Yijing*. For example, in Chinese computistics the winter and summer solstices are systematically considered points of equilibrium where all sorts of solar phenomena reach their mean value. Consequently, the passage of the sun through its perigee or apogee was not distinguished from the instant of the solstices. This wrong conception endured until it was stigmatized by Jesuit astronomers when they began to reform Chinese astronomy. Likewise, the equation of the center for solar motion was made of symmetrical components. All this is reflected in the structure of the mathematical terminology where an essential part of the technical vocabulary consists of couples of opposites such as *ying-suo* (excess/recess), *xiao-chang* (increase/decrease), *sun-yi* (augmentation/diminution), *zeng-jian* (addition/subtraction), *chu-mo* (beginning/end), *xian-hou* (before/after), and *jin-tui* (advance/recession); the list could be extended almost at will. Although the contexts of using these terms are varied, most are interchangeable from one computus to the other (Qu Anjing *et al.*, 1994: 33). Moreover, in a given context, a term is sometimes to be interpreted as if it meant exactly its opposite. Thus it happens that *zeng*, whose normal meaning is 'addition', has to be understood as 'subtraction' in a context pertaining to the same textual unit (*ibid.*: 263); the same is true of its opposite *jian* (subtraction). This linguistic peculiarity sharply contrasts with language systems where the Aristotelian law of identity, $A = A$, is enforced[15]. Chinese rejection of Euclid's *Elements* and more generally of logic, particularly Aristotelian, from the beginning of the 17th century (Martzloff, 1997: 112) is better understood in the light of such a characteristic.

Another noteworthy aspect of Chinese mathematical astronomy is linked to its extensive use of approximation formulas applied to all sorts of domains: e.g. proto-trigonometry, theoretical length of noon shadows, polar distances, and duration of night and day (Chen Meidong, 1992). The expression designates formulas characterized by a fixed degree of approximation, determined once and for all, and not liable as such to become more and more precise through some iterative process, even though such processes are also employed in Chinese mathematics. (The Chinese version of Horner's method for finding numerical roots of polynomials is a famous example of these). Such formulas are not often mentioned by historians of mathematical astronomy but a few Indian developments in the same direction are also described (Plofker, 1996). Whether or not a connection between these could be established remains to be proved.

Finally, it should be noted that epicyclic astronomy became a fundamental component of Chinese mathematical astronomy once astronomical treatises of Indian, Islamic or European origin were translated into Chinese at different periods of Chinese history.

* * *

The fact that Chinese mathematical astronomy was still an almost wholly unexplored territory a few decades ago tells much about the elementary and fragmentary character of our knowledge of China. While our understanding of Chinese archaic astronomy still remains largely obscure and conjectural, the recent emergence of Chinese mathematical astronomy has shifted the focus of historical research towards more recent historical periods. Our positive knowledge of Chinese astronomy has so significantly increased that it is now utterly impossible to claim that Chinese astronomy is purely observational and non-mathematical. At the same time, questions concerning the diffusion of knowledge have remained as important as before, but they have received more solid answers than those formerly propounded in the case of archaic Chinese astronomy. Much remains to be done, however. The future of the field incontestably belongs to those who will manage to master simultaneously the arcane fields of ancient mathematical astronomy and classical Chinese together with one or several other ancient languages such as Persian, Arabic or Sanskrit, not to mention Japanese, Korean or Manchu and even Greek or Latin, Latin being particularly important for the study of contacts between China and Europe from the beginning of the 17th century.

NOTES

[1] 'Computus' comes from a Latin word, from the late period, meaning, 'to compute'. Practically it is exclusively used in the restricted meaning of a system or set of rules to be followed in order to construct the Christian calendar and includes such notions as epact (or the age of the moon), golden number (i.e. rank of a year within the cycle of 19 years relating to intercalary months), solar cycle, computation of the date of Easter and so on. More broadly, we apply the term computus to all sorts of calendrical and astronomical calculations. We use computus and computistics not only to refer to the mathematics underlying the making of the calendar but also to the mathematics of astronomy. In this essay, 'computitistics' designates what historians often call 'mathematical astronomers'.

[2] For example, Gaubil (1729–32, t.2, 203) clearly mentions that the *Astronomia Danica* of the Danish astronomer Longomontanus (1562–1647), one of Tycho Brahe's disciples, was partly translated into Chinese at the occasion of the Chinese reform of the calendar of 1644. (Cf. Hashimoto, 1987 and 1988).

[3] Tokyo University, manuscripts T 30/95, T 30/99 and T 30/102.

[4] Historians do not agree on the exact number of reforms. Not all of them are precisely documented in the sources.

[5] In a mathematical context these functions are more often called piecewise functions.

[6] The Church was not hostile to Copernican mathematical models provided that they were presented as fictitious hypotheses and not as indisputable truths.

[7] (Dershowitz and Reingold, 1997: 16 and 184). Some mathematical properties of the sexagenary cycle and of more general simultaneous cycles are also envisioned in the same work.

[8] For a valuable description of the very numerous specific elements of the traditional Chinese calendar (definition of the beginning of the lunar year, names of the months and their numerous variations, dynastic eras, chronological and historical peculiarities, conversion between the Chinese and Christian calendar, etc., cf. Hoang, 1885a and 1885b.

[9] *Lüli zhi* means 'Treatise on the Calendar and Musical Tubes'.

[10] *Sifen* means 'one quarter' and refers to the fact that the duration of the year in the *Sifen* computus is equal to $365\frac{1}{4}$ days, a value identical to that of the Julian calendar.

[11] Note that is necessary to add 1 to $\lfloor (at \bmod 60b)/b \rfloor$ in order to get a result between 1 and 60.

[12] Adapted from Nakayama, 1982.

[13] From Martzloff, 1999 (research in progress).

[14] One of the most famous Chinese computi. Its name, *Dayan* (lit. 'the great development') is the appellation of a divinatory method of the *Yijing*, the famous *Book of Changes*.

[15] No contradiction arises once noted that in the Chinese context A implicitly depends on time.

BIBLIOGRAPHY

(1) Chinese official histories are quoted from the critical edition of the text published in Beijing by Zhonghua Shuju starting in 1974. For example (*Xin Tang Shu*, ch. 28: 648) refers to chapter 28 of the *New Tang History*, p. 648.

(2) Quotations from the *Siku quanshu* (Complete Library of the Four Treasuries) are based on the *Wenyuange* (*WYG*) reproduction of the text, published in Taipei under the title *Wenyuange Siku quanshu* in 1986. On the history of the very famous *Siku* collection, cf. Guy, R. Kent, *The Emperor's Four Treasuries, Scholars and the State in the Late Ch'ien-lung Era.* Cambridge, Massachusetts and London: 1987.

(3) The most representative articles on the history of astronomy published in China are generally published in the quarterly journal *Ziran kexue shi yanjiu* (Studies in the History of Natural Sciences) edited under the auspices of the Academia Sinica, Beijing. From time to time the astronomical journal *Zijinshan tianwentai taikan* (Publications of the Purple Mountain Observatory) also contains articles on the history of astronomy.

(4) Many other relevant publications could be listed, for example Chinese provincial universities' journals. More generally, the range of publications of potential interest has a very broad spectrum and is in no way limited to Chinese publications, astronomical or not.

Ang Tian-Se. 'The use of interpolation techniques in the Chinese calendar.' *Oriens Extremus* 23(2): 135–151, 1976.

Ang Tian-Se. *I-Hsing (683–727 AD): His Life and Scientific Work.* Ph.D. dissertation. Kuala Lumpur: University of Malaya, 1979.

Anon. *Zhongguo tianwenxue shi* (A History of Chinese Astronomy). Beijing: Kexue Chubanshe, 1981.

Anon. *Zhongguo gudai tianwen wenwu* (Collected Papers on Astronomical Cultural Artifacts). Beijing: Wenwu Chubanshe, 1989.

Bazin, Louis. *Les systèmes chronologiques dans le monde turc ancien.* Budapest and Paris: Akadémiai Kiadó and Éditions du C.N.R.S., 1991.

Bodde, Derk. *Festivals in Classical China: New Year and Other Annual Observances During the Han Dynasty, 206 B.C.–A.D. 220.* Princeton: Princeton University Press and the Chinese University of Hong Kong, 1975.

Butzer, Paul, and Dieter Lohrmann. *Science in Western and Eastern Civilization in Carolingian Times.* Basel: Birkhäuser, 1993.

Chen Jiujin, Nha Il-Seong, and F. R. Stephenson, eds. *Oriental Astronomy from Guo Shoujing to King Sejong.* Seoul, forthcoming.

Chen Meidong. 'Zhongguo gudai youguan libiao jiqi suanfa de gongshihua' [Ancient Chinese astronomical tables and the formulation of their content by means of formulas]. In *International Conference on the History of Science in China, 1987 Kyoto Symposium Proceedings*, Keiji Yamada and Tan Tanaka, eds. Kyoto: Jinbun Kagaku Kenkyūjo, 1992.

Chen Meidong. *Gu li xin tan* (New Research into Ancient Chinese Computistics). Shenyang: Liaoning Jiaoyu Chubanshe, 1995.

Chen Zungui. *Zhongguo tianwenxue shi* (A History of Chinese Astronomy). Shanghai: Shanghai Renmin Chubanshe, 4 vols, 1980–89 (Only vol. 3, published in 1984, is cited here).

Cullen, Christopher. 'Joseph Needham on Chinese Astronomy.' *Past and Present* 87: 39–53, 1980.

Cullen, Christopher. *Astronomy and Mathematics in Ancient China: the Zhoubi suanjing*. Cambridge: Cambridge University Press, 1996.

Dai Nianzu. *Zhu Zaiyu – Ming dai de kexue he yishu juxing* (Zhu Zaiyu, a Giant of Arts and Sciences of the Ming dynasty). Beijing: Renmin Chubanshe, 1986.

Deng Wenkuan. *Dunhuang tianwen lifa wenxian jiyao* (Essential Calendrical and Astronomical Documents from Dunhuang). n.p.: Jiangsu Guji Chubanshe, 1996.

Deane, Thatcher Elliott. *The Chinese Imperial Astronomical Bureau: Form and Function of the Ming dynasty Qintianjian from 1365 to 1627*. Ph.D. dissertation, University of Washington, 1989.

Dehergne, Joseph, *SJ*. 'Gaubil, historien de l'astronomie chinoise.' *Bulletin de l'Université Aurore*, ser. 3, 6(1): 168–227, 1945.

Dehergne, Joseph, *SJ*. *Répertoire des Jésuites de Chine de 1552 à 1800*. Rome and Paris: Institutum Historicum S. I. and Letouzey & Ané, 1973.

Delambre, Jean-Baptiste. *Histoire de l'Astronomie Ancienne*. 2 vols. Paris: Ve Courcier, 1817.

Dershowitz, Nachum, and Edward M. Reingold. *Calendrical Calculations*. Cambridge: Cambridge University Press, 1997.

Ding Fubao, ZhouYunqing et al. *Sibu conglu, tianwen bian* (Quadripartite Bibliography, Astronomical Section). Beijing: Shangwu Yinshuguan, 1957.

Dull, Jack. *A Historical Introduction to the Apocryphal (Ch'an-Wei) Texts of the Han Dynasty*. Ph.D. dissertation, University of Washington, 1966.

Eberhard, Wolfram. *Sternkunde und Weltbild im alten China*. Taipei: China Materials and Research Aids Service Center, Occasional Series no. 5, 1970.

Franke, Herbert. 'Mittelmongolische Glossen in einer arabischen astronomischen Handschrift.' *Oriens* 31: 95–118, 1988.

Fritsche, H. *On Chronology and the Construction of the Calendar, with Special Regard to the Chinese Computation of Time Compared with the European*. St Petersburg: Lithographed by E. Laverentz, 1886.

Gao Pingzi. *Gao Pingzi tianwen lixue lunzhu xuan* (Selected Works of Gao Pingzi on Astronomy and Computistics). Taipei: Zhongyang Yanjiuyuan, Shuxue Yanjiusuo, 1987.

Gaubil, Antoine *SJ*. 'Histoire de l'astronomie chinoise' and 'Traité de l'astronomie chinoise.' In *Observations mathématiques, astronomiques, géographiques, chronologiques et physiques tirées des anciens livres chinois; ou faites nouvellement aux Indes et à la Chine*, E. Souciet, ed. t. 2 and 3, Paris: Chez Rollin, 1729–1732.

Gaubil, Antoine, *SJ*. *Correspondance de Pékin, 1722–1759*. Geneva: Librairie Droz, 1970.

Golvers, Noel, ed. *The* Astronomia Europea *of Ferdinand Verbiest SJ (Dillingen, 1687), Text, Translation and Commentaries*. Jointly published by the Institut Monumenta Serica, Sankt Augustin, and Ferdinand Verbiest Foundation, Leuven. Nettetal: Steyler Verlag, 1993.

Grafton, Anthony. *Joseph Scaliger: A Study in the History of Classical Scholarship*, vol. 2, *Historical Chronology*. Oxford: Clarendon Press, 1993.

Grafton, Anthony. *Defenders of the Text: the Traditions of Scholarship in an Age of Science, 1450–1800*. Cambridge, Massachusetts and London: Harvard University Press, 1991.

Hashimoto, Keizo. 'Seido no shisō to dentō Chūgoku no tenmongaku' (Traditional Chinese astronomy and the Notion of Precision). *Kansai daigaku, shakai gakubu kiyō* 11(1): 93–114, 1979.

Hashimoto, Keizo. 'Longomontanus's *Astronomia Danica* in China.' *Journal for the History of Astronomy* 18: 95–110, 1987.

Hashimoto, Keizo. *Hsü Kuang-ch'i and Astronomical Reform: The Process of the Chinese Acceptance of Western Astronomy 1629–1639*. Osaka: Kansai University Press, 1988.

Ho Peng-Yoke. *Modern Scholarship on the History of Chinese Astronomy*. Canberra: The Australian National University, Occasional Paper no. 16, Faculty of Asian Studies, 1977.

Hoang, P. *De Calendario Sinico Variae Notiones, Calendarii Sinici et Europaei Concordentia*. Zi-Ka-Wei: Ex Typographia Missionis Catholicae, 1885a.

Hoang, P. *A Notice of the Chinese Calendar and a Concordance with the European Calendar*. Zi-ka-wei: Printing Office of the Catholic Mission, 1885b.

Hopkirk, Peter. *Foreign Devils on the Silk Road*. Oxford: Oxford University Press, 1982.

Huang Yilong. 'Dunhuang ben juzhu liri xintan.' (A New Study of Calendars with Hemerological Annotations, from Dunhuang). *Xin Shixue* 3(4): 1–56, 1992.

Jiang Xiaoyuan. 'The solar motion theories of Babylon and ancient China – a new lead for the relation between Babylonian and Chinese astronomy.' *Vistas in Astronomy* 31: 829–832, 1988.

Jiang Xiaoyuan. *Tianxue zhenyuan* (The True Origin of Chinese Astronomy). Shenyang: Liaoning Jiaoyu Chubanshe, 1991.

Kennedy, Edward S. and Jan P. Hogendijk. 'Two tables from an Arabic astronomical handbook for the Mongol Viceroy of Tibet.' In *A Scientific Humanist: Studies in Memory of Abraham Sachs*, Erle Leichty and Maria de J. Ellis, eds. Philadelphia: University of Pennsylvania Museum, 1988, pp. 233–242.

Laplace, Pierre-Simon. 'Mémoire sur la diminution de l'obliquité de l'écliptique qui résulte des observations anciennes.' In *Oeuvres Complètes de Laplace publiées sous les auspices de l'Académie des Sciences, par MM. les secrétaires perpétuels*. Tome 13. Paris: Gauthier-Villars, 1904, pp. 44–70.

Li Yan. *Zhong suanjia de neichafa yanjiu* (Researches into the Interpolation Techniques of Chinese Mathematicians). Beijing: Kexue Chubanshe, 1957.

Libbrecht, Ulrich. *Chinese Mathematics in the Thirteenth Century: The* Shu-shu chiu-chang *of Ch'in Chiu-shao*. Cambridge, Massachusetts: MIT Press, 1973.

Liu, Jinxi and Zhao Dengqiu. *Zhongguo gudai tianwenxue shilüe* (A Concise History of Chinese Astronomy). Shijiazhuang: Hebei Kexue Jishu Chubanshe, 1990.

Mackerras, Colin. *Western Images of China*. Oxford: Oxford University Press, 1991.

Maeyama, Y. 'On the astronomical data of ancient China (*ca.* $-100 \sim +200$): A numerical analysis.' *Archives Internationales d'Histoire des Sciences*. 25 (97): 247–276, 1975 and 26(98): 27–58, 1976.

Martzloff, Jean-Claude. 'Space and time in Chinese texts of astronomy and of mathematical astronomy in the seventeenth and eighteenth centuries.' *Chinese Science* 11: 66–92, 1993–94.

Martzloff, Jean-Claude. *A History of Chinese Mathematics*. Berlin, Heidelberg: Springer, 1997.

Martzloff, Jean-Claude. 'Jean-François Foucquet *SJ* (1665–1741) and his aborted proposal to reform Chinese astronomy based on recent research from the Académie des Sciences.' Unpublished communication presented to the *Neuvième Colloque International de Sinologie de Chantilly, Paris, 6–9 Septembre 1998*. 1998a.

Martzloff, Jean-Claude. 'Notes on planetary theories in Giacomo Rho's *Wuwei lizhi*.' In *Western Learning and Christianity in China, The Contribution and Impact of Johann Adam Schall von Bell, SJ (1592–1666)*, Roman Malek, S.V.D., ed. 2 vols. Sankt Augustin: China-Zentrum and the Monumenta Serica Institute, vol. 1, 1998b, pp. 591–616.

Martzloff, Jean-Claude. 'Les sources chinoises des manuscrits astronomiques de Seki Takakazu (?–1708).' *Daruma, Revue Internationale d'Études Japonaises* 4: 63–78, 1998c.

Maspero, Henri. 'Les instruments astronomiques des Chinois au temps des Han.' *Mélanges Chinois et Bouddhiques* 6: 183–370, 1939.

Mercier, Raymond. 'The Greek 'Persian Syntaxis' and the zīj-i īlkhānī.' *Archives Internationales d'Histoire des Sciences* 34(112): 35–60, 1984.

Mission Catholique des Lazaristes à Pékin. *Catalogue de la Bibliothèque du Pé-T'ang*. Pékin: Imprimerie des Lazaristes, 1949.

Nakayama, Shigeru. '*Futen reki* no tenmongaku shiteki ichi.' (The position of the *Futian li* in the history of astronomy). *Kagakushi* 71: 120–122, 1964.

Nakayama, Shigeru. 'Japanese activities in the history of astronomy during the early half of 1960.' *Japanese Studies in the History of Science* 4: 1–19, 1966.

Nakayama, Shigeru. *A History of Japanese Astronomy: Chinese Background and Western Impact*. Cambridge, Massachusetts: Harvard University Press, 1969.

Nakayama, Shigeru. 'Shōchōhō no kenkyū – Tōzai kansoku gijutsu no hikaku –' (Variation of tropical year length in Far Eastern astronomy and its observational basis compared with

Western techniques). In *Explorations in the History of Science and Technology in China*, Li Guohao, et al, eds. Shanghai: Shanghai Chinese Classics Publishing House, 1982, pp. 155–183.

Needham, Joseph. *Science and Civilisation in China*. vol. 3, Mathematics and the Sciences of the Heavens and the Earth. Cambridge: Cambridge University Press, 1959.

Petersen, Olaf. *A Survey of the Almagest*. Odense, Denmark: Odense University Press, 1974.

Plofker, Kim. 'How to appreciate Indian techniques for deriving mathematical formulas' In *L'Europe Mathématique: mythes, histoires, identités*. Catherine Goldstein, *et al.*, eds. Paris: Éditions de la Maison des Sciences de l'Homme, 1996, pp. 54–65.

Qu Anjing. 'Zhongguo gudai *lifa* zhong de shangyuan jinian jisuan.' (The calculation of the number of years elapsed since the Superior Epoch in the traditional Chinese calendar). *Shuxue shi yanjiu wenji* 1: 24–36, 1990.

Qu Anjing *et al. Zhongguo gudai shuli tianwenxue tanxi* (An Analysis of Ancient Chinese Mathematical Astronomy). Xi'an: Xibei Daxue Chubanshe, 1994.

Ren Jiyu, ed. *Zhongguo Kexue jishu dianji tonghui, tianwen juan* (Collected Chinese Scientific and Technical Works, Section on Astronomy. 8 vols. n.p.: Henan Jiaoyu Chubanshe, 1998.

Shi, Yunli. 'Gudai Chaoxian xuezhe de *Shoushi* yanjiu' (Researches into the *Shoushi* computus by Korean scholars). *Ziran kexue shi yanjiu* 17(4): 312–321, 1998.

Sivin, Nathan. Cosmos and Computation in Early Chinese Mathematical Astronomy. Leiden: E.J. Brill, 1969.

Sivin, Nathan. 'On the limits of empirical knowledge in Chinese and Western science.' In *Rationality in Question: On Eastern and Western Views of Rationality*. Schlomo Biderman and Ben-Ami Scharfstein, eds. Leiden: E.J. Brill, 1989, pp. 165–189.

Sivin, Nathan. 'Chinese archaeoastronomy: between two words.' In *World Archaeoastronomy: Selected Papers from the Second Oxford International Conference on Archaeoastronomy Held at Merida, Yucatan, Mexico, 13–17 January 1986*, Anthony F. Aveni, ed. Cambridge: Cambridge University Press, 1989, pp. 55–64.

Sivin, Nathan. 'Copernicus in China.' In *Science in Ancient China, Researches and Reflections*. London: Variorum, 1995, pp. 1–53.

Smith, Richard J. *Fortune-tellers and Philosophers, Divination in Traditional Chinese Society*. Boulder, San Francisco, Oxford: Westview Press, 1991.

Swarup, G., A. K. Bag, and K. S. Shula, eds. *History of Oriental Astronomy: Proceedings of an International Astronomical Union Colloquium no. 91, New Delhi, India, 13–16 November 1985*.

Swerdlow, N. M. *The Babylonian Theory of the Planets*. Princeton, New Jersey: Princeton University Press, 1998.

van Dalen, Benno. 'Table of Planetary Latitudes in the *Huihui li – an Analysis*.' Unpublished communication presented at the Eighth International Conference on the History of Chinese Science, 26–31 August 1996, Seoul National University, Korea.

Wang Rongbin. 'Liu Zhuo *Huangji li* chazhifa de goujian yuanli.' (The structure of the interpolation techniques in Liu Zhuo's *Huangji* computus). *Ziran kexue shi yanjiu* 13(4): 293–304, 1994.

Wang Wenbo, *et al. Tiangan dizhi jili yu yuce* (Calendrical notations and previsions using the sexagenary cycle of trunks and branches). Beijing: Shiyou Gongye Chubanshe, 1993.

Weschler, Howard J. *Offerings of Jade and Silk: Ritual and Symbol in the Legitimation of the T'ang Dynasty*. New Haven and London: Yale University Press, 1985.

Westman, Robert S. 'The Copernican and the churches.' In *God and Nature, Historical Essays on the Encounter between Christianity and Science*, David C. Lindberg and Ronald L. Numbers, eds. Berkeley: University of California Press, 1986, pp. 76–113.

Witek, John W., *SJ*, ed. *Ferdinand Verbiest (1623–1688), Jesuit Missionary, Scientist, Engineer and Diplomat*. Nettetal: Steyler Verlag, 1994.

Wu Wenjun, ed. *Qin Jiushao yu* Shushu jiuzhang. (Qin Jiushao and the *Shushu jiuzhang*). Beijing: Beijing Shifan Daxue Chubanshe, 1987.

Yabuuti, Kiyoshi [Yabuuchi, Kiyoshi]. 'Astronomical tables in China from the Han to the T'ang Dynasty.' In *Chūgokū chūsei kagaku gijutsu shi no kenkyū* (Research into Chinese Mediaeval Sciences and Techniques), Kiyoshi Yabuuti, ed. Tokyo: Kadokaw shoten, 1963, pp. 445–492.

Yabuuti, Kiyoshi [Yabuuchi, Kiyoshi]. *Chūgoku no tenmon rekihō* (Chinese Computistics). Tokyo: Heibonsha, 1969.

Yabuuti, Kiyoshi [Yabuuchi, Kiyoshi]. 'The influence of Islamic astronomy in China.' In *From Deferent to Equant: A Volume of Studies in the History of Science in the Ancient and Medieval Near East in Honor of E.S. Kennedy*, David King and George Saliba, eds. *Annals of the New York Academy of Sciences* 500: 1987, pp. 547–559.

Yabuuti, Kiyoshi [Yabuuchi, Kiyoshi], trans. by Benno van Dalen. 'Islamic astronomy in China during the Yuan and Ming Dynasties.' *Historia Scientiarum* 7(1): 11–43, 1997.

Yabuuti, Kiyoshi [Yabuuchi, Kiyoshi]. Zotei Zui *Tō rekihō shi no kenkyū* (Research into the History of the Computus of the Sui and Tang Dynasty, revised edition). Kyoto: Rinsen Shoten, 1997.

Yan Dunjie. 'Song, Jin, Yuan *lifa* zhong de shuxue zhishi.' (Mathematical notions in the calendrical treatises of the Song, Jin and Yuan Dynasties) In *Song-Yuan shuxue shi lunwen ji* (Collected Papers on the History of Song and Yuan Mathematics), Baocong Qian, ed. Beijing: Kexue Chubanshe, 1966, pp. 210–224.

Yan Dunjie. 'Zhongguo gudai shuli tianwenxue de tedian.' (Characteristics of ancient Chinese mathematical astronomy). *Keji shi wenji* 1: 1–4, 1978.

Yan Dunjie. 'Shi *sifen li*.' (The *Sifen* computus explained). In *Zhongguo gudai tianwen wenwu lunji* (Collected Articles on Ancient Chinese Astronomy and Cultural Artifacts). Beijing: Wenwu Chubanshe, pp. 104–110.

Zhang Peiyu. '*Shoushi li* dingshuo richan ji lishu tuibu.' (The mathematical determination of the true new moons, the motion of the sun and other calendrical elements in the *Shoushi* computus). *Zhongguo tianwenxue shi wenji* 4: 77–103, 1986.

Zhang Peiyu *et al.* '*Xuanming li* dingshuo jisuan he lishu yanjiu.' (Research into the computation of the true new moons in the *Xuanming* computus). *Zijinshan tianwentai taikan* 11(2): 121–155, 1992.

Zhu Wenxin. *Lifa tongzhi* (A General Treatise on Chinese Computistics). Shanghai: Shangwu Yinshuguan, 1934.

Zürcher, Erik, Nicolas Standaert *SJ* and Adrianus Dudink. *Bibliography of the Jesuit Mission in China ca. 1580 – ca. 1530.* Leiden: Leiden University, Centre of Non-Western Studies, 1991.

T. K. PUTTASWAMY

THE MATHEMATICAL ACCOMPLISHMENTS OF ANCIENT INDIAN MATHEMATICIANS

INTRODUCTION

The history of any remote ancient civilization is often surrounded by mystery and controversy. Even now, Western scholars have not given enough credit to the work of some of the ancient Indian mathematicians, particularly Brahmagupta and Bhāskara II, despite the fact that their entire works were translated into English by the British Sanskrit scholar Henry Thomas Colebrooke in the early part of the nineteenth century. The purpose of this article is to set forth the facts and leave the reader to interpret and assess them according to his own understanding of the subject. A sincere attempt has been made to do so. Due to limitations of space, this article discusses only some of the mathematical accomplishments of only a few outstanding mathematicians of ancient India.

THE DECIMAL SYSTEM OF NUMERATION

The most important mathematical contribution of ancient India is the invention of the decimal system of numeration, including the number zero. The unique feature of this system is the use of nine digits and a symbol zero to represent all the integral numbers by assigning a place value to the digits. In 1912, Professor G. B. Halstead remarked, 'The importance of the creation of the zero mark can never be exaggerated. This giving to airy nothing, not merely a local habitation and a name, a picture, a symbol but helpful power, is the characteristic of the Hindu race whence it sprang. It is like nirvana into dynamics. No single mathematical creation has been more potent for the general on-go of intelligence and power' (p. 20). A few years later (1926), he showed that zero existed in India before 200 BC.

Two ancient Sanskrit texts, *Yajurveda Saṃhitā* and *Taitireeya Saṃhitā*, provided lists of numerical denominations. In *Valmiki Ramayana*, a spy of the evil King Ravana narrates to his king the exact strength of the army of his enemy, Rama. He explains the numeration system employed and states that Rama's army is estimated to be $10^{10} + 10^{12} + 10^{20} + 10^{24} + 10^{30} + 10^{34} + 10^{40} + 10^{44} +$

H. Selin (ed.), *Mathematics Across Cultures: The History of Non-Western Mathematics*, 409–422.

$10^{52} + 10^{57} + 10^{62} + 5$. Regarding the ages of these works, according to B. B. Datta (1962: 97), the Arab historian Abu'l Ḥassan Al-Mas'ūdī (943 AD) writes, 'A congress of sages at the command of the creator Brahma invented the nine digits and also their astronomy and other sciences.' Writers on the Mohenjo Daro and Harappa civilizations (3000 BC) refer to the decimal system of numeration found in these excavations. The Jaina religious works *Sūrya Prajñapti*, *Jamboo Dwipa Prajñapti*, *Sthānāga Sūtra*, *Uttārādhyayana Sūtra*, *Bhagavati Sūtra* and *Anuyoga Dwāra Sūtra*, dating from 500 to 100 BC use large numbers in the decimal system. In the Buddhist work *Lalita Vistara* of the first century BC, Buddha explains to a mathematician named Arjuna the system of numerals in multiples of 100, starting from 10^7 to 10^{63}.

SULBASŪTRAS

The root meaning of the word *sulv* is to measure, and geometry in ancient India came to be known by the name *sulba* or *sulva*. Sulbasūtras mean the geometric principles of geometry. Only seven Sulbasūtras are extant: *Bodhāyana*, *Āpasthamba*, *Kātyāyana*, *Mānava*, *Maitrāyana*, *Varāha* and *Vidhūla*, named for the sages who wrote them. The sulbas are not mathematical treatises; they are only adjuncts to certain Hindu religious works. They were written between 800 and 500 BC.

The sulbas contain a large number of geometric constructions for squares, rectangles, parallelograms and trapezia. They described how to construct

(a) a square *n* times in area to a given square;
(b) a square of area equal to the sum of the two squares;
(c) a square whose whole area is equal to the difference of two squares;
(d) a square equal to a given rectangle;
(e) a triangle equal to a rectangle;
(f) a triangle equal to a rhombus; and
(g) a square equal to the sum of two triangles or two pentagons.

Several theorems are proven in the sulbas:

(a) The diagonal of a rectangle divides it into equal parts.
(b) The diagonals of a rectangle bisecting each other and opposite areas are equal.
(c) The perpendicular through the vertex of an isosceles triangle on the base divides the triangle into equal halves.
(d) A rectangle and a parallelogram on the same base and between the parallels are equal in area.
(e) The diagonals of a rhombus bisect each other at right angles.

There is also the famous theorem named after Pythagoras. The Pythagorean theorem occurs in the sulbas in the following form: The diagonal of a rectangle gives an area equal to the sum of the areas given by its length and breadth. The proof of the theorem is given as follows:

In the figure below, points are taken on the sides of the square ABCD such

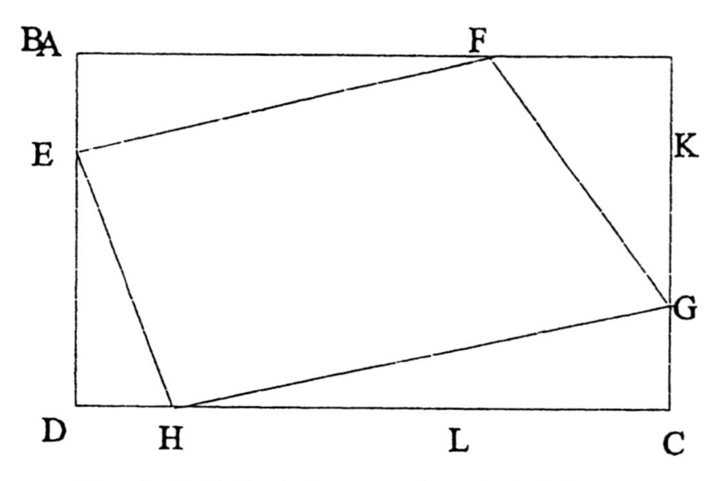

Figure 1 Right triangle theorem as shown in the *Sulbasūtras.*

that

BF = BK = CG = CL = DH = AE

Now sq ABCD = sq on DL + sq on BF + 4

$\triangle$AEF = $AF^2 + AE^2 + 4\ \triangle AEF$

EFGD is a square and sq ABCD = $EF^2 + 4\ \triangle AEF$.
Hence

$$AF^2 + AE^2 = EF^2$$

Pythagoras is believed to have lived from about 572 BC to 501 BC, and so the sulbas precede him. The German historian Hankel in *Zür Geschichte der matematik* (Leipzig, 1874: 98), is of the opinion that the above proof is more Indian in style than Greek. The well-known proof given by Euclid is his own.

A remarkable approximation to $\sqrt{2}$ occurs in three of the sulbas, *Bodhayana*, *Āpasthamba* and *Kātyāyana*, namely

$$\sqrt{2} \approx 1 + \frac{1}{3} + \frac{1}{34} - \frac{1}{3 \cdot 4 \cdot 34}.$$

This gives $\sqrt{2} = 1.4142156\ldots$, whereas the exact value is 1.414213 .... The approximation is thus correct to five decimal places. The problem arises in the construction of a square double a given square in area. Nowhere in the sulbas is it mentioned how the above approximation was arrived at. B. B. Dutta (1932) has provided an elegant proof.

The *Mānava Sūtra* gives the following:

$$40^2 + 40^2 \approx 56^2, \qquad \text{i.e. } \sqrt{2} \approx \frac{7}{5}$$

$$4^2 + 4^2 \approx [5\tfrac{2}{3}]^2, \qquad \text{i.e. } \sqrt{2} \approx \frac{17}{12}$$

$$36^2 + 90^2 \approx 97^2, \qquad \text{i.e. } \sqrt{2} \approx \frac{577}{408}$$

The very fact that ancient Indians knew that $\sqrt{2}$, $\sqrt{61}$, etc. could not be exactly determined leads to the concept of irrational numbers, although a formal mathematical definition and theory were introduced by Cantor, Dedekind and Weistrass centuries later.

THE BAKSHĀLĪ MANUSCRIPT (*ca.* 200–400 AD)

The Bakshālī Manuscript is the name given to the oldest extant manuscript in Indian mathematics. Only about 70 mutilated birch barks still exist, the greater portion of the manuscript having been lost. It deals mostly with arithmetic and algebra problems, with just a few problems in geometry.

Some indeterminate equations, which are not trivial, are solved in the Bakshālī work. We will cite two problems here:

(a) To solve in integers $\sqrt{x+a} = l$ and $\sqrt{x-b} = m$, the solution provided is: $l^2 - m^2 = a + b$. So we let $l + m = \frac{(a+b)}{z}$ and $l - m = z$ where $z \neq 0$. Then

$$x = \tfrac{1}{2}\left[\tfrac{a+b}{z} - z\right]^2 + b,$$

where z is an arbitrary non-zero number.

(b) To solve the indeterminate equation $xy - bx - cy - d = 0$, the solution is:

We have $(x - c)(y - b) = bc + d$. Let $y - b = z$, with $z \neq 0$, then

$$x - c = \tfrac{bc+d}{z} \qquad \text{and} \qquad x = \tfrac{bc+d}{z} + c.$$

A mathematical contribution of great historical importance in the Bakshālī manuscript is the formula for computing the square root of non-square number.

$$\sqrt{x} = \sqrt{a^2 + b} \approx a + \tfrac{b}{2a} - \frac{(\frac{b}{2a})^2}{2(a + \frac{b}{2a})}$$

However, there was no mention of how this formula was derived. M. N. Channabosappa (1976) has given a plausible method, which could have been employed by the author of the Bakshālī manuscript. His solution is based on the principle of iteration. Taking Heron's formula as the first approximation, he proves that the second approximation is the Bakshālī formula.

This formula is evidently a refinement of $\sqrt{x} = \sqrt{a^2 + b} = a + (\frac{b}{2a})$, which is known as Heron's formula after the Greek mathematician who lived in the second half of the first century AD

We find the following applications of the formula in the Bakshālī manuscript:

$$\sqrt{41} = \sqrt{6^2 + 5} \approx 6 + \tfrac{5}{12} - \frac{(\frac{5}{12})^2}{2\,(6 + \frac{5}{12})}$$

and

$$\sqrt{105} = \sqrt{10^2 + 5} \approx 10 + \tfrac{5}{20} - \frac{(\frac{5}{20})^2}{2\,(10 + \frac{5}{20})}$$

Similar solutions are proposed for $\sqrt{481}$, $\sqrt{889}$, and $\sqrt{339009}$.

A person travels *a yojanas* in the first day and *b yojanas* more on each successive day. Another who travels at the uniform rate of *c yojanas* per day has a head start of *t* days. When will the first man overtake the second?

If x is the number of days after which the first man overtakes the second, we have $(t + x)c = a + (a + b) + (a + 2b) + \ldots + x$ terms so,

$$(t + x)c = x[a + (\tfrac{x-1}{2})b]$$

or $bx^2 - [2(c - a) - b]x = 2tc$.

Hence

$$x = \frac{2(c-a) + b + \sqrt{[2(c-a) + b]^2 + 8btc}}{2b}.$$

This agrees exactly with the solution given in the Bakshālī treatise. This shows that the person who wrote the Bakshālī manuscript not only knew how to solve quadratic equations but also had a very good knowledge of arithmetic progression.

ĀRYABHAṬA I

Āryabhaṭa I was born in 476 AD and wrote his famous work *Āryabhaṭīya* in 499 AD. In it, he expounds the method of finding the square roots and cube roots of large numbers. W. E. Clark (1930) translated his work into English. The *Āryabhaṭīya* also contains three sections: *Ganiṭa* (mathematics), *Kāla Kriyā* (time reckoning) and *Gola* (spherical astronomy).

We find in Jaina mathematics the extraction of square roots of very large numbers, so Āryabhaṭa's method in this case was not his own. The actual working rule is not given in Jaina mathematics and Āryabhaṭa gives it in the way we do it now. There is no evidence of the method for extracting cube roots having been known earlier than Āryabhaṭa I. His rule is explained in Clark (1930), as applied to finding the cube root of 1860867.

The following occur for the first time in Āryabhaṭa's work:

- $1^2 + 2^2 + \ldots + n^2 = \frac{1}{6}n(n + 1)(2n + 1)$
- $1^3 + 2^3 + \ldots + n^3 = (1 + 2 + 3 + \ldots + n)^2 = \frac{1}{4}n^2(n + 1)^2$
- $1 + (1 + 2) + (1 + 2 + 3) + \ldots$ to n terms $= \frac{1}{6}n(n + 1)(n + 2)$

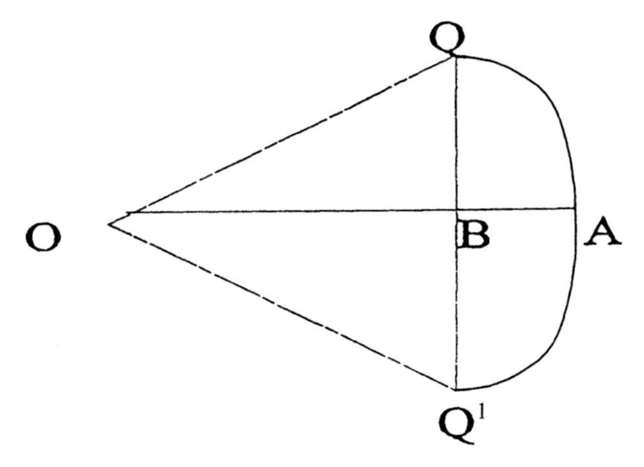

Figure 2 The Hindu meaning of sine.

The approximation of $\pi = 3.1416$ was first given by Āryabhaṭa by saying that the circumference of a circle whose diameter is 20000 is 62832. Regarding the construction of a sine table, it is not clear whether Āryabhaṭa I constructed this table himself or whether he borrowed it from the *Sūryasiddhantā*, which was available in his time.

We will now take up the Hindu meaning of sine as given in Srinivasaiengar (1967). In Figure 2, QAQ^1 is an arc of a circle with center 0. Let A be the mid-point of the arc QAQ^1. Because of its shape, QAQ^1 is called the bow, and chord QBQ^1 is called the bowstring or rope, *jya* in Sanskrit. As time passed, BQ, which is half this *jya*, itself came to be called *jya*. This *jya* is the Hindu sine. If $A\hat{O}C = \sigma$, $jya\ \sigma = BQ = R \sin \sigma$, where R is the radius of the circle. This means that the Hindu sine is the modern sine multiplied by the radius R. It was common practice to use a specific value of R.

According to Srinivasaiengar, an alternative word to *jya, jiva* became *jiba* in the hands of the Arabs and later it became *jiab*. There is an Arabic word with similar pronunciation which means heart. In the hands of the Romans, these two words were interchanged by mistake and the *jya* became *sinus* (heart). In this way, the word *jya* was transformed into the word sine.

The solution of the equation $by - ax = c$ for x and y in positive integers, where a, b, c are known integers is called *kuṭṭaka* in Hindu mathematics. Literally, *kuṭṭaka* means 'pulverizer'; it got this name because of the process of continued division that was adopted for the solution. In the West, this equation has been given the name 'Diophantine equation' after the Greek mathematician Diaphantus who is supposed to have lived in the latter part of the third century AD.

The first Indian mathematician who tackled this problem was Āryabhaṭa I. He did it in the following way: It is required to find an integer N which when divided by a leaves a remainder $\wedge_1$ and when divided by b leaves a remainder $\wedge_2$. Hence, $N = ax + \wedge_1 = by + \wedge_2$ or $by - ax = c$ where $c = \wedge_1 - \wedge_2$.

BRAHMAGUPTA

Brahmagupta was born in 598 AD. He wrote his celebrated *Brāhmasphuṭasiddhānta* in his 30th year. H. T. Colebrooke (1765–1837) translated the works of Brahmagupta and Bhāskara II into English.

Brahmagupta was interested in the construction of right-angled triangles whose sides are rational numbers and rational cyclic quadrilaterals. Brahmagupta gave the general solution $a = 2mn$, $b = m^2 - n^2$ and $c = m^2 + n^2$, where m and n are unequal rational numbers. He also presented some modifications of this problem:

(a) For the problem of constructing a right-angled triangle with rational sides, given a side a, Brahmagupta's solution is:

$$a, \frac{1}{2}\left(\frac{a^2}{n} - n\right), \frac{1}{2}\left(\frac{a^2}{n} + n\right),$$

where n is a rational number different from zero. Mahāvīra (850 AD) and Bhāskara give the solution

$$a, \frac{2ma}{m^2 - 1}, \frac{m^2 + 1}{m^2 - 1}a,$$

where m is any rational number other than ± 1. Karavindaswami (B. B. Dutta, 1932) gives

$$a, \frac{m^2 + 2m}{2m + 2}a, \frac{m^2 + 2m + 2}{2m + 2},$$

where m is any rational number other than -1. It should be observed that these solutions can easily be transformed. If we let $n = \frac{m-1}{m+1}a$ in Brahmagupta's result, we obtain the Mahāvīra–Bhāskara result, while if we let $n = \frac{a}{m+1}$ in Brahmagupta's result, we get Karavindaswami's result.

(b) Given the hypotenuse c, construct a rational right-angled triangle. Brahmagupta does not tackle this problem. Mahāvīra gives the solution:

$$c, \frac{2mnc}{m^2 + n^2}, \frac{m^2 - n^2}{m^2 + n^2}c,$$

whereas Bhāskara gives the solution as

$$c, \frac{2qxc}{q^2 + 1}, \frac{q^2 - 1}{q^2 + 1}c$$

which readily follows Mahāvīra's solution by putting $q = m/n$. These solutions were later attributed to the European mathematicians Leonardo Fibonacci (1202 AD) and François Vieta (1580 AD).

There are a couple of other results regarding the cyclic quadrilaterals which are known as Brahmagupta's theorems:

(a) The area of a cyclic quadrilateral with sides a, b, c, d is

$$\sqrt{(s-a)(s-b)(s-c)(s-d)},$$

where $2s = a + b + c + d$.

(b) If a, b, c, d in order are sides of a cyclic quadrilateral, then the diagonals are of lengths

$$\sqrt{\frac{bc+ad}{ab+cd}(ac+bd)} \quad \text{and} \quad \sqrt{\frac{ab+cd}{bc+ad}(ac+bd)}.$$

The most outstanding contribution of Brahmagupta to mathematics is his solution of *vargaprakṛiti*, which means the equation of the multiplied square. The word *prakṛiti* here means the coefficient and refers to the indeterminate equation $Nx^2 + 1 = y^2$, where N is a positive integer. This equation, now known as Pell's equation, was almost solved by Brahmagupta in 620 AD, and the complete solution was given by Bhāskara II in 1150 AD. The entire theory for the determination of the solution was expounded by Lagrange in 1767 AD and depends on the theory of continued fractions, which have no bearing on the ancient Indian method.

Brahmagupta's famous theorem

For a chosen value of c and c^1, if (a, b) and (a^1, b^1) are the sets of solutions of the equation $Nx^2 + c = y^2$ and $Nx^2 + c^1 = y^2$ respectively, then $x = ab^1 \pm a^1b$, $y = bb^1 \pm Naa^1$ are the solutions of $Nx^2 + cc^1 = y^2$. The result is called *Samāsa* or the Principle of Composition. Brahmagupta in his famous work *Brāhmasphuṭasiddhānta* (vxii, 64–65, 628 AD), solves this theorem. With its help, he almost solves the indeterminate equation $Nx^2 + 1 = y^2$. By choosing $c = c^1$ in this theorem, he shows first that $x = \frac{2ab}{c}$ and $y = \frac{(b^2 + Na^2)}{c}$ are the solutions of $Nx^2 + 1 = y^2$ and then proves that the equation $Nx^2 + c = y^2$ for $c = \pm 1, \pm 2, \pm 4$ will have integral solutions. Once one set of solutions has been determined, an infinite number of sets of solutions can be obtained by repeated application of the *samāsa* process. This is Brahmagupta's method of solving *vargaprakṛiti* (Pell's equation) – a remarkable accomplishment when we realize that this was done in 628 AD.

MAHĀVĪRA

The most famous mathematician in ancient India during the 9th century AD was Mahāvīra. He was a Jain whose work, *Gaṇitasāra Saṅgraha*, dealt with arithmetic, algebra and geometry.

Arithmetic

The following products resemble a garland, i.e., they give the same numbers whether read from left to right or from right to left.

$$139 \times 109 = 15151$$
$$27994681 \times 441 = 12345654321$$
$$12345679 \times 9 = 111111111$$
$$333333666667 \times 33 = 11000011000011$$
$$14287143 \times 7 = 100010001$$
$$142857143 \times 7 = 1000000001$$
$$152207 \times 73 = 11111111$$

Mahāvīra called these garland numbers.

Algebra

The subject of permutations and combinations owes its origin to the ancient Jains. Mahāvīra was the world's first mathematician to give the formula

$$nc_r = \left(\frac{n}{r}\right) = \frac{n(n-1)(n-2)\dots(n+r+1)}{1\cdot 2\cdot 3\dots r}.$$

Perhaps the most interesting parts of Mahāvīra's work are his techniques of obtaining unit fractions for any fraction. A unit fraction is one whose numeration is 1. We will give a few examples here.

(a) $1 = \frac{1}{2} + \frac{1}{3} + \frac{1}{3^2} + \frac{1}{3^3} + \dots \frac{1}{3^{n-2}} + \frac{1}{2} + \frac{1}{3^{n-2}}$

(b) To express 1 as the sum of odd unit fractions:

$$1 = \frac{1}{2\cdot 3\cdot \frac{1}{2}} + \frac{1}{3\cdot 4\cdot \frac{1}{2}} + \dots \frac{1}{(2n-1)2n\cdot \frac{1}{2}} + \frac{1}{2n\cdot \frac{1}{2}}$$

(c) To express a given unit fraction as the sum of $\wedge$ fractions with numerators $a_1, a_2 \dots a_\wedge$ respectively:

$$\frac{1}{n} = \frac{a_1}{n(n+a_1)} + \frac{a_2}{(n+a_1)(n+a_1+a_2)} + \dots + \frac{a_{\wedge-1}}{(n+a_1+a_2+\dots+a_{\wedge-2})(n+a_1+a_2+a_{\wedge-1})} + \frac{a_\wedge}{a_\wedge(n+a_1+a_2+\dots+a_{\wedge-1})}$$

Geometry

Mahāvīra's work also consists of constructions of rectangles and cyclic quadrilaterals. Here are some of the problems he tackled:

(a) To construct a cyclic quadrilateral having a given area;
(b) To construct a cyclic quadrilateral so that its circumdiameter equals d;
(c) To find a rectangle whose area is numerically a multiple of the perimeter or diagonal or in general, a linear combination of the sides and the diagonal.

BHĀSKARA II

Bhāskara wrote his famous work, *Siddhānta Śiromaṇi* in the year 1150 AD. It is divided into four parts: *Lilāvati*, *Bījagaṇita* (algebra), *Golādhyāya* (spherics), and *Grahagaṇita* (planetary mathematics). *Lilāvati* is essentially a work on arithmetic, whereas the third and the fourth parts deal with astronomy. He treats zero and infinity in his first chapter on algebra. He states that

$$\infty \pm K = \infty.$$

He also states that the surface area of a sphere is equal to 4 times the area of a circle and the volume of sphere equals the area $\times 2/3$ diameter.

Bhāskara gives the name *Aṅkapāśa* to the subject of permutations and combinations. His theorem states that the number of permutations of $\wedge$ things of which K_1 are one kind, K_2 another kind, etc. is $\wedge!/K_1!K_2! \ldots$.

Newton (1643–1727) and Leibnitz (1646–1716) are credited as the cofounders of differential and integral calculus. However, the rudiments of integral calculus had been known in a vague way and applied to the determination of areas and volumes by the ancient Greeks. We find the method of summation as viewed in integral calculus from Archimedes to Kepler. According to B. B. Dutta (1962), Bhāskara II determined the area and volume of the sphere by similar methods.

Bhāskara II gave the first example of a differential coefficient. To compute accurately the daily motion of a planet, he introduced in *Grahaganita* the concept of a *tātkālika* (instantaneous) method of dividing the day into a large number of small intervals and comparing the positions of a planet at the end of successive intervals. *Tātkālika gati* is essentially the instantaneous motion of the planet. If x and x^1 are the mean anomalies of the planet at the end of consecutive intervals, Bhāskara II writes that $\sin x^1 - \sin x = (x^1 - x)\cos x$, which is to say that $\delta(\sin x) = \cos x \, \delta x$.

Bhāskara II went much further into differential calculus. He stated that when the planet's motion is an extremum, then the fruit of the motion is absent (i.e., the motion is stationary), and at the commencement and end of the retrograde motion, the apparent motion of the planet vanishes. This is equivalent to stating that the differential coefficient vanishes at the extremum of the function.

Contributions in trigonometry

In Bhaskara's work *Golādhyāya*, we see the following formula:

$$\sin(A \pm B) = \sin A \cos B \pm \cos A \sin B$$

$$\sin\left(\frac{A-B}{2}\right) = \frac{[(\sin A + \sin B)^2 + (\cos A - \cos B)^2]^{\frac{1}{2}}}{2}$$

$$\sin 18^\circ = \frac{\sqrt{5}-1}{4} R$$

$$\sin 36^\circ = \sqrt{\frac{5R^2 - \sqrt{5R^4}}{8}}$$

where R is the radius of the circle.

Bhāskara II also provides a complete solution of *vargaprakṛiti* (Pell's equation) by a method which he calls *cakravāla* or the cyclic process. We can find a and b such that $Na^2 + c = b^2$ for a suitable c. We also have

$$N1^2 + (m^2 - N) = m^2.$$

Applying samāsa between (abc) and $(1mm^2 - N)$, we get

$$N\left(\frac{am+b}{c}\right)^2 + \frac{m^2 - N}{c} = \left(\frac{bm+Na}{c}\right)^2.$$

By the kuṭṭaka method, we choose m so that $am + b$ is divisible by c and so that $m^2 - N$ is numerically small. Letting $\frac{am+b}{c} = a_1$, $\frac{m^2-N}{c} = c_1$ and $\frac{bm+Na}{c} = b_1$, we have the following two theorems of Bhāskara II. Theorem 1 states that when a_1 is an integer, so are b_1 and c_1. Theorem 2 holds that after a finite number of repetitions of samāsa, we obtain a solution (A, B) of the equation $Nx^2 + c = y^2$, where $c = \pm 1$ or ± 2 or ± 4. Having gotten this solution, Brahmagupta's method will give an integral solution of the equation $Nx^2 + 1 = y^2$.

We can find the integral solutions of $61x^2 + 1 = y^2$ using the Brahmagupta–Bhāskara method. This is one of the examples given by Bhāskara II in his *Bījagaṇita* (Algebra) in 1150 AD that is of great historical interest. This was one of the examples proposed by Fermat (1601–1665) to Frenicle in a letter of February 1657. There is no evidence to show that Fermat knew about the work of Brahmagupta–Bhāskara II, particularly in view of the fact that Colebrook's English translation of Bhaskara's *Bījagaṇita* was published in 1817. Euler solved this in 1732.

Using the Brahmagupta–Bhāskara method, we choose $61 \cdot 1^2 + 3 = 8^2$, so that $a = 1$, $c = 3$ and $b = 8$. We choose m so that $\frac{m+8}{3}$ is an integer such that $\frac{m^2-61}{3}$ is small. We take $m = 7$. So $a_1 = 5$, $b_1 = 39$ and $c_1 = -4$. Hence $61 \cdot 5^2 - 4 = 39^2$. So, $61(\frac{5}{2})^2 - 1 = (\frac{39}{2})^2$, which shows that $(\frac{5}{2}, \frac{39}{2})$ is a solution of $61x^2 - 1 = y^2$. Performing samāsa between $(\frac{5}{2}\frac{39}{2} - 1)$ and $(\frac{5}{2}\frac{39}{2} - 1)$, we obtain $x = \frac{195}{2}$, $y = \frac{1523}{2}$, $c = 1$ which means that $(\frac{195}{2}, \frac{1523}{2})$ are the solutions of $61x^2 + 1 = y^2$.

Again, we perform samāsa between $(\frac{195}{2}\frac{1523}{2}\ 1)$ and $(\frac{5}{2}\frac{39}{2} - 1)$ obtaining

$$x = 3805,\ y = 29718,\ c = -1$$

which means that $(3805, 29718)$ are the solutions of $61x^2 - 1 = y^2$. Finally performing samāsa on $(3805\ 29217\ -1)$ with itself gives $x = 226153980$, $y = 1766319049$, $c = 1$ as the solution (smallest) of $61x^2 + 1 = y^2$.

The Brahmagupta–Bhāskara II method yields the solution in a few easy steps, whereas the modern method of continued fractions due to Lagrange requires the 22nd convergent of the simple continued fraction for $\sqrt{61}$.

The following example from Nārāyaṇa, who wrote *Bījagaṇita* (Algebra) in 1350 AD, is another good example of the Brahmagupta–Bhāskara II method. To find the integral solutions of $97x^2 + 1 = y^2$, take $97 \cdot 1^2 + 3 = 10^2$, so that $a = 1$, $b = 10$ and $c = 3$. We choose m so that $\frac{m+10}{3}$ is an integer such that $\frac{m^2-97}{3}$ is numerically small. We take $m = 11$. So $a_1 = 7$, $b_1 = 69$ and $c_1 = 8$. Hence we get $97 \cdot 7^2 + 8 = 69^2$.

We choose m so that $\frac{7m+69}{8}$ is an integer such that $\frac{m^2-97}{8}$ is numerically small. We take $m=3$. Hence we get $a_1=20$, $b_1=197$ and $c_1=9$. Thus we get $97 \cdot 20^2+9=197^2$. We choose m so that $\frac{20m+197}{9}$ is an integer such that $\frac{m^2-97}{9}$ is numerically small. We take $m=14$ so that $a_2=53$, $b_2=522$ and $c_2=11$. We choose m so that $\frac{53m+522}{11}$ is an integer such that $\frac{m^2-97}{11}$ is numerically small. We take $m=8$, so that $a_3=86$, $b_3=847$ and $c_3=-3$. Hence we get $97(86)^2-3=(847)^2$. We choose m so that $\frac{86m+847}{3}$ is an integer such that $\frac{m^2-97}{3}$ is small. We take $m=10$. Hence we get $97(569)^2-1=(5604)^2$. The samāsa between $(569\ 5604\ -1)$ with itself yields $x=6377352$, $y=62809633$ as integral solutions of $97x^2+1=y^2$.

Having made contributions to the solution of *vargaprakṛiti* Bhāskara II later treated the solution of more general equations of second degree. Realizing when they lived and the general standard of mathematics prevailing then, the brilliance and ingenuity that Brahmagupta and Bhāskara II exhibited should place them in the front line of world class mathematicians.

INDIAN MATHEMATICS AFTER BHĀSKARA II

After Bhāskara II, for several centuries, we come across only a few commentaries, except in Kerala in southwest India. C. M. Whish (1835) refers to four works in Kerala, *Karanapaddhati, Tantra-Sangraham, Yuktibhāṣā* and *Sadranthamala. Karanapaddhati* has been published in the Trivandram Sanskrit Series (No. 20). The author preferred to remain anonymous except that he is a Somayajin. Raja Rajaverma (1949) is of the opinion that the date of this work is between 1375 AD and 1475 AD. From a verse taken from Govinda Bhatta's *Gaṇitasūchika Granṭha*, the date is believed to be 1430 AD.

The series $\arctan t = t - (\frac{t^3}{3}) + (\frac{t^5}{5}) - \ldots\ t \leq 1$ is called the Gregory Series after James Gregory (1638–1675) of Scotland, who obtained his result in 1671 AD. If we accept the date of *Karanapaddhati* as some time in the 15th century, it means that this series and a number of transformations had been discovered in India two centuries before Gregory. According to C. T. Rajagopal and K. Mukand Murar (1944), Nīlakaṇṭha wrote *Tantrasaṅgraha* in 1502 AD. This book also gives particular cases of the Gregory Series corresponding to $t=1$ and $t=1/\sqrt{3}$, and goes much deeper by giving excellent approximations of π and series which converge very rapidly. The *Sandratnamala* is a treatise on astronomy written by Sankara Varman in about 1530 AD. The *Yuktibhāṣā* was written by Jyeṣṭadeva who lived between 1475 and 1575 AD. The uniqueness of this work is that it gives proofs for all the results it states. One such result is Gregory's Series, for which a simple and straightforward proof has been provided,

$$\lim n \to \infty \frac{1^P + 2^P + \ldots + (n-1)^P}{n^{P+1}} = \frac{1}{p+1},$$

which is not elementary.

The political upheavals due to frequent foreign invasions that began in the subcontinent after Bhāskara II may have been contributed to the barrenness of mathematical activity during that time. Ancient India, which stood in the forefront of mathematical knowledge and research, sank into a state of almost inactivity except in Kerala in the remote southwest corner, until the early part of the twentieth century, when Srinivasa Ramanujan dazzled the mathematical community with his brilliance and ingenuity.

ACKNOWLEDGMENT

In the preparation of this article, I acknowledge my indebtedness to the books and research papers listed in the bibliography. I hope to publish a detailed comprehensive book on the history and the mathematical accomplishments of the ancient Indian mathematicians from 800 BC to the seventeenth century, arranged chronologically. The publication of this envisaged book should satisfy a long felt need.

BIBLIOGRAPHY

Balagangadharan. 'A consolidated list of Hindu mathematical works.' *Mathematics Student* 15: 59–69, 1947.

Bannerji, H. C. *Colebrooke's Translation of Lilāvati.* Calcutta: The Book Company Limited, 1927.

Bapudeva, Sastri and Pandit Chandradeva, eds. *Siddhānta Śiromaṇi of Bhāskara II.* Benares: n.p., 1891.

Bell, E. T. *Development of Mathematics.* New York: McGraw Hill Book Company, 1945.

Bose, D. M., S. N. Sen and B. V. Subbarayappa. *A Concise History of Science in India.* Delhi: Indian National Science Academy, 1971.

Channabasappa, M. N. 'A note on Colebrooke's translation of a stanza from Bhaskara's Lilāvati.' *Indian Journal of History of Science* 9(2): 221–223, 1974.

Channabasappa, M. N. 'On the square root formula in Bakshālī manuscript.' *Indian Journal of History of Science* 2(2): 112–124, 1976.

Clark, W. E. *The Āryabhaṭīya, Translated with Notes.* Chicago: University of Chicago Press, 1930.

Colebrooke, H. T. *Algebra with Arithmetic and Mensuration from the Sanskrit of Brahmagupta and Bhāskara II.* London: John Murray, 1817.

Dickson, L. E. 'Rational triangles and quadrilaterals.' *American Mathematical Monthly* 28: 244–250, 1921.

Dickson, L. E. *History of Theory of Numbers*, vol. 2. New York: G. E. Stechert and Co., 1934.

Dutta, B. B. 'On Mahāvīra's solution of rational triangles and quadrilaterals.' *Bulletin of Calcutta Mathematical Society* 20: 267–294, 1928–29.

Dutta, B. B. 'The scope and limitations of Hindu Gaṇita.' *Indian Historical Quarterly* 5: 479–512, 1929.

Dutta, B. B. 'The Jaina school of mathematics.' *Bulletin of Calcutta Mathematical Society* 21: 115–145, 1929.

Dutta, B. B. *The Science of Sulvas.* Calcutta: Calcutta University Press, 1932.

Dutta, B. B. 'On the origin of Hindu terms for root.' *American Mathematical Monthly* 38: 371–376, 1931.

Dutta, B. B. 'Early literacy evidence for the use of zero in India.' *American Mathematical Monthly* 38: 569, 1931.

Dutta, B. B. 'Elder Āryabhaṭa's rule for the solution of indeterminate equations of first degree.' *Bulletin of Calcutta Mathematical Society* 24: 35–53, 1932.

Dutta, B. B. 'The Bakshālī mathematics.' *Bulletin of Calcutta Mathematical Society* 21: 1–66, 1929.

Dutta, B. B. and A. V. Singh. *History of Hindu Mathematics*, vol. 1 and 2. Bombay: Asia Publishing House, 1962.

Eves, H. *An Introduction to the History of Mathematics*. New York: Holt, Reinhart and Winston, 1964.
Ganapathri, Sastry. *Tantra Samuccya of Nārāyaṇa with Commentary*. Trivandrum: Trivandrum Sanskrit Series, 68.
Ganguly, S. K. 'Bhāskara Cārya and simultaneous indeterminate equations of first degree.' *Bulletin of Calcutta Mathematical Society* 18: 89–98, 1916.
Ganguly, S. K. 'Notes of Āryabhaṭa I.' *Journal of Bihar and Orissa Research Society* 12: 89–99, 1926.
Ganguly, S. K. 'Indian's contribution to the theory of indeterminate equations of first degree.' *Journal of the Indian Mathematical Society* 19: 110–121 and 129–142, 1931.
Halstead, G. B. *On the Foundation and Techniques of Arithmetic*. Chicago: Open Court, 1912.
Halstead, G. B. 'Evidence of early use of zero in India.' *American Mathematical Monthly* 33: 449–454, 1926.
Krishnaswamy Ayyangar, A. A. 'Bhāskara II and Samslishta Kuṭṭaka.' *Journal of Indian Mathematical Society* 18: 232–245, 1929.
Kuppanna, Sastri, ed. *Mahābhāskariya of Bhāskara I with Bhāsya of Govindaswami*. Madras: Madras Government Oriental Manuscript Library, 1957.
Lancelot, H. *Mathematics for the Millions*. London: George Allen and Unwin Limited, 1957.
Mazumdar, N. K. 'Āryabhaṭa's rule in relation to indeterminate equations of first degree.' *Bulletin of Calcutta Mathematical Society* 3: 2–9, 1912.
Mazumdar, N. K. 'The Mānava Sulva Sutra by Manu.' *Journal of the Department of Letters in the University of Calcutta* 8, 1922.
Nayar, S. K. *Karana Paddhati*. Madras: Madras Government Publications, 1956.
Pillai, S. K., ed. *Tantrasaṅgraha of Nīlakaṇṭha, with Commentary of T. Sankara Variyar*. Trivandrum Sanskrit Series, 188, 1958.
Rajagopal, C. T. and K. Mukund Murar. 'Hindu quadrature of the circle.' *Journal of Royal Asiatic Society* (Bombay Branch) 20: 66–82, 1944.
Rangacharya, M., ed. *Gaṇita sara Samgraha of Mahāvīra, with Translation and Notes*. Madras: Madras Government Publications, 1912.
Sambasiva, Sastry K., ed. *Āryabhatiya, with Bhāsya of Nilkantha*. Trivandrum: Trivandrum Sanskrit Series, 68, 1930.
Saraswatiamma, T. A. *Geometry in Ancient and Medieval India*. Delhi: Motilal Banarsidas Publishers, 1979.
Sengupta, P. C. 'Brahmagupta on interpolation.' *Bulletin of Calcutta Mathematical Society* 23: 123–128, 1931.
Sharma, V., ed. *Sulvasūtra of Katyayana*. Benaras: Achyuta Granthamala, 1928.
Shukla, K. S., ed. *Patigaṇita of Sridharacarya*. Lucknow: Lucknow University Astronomical and Mathematical Text Series 2, 1959.
Smith, D. E. *History of Mathematics*. New York: Dover Publications, 1958.
Srinivasachar, D. and S. Narasimhachar, eds. *Apasthamba Sulbasūtras*. Mysore Oriental Library Publications, 73, 1931.
Srinivasaiengar, C. N. *History of Ancient Indian Mathematics*. Calcutta: The World Press Private Limited, 1967.
Sudhakara, Dvivedi, ed. *Trisatika of Sridharacārya*. Benaras: Chowkhamba Sanskrit Book Depôt, 1899.
Thampuran, R. and A. R. Akileswara Iyer. *Yuktibhasa, Part I with Notes*. Trichur: Mangalodayam Limited, 1952.
Thibaut, G., ed. *The Sulva Sutra of Bodhayana, with Commentary of Dvarakanatha Yajavan*. Benaras: The Pandit Series 9 and 10, 1874–1875.
Thibaut, G. *Mathematics in the Making in Ancient India*. Calcutta: K. P. Bagchi and Co., 1984.
Upadye A. N. and H. Jain, eds. *Jambudvipa Prajnapti Samgraha*. Sholapur: Jaina Samkriti Samraksak Sangh Press, 1958.
Varma Raja Raja, V. *History of Sanskrit Literature in Kerala*. Trivandrum: Kamalalaya Book Depot, 1949.
Whish, C. M. 'On the Hindu quadrature of a circle.' *Transactions of Royal Asiatic Society of Great Britain and Ireland* 3: 509–523, 1835.

JOCHI SHIGERU

THE DAWN OF *WASAN* (JAPANESE MATHEMATICS)

Wasan, Japanese arithmetic, is an amazing mathematical system. Most people think that there is only one kind of mathematics, just as some philosophies will have us believe that there is only one truth. In fact the Japanese have their own special system which is unique to their culture. We used to believe that modern mathematics was the only type, but we now know that there are different varieties, each with its own value.

The Japanese called mathematics *sugaku* (mathematical science). But in the Edo era (1603–1867), mathematics was called *san* (arithmetic or counting), not adding the *wa* (Japanese). Sometimes the term *san* was added to *jutsu* (method), and was called *sanjutsu*, or it was added to *gaku* (science) and called *sangaku*. Japanese mathematicians at that time also thought their mathematics was the only one because there was national isolation and little information about foreign mathematical works. After Western mathematics was introduced, historians of mathematics developed the concept of *wasan*, which meant the mathematics of the Yamato (great Japanese) race.

THE ERA OF JAPANESE MATHEMATICS

Historians of mathematics now study mathematics in relation to culture. They call Japanese traditional mathematics *wasan* and call Western mathematics *yosan* or *seisan*. Wasan is the traditional mathematical system developed to administrate an agrarian society; Western mathematics was introduced into Japan to establish an industrial society in the 19th century. In 1877, the Mathematical Society of Japan was established. This marked the beginning of the modern era in the history of Japanese mathematics, and the word mathematics was translated as *sugaku*.

There are three eras in the history of Japanese mathematics. The last one, the modern era, is the Sugaku era. At this time, Japanese mathematicians used Western mathematics, although they also copied Chinese mathematics. We call traditional mathematics before the Meiji era *wasan*. However, there are really three types of wasan, as can be seen in the following table:

H. Selin (ed.), Mathematics Across Cultures: The History of Non-Western Mathematics, 423–455.

Table 1 The eras of Japanese mathematics

		(Ritsuryo)
	Ritsuryo–Kakushiki age	731
Wasan (wide sense) age		(Kakushiki)
		14th century
		(Before Seki)
	Wasan (semi wide sense) age	1674
		Wasan (narrow sense)
Sugakus (or Yosan) age		1877–

It is very hard to compare Japanese mathematics with modern mathematics unless we use a historical point of view to analyze it. Wasan flourished from the 17th to the 19th centuries. That gives us only two hundred years on which to base our study, a short period of time for an analysis.

As we said, the Japanese called mathematics *sugaku* (*shu xue* in Chinese). The term *shu xue* has existed since the Song dynasty in 1109. According to material in the *Song Shi*,[1] the Secretary, Zhang Fengchang, wished to decorate mathematicians in their old age, but the counselor of personnel, Wu Shi, opposed it. Wu Shi first used the word *shu xue*. We have few historical materials from that time, but it was the age when the *tianyuanshu* method (*tengen jutsu* in Japanese, celestial element method) and the method of solving higher degree equations, a new type of Chinese mathematics, was developed.[2] The Chinese mathematician Qin Jiushao also used the word in the preface of his book, the *Shu Shu Jiu Zhang* (*Susho Kyusho* in Japanese). After that it was not used commonly until modern times.

In this essay, we shall use *san* (*suan* in Chinese) to denote the Asian system of mathematics and *sugaku* to refer to Western mathematics. There are many differences between the two, but we will offer no value judgments as to which is better. Since people could calculate area and cubic content, and they knew how to use a matrix to answer questions, some scholars go so far as to say that the Japanese invented calculus. However, even though they did use matrices, it was not the same as differential and integral calculus.

One reason that the Japanese might have a feeling of inferiority about their own mathematics is that it has not been taught in school; Western mathematics has been the standard of the curriculum. Now Japanese mathematics is also taught in elementary schools; it has been called *sansu* (counting numbers) since 1941.[3] The original meaning of *san* was literally operation of bamboo (counting rods). Since the emphasis of Eastern mathematics is on calculation, and since the abacus was the popular calculating tool ever since the Ming dynasty in China, it was easy to name mathematics after calculating tools.

JAPANESE MATHEMATICS IN THE RITSURYO–KAKUSHIKI AGE

Japanese society was a tribal society before the Ritsuryo age. At that time they tried to establish a centralized government like China and developed criminal and administrative law. The Japanese borrowed freely from northern Chinese culture. They did not make a conscious choice to introduce mathematics; mathematics was just part of the culture. Japanese mathematicians needed Chinese mathematics to build cities, temples, statues of the Buddha, riparian works, and to make the calendar, which is one of the most important things in an agricultural society.

The counting tool of the northern Chinese culture was the counting rod, *chou* in Chinese and *Chu* in Japanese.[4] The counting rods system in China began in the Han dynasty. The oldest record of counting rods in Japan was in the text of the *Taiho Ritsuryo* (Laws in the Taiho Period) in 701. In Chapter 2 it said, 'The bureaucracies in capital and local authorities must receive taxes face-to-face with taxpayers. And they count taxes by counting rods ...' (Inoue, 1976: 407).

Bureaucrats of the Ritsuryo age studied counting rods because they were important for their work. They always put counting rods into a *sanbukuro* (carrier bag for counting rods). Emperor Shomu Tenno (701–756, r. 724–749) presented counting rods to the Todaiji temple after his death. Generally, they were made from bamboo, with a diameter of 1 *fen* and a length of 6 *cun*.[5] Red rods represented plus, and black ones were minus. This coloring system is the same as Chinese counting rods and the opposite of Western account books.

Specialists studied the counting rods system at the Daigakuryo (university). Mathematical education in Japan was very similar to the way it was during the Tang dynasty in China. There were two professors of mathematics (*San no Hakase*), because the Daigakuryo had two mathematical classes. An assistant (*zhujiao*) helped professors in China, but there was no equivalent post in Japan. The official rank of professors was higher in Japan than in China, where it was the lowest rank.

Students were from 13 to 16 years old, which was a little younger than Chinese students, who were aged 14 to 19. Moreover, their ranks were much higher. In China, students were lower government officials, but Japanese students were noble's sons. The graduates attained a higher grade in the government ranking.

The textbooks were very similar. The *Rokusho* and the *Sankai* were used in Korea, too; the *Kuji* was the original Japanese book. However, the level of the *Kuji* was probably not so high.

The textbook for operating counting rods was the *Songzi Suan Jing* (Master Sun's Mathematical Manual). Because this book explained the counting rods system in detail, it was listed at the top of the *Gakuryo* (law for education). Mathematics books were difficult to study because they were written in Chinese. Students studied one book for two hundred days (Jochi, 1987); thus it took three or four years to graduate. There were two classes, one studying the *Jiu Zhang Suan Shu* and six other books, the other studying the *Rokusho* and the *Zhui Shu*. Since the *Zhui Shu* was difficult even for Chinese scholars, eventually it was dropped from the curriculum.

Students took examinations in order to graduate, but the examination was not a good barometer of mathematical achievement. Students had to learn mathematical textbooks by heart. There were three questions from the *Jiu*

Table 2 Comparing Chinese and Japanese mathematical systems

	China (Tang)	Japan
Professor	*Suanxue Boshi* (2)	*San no Hakase* (2)
Assistant	*zhujiao*	
Students	14–19 years old	13–16 years old
Rank	Less than 8th grade	Noble's sons (up to 5th grade)
Graduate rank	30th rank	27th rank (1st class) 28th rank (2nd class)
Text books group A	*Zhoubi Suanjing* *Jiuzhang Suanshu* *Haidao Suanjing* *Sunzi Suanjing* *Wucao Suanjing* *Zhang Qiujian Suanjing* *Xia Houyang Suanjing* Wujing Suanshu	*Shuhi Sankei* *Kyusho Sanjutu* *Kaito Sankei* *Sonshi Sankei* *Goso Sankei* *Kuji* *Sankai Jusa*
Group B	*Qigu Suanjing* *Zhui Shu*	*Tetsujutu* *Rokusho*
Basic text	*Shushu Jiyi* *Sandengshu*	

Zhang Suan Shu and the *Rokusho* and one question from each of the other books. Students had to pass at least one question from the two main books. Graduates of the mathematical course could be *Sanshi* (government accountants). If they had good results in mathematics, they could take good government posts. Some graduate students went on to be professors. Because the official ranks of mathematicians were not so high, there were few records in official histories. There are records that show that some students became bureaucrats in *Genbaryo* (The Bureau of Buddhism and Diplomacy) and governors (Otake, 1983).

After the Heian period (794–1192), Japanese statesmen made new laws called *Kakushiki.* Professor of Mathematics became a hereditary title, so that only the Ozuki and Miyoshi families could become professors of mathematics. The idea of *Ritsuryo* was to centralize power and deny hereditary titles. But mathematics, which ought to have been based on a merit system, also became part of the hereditary system. These families then changed the law for examinations in 731, so that students of mathematics had to study the *Zhou Bi Suan Jing.* The *Zhou Bi Suan Jing* was a Chinese astronomy and philosophy book, based on the theory of *gaitian lun* (heaven is a round canopy covering the earth). People without a good understanding of Chinese or Chinese culture would find it very difficult to understand this book. The Oduki and Miyoshi families came from the continent and could speak Chinese much better than the native Japanese. This made it possible for families from the continent to keep the mathematical post for themselves.

JAPANESE MATHEMATICS IN THE WASAN AGE, PRE-SEKI ERA

Science is never the creation of only one person. But often one person gets all the credit for the achievements of his associates. This is the case with Seki Takakazu. Actually, there were many excellent Japanese mathematicians who set the stage for Seki Takakazu's great achievements.

Mathematics was very popular in the Edo era. Each village leader had at least one or two Japanese mathematical books, and the method of division and multiplication was handed down all over the country through private elementary schools. Some unique ways were used to do this. I will introduce two.

The four place decimal place system

Generally, wasan was the mathematics of the Edo period (1603–1867). Japanese mathematics existed in the Ritsuryo period, but it was only an imitation of Chinese mathematics. At the end of the Kamakura period (1192–1333), Japanese culture underwent many changes. The Mongolian and Chinese empires attacked Japan, but Samurai succeeded in defending the country. After this, Chinese culture was no longer a source of respect, and the focus shifted to maintaining a Japanese culture. Mathematics was part of this shift of focus.

Let us begin with the numbering system. Asian countries used the decimal place system. The Japanese basically used the decimal place system too, but they used a new number every four places (Oys, 1980: 85). The Chinese used three types of systems, and the most popular one was an eight decimal place system. Britain and America used to use different place systems, although now they are quite similar.

The Japanese system is as follows:

Main system: *Man* (10^4), *Oku* (10^8), *Cho* (10^{12}), *kei* (10^{16}), *Gai* (10^{20}), *Jo* (10^{24}), *Jou* (10^{28}), *Kou* (10^{32}), *Kan* (10^{36}), *Sei* (10^{40}), *Sai* (10^{44}), *Goku* (10^{48}), *Gogasha* (10^{52}), *Asougi* (10^{56}), *Nayuta* (10^{60}), *Fukashigi* (10^{64}), *Muryo-taisu* (10^{68}).

Sub-system: *Ju* (10), *Hyaku* (100), *Sen* (1000).

For example, the *shi wan yi* in the old Chinese system was *juccho* in the Japanese system for 10^{13}. All of the names are the same as Chinese ones, but the system is not the same. Today, the Chinese have started to use the Japanese system. At any rate, we can date the epoch of the wasan period from the end of the Kamakura era.

The Japanese abacus

In the Chinese abacus two beans are on the top and five are on the bottom in one reed, and a bar divides the top and bottom beans. The top beans represent fives and the bottom beans represent ones. There are 15 numbers in one reed, so this calculator is suitable for a hexadecimal system.

In the Edo period, the Japanese (and Chinese) used the decimal place system more than the hexadecimal system, so Japanese mathematicians developed an abacus of 1–5 beans, one top bean and five bottom beans. After World War II, they used a 1–4 bean abacus. This type was much lighter than the 2–5 bean abacus, so it was more useful for everyday life. After the Chinese abacus (which is still in use on the Chinese mainland today) was introduced into Japan in the 16th century, the Japanese mathematicians made a new abacus.

Idai Keisho

Japanese mathematics changed because of the textbooks used. In them, the answers to questions were not printed on the final page, so the problems could not be done as 'homework'. The mathematicians called those questions 'predilection' instead of 'homework'. The system was established by Endo Toshisada (1843–1915), who was a scholar of Japanese mathematics (Endo, 1915). After the readers mastered the contents of the book, a new solution for the questions could be created and could challenge the original answers. The readers might become the writers if they found a new way to deal with the questions.

The *Jinkoki* (Permanent Mathematics, first edition 1627) was published in the age of Kan'ei (1624–1644) and was written by Yoshida Mitsuyoshi (1598–1672). He picked the questions from the abacus calculation book *Suan Fa Tong Zong*, and this book became the most popular mathematical treatise at that time. After that some good mathematical books were published which competed with Yoshida's, so he devised a new method, the *idai keisho*, in his 1641 edition. Yoshida's questions had been asked in the *Sanyo Roku*, *Empo Shikan-ki*, and the *Sampo Ketsugisho*. Once these books began to ask *idai* questions, the *idai keisho* system began.

The series from the *Sampo Ketsugisho* was quite important. Those idai led to the *Sampo Kongenki*, which was the third generation. The fourth generation was the *Kokon Sampo-ki*, and the fifth and last was the *Hatsubi Sampo* by Seki Takakazu. The idai of the later generations were more difficult, almost always involving discussions about the *tianyuanshu* (*tengen jyutsu* in Japanese) method. Japanese mathematicians learned the *tengen jyutsu* by solving the idai questions.

Seki's great discovery was in solving an equation of 1458 degrees. Chinese algebra could solve any order of higher equations theoretically, but there was in fact a limit in the tengen jyutsu method; a new method was looked for. After the homework method which Seki Takakazu had invented, other methods gradually appeared. The *tenzan jyutsu* method meant that you could use some marks instead of using equations. All you had to do was put down an unknown number and calculate the answer. It was unnecessary to use a new method whenever you tried to solve a new homework problem. The real age of Japanese mathematics started after such changes happened.

Sangaku Hono (mathematical votive picture tablets[6])

Mathematical *Ema*[7] was the other reason for the popularity of wasan. It is a unique Japanese religious custom, a way to ask the gods to solve questions.

Generally, the standard form of the votive tablets was to write the questions, then the answers and finally the methods down in order. The length of the tablet was about 2 or 3 meter lengths bigger than the one which is used now in Japan. Most questions were about geometry, but the calculation of area and volume were also performed. The oldest votive broadsheet, written by Murayama Yoshishige in about 1683, is preserved in the Hoshinomya Shinto shrine, Sano city, Tochigi Prefecture. This was very rare, since Kyoto and Osaka were the centers of mathematics and culture before the age of Genroku (1688–1704). The broadsheet, which was presented to the Yasaka Shinto shrine in Kyoto in 1692, was representative of the first stage of the votive tablet. Because the center of mathematics moved to Edo after the reign of Kasei (1804–1830), votive tablets were used widely from the East to the North.

Fujita Sadasuke (or Teishi, 1734–1807) used this custom to popularize his own school and published a book called *Shinheki Sampo*. Even after the Meiji era, some votive broadsheets were preserved in the North and East. These were not only about belief; to Japanese mathematicians, they were a means to present their achievements to the public. Aida Yasuaki (1747–1817) and Fujita disputed about the meaning of the votive tablets. Aida wished to be Fujita's student and admired his knowledge of mathematics; Fujita in turn regarded Aida as a capable person. For his first task, Aida was instructed to correct the errors of the broadsheet in his *sangaku*. He misunderstood the instructions from Fujita and thought that his teacher did not think highly of him at all. So, he just published all the errors from Fujita Teishi's books, and that was the reason they had a falling out.

The styles of Seki Takakazu and Aida Yasuaki advanced mathematical research and made great progress in the history of Japanese mathematics. At the end of their era, the general standard advanced higher because of their efforts. The votive broadsheet perfectly fulfilled its mission to present or promote new results of research. Endo Toshisada (1843–1915) has gathered data on Japanese mathematical votive broadsheets. Research on *sangaku* marked the start of studying Japanese mathematics and is one of the most active fields for history of mathematics now.

THE DAWN OF WASAN

The most popular mathematician in Japan is Seki Takakazu (or Seki Kowa, 1642?–1708). However, wasan was not only Seki's work; it has a history of several hundred years. Takebe Katahiro (1664–1739), his student, created more excellent and more complete works than his teacher. We say that Seki Takakazu was the father of wasan because he made the model for coming generations. Seki founded the Japanese algebra system, which was named *tenzan jutsu* by Matsunaga Yoshisuke (1690?–1744), a mathematician of Seki's school.

Seki Takakazu was usually called Seki Kowa,[8] because the Japanese read one Chinese character in two or more ways. Generally, they read the name of great men using the *on* reading, which is the reading in Chinese. Takakazu is the *kun* reading and Kowa is the *on* reading. Seki Takakazu refers to himself as Takakazu, but his students and Japanese mathematicians in later generations called him Kowa. Because Takakazu is a kind of pen name, it was written on books using Chinese characters. Seki's official name is Shinsuke, which is on an official document. Most of his works were written in Chinese characters, and his paradigm must have been traditional Chinese mathematics, as in the *Jiu Zhang Suan Shu* (Nine Chapters on the Mathematical Arts). Seki applied Chinese mathematics in the Song and Yuan dynasties.

Biography of Seki Kowa (Takakazu)

Seki Takakazu was a Samurai warrior, a landlord of a 250-person village (which later became 300 *koku*[9]). He was born about 1640, probably in Fujioka, Gumma or Edo (now Tokyo). His father, Uchiyama Nagaakira, worked in Edo after 1639 and in Fujioka before then. Some historians said that he was born in 1642, the same year as Sir Isaac Newton, and in Fujioka, but we cannot find historical evidence supporting this opinion. Since Seki Takakazu's son-in-law, Shinshichiro, lost his position owing to his gambling activities, we cannot fix even his birth year exactly, as the family records were lost with the loss of position. Probably he was born between 1637 and 1642. Seki Takakazu was the second son of Uchiyama, so we do not know the name of his mother. Then he became a son-in-law of Seki Gorozaemon and succeeded to the Seki family. In 1678, he became an auditor of the Shogun's family and then became a *hatamoto* (guard), a landlord of the Shogun government.

Seki studied mathematics under Takahara Yoshitane. Takahara Yoshitane was a disciple of Mori Shigeyoshi, who wrote the *Warizansho*, the second mathematical book in Japanese. But his mathematical works were little known, and Seki taught himself mathematics using famous texts of Chinese and Japanese mathematicians, such as the *Yang Hui Suan Fa* (1275), and Isomura Yoshinori's (1640?–1710) *Sampo Ketsugi Sho*. These books described the method of solving equations of powers two and three. It is quite easy to solve two and three degree equations using geometric models.[10] Seki hand-copied the *Yang Hui Suan Fa*, and the recopied book was kept at the Koju Bunko in Shimminato, Toyama, Japan. Seki hand-copied a Korean edition of this book in 1661. This manuscript is no longer extant but Ishiguro Nobuyoshi (1760–1836), a mathematician of the Seki School, copied it by hand and his manuscript remains.

Founding the Seki-ryu school

Seki died in Edo (now Tokyo), on December 5, 1708 (October 24, 1708, in our calendar), but his students kept his school going.

The system of mathematical licensing began after Seki's works had been used widely. There were five steps: *kendai* (clear problem), *indai* (hide problem), *fukudai* (secret problem), *betsuden* (special teaching) and *inka* (admitted to the certificate). Seki did not establish this system; it was probably started by Yamaji Nushizumi (1704–1772).

Table 3 Mathematicians in the Edo era

	1st	2nd	3rd	4th	5th
				Arima Yoriyuki	
		Kurushima Yoshihiro	(1714–1783)	(1714–1783)	Kusaka Makoto
		(?–1757)		Ajima Naonobu	(1764–1839)
Seki Takakazu	Araki Murahide	Matsunaga Yoshisuke	Yamaji Nushizumi	(1739–1798)	Sakabe Kohan
(1642–1708?)	(1640–1718)	(?–1744)	(1704–1772)	Toita Yasusuke	(1759–1824)
	Takebe Katahiro	Nakane Genkei		(1708–1784)	
	(1664–1739)	(1662–1733)		Fujita Sadasuke	
		│		(1734–1807)	
		└──────	* — * — * —	Honda Toshiaki —	Aida Yasuaki
				(1744–1821)	(1747–1817)

Seki's tenzan-jutsu (Seki called it *bosho-ho*) did not explain division clearly; in this sense it was incomplete. Matsunaga Yoshisuke (?–1744) developed the dividing system and added Seki's method to the tenzan-jutsu. Both before and after Seki, there were many other outstanding Japanese mathematicians. We show some of the main mathematicians in Table 3 (Shimodaira, 1965–70, vol. 1: 182).

Similarity between Chinese mathematics and Seki's works

Seki amended works of other mathematicians and founded the Seki-ryu school. He is generally considered the founder of Japanese mathematics.

According to Hirayama Akira's (1904–1998) studies, (1959: 175–7), mathematics can be classified into 15 categories, as follows:

(a) *bosho-ho* and *endan-jutsu* (Japanese algebra system, *tenzan-jutsu*)
(b) solution of higher degree equations
(c) properties (e.g., number of solutions) of higher degree equations
(d) infinite series
(e) *reyaku-jutsu* (approximate value of fractions)
(f) *senkan-jutsu* (indeterminate equations)
(g) *shosa-ho* (method of interpolation)
(h) obtained Bernoulli numbers by *ruisai shosa-ho*
(i) computing area of polygons
(j) *enri* (principle of the circle)
(k) obtained Newton's formula by *kyusho*
(l) computing area of rings
(m) conic curves line
(n) *hojin* (magic squares) and *enjin* (magic circles)
(o) *mamakodate* (Josephus question) and *metuke-ji* (game of finding a Chinese character)

Even though Seki's works were great, most of the subjects he took up were not his original ideas, being typical works of Chinese mathematics in the Song and Yuan dynasties. These are, according to Li Yan's (1892–1963) studies (1937), shown in Table 4.

Table 4 Classification of Chinese mathematical works and important books in the Song and Yuan dynasties

(a) *Chengchu Kejue* 乘除歌訣 (verses for multiplication and division)
Yang Hui Suan Fa
Suan Fa Tong Zong

(b) *Cong Huang Tu Shuo* 縦横図説 (magic squares)
Yang Hui Suan Fa
Suan Fa Tong Zong

(c) *Shu Lun* 数論 (number theorems)
Shu Shu Jiu Zhang

(d) *Ji-Shu Lun* 級数論 (series)
Meng Xi Bi Duan
Yang Hui Suan Fa
Si Yuan Yu Jian

(e) *Fangcheng Lun* 方程論 (higher degree equations)
Ce Yuan Hai Jing 測円海鏡
Shu Shu Jiu Zhang
Suan Xue Qi Meng 算学啓蒙

(f) *He Yuan Shu* 割円術 (the method of dividing a circle)
Shou Shi Li 授時暦

Considering to which categories Seki's work belongs in Hirayama's table, we see that only *a* and *n* are Seki's original subjects; the others belong to Chinese subjects in the Song and Yuan dynasties, as illustrated in Table 5.

Now, let us consider two of Seki's works, one about solving higher degree equations and the other about magic squares. The former is one of the most important matters from the view of modern mathematics. The latter is very different from modern mathematics. We will consider three works in two categories and will try to consider the characteristics of Japanese mathematics.

Table 5 Similarity between Chinese mathematics and Seki's work

Categories	Seki's work
(a) Verse	–
(b) Magic squares	n,
(c) Indeterminate equations	e, f, o
(d) Series	d, g, h, k,
(e) Higher degree equations	b, c,
(f) Circles	i, j, l

Seki's mathematical works

Tenzan-jutsu – Japanese algebra method

Japanese mathematics was based on Chinese mathematics until Seki improved on Chinese mathematics and founded the Japanese mathematics system.

Seki calculated on paper. Traditional Chinese mathematics used counting tools; they had no algebraic symbols. Since the position of the rods (or beads of *suanpan*) indicated the numbers and the power of unknown numbers, this system could only operate with one unknown number. Seki's tenzan-jutsu could operate with unknown formulae, making it possible to solve complicated problems.

There are two parts to equations in Eastern mathematics: making the equations and solving them. The solving method was created in southern China in the Song and Yuan dynasties, in Qin Jiushao's *Shu Shu Jiu Zhang* by (1247). The *Yang Hui Suan Fa* also made some important contributions. The method of making equations was developed in northern China. It is called the *tianyuanshu* (celestial element method, Chinese algebra system). The first examples were in the *Ce Yuan Hai Jing*, although historians believe it was probably invented earlier. The tianyuanshu method was introduced into Japan in Zhu Shijie's *Suan Xue Qi Meng*. This system was for the *suanchou* (counting rods; *sangi* in Japanese) in the Song and Yuan dynasties. In the Ming dynasty, however, new counting tools, *suanpan* (abacus, *soroban* in Japanese), became popular, because mathematicians could not understand the tianyuanshu method. Japanese mathematicians had to teach the tianyuanshu method using a Korean edition of the *Suan Xue Qi Meng*. Sawaguchi Kazuyuki became the first Japanese mathematician to master the tengen-jutsu[11] method.

But the tianyuanshu method had its limits. This method used counting rods and indicated powers of unknown numbers by places, as we use the decimal place system. One rod sometimes indicated x, and sometimes x^2 if the place were not the same. Therefore it is possible to indicate $\frac{1}{x}$, but impossible to indicate even such an easy formula as $\frac{1}{x-1}$. Zhu Shijie wrote the *Si Yuan Yu Jian* and tried to use four unknown numbers, but it was not successful because he also used counting rods.

Seki called his method *endan-jutsu* or *bosho-ho*. Matsunaga Yoshisuke (1693–1744), a mathematician of the Seki-ryu school, renamed it tenzan-jutsu. Seki indicated one unknown number using one Chinese character and created symbols for calculations and powers. Because in the Chinese language one syllable has one notion and one Chinese character, the language itself is rather algebraic. But Japanese and many Western languages are not like this. As one word has several syllables, Japanese mathematicians had to use a special symbol for algebra. Seki's symbols are quite similar to the counting rods system, but he used one Chinese character to indicate an unknown number. This enabled Japanese mathematicians to solve equations for some unknown numbers. Seki used the sexagenary cycle.

Seki's *bosho-ho* is as follows (Murata, 1981: 125):

<table>
<tr><td>$a+b$</td><td>| ko</td><td>| otsu</td></tr>
<tr><td>$2a+b$</td><td>|| ko</td><td>| otsu</td></tr>
<tr><td>$ab-c$</td><td>| ko otsu</td><td>∤ hei</td></tr>
<tr><td>x^2</td><td>| ko beki</td><td></td></tr>
<tr><td>x^3</td><td>| ko san</td><td></td></tr>
<tr><td>root x</td><td>| ko sho</td><td></td></tr>
<tr><td>$\frac{a}{b}$ (after Matsunaga)</td><td>otsu | ko</td><td></td></tr>
</table>

That is, *ko* (a), *otsu* (b), *hei* (c), *beki* (power 2), *sho* (root), *san* (three, power 3) the same as after *shi* (four, power 4) and so on. Seki did not describe a division system, which was the limit of what the counting rods could do. But it was possible to apply his method to a division system. Matsunaga Yoshisuke (1690?–1744) a mathematician of Seki's school, developed the division system.

Bernoulli numbers – application of Japanese algebra

In 1712, one year before Bernoulli's *Ars Conjectandi* (1713), Seki obtained Bernoulli numbers using a system called *ruisai shosa-ho*. It is a characteristic example of his mathematics.

Eastern mathematicians used a form of what we now call Pascal's triangle since the *Yang Hui Suan Fa*.

Using Table 6, Seki computed to higher powers and obtained the fraction numbers of the last line of Table 2.

$$\lambda_0 = 1, \qquad \lambda_1 = \frac{1}{2}, \qquad \lambda_2 = \frac{1}{6}, \qquad \lambda_3 = 0, \ldots$$

These numbers fulfill the following formula:

$$\sum_{i=1}^{n} i^{\sigma-1} = \frac{1}{p} \sum_{i=0}^{p-1} \left(\lambda_1 \binom{p}{1} n^{\sigma-1} \right).$$

Table 6 Shosa-ho's table, Chapter 1 of the *Katsuyo Sampo* (1712)

	1	1											
x^2	1	2	(1)										
x^3	1	3	3	(1)									
x^4	1	4	6	4	(1)								
x^5	1	5	10	10	5	(1)							
x^6	1	6	15	20	15	6	(1)						
x^7	1	7	21	35	35	21	7	(1)					
x^8	1	8	28	56	70	56	28	8	(1)				
x^9	1	9	36	84	126	126	84	36	9	(1)			
x^{10}	1	10	45	120	210	242	210	120	45	10	(1)		
x^{11}	1	11	55	165	330	462	462	330	165	55	11	(1)	
x^{12}	1	12	66	220	495	792	924	792	495	220	66	12	(1)
power	1	2	3	4	5	6	7	8	9	10	11	12	13
(+1)	1												
	1	$\frac{1}{2}$	$\frac{1}{6}$	0	$-\frac{1}{30}$	0	$-\frac{1}{42}$	0	$-\frac{1}{30}$	0	$\frac{5}{66}$	0	

These are Bernoulli's numbers. Seki computed much larger numbers than other mathematicians had before and obtained general rules by the inductive method. He and mathematicians of his school worked this way habitually.

MAGIC SQUARES

Magic squares as mathematics

Seki and his school studied the *Yang Hui Suan Fa*, but other Japanese mathematicians did not have access to it. The *Yang Hui Suan Fa* is the best work for studying magic squares. In this section, I will analyze both the works of Yang Hui and Seki Takakazu and will consider how Seki Takakazu applied the works of Yang Hui to his work.

Magic squares involve arranging integers in a square grid such that the sums of individual rows and columns are the same. Also all the integers from 1 to n^2 (we call it an n degree magic square) are used uniquely. For example, Figure 1 is the simplest magic square, and the order is three, so we call Figure 1 a three degree magic square. Figure 2 is a four degree magic square, which is in the picture 'Melancholia' drawn by Albrecht Durer (1471–1528) in 1514.

Today, the study of magic squares is not regarded as a subject of mathematics, but many earlier mathematicians in China and Japan studied it. Fujita Sadasuke (1734–1807) classified Japanese mathematics into three categories (1781): *Yo no Yo* (necessarily for necessarily) *Muyo no Yo* (unnecessarily for necessarily) and *Muyo no Muyo* (unnecessarily for unnecessarily). Fujita believed that *Yo no Yo* was mathematics in the modern sense, from daily calculation to astronomical calculation. And magic squares belonged to *Muyo no Yo*. This kind of mathematics was not practical, but it was quite helpful to develop mathematical thinking. *Muyo no Muyo* was more of an abstract art form. He suggested studying *Muyo no Yo*, but reproved studying *Muyo no Muyo*.

In China, mathematicians studied magic squares up to the Ming dynasty. After the Qing dynasty, some characteristics of Chinese mathematics changed. Mei Juecheng (1681–1763), for example, did not admit that magic squares were a part of mathematics.[12] Japanese mathematicians, however, considered the Chinese method of making magic squares science and applied squares from

4	9	2
3	5	7
8	1	6

Figure 1 Luo Shu.

16	3	2	13
5	10	11	8
9	6	7	12
4	15	14	1

Figure 2 Melancholia.

China to their own works. In this section, I am going to trace the development of studies of magic squares.

In magic squares the sums of individual rows and columns are the same. Even with only this condition, magic squares are mysterious. Moreover the simplest magic square in China, *Luo shu* embodied Chinese philosophy.[13]

(a) Odd numbers are on the edge; even numbers are in the corners.
(b) Sums of symmetrical positions are the same, therefore the sums of rows and columns are the same.
(c) This order is the Principle of Mutual Conquest of the Five Phases (Ho, 1991).

As magic squares are quite mysterious, people thought that they had special powers and drew them on the walls of their houses for exorcism.

Western mathematicians can make odd number magic squares easily, but it is difficult to make even number magic squares. However, Seki developed a method of making magic squares of any order.

In Japan, the oldest evidence of magic squares is probably in the *Kuchizusami* in 970, a 3rd degree magic square (Noguchi, 1991). Because the *Shu Shu Ji Yi*, the Chinese mathematics textbook, described this magic square, it was familiar to Japanese scholars. But Japanese mathematicians could not make the larger magic squares. One of the oldest studies in the Edo era is the *Kigu Hosu*.[14] The author of this book designed magic squares from the 3rd to the 16th degree. He made 3rd degree magic squares using Yang Hui's method. He made a core magic square, arranged the outer stratiform pattern, and the sums of pairs in symmetrical positions (or diagonal positions) were the same value. Yang Hui developed this method and Cheng Dawei commented on it. Let us call it the 'stratiform pair method'. This method was very popular with Japanese mathematicians.

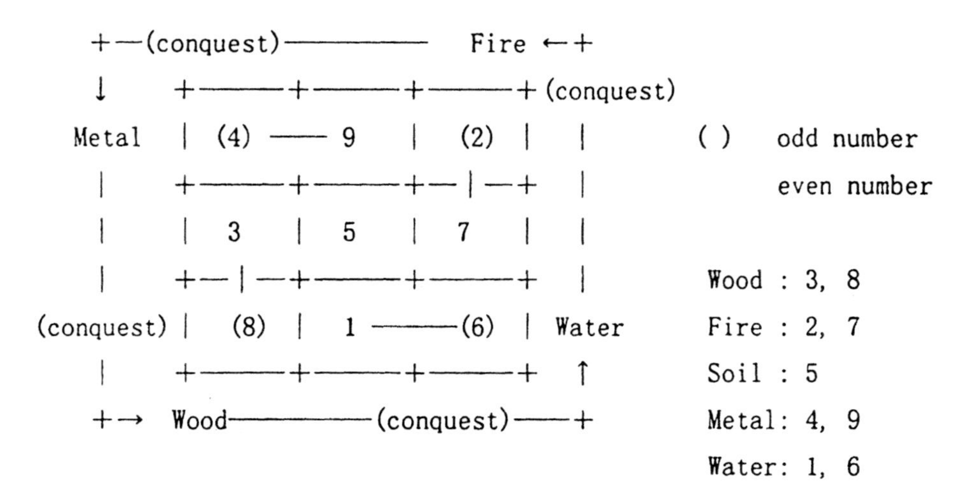

Figure 3 An unusual magic square.

Figure 4 Magic square writing on the door of an old house, Ponghu, Taiwan.

For example, a 5th degree magic square was made from the *Luo Shu* (Figure 5). The author of *Kigu Hosu* did not describe the details of making it, but the method must have been: (1) Add eight to each item of *Luo shu* (Figure 6). (2) Make the outer stratum so that the sum of each pair is 26 (Figure 7).

The 5th degree magic square of Yan Hui is the *Wu Wu Yin Tu* (Negative Five by Five Figure). It is made with numbers from 9 to 33; all other magic squares are made by using numbers from 1 upwards. Of course if each number of a magic square is increased by the same value, a new magic square is created. It is a simple principle, but it is easy to make mistakes unless care is taken. Notice that the value is 8, the same value as in Figure 6. At that time, the stratiform pair method was popular among Japanese mathematicians. Shimada

4	9	2
3	5	7
8	1	6

Figure 5 Luo Shu.

12	17	10
11	13	15
16	9	14

Figure 6 Luo Shu.

4	19	21	1	20
8	12	17	10	18
23	11	13	15	3
24	16	9	14	2
6	7	5	25	22

Figure 7 Luo Shu.

Sadatsugu, however, could not find a general method to arrange pairs; his method was by trial and error. Japanese mathematicians probably knew about 'compound magic squares', so they tried to construct magic squares which could not be made using that method, i.e., magic squares of prime number degree. The typical pattern was to make 19 degree magic squares.

This question was posed in the *Sampo Ketsugi-sho* in 1659, and the first answer was given in Muramatsu Shigekiyo's (d. 1695) *Sanso* in 1663. The magic square is shown in Figure 8 (Mikami, 1917: 16); it was an *idai keisho*.

It is certain that he used the 'stratiform pair method' to make this magic square because the core magic square is the *Luo shu* with 176 added (Figure 9).

Seki's magic squares

Before Seki, magic squares were usually called *narabemono*. Another name, *rakusho* in Japanese (*luo shu* in Chinese), was introduced into Japan by the *Suan Fa Tong Zong* and the *Yang Hui Suan Fa*. Seki gave magic squares a new name, *hojin* (square formation). I think that the *Luo shu* in the Korean edition of the *Yang Hui Suan Fa* was mistaken. Since Seki did not want to use the name *luo shu*, he invented other names for magic squares.

In China, the term *fang chen* (square formation) was never used. But there is a term similar to it in the *Shu Shu Jiu Zhang*, *fang bian rui chen* (*zhen*; changing formation from square to sharp triangle), which is the title of question 2 of chapter 15 (Ren, 1993, vol. 2: 376–8). It does not mean magic squares, so it is difficult to conclude that Seki used the *Shu Shu Jiu Zhang*'s term. But is it an accident that Seki's unique term, which was never used in both China and Japan, is very similar to Qin Jiushao's?

Seki's methods for magic squares: stratiform pair methods

Seki applied the traditional method, after the *Yang Hui Suan Fa*. But he also classified magic squares into three categories instead of two:

(a) *ki hojin* (odd number degree magic squares),

359	2	5	7	8	10	11	13	14	326	328	331	332	334	335	339	341	343	1
358	323	296	41	42	43	45	47	48	49	294	299	300	301	302	304	306	37	4
356	52	291	70	72	75	76	79	80	266	267	269	272	273	276	280	69	310	6
353	53	94	263	98	100	102	103	105	242	244	246	249	250	254	97	268	309	9
350	54	92	261	222	231	230	229	228	227	129	127	125	121	122	101	270	308	12
347	55	91	258	123	171	176	169	216	221	214	153	158	151	239	104	271	307	15
346	57	88	256	124	170	172	174	215	217	219	151	154	156	238	106	274	305	16
345	59	87	255	126	175	168	173	220	213	218	157	150	155	236	107	275	303	17
35	64	85	119	128	162	167	160	180	185	178	198	203	196	234	243	277	298	367
33	65	84	117	130	161	163	165	179	181	183	197	199	201	232	245	278	297	329
32	67	83	115	223	166	159	164	184	177	182	202	195	200	139	247	279	295	330
29	311	289	114	224	207	212	205	144	149	142	189	194	187	138	248	73	51	333
26	312	288	111	225	206	208	210	143	145	147	188	190	192	137	251	74	50	336
25	316	285	110	226	211	204	209	148	141	146	193	186	191	136	252	77	46	337
24	318	284	109	240	131	132	133	134	135	233	235	237	241	140	253	78	44	338
22	322	281	265	264	262	260	259	257	120	118	116	113	112	108	99	81	40	340
20	324	293	292	290	287	286	283	282	96	95	93	90	89	86	82	71	38	342
18	325	66	321	320	319	317	315	314	313	68	63	62	61	60	58	56	39	344
361	360	357	355	354	352	351	349	348	36	34	31	30	28	27	23	21	19	3

Figure 8 An idai keisho.

180	185	178
179	181	183
184	177	182

4	9	2
3	5	7
8	1	6

Figure 9 Stratiform pair method.

(b) *tan-gu hojin* (oddly-even number $(4n-2)$ degree magic squares), and
(c) *so-gu hojin* (doubly-even number $(4n)$ degree magic squares).

ODD NUMBER DEGREE MAGIC SQUARES

Seki computed how many items needed to be added to make the next larger magic square, then he arranged *hyosu* (front numbers) and *risu* (back numbers) around the outer circumference of the new magic square symmetrically. The method of arrangement for odd degree magic squares is as follows.

The degree is to be $n = 2m + 1$. We will demonstrate with an eleven degree magic square, therefore the value of *m* (his term was *ko dan su* (number of A, see Figure 10) is 5. The direction of the row is from right to left normally.

1. First, arrange *hyosu* (front numbers) as in Figure 10.
   (a) Put 1 next to the top-right corner, arrange *m* numbers, from 1 to *m* from here. The direction is clockwise, arriving at the corner below.
   (b) Arrange $(m-1)$ numbers; his term was *Otsu Dan Su* (number of B, from $(m+1)$ to $(2m-1)$), from the third position on the top row to the left.
   (c) Arrange $(m+1)$ numbers; his term was *Hei Dan Su* (number of C, from $2m$ to $3m$), from the next position of A downwards.
   (d) Arrange numbers, *Tei Dan Su* (number of D, from $(3m+1)$ to $(4m)$), from the next position from B to the left (Hirayama, 1974: 199–200).
2. Second, arrange *risu* (back numbers) symmetrically with the front numbers; the corner numbers are arranged diagonally.
3. Third, interchange some items, according to this method:
   (a) Interchange columns from 1 symmetrically.
   (b) Interchange similar rows (Hirayama, 1974: 200).

We do not know how he discovered this interchanging method, but Figure 11 is certainly a perimeter magic square.

DOUBLY-EVEN DEGREE MAGIC SQUARES

So gu ho (doubly-even degree magic squares) were also obtained by a similar method. We will explain the case of a twelve degree magic square, $n = 4m$, so *m* is 3 (see Figure 12).

		m (D)									
20	19	18	17	16	9	8	7	6	1	2	m
119										3	(A)
118					(B)	m - 1				4	
117										5	
112										10	
111										11	
110										12	(C)
109										13	m + 1
108										14	
107										15	
120	103	104	105	106	113	114	115	116	121	102	

Figure 10 An eleven degree magic square.

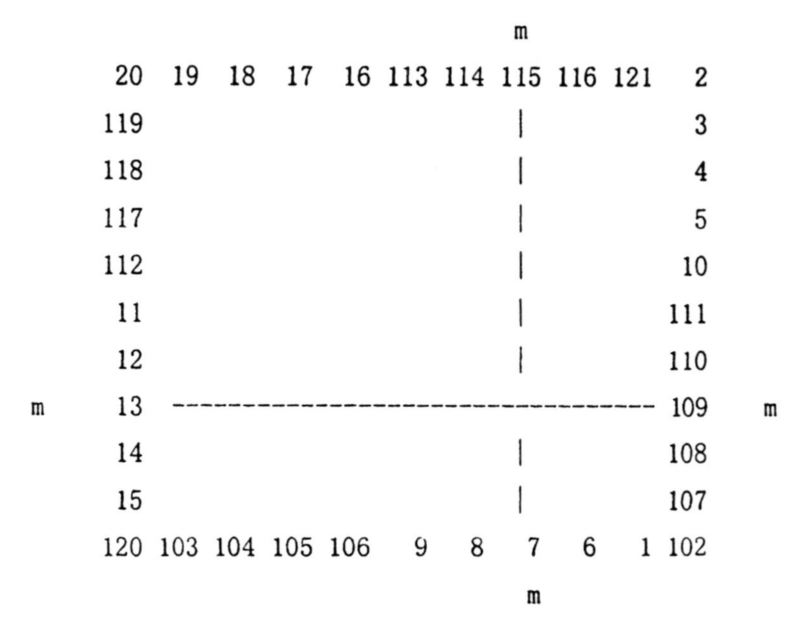

Figure 11 A perimeter magic square.

> Change $(4m-2)$ numbers (*ko dan su*) (A, from 1 to $(4m-2)$), from the third position of the top row to the left. Change two numbers (*otsu dan su*) (B, $(4m-1)$ and $4m$), from the top-right corner to the left. Change $(4m-2)$ numbers (*hei dan su*) (C, from $(4m+1)$ to $(8m-2)$), from just below the top-right corner downwards (Hirayama, 1974: 200).

Interchange within the columns:

> Interchange the two numbers in the fourth and fifth positions from the top-right corner. Do not interchange the next two numbers. Interchange the next two, but do not interchange the next two numbers. Continue interchanging in the same way (Hirayama, 1974: 200).

Interchange within the rows:

> Interchange the two numbers from the top-right corner. Do not interchange the next two numbers, then interchange the next two. Do not interchange the next two numbers, then again interchange the next two. Interchange in this same way until the third position from the bottom-right corner is reached (Hirayama, 1974: 200).

ODDLY-EVEN DEGREE MAGIC SQUARES

Tan Gu Ho (oddly-even degree magic squares) were also obtained by a similar method. We will explain the case of a fourteen degree magic square, $n = 4m + 2$, so m is 3 (see Figures 14 and 15).

			4m − 2							2		
10	9	8	7	6	5	4	3	2	1	12	11	
132	2		2		2		2		2	2	13	
131											14	
130										2	15	
129											16	
128										2	17	
127											18	4m −2
126										2	19	
125										2	20	
124											21	
123											22	
134	136	137	138	139	140	141	142	143	144	133	135	

Figure 12 Doubly-even degree magic square.

11	9	8	138	139	5	4	142	143	1	12	10
13											132
131											14
130											15
16											129
17											128
127											18
126											19
20											125
21											124
123											22
134	136	137	7	6	140	141	3	2	144	133	135

Figure 13 A twelve degree magic square.

					4m	(A)					1	(C)		
12	11	10	9	8	7	6	5	4	3	2	1	25	13	
183			2		2		2		2		3	1	14	
182													15	
181												2	16	
180													17	
179												2	18	
178													19	4m
177												2	20	(B)
176												2	21	
175													22	
174													23	
173													24	
171												1	26	1 (D)
184	186	187	188	189	190	191	192	193	194	195	196	172	185	

Figure 14 Oddly-even degree magic square.

> Change $4m$ numbers (*ko dan su*) (A, from 1 to $4m$) from the third position on the top row to the right. Change $4m$ numbers, (*otsu dan su*) (B, from $(4m+1)$ to $8m$) from the top-right corner downwards until the third position from the bottom-right corner. Then change one number (*hei dan su*) (C, the number $(8m+1)$) left of the top-right corner; change one number (*tei dan su*) (D, $(8m+2)$) above the right-bottom corner (Hirayama, 1974: 200).

Interchange by columns:

> Interchange three numbers from the top-right corner. Do not interchange the next two numbers, then interchange the next two. Do not interchange the next two numbers, then interchange the next two. Continue until the third position from the top-left corner is reached (Hirayama, 1974: 200).

Interchange by rows:

> Interchange in one row below the top-right corner (2nd row). Do not interchange the next two numbers, then interchange the next two. Do not interchange the next two numbers, the interchange the next two. Continue interchanging until the second to last row (Hirayama, 1974: 200).

12	11	187	188	8	7	191	192	4	3	195	196	172	13
14		2		2		2		2			3	1	183
182													15
181												2	16
17													180
18												2	179
178													19
177												2	20
21												2	176
22													175
174													23
173													24
26												1	171
184	186	10	9	189	190	6	5	193	194	2	1	25	185

Figure 15 Oddly-even degree magic square.

AFTER SEKI TAKAKAZU

In Japan

Seki's method was based on calculation. The magic square is a geometrical matter at first glance, but he solved the problems by calculation. He probably used a particular calculating tool such as the abacus. Seki Takakazu could construct any degree magic square by applying the stratiform pair method. The *Yang Hui Suan Fa* and some Japanese mathematicians' methods influenced him, so his method was not totally original. However, he was the first to discover this method of arrangement.

In 1683, Tanaka Yoshizane (1651–1719) wrote the *Rakusho Kigan* (Mirror of Luo shuon Turtle's Shell) (Kato, 1956, Zatsuron vol. 3: 266–71) in which he found another method of arrangement, which builds the magic square from the outer circumference in to the inner circumference. Most of Tanaka's terms were influenced by the study of the *Yijing* (*I Ching, Book of Changes*), but his method was also a traditional method in Japan.

Japanese mathematicians advanced step by step, and this was a most important advance. The study of magic squares as 'normal science' had been transferred from culture to culture, i.e., from China to Japan. During the Edo period in Japan, mathematicians continued to study magic squares. We have evidence from Isomura Yoshinori's *Sampto Ketsugi Sho* (1659), in which he makes a 19 degree magic square, to Gokayu Yasumoto's *Sampo Semmon Sho* (1840), which uses a diagonal method.[15]

After Seki Takakazu created his method for composing magic squares in the 18th century, later mathematicians were able to study more advanced methods of making them. Tanaka Matsunaga and Uchida studied magic squares in a similar way to Seki's, i.e., classifying magic squares into three categories, then considering separate methods of composition.

Another important method is Takebe Katahiro's (1664–1739), a student of Seki's. His method, for example in the case of a seven degree magic square, is:

1. Arrange a natural square (Figure 16-a).
2. Four lines or rows are already the same as the lines in a magic square. These are the middle column, the middle row and the two diagonal lines. The diagonal line from top-right to bottom-left is called *usha* (right diagonal) and the other is called *sasha* (left diagonal).
3. Make the original middle row the new *usha*, the original *usha* the new middle column, the original middle column the new *sasha*, and the original *sasha* the new middle row (Figure 16-b).
4. Reverse the middle column and row, and interchange the *usha* and the *sasha* (Figure 16-c).

43	36	29	22	15	8	1
44	37	30	23	16	9	2
45	38	31	24	17	10	3
46	39	22	25	18	11	4
47	40	33	26	19	12	5
48	41	34	27	20	13	6
49	42	35	28	21	14	7

a

22			1			4
	23		9		11	
		24	17	18		
43	37	31	25	19	13	7
		22	33	26		
	39		41		27	
46			49			28

b

4			49			28
	11		41		27	
		18	33	26		
7	13	19	25	31	37	43
		24	17	32		
	23		9		39	
22			1			46

c

	+42	+28	+14		−14	−28	−42
−6	4	<36>	(29)	49	(15)	< 8>	28
−4	<44>	11	(30)	41	(16)	27	< 2>
−2	<45>	(38)	18	33	26	(10)	< 3>
	7	13	19	25	31	37	43
+2	<47>	(40)	24	17	32	(12)	< 5>
+4	<48>	23	(34)	9	(20)	39	< 6>
+8	22	<42>	(35)	1	(21)	<14>	46

d

Figure 16 Wasan.

5. There are some irregularities (Figure 16-d), but these can be corrected by interchanging one item in each column and row. For example, interchange 15 and 29, or 21 and 35 for correcting the third and fifth row.

Later Japanese mathematicians such as Murai Chuzen (1708–1797), Gokayu Yasumoto (fl. 1840) and Sato Mototatsu (1819–1896) also applied Takebe Katahiro's method.

Magic squares in China

In the Qing dynasty, magic squares were not considered part of the study of mathematics, so the level of studying magic squares was the same as it was before the Ming dynasty. It was an age in which modern western science was introduced into China, so that studies of magic squares were less influential. I suspect it was no accident that Mao Jin's manuscript of the *Yang Hui Suan Fa* omitted the part concerning magic squares. One reason for this was that there were no systematic schools for mathematics in China (Hua Yinchun, 1987: 85–6). In contrast, there were many schools called *terakoya* (elementary schools) in Japan, and there was fierce competition among them. Most Japanese mathematicians were teachers of mathematics, and magic squares were a good advertisement for their schools.

THE BIRTH OF SUGAKU

The destruction of wasan

In the 19th century, the Industrial Revolution took place on both shores of the Atlantic Ocean. Following that, Western European culture swept all over the world. Western mathematics arose from the differential and integral calculus that was based on physics and was closely related with natural science. How to assimilate western knowledge was a burning issue for Japan; otherwise Japan opened itself to the possibility of being colonized. The Japanese imported science from the West instead of from China. The first generation of Sugaku, most of them Army and Navy officers, studied only Western mathematics (Ogura, 1940: 137). They especially emphasized chemistry, physics, and engineering (especially military engineering).

One of the first Western mathematical books, *Yozan Yoho*, had quite interesting questions. Yanagawa Shunsan (1832–1870) asked a question about a battle ship: One battle ship (line ship) in Germany has 80 guns and 2000 crew members. One cruiser (frigate) has 36 guns and 600 crew. How many guns and crew members do five battle ships and ten cruisers have?

Kondo Makoto (1831–1886) was a professor from the navy. Tsukamoto Akitake (1833–1885) edited one of the first mathematical textbooks in 1869, and he was a professor from the army. Sasaki Tsunachika (?), another professor from the army, also edited mathematical books in 1871. In 1872, the Meiji government decided to import Western mathematics. Many students went abroad to study. A characteristic example was Kikuchi Dairoku (1855–1917) who went to England and studied mathematics in the University of Cambridge and obtained first class honors.

In 1877, some of this generation and some wasan mathematicians established the Tokyo Sugaku Kaisha (Tokyo Mathematical Society). Then, in 1882, the name *sugaku* was chosen in a vote. The last generation of wasan mathematicians and the president of the Tokyo Sugaku Kaisha, Okamoto Noribumi (1847–1931) insisted that it should be changed to *sugaku*, although Kikuchi wanted *surigaku* (mathematical science). Kikuchi reorganized the Tokyo Sugaku Kaisha, and established a new academic society, Tokyo Sugaku-Butsurigaku Kai[16] (Tokyo Mathematical and Physical Society) in 1885.

JAPANESE MATHEMATICS AS AN OBJECT OF STUDY

The mathematician Endo Toshisada's (1843–1915) book, *Dainihon Sugakushi* (1896), was a great work, which was not built on any references or textbooks at that time. This work was influential in making sure that most parts of Japanese mathematics had a chance to succeed to the next generation. At the same time, Japanese mathematics became an object of historical study. Endo decided to gather all the historical materials into the style of a chronological table. Since there were no references or textbooks for editing his work, it is not strange that there were some mistakes found by mathematicians in the following generation. Mikami Yoshio (1875–1950) revived the book, changed the name to *Zoshu Nihon Sugakushi* (Endo, 1918) and published it even though there were some errors. This book has become the bible of Japanese mathematics to all modern historians.

In the last years of the Meiji age, Japan won several wars and there was a great nationalistic fervor. Japanese mathematics became part of this, and Kikuchi began the work of editing wasan. This was a great irony, since he was one of its destroyers. But when he was the principal of Nihon Gakushiin (the Japan Academy), he became a historian and developed an interest in the older mathematical art.

The Nihon Gakushiin published a series called *Meijizen Nihon Kagakushi* (History of Japanese Science before the Meiji Era). The books were published in the 2600th year of the Japanese emperor, 1940, under the editorship of Fujiwara Shozaburo (1881–1946). They are the most complete collection of books on wasan.

* * *

In this chapter, we introduced the history of Japanese mathematics. Since the mathematics was done at least 100 years ago, it might seem quite simple from the view of modern mathematics. But it is not enough to see it from only a modern perspective. If that were so, all culture which is not directly connected with the present would have no value. We did not introduce only pure mathematical matter but also magic squares, which blend mathematics and superstition. Most of the books about wasan are collected in Tohoku University, the Japan Academy, and the Wasan Institute.[17] The study of wasan is not necessarily part of the official academy; often it is seen as folklore. But it is an important part of a movement in academia to bring many parts of the world together and value them equally.

NOTES

[1] See chapter 105 (Vol. 7: 351) and chapter 347 (vol. 8: 1241).

[2] Liu Yi (11c) developed the method of solving higher degree equations about 1019, at least before 1113. Zhongwai, 1986: 247–8.

[3] It was called *sanjutsu* (arithmetic) from 1872 to 1941 (Hiraoka, 1992: 201–6).

[4] Japanese called counting rods *Chu* in the Ritsuryo age, and *San*, *Sanko*, and then *Sangi* in the Wasan age.

[5] The *Shuo Wen* and vol. 1 of Lüli-zhi, *Han Shu*. Some counting rods were unearthed at Han dynasty tombs, for example, Qianyang, Xianxi in 1971, Zhangjiashan (186 BC), Jiangling, Hubei in 1983. The lengths are 12.6–13.8 cm; the diameters are 0.2–0.4 cm. These counting rods were in a silk bag.

[6] Some scholars translate this as 'temple geometry', but this custom was in the Shinto shrine, not in the Buddhist temple.

[7] Horses are holy animals in Shinto, and believers offer votive pictures of horses.

[8] Mikami Yoshio (1875–1950) usually called him 'Kowa' in English, but Mikami used both in the preface of his book (Mikami, 1913; 1961; 1971).

[9] *Koku* is a unit of volume, about 180 liters. The samurai's salary was given in rice, and his land was counted by *koku*. But the product (and the expenses) of one person is close to one *koku*. Therefore I translated one *koku* as the product of one person.

[10] The *Jiu Zhang Suan Shu* already solved two degree equations and the *Qi Gu Suan Jing* (Continuation of Ancient Mathematics) by Wang Xiaotong (620), solved three degree equations.

[11] This method was called *li tian yuan yi shu* (setting up *tian yuan* method) in China, and also in Japan. Then Matusnaga Yoshisuke named it *tengenjutsu* (*tian yuan shu* in Chinese).

[12] The original book (*Suan Fa Tong Zong*) drew *He Tu* and *Luo Shu* in the initial volumes. They seemed to be using mathematics, because they used numbers. However they were used for fortune-telling, and all books about divination commented on them. Since they were not useful for arithmetic, they were omitted in this book.

[13] In Zhang Huang's (fl. 1562) explanation, chapter 1 of the *Tu Shu Bian* (Encyclopaedia of Maps and Books), there are two more restrictions: In *Luo Shu*, *yang* (odd numbers) are on four sides, and *yin* (even numbers) are on four squares. Five is the pivotal number, and the sums of symmetrical numbers (about the pivot) are ten, excepting the number at the center. One-nine, three-seven, two-eight and four-six are all sums of verticals and horizontals. According to the idea of *wuxing* (five phases), however, this order is the reverse of heaven, i.e. one-sixth of water conquers two-sevenths of fire; two-sevenths of fire conquers four-ninths of metal; four-ninths of metal conquers three-eighths of wood; three-eighths of wood conquers five of soil in the center, moving in a counter-clockwise direction as in heaven (SKQS, vol. 968: 12).

[14] Another name is *Narabemono-jutsu* (Arranged Matter) written about 1653, which was not published and is in a manuscript preserved at the Nihon Gakushiin (Japan Academy). The author is unknown, and I suggest it was written by Shimada Sadatsugu (1608–1680) (Jochi, 1993: 116 and 146).

[15] Mikami, 1917, provides a comprehensive list of studies of magic squares by Japanese mathematicians.

[16] It has been the Nihon Sugaku Kai (Mathematical Society of Japan) since 1946. It was the Nihon Sugaku Butsurigaku Kai from 1927 to 1946.

[17] This institute was established in 1997 using books from Shimodaira Kazuo (1928–1994), former president of the History of Mathematics Society in Japan.

BIBLIOGRAPHY

ABBREVIATIONS

DNK	*Dai Nihon Komonjo*
ESWS	*Edo Shoki Wasan Sensho*, see Shimodaira Kazuo *et al.*, eds., 1990–
GXJBCS	*Guo Xue Ji-Ben Cong-Shu* by SWYSG
KXCBS	Kexue Chubanshe (Science Press)
SKQS	*Si Ku Quan Shu*
SWYSG	Shangwu Yinshuguan (Commercial Press)
STZ	*Seki Takakazu Zenshu*, see Hirayama *et al.*, eds., 1974
SJSS	*Suan Jing Shi Shu*, see Qian Baocong, ed., 1963
MZNSS	*Meiji-Zen Nihon Sugaku-shi*, see Nihon Gakushiin, ed., 1954
WHM	*Wasan no Hojin Mondai*, see Mikami Yoshio, 1917
ZHSJ	Zhonghua Shuju (China Press)
ZGR	*Zoku Gunsho Ruiju*

(1) Original Books (by title, before 1867 in Japan and China)
Title (English). Author. Place (country), Year, Reference.

Araki Murahide Sadan (Talks of Araki Murahide). Araki, Murahide. Japan, 1640–1718?. Hirayama Akira, 1959. MZNSS.

Burin Inkenroku (Anecdotes of Mathematicians). Saito, Yajin. Japan, 1738. MS kept at Kokuritsu Kobunsho-kan. MZNSS. Shimodaira Kazuo, 1965–1870.

Byodai Meichi no Ho (Methods of Correcting Failures as Questions). Seki, Takakazu. Japan, 1685. STZ, 1974.

Daijutsu Bengi no Ho (Methods of Discriminant). Seki, Takakazu. Japan, 1685. STZ.

Dokaisho (Mathematical Introductions for Pupils). Nozawa, Sadanaga. Japan, 1664.

Empo Shikan-ki (Four Chapters Book for Methods of the Circle). Hatsusaka, Shigeharu. Japan, 1657. ESWS.

Fuddankai To-jutsu (Answers and Methods of the Sampo Fuddankai). Seki, Takakazu. Japan, 1674. STZ.

Fukyu Tetsu-jutsu (Sir Takebe's Bound Methods). Takaebe, Katahiro. Japan, 1722. See Tetsu-jutsu Sankyo.

Hai Dao Suan Jing (Sea Island Mathematical Manual). Liu, Hui. China, 263?. Bai Shangshu, ed. 1983. SJSS.

Happo Ryakketsu (Short Explanations of Eight Items). Seki, Takakazu. Japan, 1680. STZ.

Hatsubi Sampo (Mathematical Methods for Finding Details). Seki, Takakazu. Japan, 1674. STZ.

Hatsubi Sampo Endan Genkai (Introduction of Seki's Hatsubi Sampo). Takebe, Katahiro. Japan, 1685. STZ.

Ho-Enjin no Ho (Methods of Magic Squares and Magic Circles). Aida, Yasuaki. Japan, 1747–1817. WHM. MZNSS.

Hojin Henkan no Jutsu (Transforming Technique of Magic Squares). Aida, Yasuaki. Japan, 1747–1817. WHM. MZNSS.

Hojin no Ho, Ensan no Ho (Methods of Magic Squares and Methods of Magic Circles). Seki, Takakazu. Japan, 1683. STZ.

Hojin Shin-jutsu (New Methods of Magic Squares). Takebe, Katahiro. Japan, 1760. WHM. MZNSS.

Hojin Shin-jutsu (New Methods of Magic Squares). Matsunaga, Yoshisuke. Japan, d. 1744. WHM. MZNSS.

Inki Sanka (Mathematical Songs for Multiplication and Division). Imamura, Chisho. Japan, 1640. ESWS.

Iroha Jiruisho (Iroha Dictionary). Tachibana-no, Tadakane. Japan, 1177?–13c?. DNK.

Jinko-ki (Permanent Mathematics). Yoshida, Mitsuyoshi. Japan, 1627. ESWS.

Jiu Zhang Suan Shu (Nine Chapters on the Mathematical Arts). ?. China, 1c?. Bai Shangshu, 1983. SJSS.

Jugairoku (Record of Formulae). Imamura, Chisho. Japan, 1639.

Juji Hatsumei (Comments on the Works and Days Calendar). Seki, Takakazu. Japan, 1680. STZ.

Jujireki-kyo Rissei no Ho (Methods of Manual Tables of the Works and Days Calendar) Seki, Takakazu. Japan, 1681. STZ.

Kai Fukudai no Ho (Methods of Solving Secret Questions). Seki, Takakazu. Japan, 1683. STZ.

Kaiho Sanshiki (Formulae for Solving Higher Degree Equations). Seki, Takakazu. Japan, 1685. STZ.

Kai Indai no Ho (Methods of Solving Concealed Questions). Seki, Takakazu. Japan, 1685. STZ.

Kai Kendai no Ho (Methods of Solving Findable Questions). Seki, Takakazu. Japan, 1685. STZ.

Kaizan-ki (Improved Mathematical Record). Yamada, Masashige. Japan, 1659.

Kanja Otogizoshi (Mathematicians' Fairy Tales). Nakane, Hikodate. Japan, 1743. MZNSS. WHM.

Katsuyo Sampo (Essential Points of Mathematics). Seki, Takakazu. 1712. Japan. STZ.

Kemmon Zakki (Notes of Experience). Sosho. Japan, about 1466. ZGR.

Kenki Sampo (Counterargument for Saji's Books). Takebe, Katahiro. Japan, 1683.

Ketsugi-Sho To-jutsu (Answers and Methods of the Sampo Ketsugi-Sho). Seki, Takakazu Japan, 1672?. STZ.

Kigu Hosu (Odd and Even Squares). Shimada, Sadatsugu?. Japan, about 1653. MS kept at Waseda University Library.

Kiku Yomei Sampo (Essential Mathematical Methods of Measures). Seki, Takakazu?. Japan, 1711. STZ.

Kokon Sampo-ki (Mathematics of All Ages). Sawaguchi, Kazuyuki. Japan, 1671. MS kept at Nihon Gakushiin Library MZNSS. ESWS.

Ko Ko Gen Sho (Manuscript on the Sides of Right Triangles). Hoshino, Sanenobu. Japan, 1672. MS kept at Nihon Gakushiin Library MZNSS. WHM.

Kuchizusami (Humming). Minamoto-no, Tamenori. Japan, 970. ZGR.

Ku-shi Iko (Posthumous Manuscripts of Mr. Kurushima). Kurushima, Yoshita. Japan, 1755. MZNSS. WHM.

Meng Xi Bi Duan (Dream Pool Essays) Sheng Guo. China, 1086–93?.

Ni Chu Reki (Two Hand Almanac). ?. Japan, 14c. ZGR.

Nihon-Koku Kenzaisho Mokuroku (Catalogue of Books Seen in Japan). Fujiwara-no, Sukeyo Japan, 891.

Qi Gu Suan Jing (Continuation of Ancient Mathematics). Wang, Xiaotong. China, 620. SJSS.

Rakusho Kigan (Mirror of 'Rakusho' (*luo shu*) on the Turtle). Tanaka, Yoshizane. Japan, 1683. MZNSS. WHM.

Renchu-sho (Records of the Court). Fujiwara-no, Suketaka. Japan, 12c. Kaitei Shiseki Shuran, 1739.

Ritsu Hojin (Magic Cubes). Kurushima, Yoshita. Japan, 1755. MZNSS. WHM.

Sampo Doshi Mon (Mathematical Methods of Pupils' Questions). Murai, Chuzen. Japan, 1784. MZNSS. WHM.

Sampo Fuddankai (Mathematical Methods of Not Bothering to Correct). Murase, Yoshieki Japan, 1673. MZNSS.

Sampo Ketsugi Sho (Solving Mathematical Questions). Isomura, Yoshinori. Japan, 1659. 2nd ed. *Kashiragaki Sampo Ketsugi Sho* (1661). Solving Mathematical Questions with Author's Comment, 1661.

Sampo Kongen-ki (Origin of Mathematics). Sato, Masaoki. Japan, 1669. WHM. MZNSS.

Sampo Kokon Tsuran (Japanese Mathematics in All Ages). Aida, Yasuaki. Japan, 1795.

Sampo Semmon Sho (Manuscript of Mathematical Methods of Easy Questions). Gokayu, Yasumoto. Japan, 1840. WHM. MZNSS.

Sampo Shigen-ki (Record of Origin of Mathematics). Maeda Kenjo. Japan, 1673. MZNSS.

Sampo Tensho-ho Shinan (Instruction of Aida's Algebra). Aida, Yasuaki. Japan, 1810. Fujii ed., 1997.

Sandatsu no Ho, Kempu no Ho (Methods of Solving Josephus Questions, Methods of the Check of Sign). Seki, Takakazu. Japan, 1683. STZ.

Sangaku Keimo Genkai Taisei (Comments for Zhu's Suanxue Qimeng). Takebe, Katahiro Japan, 1690. MZNSS.

Sanso (Cutting Board for Mathematics). Muramatsu, Shigekiyo. Japan, 1663. Sato Ken'ichi, ed., 1987.

Sanyo-ki (Calculation Manual). ?. Japan, 16c?. ESWS.

Sekiryu Go Hojin Hensu Jutsuro narabini Sukai (Transformation Methods and Comments of Five Degree Magic Squares at Seki's School). Yamaji, Nushizumi. Japan, 1771. WHM. MZNSS.

Seki Teisho (Seki's Amendments). Seki, Takakazu. Japan, 1686. STZ.

Shi Hojin Hensu (Variables of Four Degree Magic Squares). Ishiguro Nobuyoshi. Japan, 1760–1836?. WHM.

Shinheki Sampo (Mathematical Method on the Holly Wall). Fujita, Sadasuke. Japan, 1789. MZNSS.

Shokambumono (Many Materials for Calculation). Momokawa, Jihei. Japan, 1622. ESWS.

Shoyaku no Ho (Methods of Reduction). Seki Takakazu. Japan, 1683. See *Katsuyo Sampo*.

Shu Du Yan (Generalizations on Numbers). Fang, Zhongtong. China, 1661. ZSSRC.

Shu Shu Jiu Zhang (Mathematical Treatise in Nine Sections). Qin, Jiushao. China, 1247. SKQS. GXJBCS, 1937. Libbrecht, Ulrich. 1973. ZKZDTH.

Shu Shu Ji Yi (Memoir on some Traditions of Mathematical Art). Zhen, Luan. China, 6c. SJSS.

Si Ku Quan Shu (Complete Works of the Four Categories). Yu, Meizhong, Tongjun Liu *et al.*, eds. China, 1782. SWYSG (Taiwan), 1984–1988 (Wenyange version).

Si Yuan Yu Jian (Precious Mirror of the Four Elements). Zhu, Shijie. China, 1303. KXJBCS, 1937.

Suan Fa Tong Zong (Systematic Treatise on Arithmetic). Cheng, Dawei. China, 1592. ZKZDTH.

Suan Jing Shi Shu (The Ten Mathematical Manuals). Dai, Zhen, ed. China, 1774. Qian, Baocong, ed., 1963.

Suan Xue Qi Meng (Introduction to Mathematics Studies). Zhu, Shijie. China, 1299. Ruan Yuan reprinted edition, 1839. Kodama, 1966.

Sun-zi Suan Jing (Master Sun's Mathematical Manual). Sunzi. China, about 400. SJSS.

Taisei Sankyo (Complete Mathematical Manual). Seki, Takakazu, Kataaki Takebe and Katahiro Takebe. Japan, 1710. STZ.

Tan'i Sampo (Searching of Bringing-up in Mathematics). Kemmochi, Masayuki. Japan, 1840. MZNSS.

Tetsu-jutsu Sankyo (Mathematical Manual of Bound Methods). Takaebe Katahiro. Japan, 1722. MZNSS. See Fukyu Tetsu-jutsu.

Tian Mu Bi Lei Cheng Chu Jie Fa (Practical Rules of Arithmetic for Surveying). Yang, Hui. China, 1275. See *Yang Hui Suan Fa*.

Toyoshima Sankyo Hyorin (Comments of Toyoshima's Mathematical Manual). Aida, Yasuaki Japan, 1804. MZNSS.

Tu Shu Bian (Encyclopaedia of Maps and Books). Zhang, Huang. China, 1562. SKQS.

Warizan Sho (Division Book). Mori, Shigeyoshi. Japan, 1622. ESWS.

Wu Cao Suan Jing (Mathematical Manual of the Five Government Departments). Zhen, Luan China, 6c. SJSS.

Wu Jing Suan Shu (Arithmetic in the Five Classics). Zhen, Luan. China, 6c. SJSS.

Xia-Huo Yang Suan Jing (Xiahuo Yang's Mathematical Manual). Han, Yan. China, 500? SJSS.

Xu Chou Ren Zhuan (Biographies of Mathematicians and Astronomers, part 2). Luo, Shilin China, 1840. GXJBCS, 1935.

Xu Gu Zhai Qi Suan Fa (Continuation of Ancient Mathematical Methods for Elucidating the Strange [Properties of Numbers]). Yang, Hui. China, 1275. See *Yang Hui Suan Fa*.

Yang Hui Suan Fa (Yang Hui's Method of Computation). Yang, Hui. China, 1275. Kodama Akio, 1966. Lam Lay-Yong, 1977.

Yi Jing (The Book of Changes) ?. China, Zhou dynasty?. Wilhelm (tr.), 1950. SKQS.

Yozan Yoho (Methods for Western Mathematics). Yanagawa, Shunsan. Japan, 1857. Ogura, 1940.

Zhang Qiu-Jian Suan Jing (Zhang Qiujian's Mathematical Manual). Zhang, Qiujian. China, 466. SJSS.

Zhou Bi Suan Jing (The Arithmetical Classic of the Gnomon and the Circular Paths of Heaven). China, B.C. 1c?. SJSS.

Zhui Shu (Bound Methods). Zu, Chongzhi. China, about 463. no version extant.

Zoku Gunsho Ruiju (Encyclopaedia of Wagaku Kodansho, part 2). Haniho, Kiichi, ed. Japan since 1793. Tokyo: Hobunsha. 1925, 3rd ed. 1974.

After 1868

Abe, Rakuho. 'Yoki Sampo ni Aru Hojin' (Magic squares in the Yang Hui Suan Fa). *Sugaku-shi Kenkyu* 70: 11–32, 1976.

Bai, Shangshu. *Jiu Zhang Suan Shu Zhu Shi* (Comments of Nine Chapters on the Mathematical Arts). Beijing: KXCBS, 1983.

Endo, Toshisada. *Nihon Sugaku-shi* (History of Mathematics in Japan). Tokyo: Iwanami Shoten, 1896.

Endo, Toshisada. *Zoshu Nihon Sugaku-shi* (History of Mathematics in Japan, enlarged ed.). Tokyo: Koseisha Koseikaku, 1918; reprinted 1960 and 1981.

Fujiwara, Shozaburo. *Nihon Sugaku Shiyo* (Essences of the History of Japanese Mathematics). Tokyo: Hobunkan, 1952.

Fujiwara, Shozaburo, 1954–60. *See* Nihon Gakushiin.

Hagino, Kougo. *Nihon Sugaku-shi Bunken Soran* (Bibliography of the History of Japanese Mathematics). Vol. 1 of Oya, Shin'ichi, 1963–64,

Hayashi, Tsuruichi. *Hayashi Tsuruichi Hakushi Wasan Kenkyu Shuroku* (Complete Studies of Japanese Mathematics by Dr. Hayashi Tsuruichi). 2 vols. Tokyo: Tokyo Kaiseikan, 1937.

Hiraoka, Tadashi, ed. *Shogakko Sansu-ka Kyoiku no Kenkyu* (Study of Mathematical Education in Elementary School). Tokyo: Kempakusha, 1992.

Hirayama, Akira. *Seki Takakazu* (The Studies of Seki Takakazu). Tokyo: Koseisha Koseikaku, 1959.

Hirayama, Akira and Rakuho Abe. *Hojin no Kenkyu* (Studies of Magic Squares). Osaka: Osaka Kyoiku Tosho, 1983.

Hirayama, Akira, Kazuo Shimodaira and Hideo Horose, eds. *Seki Takakazu Zenshu* (Takakazu Seki's Collected Works, Edited with Explanations). Osaka: Osaka Kyoiku Tosho, 1974.

Ho, Peng-Yoke. 'Magic squares in East and West.' *Papers on Far Eastern History, Australian National University, Department of Eastern History* 8: 115–41, 1973.

Hong, Wansheng, ed. *Tantian San You* (Talking about Three Friends). Taipei: Mingwen Shuju, 1993.

Horiuchi, Annick. *Les Mathématiques Japonaises a l'époque d'Edo.* Paris: Vrin, 1994.

Hosoi, So. *Wasan Shiso no Tokushitsu* (Specificity of Japanese Mathematical Thought). Tokyo: Kyoritsusha, 1941.

Hua, Yinchun. *Zhongguo Zhusuan Shi Gao* (Manuscript on the History of the Abacus in China). Beijing: Zhongguo Caizheng Jingji Chubanshe, 1987.

Inoue Mitsusada *et al.*, eds. *Ritsuryo*. Tokyo: Iwanami Shoten. 1976; reprinted 1985.

Jeon, Sang-woon. *Science and Technology in Korea.* Cambridge, Massachusetts: The MIT Press, 1974.

Jochi, Shigeru. 'Ritsuryo-ki no Sugaku Kyoiku' (Mathematics teaching in Japan from 8c to 12c). *Sugaku-shi Kenkyu* 112: 13–21, 1987.

Jochi, Shigeru. 'Nitchu no Hoteiron Saiko' (A reconsideration of higher degree equations in China and Japan). *Sugaku-shi Kenkyu* 128: 26–35, 1991a.

Jochi, Shigeru. *The Influence of Chinese Mathematical Arts on Seki Kowa.* Ph.D dissertation, University of London, 1993.

Jochi, Shigeru. 'Mohitotsuno Tengen-jutsu; Daien Kyuitsu-jutsu' (Another 'celestial element method'; the technique of acquiring one in Dayan). *Sugakushi Kenkyu* 148: 1–12, 1996a.

Kanda, Shigeru. 'Genna-ban no Ryukoku Daigaku Bon Sanyo-ki; Nihon de Ichiban Furui Kampon Wasansho' (Sanyo-ki printed in the Genna period (1615–1623). Collected at Ryukoku University. The oldest printed mathematical book in Japan). *Sugaku-shi Kenkyu* 37: 48–54, 1968.

Kato, Heizaemon. *Wasan no Kenkyu* (The Study of Wasan). 5 vols. (*Zatsuron* (Sundry) 3 vols. *Hoteishiki-ron* (Equation Theorem), *Seisu-ron* (Numbers Theorem)) Tokyo: Nippon Gakujutsu Shinko Kai (Japanese Academy Press), 1956–64.

Kato, Heizaemon. *Sansei Seki Takakazu no Gyoseki.* (Achievements of the Mathematician Seki Takakazu). Tokyo: Maki Shoten, 1972.

Kim, Yong-woon, and Yong-guk Kim. *Kankoku Sugaku-shi* (History of Mathematics in Korea). Tokyo: Maki Shoten, 1978.

Kodama, Akio. *15 Seiki no Chosen-Kan Do-Katsuji-Ban Sugaku-sho* (Mathematical Books Printed in Korea in the 15c by Copper Printing Type). Tokyo: private publication, 1966.

Kodama, Akio. *16 Seiki Matsu Min-kan no Shuzan-sho* (Abacus Books Printed in the Ming Dynasty at the end of the 16c). Tokyo: Fuji Tanki Daigaku Shuppanbu, 1975.

Lam, Lay-Yong. *A Critical Study of Yang Hui Suan Fa.* Singapore: Singapore University Press, 1977.

Li, Di. *Zhongguo Shuxue-shi Jianbian* (Brief History of Mathematics in China). Shenyang: Liaoning Renming Chubanshe, 1984.

Li, Yan. *Zhongguo Suanxue-shi* (History of Chinese Mathematics). 1st and 2nd editions, Shanghai: SWYSG, 1937 and 1938; 3rd edition. Shanghai: Shanghai Shudian, 1984. 4th edition, Taipei: TSWYSG, 1990.

Li, Yan. *Zhongsuan-shi Lun Cong* (Collected Theses on the History of Chinese Mathematics). 4 vols. Shanghai: SWYSG. 2nd Series. 5 vols. Beijing: KSCBS, 1933–47; 1954–5.

Libbrecht, Ulrich. *Chinese Mathematics in the Thirteenth Century.* Cambridge, Massachusetts and London: The MIT Press, 1973.

Martzloff, Jean-Claude. *Histoire des mathématiques chinoises.* Paris: Masson, 1988.

Matsuzaki, Toshio, ed. *Saiho Sampo Kokon Tsuran (1795).* Tokyo: Tsukuba Shorin, 1996.

Mikami, Yoshio. *The Development of Mathematics in China and Japan.* Leipzig: Teubner, 1913; 2nd ed. New York: Chelsea Publishing Co., 1974.

Mikami, Yoshio. *Wasan no Hojin Mondai* (Problems of the Magic Square in Japanese Mathematics). Tokyo: Teikoku Gakushiin, 1917.

Mikami, Yoshio. *Bunka-shi jo yori Mitaru Nihon no Sugaku* (Japanese Mathematics from the Viewpoint of the History of Culture). Tokyo: Sogensha, 1922, 1947, 1984, 1999.

Mikami, Yoshio. 'Seki Kowa no Gyoseki to Keihan no Sanka narabini Shina no Sampo tono Kankei oyobi Hikaku' (The achievements of Seki Kowa and his relations with mathematicians of Osaka and Kyoto and with Chinese mathematics). *Toyo Gakuho* 20: 217–249, 20: 543–566, 21: 45–65, 21: 352–372, 21: 557–575, 22: 54–99. *Sugaku-shi Kenkyu* 22: 1–51, 23: 53–109, 1932–4; 1944.

Murata, Tamotsu. *Nihon no Sugaku, Seiyo no* Sugaku (Japanese Mathematics, Occidental Mathematics). Tokyo: Chuo Koron-sha, 1981.

Nakayama, Shigeru. *Kinsei Nihon-no Kagaku Shiso* (Scientific Thought of the Modern Era in Japan). Tokyo: Kodansha, 1993.

Needham, Joseph. *Science and Civilisation in China.* Cambridge: Cambridge University Press, 1954–.

Nihon Gakushiin (Japanese Academy), ed. (Fujiwara Shozaburo). *Meiji-zen Nihon Sugaku-shi* (History of Pre-Meiji Japanese Mathematics). 5 vols. Tokyo: Iwanami Shoten. 2nd ed. Tokyo: Noma Kagaku Igaku Kenkyu Shiryokan, 1954–60; 1979.

Noguchi, Taisuke. 'Shosekini Miru Sugaku Yugi' (Recreational mathematics). *Sugaku-shi Kenkyu* 130: 13–26, 1991.

Ogawa, Tsukane. *Kinsei Nihon Sugaku ni okeru Enri no Hoga to sono Tokushitsu* (Germination and Characteristics of the Principle of the Circle in the Edo Period, Japan). Ph.D. thesis, University of Tokyo, 1999.

Ogura, Kinnosuke. *Sugaku-shi Kenkyu* (Study of History of Mathematics). 2 vols. Tokyo: Iwanami Shoten, 1935–48.

Ogura, Kinnosuke. *Nihon no Sugaku* (Japanese Mathematics). 2nd ed. Tokyo: Iwanami Shoten, 1964. English trans. by Norio Ise. *Wasan: Japanese Mathematics.* Tokyo: Kodansha, 1940; 1964, 1993.

Ogura, Kinnosuke. *Chugoku Nihon no Sugaku* (Mathematics in China and Japan). Tokyo: Iwanami Shoten, 1978.

Okamoto, Noribumi, ed. *Wasan Tosho Mokuroku* (Catalogue of Japanese Mathematical Books). Kyoto: Rinsen Shoten, 1932; 1981.

Otake, Shigeo. 'Kodai Ritsuryo Seika no San Reki Ka no kan'i ni Tsuite' (On the official rank of arithmetician and astronomer in ancient Japan). *Sugaku-shi Kenkyu* 97: 1–16, 1983.

Oya, Shin'ichi, ed. *Sugakushi Kenkyu Sosho.* (Studies of the History of Mathematics). 7 vols. Tokyo: Fuji Tanki Daigaku Shuppanbu, 1963–64.

Oya, Shin'ichi. *Wasan Izen* (Before Wasan). Tokyo: Chuo Koron-sha, 1980.

Qian, Baocong, ed. *Suan Jing Shi Shu* (The Ten Mathematical Manuals). Beijing: CHSJ, 1963.

Qian, Baocong. *Zhongguo Shuxue-shi* (History of Mathematics in China). Beijing: KXCBS, 1964.

Qian, Baocong, ed. *Song Yuan Shuxue-shi Lunwen-Ji* (Collected Theses on the History of Chinese Mathematics in the Song and Yuan Dynasties). Beijing: KXCBS, 1966 and 1985.

Ren, Jiyu *et al.*, eds. *Zhongguo Kexue Zhishu Dianji Tong Hui* (Classics of Science and Technology in China) (Mathematics; 5 vols.). Ji'nan: Henan Jiaoyu Chubanshe, 1993.

Sasaki, Chikara. 'Asian mathematics from traditional to modern.' *Historia Scientiarum*, ser. 2, 4: 69–77, 1994.

Sawada, Goichi. *Nara Cho Jidai no Minsei Keizai no Suteki Kenkyu* (Mathematical Studies for Civil Administration and Economy in the Nara Dynasty). Tokyo: Fujibo, 1927.

Sawada, Goichi. *Nihon Sugaku-shi Kowa* (Lecture on the History of Japanese Mathematics). Tokyo: Toe Shoin, 1928.

Shimodaira, Kazuo. *Wasan no Rekishi* (History of Japanese Mathematics in the Edo Period). 2 vols. Tokyo: Fuji Tankidaigaku Shuppanbu, 1965–70.

Shimodaira, Kazuo *et al.*, eds. *Edo Shoki Wasan Sensho* (Selected Collection of Japanese Mathematical Arts in the Early Edo Period), 5 vols. Tokyo: Kenseisha, 1990–1998.

Suzuki, Hisao. *Shuzan no Rekishi* (History of the Abacus). Tokyo: Fuji Tankidaigaku Shuppanbu, 1964.

Tokyo Daigaku Shiryo Hensanjo, eds. *Dainihon Komonjo* (Great Collection for the History of Japan). Tokyo: Tokyo University Press, 1901, 1968, 1998.

Toya, Seiichi. 'Isomura Yoshinori no Hojin Sakusei no Kangaekata' (A way of thinking of designing magic squares by Isomura Yoshinori). *Fuji Ronso* 32(1): 19–33, 1987.

Wang, Qingxiao (O Seisho). *Sangi wo Koeta Otoko* (The Man Behind the Counting Rods). Tokyo: Toyo Shoten, 1999.

Wu, Wenjun, ed. *Zhongguo Shuxue-shi Lunwen Ji* (Collected Theses on the History of Mathematics in China), 3 vols. Ji'nan: Shandong Jiaoyu Chubanshe, 1985–87.

Xiong, Jisheng. 'Yang Hui Wu-Wu-Tu Qian Shi' (Analysis of Yang Hui's five degree magic square). *Shuxue Tongxun* 9: 22–6, 1955.

Yabuuchi, Kiyoshi. *Chugoku no Sugaku* (Mathematics in China). Tokyo: Iwanami Shoten, 1974.

Yabuuchi, Kiyoshi, ed. *Chugoku Temmongaku*-Sugaku *Shu* (Materials on Astronomy and Mathematics in China). Tokyo: Asahi Shuppansha, 1980.

Yamazaki, Yoemon. *Jinko-ki no Kenkyu; Zuroku-hen* (Studies of 'Permanent Mathematics'; Diagrams). Tokyo: Morikita Shuppan, 1966.

Yan, Dunjie. 1964. 'Song Yang Hui Suan-Shu Kao' (Study of Yan Hui's Mathematical Books in the Song Dynasty). In Qian, Baocong, ed. *Song Yuan Shuxue-shi Lunwen-Ji* (Collected Theses on the History of Chinese Mathematics in the Song and Yuan Dynasties). Beijing: KXCBS, 1966 and 1985.

Yang, Tsui-hua and Yi-Long Huang, eds. *Jindai Zhongguo Keji-shi Lunji* (Science and Technology in Modern China). Taipei: Zhongyan Yanjiuyuan Jindaishi Yanjiusuo, Qinghua Daxue Lishi Yanjiusuo, 1991.

Yoshida, Yoichi. *Rei no Hakken* (Discovery of Zero). Tokyo: Iwanami Shoten, 1969.

Zhong Wai Shuxue Jian Shi Bianxie-zu, ed. (Edition Committee on History of Mathematics). *Zhongguo Shuxue Jian Shi* (Brief History of Mathematics in China). Ji'nan: Shangdong Jiaoyu Chubanshe, 1986.

KIM, SOO HWAN

DEVELOPMENT OF MATERIALS FOR ETHNOMATHEMATICS IN KOREA

D'Ambrosio (1985) used the expression 'ethnomathematics' to refer to forms of mathematics that vary as a consequence of being embedded in cultural activities whose purpose is other than 'doing mathematics'. Ethnomathematics has the goal of broadening the history of mathematics to one that has a multicultural, global perspective. It involves the study and presentation of mathematical ideas of traditional peoples. This broadening of perspective to include other cultures has the associated effect of enlarging the history of mathematics from dealing primarily with the Western professional class called mathematicians to involving all sorts of people (Ascher, 1991).

There are two ways of resuscitating ethnomathematics in the process of teaching and learning mathematics. One is to use a creative way of thinking and unique ideas of ethnomathematics in one's own cultural area; the other is through other cultural areas. In order to develop mathematical acculturation, it is important for us to develop materials for ethnomathematics in both the person's own and other cultural areas.

It has long been common practice to divide mathematics into 'pure' and 'applied' categories. Shirley (1995) also makes a significant cross-categorization, dividing mathematics into 'formal' and 'informal'. The result is a two-by-two grid of four cells labeled academic, technical, recreational, and everyday mathematics.

There is a tendency to make the mathematics classroom a place for a small number of the elite as a result of attaching great importance to mathematical structure in academic mathematics. For a while, mathematics educators concentrated on everyday mathematics. Considering that we should provide students with all kinds of mathematics to cope with a rapidly changing society, the dichotomous categorization of school mathematics should be reconsidered. It is desirable for us to classify mathematics into six categories: academic, technical, school, interdisciplinary, recreational, and everyday mathematics on a three-by-two grid of six cells as in Table 1.

I am very interested in school mathematics and interdisciplinary mathematics for students' mathematical acculturation. Bishop (1988) argues that the mathe-

H. Selin (ed.), Mathematics Across Cultures: The History of Non-Western Mathematics, 455–465.

Table 1 Six categories of mathematics

	Type of Mathematics	
Level of Mathematical Culture	Pure	Applied
Technical	Academic	Technical
Formal	School	Interdisciplinary
Informal	Recreational	Everyday

matical enculturation curriculum should contain the symbolic component, which is concept-based; the cultural component, which is investigation-based; and the social component, which is project-based.

Mathematical knowledge is believed to develop by invention rather than discovery, since it has the characteristics of both physical abstraction based on reality and reflective abstraction based on possibility. And there can be excellent mathematical ideas in ethnomathematics. Therefore, it is included as an important element of mathematical culture from the viewpoint of reviewing the old and learning the new. Furthermore, it is necessary for us to make everyday cognition an object to explore mathematics, broadening the scope of mathematical concepts.

In sum, the originality of ethnomathematics and the practicality of everyday mathematics should be embodied in the process of teaching and learning mathematics to develop a valuable mathematical culture including counting, locating, measuring, designing, playing, and explaining.

MATERIALS FOR ETHNOMATHEMATICS IN KOREA

This part of the essay presents a series of games and situations in which mathematics plays a part. I have suggested problems that students can tackle to understand ethnomathematics better.

Wooden die for drinking game

This game is made of a wooden tetradecahedron which was unearthed at Anapzi, the site of a pond in the royal gardens in the era of Unified Silla, around 674 AD. Archaeologists suggest that it was used as a method of drinking rice wine and giving zest to a banquet when a lot of people came together for a ceremonial feast. It has 6 facets of 2.5 cm × 2.5 cm regular squares and 8 facets of 2.5 cm × 0.8 cm hexagons. It is geometrically harmonized with a height of 4.8 cm. One can make a tetradecahedron by cutting 8 vertices from a cube not to be pointed, and one can also cut six vertices from an octahedron. Park and Lee (1987) argued that 14 surfaces had the same probability. The contents written on the 14 faces are as follow:

- dance silently
- stay motionless regardless of persons' rushing upon oneself

- giving a loud laugh after drinking
- everybody hit on the nose
- singing and drinking for oneself
- drinking three cups of wine at once
- drink in one gulp with bending one's forearm
- staying motionless regardless of people tickling your face
- having someone sing a song
- sing a tune for 12 months
- sing a special song for oneself
- compose a poem
- throw out one cup if you have two cups of wine
- not throwing away an ugly object

If you are teaching Korean mathematics, you can give students two problems. You can ask them to draw a development figure, and make the tetradecahedron Wooden Die for Drinking Game or have them calculate the areas of 14 surfaces, and experiment on the probability of 14 surfaces.

Figure 1 Wooden die for drinking game (Ko, 1996: 81). Used with permission.

Investigation by game: primitive counting sticks

Origin: Counting sticks were once used to count in Korea. The game using the primitive counting sticks seems to be simple, but careful observation and delicate hands are required to perform it skillfully.

Number of Persons 2–5

Method

(1) Determine the order of play.
(2) The first player holds the counting sticks in one hand, sets them up on the floor of a room or on a desk, and then scatters them naturally by releasing his hold.
(3) The player picks up the sticks according to the order of play. The person goes on if he/she does not touch the other sticks, but loses if he/she does.
(4) The picked up sticks can be used to pick up the other sticks.
(5) If all the sticks are picked up, each player counts the scores of his sticks according to color, and the player whose scores highest wins.

Scores: Yellow Stick: 2 points; Red Stick: 3 points; Blue Stick: 5 points; Green Stick: 10 points; Black Stick: 20 points.

Tangram

Tangram (Figure 2) is a game in which one makes up several shapes with seven pieces; two big right-angled triangles, a middle-sized triangle, two small triangles, a square, and a parallelogram cut away from a 10 cm × 10 cm square. The game began in China about two thousand years ago and remains in use both in Korea and all over the world.

It is a very interesting and useful game requiring the power of imagination, observation, and perseverance. There are numerous shapes which can be made up by seven pieces. There is a work of art, The Seven Magicians (which means the figures which the Vega saw in a dream) on the wall in the subway station of the Kimpo International Airport. Such an example is a good teaching aid for presenting tangrams.

It includes a man who is walking, a man who is running, a water bird, a lotus blossom, 'small' in Chinese, a boat, 'six' in Chinese, a man who is reading a book while lying down, a chair, and a tower.

Tracing one's kinship at Chusok, the Korean Thanksgiving Day

Chusok falls on 15th August using a lunar calendar. For Koreans, the Full Moon Harvest Day is the most exciting holiday with a variety of festivals, traditional games, and exchanges of gifts.

Here is a problem situation that can be used with this day. Brotherhood is the 2nd degree of kinship, and the relationship between father and son is 1st degree. For example, the relationship between J. W. Kim and H. R. Kim is 2nd degree, but J. W. Kim and J. H. Kim are in 4th degree, as in Figure 3.

The student must determine the degree of the kinship between J. W. Kim and C. S. Kim. If C. S. Kim has a baby, what is the degree of kinship between J. W. Kim and C. S. Kim's baby? We can also see that J. W. Kim has no 6th degree of brothers and sisters. What is the reason for this?

A stonemason's wisdom

Most Koreans visit their ancestral graves after observing a worship service for family ancestors on Chusok morning. The price of the stone offertory table in front of a tomb depends upon the length of the diagonal line AB.

How can a stonemason measure the length of the diagonal line without knowing the Pythagorean theorem?

Mathematical ideas using the traditional calendrical system

The ten 'celestial stems' (*gan*) are such characters as 甲, 乙, 丙, 丁, 戊, 己, 庚, 辛, 壬, 癸, and the twelve horary characters (*ji*) are 子, 丑, 寅, 卯, 辰, 巳, 午, 未, 申, 酉, 戌, 亥. Combining these into a recurring sexagenary cycle gave rise to the Korean

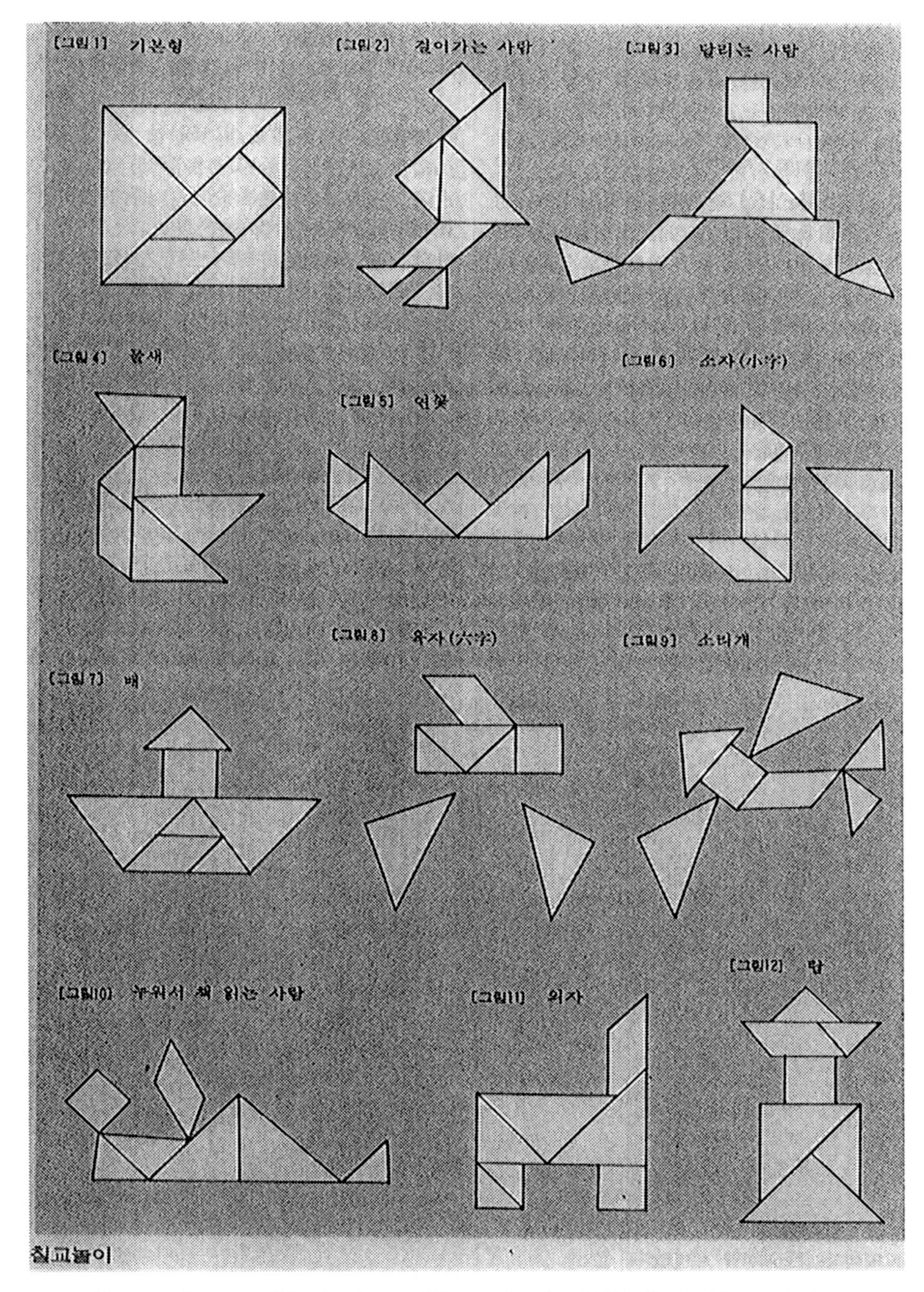

Figure 2 Tangram (The Academy of Korean Studies, 1991). Used with permission.

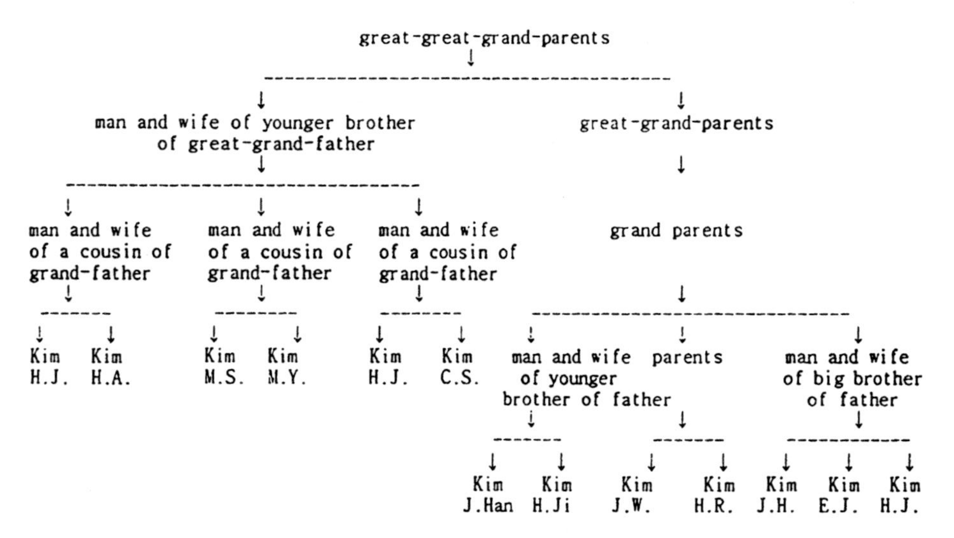

Figure 3 J. W. Kim's kinship diagram.

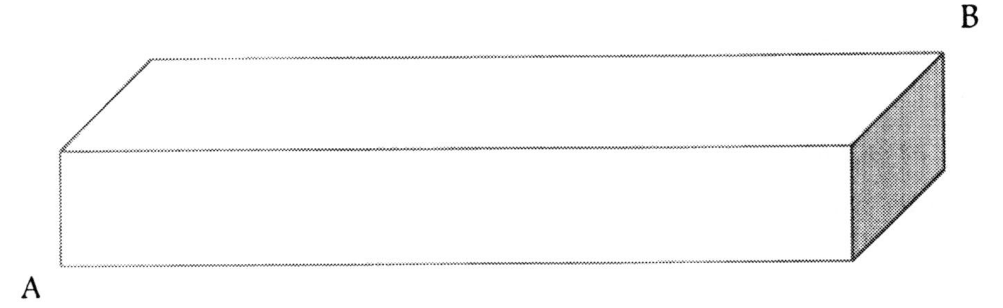

Figure 4 The stone offertory table in front of a tomb.

and the Chinese traditional calendar system as we know it (Needham, 1954). The number of the ten celestial stems is derived from the base of 10 in the decimal system, and the number of the twelve horary characters is from measuring astronomical phenomena. The numbers 10, 12, and their least common multiple 60, are the symbolic ally related to the thought of Providence as well as units of year, month, day, and time (Kim and Kim, 1982). Figure 5 shows the points of the compass in geomancy, which is the science and art of choosing an auspicious place for a home or a grave.

There are some problems we can create using Figure 5, which are useful for determining direction and time. In Figure 5, the directions of 子, 卯, 午, and 酉, are N, E, S, and W. What are the directions of the duodenary cyclical characters? For example, an azimuth angle of 丑 is N 30°E.

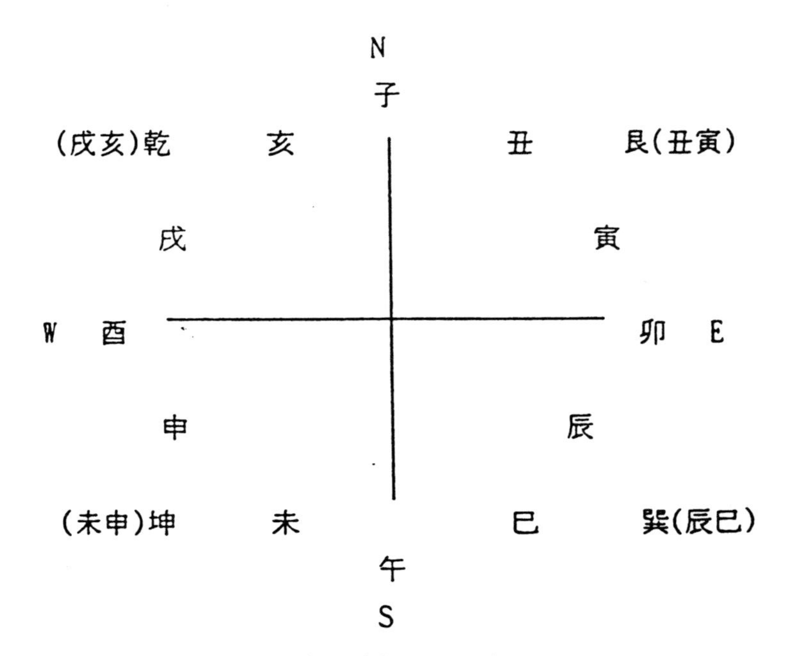

Figure 5 Points of the compass in geomancy.

The duodenary cyclical characters are used as horary characters. For example, the character 子 shows the time between 11 p.m. and 1 a.m. Ask each student to determine when he was born using the duodenary cyclical characters.

64 gua and the binary system in the Book of Changes (Yijing or I Ching)

The 64 symbols in the *Book of Changes* (*Yijing*) provided a set of abstract conceptions capable of subsuming a large number of the events and processes in the natural world. The symbols are all made up of lines (*xiao*), some full or unbroken (*yang* lines), and others broken, i.e. in two pieces with a space between (*yin* lines). By using all the possible permutations and combinations, 8 trigrams and 64 hexagrams are formed, all known as *gua* (Needham, 1954; Kim and Kim, 1982).

Here is a problem that can be used with the 64 *gua* and the binary system in the *Book of Changes*: Supposing that *yang* lines mean 1, and *yin* lines mean 0 as in Table 2, represent binary numbers and decimal numbers from 0 to 63 in Table 3 with 64 hexagrams.

Table 2 8 trigrams in the *Book of Changes* (*Yijing*)

Sex	♀	♂	♂	♀	♂	♀	♀	♂
Korean (Chinese)	곤(坤)	간(艮)	감(坎)	손(巽)	진(震)	리(離)	태(兌)	건(乾)
Romanized	Khun	Ken	Khan	Sun	Chen	Li	Tui	Chhien
Denary	0	1	2	3	4	5	6	7
Binary	000	001	010	011	100	101	110	111
Kua	☷	☶	☵	☴	☳	☲	☱	☰

Making the standardized national flag of Korea

The national flag of Korea is composed of the Great Absolute (i.e. the source of the dual principle of *yin* and *yang*) and 4 *gua*: *Qien, Li, Kan, Kun.*

We can construct the flag according to the standardized rule following Figure 6. Doing this requires mathematical concepts such as proportion, binary system, geometrical design, and measuring the area of plane figures.

There are many methods for developing materials for ethnomathematics. We can make use of literature on both the history of mathematics and on cultural heritage. There are many books, including *Great Moments in Mathematics* (Her and Oh, 1994, Korean version), *An Introduction to the History of Mathematics* (Lee and Shin, 1995, Korean version), *Annapzi* (Ko, 1996), *Castle of Suwon* (Kim, 1996), and *Traditional Folk Games* (Kim, 1997) that are appropriate to the study of ethnomathematics. We can also make use of traditional folk games. We know of more than 216 kinds of traditional games including the ones mentioned here.

Table 3 The binary numbers and decimal numbers of 64 hexagrams in the *Book of Changes* (*Yijing*)

		Upper 8-trigrams							
		- - - - - -	--- - - - -	- - --- - -	--- --- - -	- - - - ---	--- - - ---	- - --- ---	--- --- ---
Under 8-trigrams	- - - - - -	000000 0	000001 1	000010 2	000011 3	000100 4	000101 5	000110 6	000111 7
	--- - - - -	001000 8	001001 9	001010 10	001011 11	001100 12	001101 13	001110 14	001111 15
	- - --- - -	010000 16	010001 17	010010 18	010011 19	010100 20	010101 21	010110 22	010111 23
	--- --- - -	011000 24	011001 25	011010 26	011011 27	011100 28	011101 29	011110 30	011111 31
	- - - - ---	100000 32	100001 33	100010 34	100011 35	100100 36	100101 37	100110 38	100111 39
	--- - - ---	101000 40	101001 41	101010 42	101011 43	101100 44	101101 45	101110 46	101111 47
	- - --- ---	110000 48	110001 49	110010 50	110011 51	110100 52	110101 53	110110 54	110111 55
	--- --- ---	111000 56	111001 57	111010 58	111011 59	111100 60	111101 61	111110 62	111111 63

We can investigate mathematical ideas by visiting museums such as the National Folklore Museum, the National Museum of Korea, the National Museum of Kyungju, and the Sam Sung Children's Museum. There are many scientific devices including a sundial and the Wooden Die for Drinking game at the National Museum of Kyungju, and tangrams, geoboards, and counting sticks in the Sam Sung Children's Museum that students can explore.

Visiting such historic relics as Su Won Castle, Skkuram cave, and Annapzi provide opportunities to measure them to scale and explore their geometric patterns. Ethnic customs such as tracing the degree of kinship, mathematical ideas in the denary and duodenary cyclical characters, and the binary system of 64 *gua* in the *Book of Changes*, can also be investigated. Toy shops and stationery shops also give ideas for mathematical games.

* * *

In conclusion, it is important for learners to appreciate the originality of

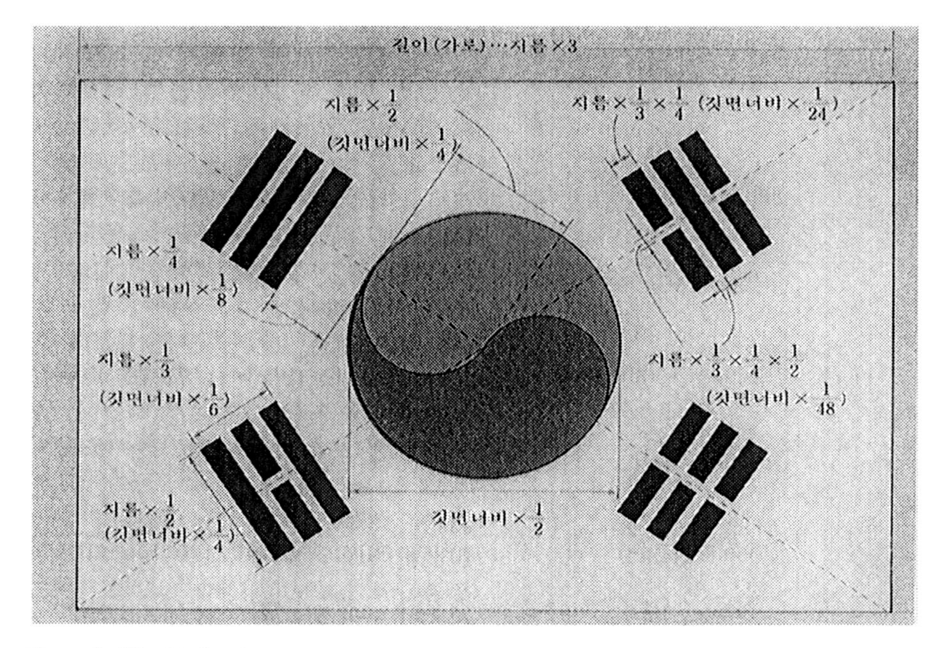

Figure 6 The National Flag of Korea (The Academy of Korean Studies, 1991). Used with permission.

ethnomathematics and the practicality of everyday mathematics in the process of teaching and learning mathematics. Therefore, we must develop available materials for mathematical acculturation. For this, we should keep in mind the following details. First, we should recognize that each culture has its own ethnomathematics and this should be integrated into school mathematics. We should remember to review the old and learn the new. Second, we should make use of various methods for developing materials for ethnomathematics such as making use of literature and traditional folk games, investigating ethnic customs, and visiting museums, historic relics, toy shops, and stationery shops. Third, it is important to explore ethnomathematics actively in familiar environments so that they can use their mathematical ability in original, creative ways. Finally, we should make learners prepare for leading a high quality of life in the future by exploring the mathematical ideas integrated in the problem situations of everyday life.

BIBLIOGRAPHY

Ascher, M. *Ethnomathematics. A Multicultural View of Mathematical Ideas.* Pacific Grove, California: Brooks/Cole Publishing Company, 1991.

Bishop, A. J. *Mathematical Enculturation – a Cultural Perspective on Mathematics Education.* Dordrecht: Kluwer Academic Publishers, 1988.

D'Ambrosio, U. *Socio-Cultural Bases for Mathematics Education.* Campinas: UNICAMP, 1985.

Her, Min and Oh Hyeyoung. *Great Moments in Mathematics* (Korean version. Originally published

Washington DC: The Mathematical Society of America, 1983). Seoul: Kyongmoonsa Publishing Co., 1994.

Kim, Dongwook. *Castle of Suwon* (Korean version). Seoul: Daewonsa Publishing Co. Ltd., 1996.

Kim, Kwangun. *A Traditional Folk Game.* (Korean version). Seoul: Daewonsa Publishing Co. Ltd., 1997.

Kim, Soohwan. 'Development of a teaching /learning model for the mathematical enculturation of elementary and secondary school students.' *Journal of the Korea Society of Mathematical Education Series D: Research in Mathematical Education* 1(2): 107–116, 1997.

Kim, Youngwoon and Kim, Youngkuk. *Korean Mathematical History.* (Korean version). Seoul: Yolhwadang Publishing Co., 1982.

Ko, Kyonghee. *Annapzi* (Korean version). Seoul: Daewonsa Publishing Co. Ltd., 1996.

Lee, Wooyoung and Shin, Hangkyuoon. *An Introduction to the History of Mathematics* (Korean version. Originally published New York: Rinehart, 1953). Seoul: Kyongmoonsa Publishing Co., 1995.

Needham, J. *Science and Civilisation in China.* Cambridge: Cambridge University Press, 1954.

Park, Hanshick and Lee, Kangsub. 'Tetradecahedron dice.' *Mathematics Education* 25(2): 19–21, 1987.

Shirley, L. 'Using ethnomathematics to find multicultural mathematical connections.' In *Connecting Mathematics Across the Curriculum, 1995 Yearbook*, P. A. House and A. F. Coxford, eds. Reston, Virginia: The National Council of Teachers of Mathematics, 1995, pp. 34–43.

The Academy of Korean Studies. *The Encyclopedia of Korean Culture.* Seoul: Samhwa Publishing Company, 1991.

Index

Printed in the United States
84333LV00001B/6/A

9 781402 002601